SOME CLASSES OF SINGULAR EQUATIONS

North-Holland Mathematical Library

VOLUME 17

NORTH-HOLLAND PUBLISHING COMPANY
AMSTERDAM · NEW YORK · OXFORD

SOME CLASSES OF SINGULAR EQUATIONS

SIEGFRIED PRÖSSDORF

Zentralinstitut für Mathematik und Mechanik der Akademie der Wissenschaften der DDR

1978

NORTH-HOLLAND PUBLISHING COMPANY
AMSTERDAM · NEW YORK · OXFORD

North-Holland ISBN for this volume: 0 7204 0501 7

A translation of:

EINIGE KLASSEN SINGULÄRER GLEICHUNGEN

Translated by:

Prof. Dr. Siegfried Dümmel, Technische Hochschule Karl-Marx-Stadt.

Published by:

NORTH-HOLLAND PUBLISHING COMPANY
AMSTERDAM · NEW YORK · OXFORD

Sole distributors for the USA and Canada :
ELSEVIER NORTH-HOLLAND INC.
52 Vanderbilt Avenue
New York, N.Y. 10017

Library of Congress Cataloging in Publication Data

Prössdorf, Siegfried.
Some classes of singular equations.
(North-Holland mathematical library; v. 17)
Translation of Einige Klassen singulärer Gleichungen.
Bibliography: p.
Includes indexes.
1. Integral equations. I. Title.
QA431.P7613 515'.45 77-16710 ISBN 0-7204-0501-7

PRINTED IN GDR

Dedicated to my teacher
S. G. Mihlin

PREFACE TO THE ENGLISH EDITION

The essential difference between the present edition of this book and the original edition lies in the last two chapters which are completely new. They essentially generalize and complete the contents of the two short appendices of the german edition.

In Chapter 10 the most important results of the preceding chapters are generalized to the case of singular equations with piecewise continuous resp. measurable bounded coefficients and to the case of open integration curves.

Chapter 11 investigates several stable projection methods for the approximate solution of singular equations of normal and non-normal type. In particular, we consider the reduction method and the collocation method. Here for the diverse concrete classes of equations and spaces, in addition to the proofs of convergence, error estimates are also given.

Most of the results stated in these two chapters were obtained in the last three years.

Furthermore, the statements of several theorems have been sharpened and the proofs simplified.

The question of the one-sided invertibility of the singular operators of non-normal type was investigated in more detail.

Finally, the bibliography and the references have been expanded.

I express my sincere thanks to Prof. Dr. S. Dümmel for the careful translation and for his co-operation during the time of the translation.

Berlin, April 1976 S. Prössdorf

PREFACE TO THE GERMAN EDITION

The basis for the development of a general theory of one-dimensional singular integral equations was given in the fundamental papers of F. NOETHER [1] on integral equations with Hilbert kernel and of N. WIENER and E. HOPF [1] on integral equations with a difference kernel on the half-line. The equations considered by NOETHER and the integral equations with Cauchy kernel which are closely related to these equations are in general called "singular integral equations". For integral equations with a difference kernel on the half-line the term "Wiener-Hopf equations" is used. In this book the concept "singular equations" contains both type of equations and also some other types.

The theory of singular equations of normal type can essentially be considered as closed. Soviet mathematicians played a decisive role in its development. A complete representation of this theory is given in the book by N. I. MUSHELIŠVILI [1] (for equations with Cauchy and Hilbert kernels and spaces of Hölder continuous functions) and in the monograph by I. C. GOHBERG and I. A. FELDMAN [1] (for Wiener-Hopf equations and some more general singular equations).

Recently in the theory of singular equations there arose an independent direction in which equations whose symbol has zeros (degenerate equations or equations of non-normal type) are investigated. Theoretical as well as practical interest in such equations increased greatly. There appeared an extensive literature in which these equations are studied from different points of view and with very different methods. Therefore, in this book I gave myself the task, for some important classes of one-dimensional singular equations of non-normal type, to build up a closed theory which includes all essential results in this field known at the present time. Singular integral equations with Cauchy or Hilbert kernel, Wiener-Hopf integral equations, their discrete analogue and several paired equations with difference kernels are of this type.

Some important questions which are closeley connected with the subject of this book could not be considered here for reasons of space. This holds especially for similarily placed (non-elliptic) multi-dimensional problems.

Recently great advances have been made in this field[1]). We also do not consider non-linear singular integral equations.

In the present book the methods of functional analysis are mainly used. We assume that the reader is familiar with the elements of the theory of linear operators in Banach spaces and also with the elements of the theory of linear topological spaces. In some parts of the book a basic knowledge of Banach algebras is necessary (roughly to the extent of the first chapter of the book by I. M. GELFAND, D. A. RAIKOV and G. E. ŠILOV [1]).

I owe my interest in the methods of functional analysis and especially in the subject of this book to my esteemed teacher Prof. S. G. MIHLIN (Leningrad). Through several discussions with Prof. I. C. GOHBERG (Kišinov) in the last few years I was stimulated to write this book. Prof. S. G. MIHLIN and Prof. I. C. GOHBERG also read a great deal of the manuscript and gave much useful advice. For this I express my sincere thanks to them.

Furthermore, I wish to thank Dr. B. SILBERMANN (Karl-Marx-Stadt) who read the whole manuscript. Finally I thank the Akademie-Verlag, Berlin, especially the subeditors for mathematics Dipl.-Math. R. HELLE and Dr. R. HÖPPNER for taking into consideration all my wishes in an understanding way and for the careful layout of the book.

Karl-Marx-Stadt, December 1972 S. PRÖSSDORF

1) The reader can find references up to 1969 in the article [15] of the author.

CONTENTS

INTRODUCTION

1. The following three types of equations which play an important role in many problems of mathematical physics are the most essential representatives of the classes of singular equations considered in this book.

a) The *Wiener-Hopf integral equation*

$$A\varphi = c\varphi(t) + \int_0^\infty k(t-s)\,\varphi(s)\,ds = f(t) \qquad (0 \leqq t < \infty)\,, \tag{0.1}$$

where c is a constant, $k(t)$ $(-\infty < t < \infty)$ an arbitrary absolutely integrable function, $f(t) \in L^p(0, \infty)$ $(1 \leqq p < \infty)$ a given function and $\varphi(t) \in L^p(0, \infty)$ the required function searched for.

b) The *discrete Wiener-Hopf equation*

$$\sum_{j=0}^{\infty} a_{k-j}\xi_j = \eta_k \qquad (k = 0, 1, \ldots)\,. \tag{0.2}$$

Here

$$\sum_{-\infty}^{\infty} |a_j| < \infty\,,$$

$\eta = \{\eta_j\}_0^\infty$ is a given vector and $\xi = \{\xi_j\}_0^\infty$ is a vector sought for in the space $l^p(1 \leqq p < \infty)$.

c) The *singular integral equation* with Cauchy kernel

$$A\varphi = a(t)\,\varphi(t) + b(t)\,S\varphi + T\varphi = f(t) \qquad (t \in \Gamma)\,, \tag{0.3}$$

where $a(t)$ and $b(t)$ are continuous functions on a closed Ljapunov curve Γ and the so-called singular integral

$$S\varphi = \frac{1}{\pi i}\int_\Gamma \frac{\varphi(\tau)}{\tau - t}\,d\tau \qquad (t \in \Gamma)$$

is understood in the sense of the Cauchy principal value.

We consider the equation (0.3) in the space $L^p(\Gamma)$ $(1 < p < \infty)$ and assume that T is a compact linear operator in this space. With the equations

(0.1)—(0.3) we associate the functions

$$A(\lambda) = c + \int_{-\infty}^{\infty} e^{i\lambda t}\, k(t)\, dt \qquad (-\infty \leqq \lambda \leqq \infty)\,, \tag{0.1'}$$

$$A(z) = \sum_{j=-\infty}^{\infty} a_j z^j \qquad (|z| = 1)\,, \tag{0.2'}$$

$$A(t, \theta) = a(t) + \theta b(t) \qquad (t \in \Gamma;\ \theta = \pm 1)\,, \tag{0.3'}$$

respectively. These functions are called the *symbol* of the corresponding equation resp. of the operator A defined by this equation. The operator A and its symbol are closely related: The operators of the form (0.3) form an algebra $\mathfrak{A}$ (the corresponding statement holds for the operators (0.1) and (0.2) perturbed by a compact summand). The symbol effects a (continuous) isomorphism of the factor algebra $\mathfrak{A}/\mathfrak{K}$ ($\mathfrak{K}$ the ideal of the compact operators in the space $L^p(\Gamma)$) onto the algebra of all functions of the form (0.3').

The numbers [1])

$$\varkappa = \frac{1}{2\pi}[\arg A(\lambda)]_{\lambda=-\infty}^{\infty}\,, \qquad \varkappa = \frac{1}{2\pi}[\arg A(z)]_{|z|=1}\,,$$

$$\varkappa = \frac{1}{2\pi}\left[\arg \frac{a(t) + b(t)}{a(t) - b(t)}\right]_\Gamma$$

are called the *index* of the corresponding symbol. The situation with respect to the solvability of the equations (0.1) and (0.2) is clarified by the following theorem.

Theorem 0.1. *For the operator A to be at least one-sided invertible it is necessary and sufficient that its symbol vanishes nowhere.*

If this condition is satisfied, then the operator A is left, right or two-sided invertible if the number $\varkappa$ is positive, negative or equal to zero, respectively.

The corresponding inverses of the operator A can be effectively constructed.

In the case $T = 0$ Theorem 0.1 remains valid also for the equation (0.3). For any compact operator T the following theorem holds.

Theorem 0.2. *Let $a(t)$, $b(t)$ be arbitrary functions on the curve Γ and $A(t, \theta) \neq \neq 0$, i.e.*

$$a^2(t) - b^2(t) \neq 0 \qquad (t \in \Gamma)\,.$$

Then

(1) *The operator A is normally solvable, i.e. the equation* (0.3) *is solvable if and only if the right-hand side $f(t)$ is orthogonal to all solutions of the adjoint homogeneous equation $A^*\psi = 0$.*

(2) *The equations $A\varphi = 0$ and $A^*\psi = 0$ have finitely many linearly independent solutions.*

---

[1]) The sign $[\ ]_\Gamma$ denotes the increment of the expression enclosed in brackets as t traverses the curve Γ once in the positive direction.

(3) *The difference of the numbers of the linearly independent solutions of the equations mentioned in* (2) (*the so-called index of the operator* A) *does not depend on the compact part* T *and is equal to* $-\varkappa$.

The statements (1)—(3) are called the *Noether theorems* for the singular integral equation (0.3).

Equations whose symbol does not satisfy the conditions of Theorem 0.1 and 0.2 are called *singular equations of non-normal type*. E.g. the Wiener-Hopf equations of the first kind (i.e. equation (0.1) with $c = 0$), which are of great practical interest, are of this type.

The operators defined by equations (0.1) to (0.3) and also the other singular operators considered in this book have a common property: They are functions of a one-sided invertible operator resp. pairs of functions of an invertible operator. E.g. we can regard the operator A defined an the space l^p by the equation (0.2) as the value of the function (0.2′) of the shift operator V, i.e.

$$A = \sum_{j=-\infty}^{\infty} a_j V^{(j)} ,$$

where this series converges in the operator norm. Here

$$V^0 = I, \qquad V^{(j)} = V^j (j = 1, 2, \ldots) , \qquad V^{(j)} = [V^{(-1)}]^{|j|} \ (j = -1, -2, \ldots) ,$$

and the shift operators V and $V^{(-1)}$ are defined by the equations

$$V\{\xi_j\}_0^\infty = \{0, \xi_0, \xi_1, \ldots\} , \qquad V^{(-1)}\{\xi_j\}_0^\infty = \{\xi_{j+1}\}_0^\infty .$$

Obviously, $V^{(-1)}$ is a left inverse of V.

2. The singular equations of non-normal type whose symbol vanishes at several points occupy a central position in this book. In this case the corresponding operator A is not normally solvable in the above-mentioned space X (X one of the spaces $L^p(0, \infty)$, l^p, $L^p(\Gamma)$). But a simple application of subsequent constructions enables us to generate Banach spaces $\overline{X}$ and $\widetilde{X}$ such that the relation

$$\overline{X} \subset X \subset \widetilde{X}$$

(in the sense of a continuous embedding of normal spaces) hold and the operator A considered as an operator from $\widetilde{X}$ into the space $\overline{X}$ has the properties formulated in no. 1.

Namely, we can prove that the singular operator A whose symbol has at most finitely many zeros (of finite order) can be represented in the form

$$A = BCD . \tag{0.4}$$

Here C is a singular operator of normal type. B, D are operators of non-normal type with extremely simple symbols, and moreover, they satisfy the following conditions:

(1) The equation $B\varphi = 0$ has only the zero solution.

(2) There exist an operator $D^{(-1)}$ defined on the whole space X with a range $\widetilde{X} = D^{(-1)}(X)$ $(\supset X)$ and a linear extension $\widetilde{D}$ of the operator D from X to the set $\widetilde{X}$ such that the relations

$$D^{(-1)}Df = f\,, \qquad \widetilde{D}D^{(-1)}f = f$$

hold for any $f \in X$. We introduce the space $\overline{X} = B(X)$. With the norms

$$||\varphi||_{\overline{X}} = ||B^{-1}\varphi||_X\,, \qquad ||\varphi||_{\widetilde{X}} = ||\widetilde{D}\varphi||_X$$

$\overline{X}$ and $\widetilde{X}$ are Banach spaces satisfying the above-mentioned conditions. Obviously, $\widetilde{D}$ and B induce an isometric isomorphism of $\widetilde{X}$ onto X and of X onto $\overline{X}$, respectively. Hence the singular equation $A\varphi = f (\varphi \in \widetilde{X})$ with a right-hand side $f \in \overline{X}$ is equivalent to the equation of normal type

$$C\psi = B^{-1}f$$

considered in the space X.

The operators B and D in the representation (0.4) are not uniquely determined. Under suitable hypotheses on the symbol we can realize the two limiting cases $B = I$ (i.e. $\overline{X} = X$) or $D = I$ (i.e. $\widetilde{X} = X$). We have to choose the factors B, D in such a way that the spaces $\overline{X}$ and $\widetilde{X}$ admit an analytic description which is as simple as possible.

3. We explain the considerations made in no. 2 by the example of the *Wiener-Hopf integral equation of the first kind* of the form (0.1) with $c = 0$. In this case the symbol defined by (0.1′) vanishes at the point at infinity. For simplicity we assume that the point $\lambda = \infty$ is the only zero of the function $A(\lambda)$ and that $A(\lambda)$ moreover has the form

$$A(\lambda) = \frac{B(\lambda)}{(\lambda + i)^m}\,,$$

Here m is a positive integer, $B(\lambda)$ a function of the form (0.1′) and $B(\lambda) \neq 0$ $(-\infty \leqq \lambda \leqq \infty)$. Choosing two *arbitrary* non-negative integers μ and ν with $\mu + \nu = m$ we can represent the function $A(\lambda)$ in the form

$$A(\lambda) = \frac{1}{(\lambda - i)^\mu}\, C(\lambda)\, \frac{1}{(\lambda + i)^\nu}\,,$$

where $C(\lambda) = \left(\dfrac{\lambda - i}{\lambda + i}\right)^\mu B(\lambda)$ again has the form (0.1′). By B and D we denote the Wiener-Hopf integral operators of the first kind with symbols $(\lambda - i)^{-\mu}$ and $(\lambda + i)^{-\nu}$, respectively. Then for the operator A we obtain the representation (0.4) from a well-known convolution theorem, and we have $B = I$ if $\mu = 0$ and $D = I$ if $\nu = 0$.

In the present case the space $\widetilde{L}^p(0, \infty)$ consists of all functions of the form

$$D^{(-1)}f = i^\nu \left(\frac{d}{dt} + 1\right)^\nu f(t) \qquad (f(t) \in L^p(0, \infty)) , \tag{0.5}$$

where the derivative $(d/dt)\, f(t)$ is to be understood in the sense of Schwartz's distribution theory.

The space $\overline{L}^p(0, \infty)$ consists of all functions $f(t)$ for which the derivatives $f^{(l)}(t)$ $(l = 0, 1, \dots , \mu - 1)$ are absolutely continuous on the positive half line $(0, \infty)$ and $f^{(l)}(t) \in L^p(0, \infty)$ $(l = 0, 1, \dots , \mu)$ holds. The norm in the space $\overline{L}^p(0, \infty)$ is equivalent to the norm

$$||f|| = \sum_{l=0}^{\mu} ||f^{(l)}||_{L^p} ,$$

and for the inverse of the operator B we have $B^{-1} = i^\mu \left(\frac{d}{dt} - 1\right)^\mu .$

By Theorem 0.1 the operator $A\colon \widetilde{L}^p(0, \infty) \to \overline{L}^p(0, \infty)$ is left invertible, right invertible or two-sided invertible if the number $\frac{1}{2\pi}[\arg C(\lambda)]_{\lambda=-\infty}^{\infty}$ is positive, negative or equal to zero, respectively.

We remark here that solutions of the form (0.5) are also of practical importance (e.g. in several problems of control theory).[1]

4. Since it is sometimes difficult to construct the inverse or the one-sided inverse of a singular operator, methods for approximate solution of singular equations are of great importance. In this book we consider the reduction method, the collocation method and some more general projection methods.

For the discrete Wiener-Hopf equation the reduction method (also called "interception method") can be described as follows: We replace the equation (0.2) by the system of $n + 1$ equations with $n + 1$ unknowns

$$\sum_{j=0}^{n} a_{k-j}\xi_j = \eta_k \qquad (k = 0, 1, \dots , n) \tag{0.6}$$

and consider the solution of this intercepted system as the approximate solution for the given system (0.2). The following theorem answers the question for the convergence of the reduction method.

Theorem 0.3. *Let the following two conditions be satisfied:*

$$1)\ A(z) \neq 0\ (|z| = 1) , \qquad 2)\ \varkappa = \operatorname{ind} A(z) = 0 .$$

Then for all sufficiently great n and for any right-hand side $\eta = \{\eta_j\}_{j=0}^{\infty} \in l^p$ $(1 \leq p < \infty)$ the system (0.6) possesses precisely one solution $\{\xi_j^{(n)}\}_{j=0}^{n}$ and

[1]) Cf. e.g. V. V. Ivanov [1], [2] (and the papers cited there).

for $n \to \infty$ the sequence

$$\xi^{(n)} = \{\xi_0^{(n)}, \xi_1^{(n)}, \dots, \xi_n^{(n)}, 0, \dots\}$$

converges in the norm of the space l^p to the solution ξ of the system (0.2).

In this case we speak of the convergence of the reduction method (0.6) in the space l^p for *all* $\eta \in l^p$.

The conditions 1) and 2) of the Theorem 0.3 are together equivalent to the invertibility of the operator A. If only the condition 1) satisfied, then A is only one-sided invertible. In this case we easily obtain an approximation method for the equation (0.2) (if (0.2) is solvable), if we take into consideration that the operator

$$B = \sum_{j=-\infty}^{\infty} a_{j+\varkappa} V^{(j)}$$

is invertible and that for the left inverse resp. the right inverse of the operator A the relations

$$A^{(-1)} = B^{-1} V^{(-\varkappa)}\ (\varkappa > 0)\,, \qquad A^{(-1)} = V^{-\varkappa} B^{-1}\ (\varkappa < 0)$$

hold.

If on the contrary the condition 1) is not satisfied, then the reduction method (0.6) no longer converges in the space $l^p (1 \leqq p < \infty)$ for all $\eta \in l^p$ for which the equation (0.1) is solvable. We can already see this for the simple system

$$\xi_k - \xi_{k+1} = \eta_k \qquad (k = 0, 1, \dots)\,, \tag{0.7}$$

whose symbol $A(z) = 1 - z^{-1}$ vanishes at the point $z = 1$. If the system (0.7) is solvable, then it has the unique solution

$$\xi_j = \sum_{k=j}^{\infty} \eta_k \qquad (j = 0, 1, \dots)$$

and the matrix A_n ot the intercepted system (0.6) here has the inverse $A_n^{-1} = \{c_{jk}\}_{j,k=0}^{n}$ with $c_{jk} = 0$ for $j > k$ and $c_{jk} = 1$ for $j \leqq k$. Thus for the unit vectors $e_n = \{0, \dots, 0, 1\} \in R^{n+1}$ we obtain the obviuos relation $\|A_n^{-1} A e_{n+1}\|_{l^p} = (1 + n)^{1/p}$. But the reduction method for the system (0.7) already converges in l^p for all $\eta \in l^{p,1}$.

Here by $l^{p,\mu} (1 \leqq p < \infty, -\infty < \mu < \infty)$ here we denote the space of all sequences $\xi = \{\xi_j\}_0^\infty$ with a finite norm

$$\|\xi\|_{p,\mu} = \left[\sum_{j=0}^{\infty} |(1 + j)^\mu \xi_j|^p\right]^{1/p}$$

and by $W^\mu (0 \leqq \mu < \infty)$ the collection of all functions of the form (0.2′) with $\sum_{j=-\infty}^{\infty} (1 + |j|)^\mu |a_j| < \infty$. Let α_j $(j = 1, \dots, r)$ be the zeros of the symbol $A(z)$

on the unit circle and the integers $m_j \geqq 0$ their multiplicities. We put

$$C(z) = A(z) \prod_{j=1}^{r} (z^{-1} - \alpha_j^{-1})^{-m_j}, \qquad D(z) = A(z) \prod_{j=1}^{r} (z - \alpha_j)^{-m_j}.$$

Then the following theorem holds.

Theorem 0.4. *Let $C(z) \in W^m$, where $m = \max(m_1, \ldots, m_r)$.*

(a) *If* ind $C(z) = 0$, *then the reduction method* (0.6) *converges in the space* l^p $(1 \leqq p < \infty)$ *for all* $\eta \in l^{p,m}$. *If* $C(z) \in W^{m+\mu}$ *and* $\eta \in l^{p,m+\mu}$ $(\mu > 0)$, *then moreover the estimate*

$$\|\xi - \xi^{(n)}\|_{l^p} = 0(n^{-\mu})$$

holds.

(b) *If* ind $D(z) = 0$, *then the reduction method* (0.6) *converges in the space* $l^{p,-m-\varepsilon(p)}$ *for any right-hand side* $\eta \in l^p$. *Here*

$$\varepsilon(p) = 0 \quad for \quad 1 < p < \infty \quad and \quad \varepsilon(p) = \varepsilon > 0 \quad for \quad p = 1.$$

If $C(z) \in W^{m+\varepsilon(p)+\mu}$ *and* $\xi \in l^{p,-m-\varepsilon(p)+\mu}$, *then moreover the estimate*

$$\|\xi - \xi^{(n)}\|_{p,-m-\varepsilon(p)} = 0(n^{-\mu})$$

holds.

Analogous results hold for the reduction and Galerkin methods for Wiener-Hopf integral equations and several paired equations, as well as for the reduction and collocation methods for singular integral equations. All these theorems follow from more general theorems on projection methods for the solution of linear operator equations resp. abstract singular equations.

5. Now some remarks about the structure of the book.

Chapter 1 contains the theory of Noether operators. This theory is stated for closed operators an Fréchet spaces.

Chapter 2, 3 and 7 contain the theory of several classes of singular equations of normal type and of the corresponding equation systems in a compact form. These parts of the book are chiefly intended as an introduction to the other chapters.

Chapters 4 to 6 and 8 to 11 form the main contents of the book. In this chapters non-normal singular integral equations, Wiener-Hopf integral equations and their discrete analogue, several paired equations with a continuous and also with a discontinuous symbol are investigated in detail. Here the different possibilities of the degeneration of the symbol at several points (zeros of integral and fractional order, of logarithmic type and others) are considered for one equation as well as for systems, of equations.

In Chapters 5, 6 and 8 the above equations, with a continuous symbol, are investigated in concrete Banach spaces. These spaces are fully described at the beginning of each of these chapters (resp. at the beginning of the corresponding seetion within a chapter). They are constructed by means of the above-mentioned method which is described in still more detail in Chapter 4.

In Chapter 9 the results of the preceding chapters are used to study of the singular equations of non-normal type in some countably normed spaces and spaces of distributions over countably normed spaces.

In Chapter 10 the most important results of the preceding chapters are generalized to the case of singular equations with piecewise continuous resp. arbitrary measurable bounded coefficients and to the case of open integration curves (at singular integral equations).

The last chapter investigates the reduction method and the collocation method for systems of discrete Wiener-Hopf equations, Wiener-Hopf integral equations and singular integral equations of normal and non-normal type. Here the general theorems on the convergence, stability and error estimate for projection methods with (in general) unbounded projections for the solution of linear equations in Banach spaces stated in the first section of Chapter 11 play an important role.

For the sake of clarity we have as a rule in the text not introduced exakt references at each theorem. These references are comipled at the end of each chapter. Hints to other results which are closely connected to the questions considered in this book can sometimes also be found here.

CHAPTER 1

NOETHER OPERATORS

The singular operators treated in this book possess the following extremely important properties in suitably chosen locally convex vector spaces: they are normally solvable and have a finite index. Today such operators are generally called Noether operators (briefly Φ-operators). For them there is a completely closed theory, at least in the case of closed operators in Banach spaces or of continuous operators in locally convex spaces.

This chapter begins with a summary of the most important concepts of functional analysis which we shall use. In Sections 1.2 and 1.3 the fundamental theorems of the theory of normally solvable and Noether operators are proved for the case of closed operators in Fréchet spaces. The last two sections deal with the method of regularization of operator equations (especially with the equivalent regularization). For applications to singular integral equations of the non-normal type the operators with an unbounded regularizor are intensively studied in Section 1.5.

In this book Fréchet spaces will appear only in Chapter 9. To understand all the other chapters it is sufficient to restrict oneself in the present chapter to the case of the Banach space.

1.1. Some Fundamental Concepts and Notations

This section contains a summary of the most important concepts and theorems of functional analysis we shall use. As references we recommend the following books:

L. V. Kantorovič and G. P. Akilov [1], A. P. Robertson and W. Robertson [1], A. Pietsch [3].

1.1.1. A linear space (vector space) X over the field K of the complex (or real) numbers is called a *linear topological space* (or a *topological vector space*) if there is defined a topology or X such that the mappings $(x, y) \to x + y$ of $X \times X$ into X and $(\lambda, x) \to \lambda x$ of $K \times X$ into X are continuous. Every (not necessarily closed) linear subset of X is called a subspace of X. A set $E \subset X$ is said to be *convex* if for arbitrary numbers $\lambda, \mu \in K$ satisfying the conditions $\lambda \geqq 0$, $\mu \geqq 0$ and $\lambda + \mu = 1$ it follows from $x \in E$ and $y \in E$ always that $\lambda x + \mu y \in E$.

A linear Hausdorff topological space X is said to be *locally convex* if there is a fundamental system $\mathfrak{U}$ of convex neighborhoods of zero in X. The quotient space (factor space) X/X_0 of a locally convex space X with respect to the subspace X_0 is also locally convex, and the sets $\varphi(U)$ ($U \in \mathfrak{U}$) form a fundamental system of convex neighborhoods of zero. Here φ is the so-called *canonical mapping* of X onto X/X_0, i.e. that mapping which, to each element $x \in X$, assigns the coset $\hat{x} = x + X_0$.

A class of locally convex spaces which is one of the most important for the application in analysis is formed by the so-called Frechet spaces. A locally convex space is called a *Fréchet space* (or F-space) if it is metrizable and complete. F-spaces are characterized by the fact that in them the topology is generated by a countably infinite system of semi-norms of the form

$$||x||_1 \leqq ||x||_2 \leqq \dots \leqq ||x||_n \leqq \dots \tag{1.1}$$

Therefore the system (1.1) is called a generating system of seminorms of the corresponding F-space. In such a space the sets $U_{p,\varepsilon} = \{x \in X: ||x||_p < \varepsilon\}$ form a fundamental system of convex neighborhoods of zero.

A *semi-norm* on a linear space X is a real valued function $||x||$ with the following properties:

1. $||x + y|| \leqq ||x|| + ||y||$.
2. $||\lambda x|| = |\lambda|\, ||x||$ for all $x, y \in X$ and $\lambda \in K$.

It is obvious that $||x|| \geqq 0$ and $||0|| = 0$. If $||x|| = 0$ implies $x = 0$, then $||x||$ is called a *norm*.

A linear space on which a norm is defined is called a *normed* or a *normable* space. A complete normed space is called a *Banach space* (or B-space). It is obvious that such a space is an F-space.

A closed subspace of an F-space is itself an F-space. The topological product and the sum of two F-spaces are also F-spaces. In both cases we obtain a generating system of semi-norms by adding the semi-norms of the corresponding spaces with equal indices. Moreover the quotient space X/X_0 of an F-space X with respect to a closed subspace $X_0 \subset X$ is an F-space. It is possible to introduce a generating system of semi-norms in X/X_0 by equations

$$||\hat{x}||_p = \inf_{x \in \hat{x}} ||x||_p \qquad (\hat{x} \in X/X_0;\ p = 1, 2, \dots)\,.$$

It is easy to see that the linear space $C^\infty[a. b]$ of all infinitely differentiable (real or complex valued) functions f on the closed interval $[a, b]$ is an F-space if we choose the system of norms

$$||f||_n = \sum_{k=0}^{n} \max_{a \leqq t \leqq b} |f^{(k)}(t)| \qquad (n = 0, 1, 2, \dots)$$

as a generating system. The same holds for the space $C^\infty(\Gamma)$. where Γ is an arbitrary closed rectifiable plane curve of the class C^∞.

The linear space $C^\infty(-\infty, \infty)$ of all infinitely differentiable functions on the real line is an F-space if we use the system of semi-norms

$$||f||_n = \sum_{k=0}^{n} \max_{-n \leqq t \leqq n} |f^{(k)}(t)| \qquad (n = 0, 1, 2, \ldots)$$

as a generating system. $C^\infty(\Gamma)$ and $C^\infty(-\infty, \infty)$ are examples of F-spaces which are not normable.

1.1.2. Let X and Y be two arbitrary locally convex spaces. We say that A is an *operator from X into Y* (or the operator A acts from X into Y) if A maps a set $D(A) \subset X$ (*domain* of A) onto a set im $A \subset Y$ (*image space* or *range* of A). In the case $X = Y$ we call A an operator on X. The set ker $A = \{x \in D(A)\colon Ax = 0\}$ is called the *kernel* of the operator A. If U is a subset of $D(A)$, by $A(U)$ we denote the collection of all elements Ax with $x \in U$.

In the following we shall consider only *linear operators,* i.e. such operators for which $D(A)$ is a linear subset of X, and moreover. the relation

$$A(\alpha x + \beta y) = \alpha Ax + \beta Ay$$

holds for all $x, y \in D(A)$ and $\alpha, \beta \in K$.

We express the statement that A is a linear operator from X into Y in the sequel by $A(X \to Y)$. Every operator $A(X \to Y)$ can be extended to an operator $\tilde{A}(X \to Y)$ with $D(\tilde{A}) = X$. Namely, we have only to put $\tilde{A}x = 0$ for all elements x of the algebraic complement of $D(A)$ in X (compare Section 1.2.3).

The operator $A(X \to Y)$ is *continuous* if for every zero neighborhood V in Y there exists a neighborhood of zero U in X with $A[U \cap D(A)] \subset V$. A is called *sequentially continuous* if for every sequence $\{x_n\} \subset D(A)$ with $x_n \to 0$ we have also $Ax_n \to 0$. The operator A is called *bounded*[1]) if it maps every bounded set $E \subset D(A)$ into a bounded set of the space Y. Here the set $E \subset X$ is called bounded if for every neighborhood of zero $U \subset X$ there is a number $\lambda > 0$ with $\lambda E \subset U$. In the case of an F-space X this is equivalent to the statement that the inequalities

$$||x||_n \leqq C_n \qquad (n = 1, 2, \ldots)$$

are fulfilled for certain constants C_n and all $x \in E$.

The operator $A(X \to Y)$ is called a *homomorphism* if for every open set $U \subset D(A)$ (with the topology induced by X in $D(A)$) $A(U)$ is open in im A (with the topology induced by Y in im A). It is obvious that a one-to-one operator $A(X \to Y)$ possesses a continuous inverse A^{-1} (of im A onto $D(A)$) if and only if A is a homomorphism.

---

[1]) Our definition of a bounded operator an topological vector spaces can befound e.g. in L. V. Kantorovič and G. P. Akilov [1], I. M. Gelfand and G. E. Šilov [2]. This concept is also used with another meaning ($A(U)$ bounded for a certain neighborhood U) of zero.

The canonical mapping of X onto a quotient space X/X_0 is both continuous and a homomorphism.

The continuity of an operator $A(X \to Y)$ implies sequential continuity and boundedness. In general, the converse is not true. Assuming that X and Y are both F-spaces. the operator $A(X \to Y)$ is continuous (or sequentially continuous) if and only if it is bounded, and hence if and only if for every natural number n there exist a natural number m_n and a constant $C_n > 0$ such that

$$||Ax||_n \leqq C_n ||x||_{m_n} \quad (x \in D(A)) .$$

For a continuous operator A from X into Y we always assume that $D(A) = X$ unless otherwise stated. We denote the set of all these continuous operators by $\mathscr{L}(X, Y)$. In the case $X = Y$ we simply write $\mathscr{L}(X)$ instead of $\mathscr{L}(X, X)$. With the operation of addition and multiplication defined in the usual way the set $\mathscr{L}(X)$ forms a non-commutative ring with unit element.

1.1.3. Let X and Y be two F-spaces. An operator A is said to be *closed* if $x_n \in D(A)$, $x_n \to x$ and $Ax_n \to y$ imply $x \in D(A)$ and $Ax = y$. If A is one-to-one, then the closedness of A implies the closedness of the inverse operator A^{-1} and conversely.

For the sequel the following two theorems are of greatest importance.

Closed Graph Theorem. *An operator $A(X \to Y)$ with $D(A) = X$ is continuous if and only if it is closed.*

Banach's Theorem on Homomorphisms. *Every continuous linear operator A from the F-space X onto the F-space Y is a homomorphism.*

From the last theorem it follows in particular that every continuous linear operator which is a one-to-one mapping of an F-space onto another F-space has a continuous inverse.

1.1.4. The linear space $X^* = \mathscr{L}(X, K)$ of all continuous linear functionals on the locally convex space X is called the *dual* (or *conjugate*) *space* of X.

On the linear space $\mathscr{L}(X, Y)$ we can introduce a locally convex topology in various ways. The most important locally convex topologies are the strong and the weak topology. A strong neighborhood of the zero functional $0 \in X^*$ is, for an arbitrary bounded set $E \subset X$ and an arbitrary number $\varepsilon > 0$, the collection of all functionals $f \in X^*$ with the property

$$\sup_{x \in E} |\langle x, f\rangle| < \varepsilon . \tag{1.2}$$

Here and in the sequel $\langle x, f\rangle = f(x)$ is the value of $f \in X^*$ in $x \in X$. The weak neighborhoods of zero in X^* are obtained if E runs in (1.2) through all finite subsets of X. Whenever we consider the dual space X^* of a locally convex space X, we assume always that X^* is equipped with one of these two norms.

Let X and Y be Banach spaces; then $\mathcal{L}(X, Y)$ is also a Banach space if the norm of the operator $A \in \mathcal{L}(X, Y)$ is defined by the relation

$$||A|| = \sup_{||x|| \leq 1} ||Ax|| \,. \tag{1.3}$$

If $X = Y$, then $\mathcal{L}(X)$ with this norm is even a Banach algebra. In particular X^* is a Banach space if X is a Banach space.

One of the fundamental theorems of the theory of locally convex spaces is the

Hahn-Banach Theorem. *Let X be a locally convex space, $X_0 \subset X$ a subspace and f_0 a continuous linear mapping into K defined on X_0. Then there exists a continuous linear functional $f \in X^*$ which coincides with f_0 on X_0.*

1.1.5. For every operator $A(X \to Y)$ with a dense domain $(\overline{D(A)} = X)$ we define *adjoint* A^* as follows. Its domain consists of all functionals $f \in Y^*$ for which $\langle Ax, f \rangle$ is a continuous linear functional in x on $D(A)$. Consequently to every $f \in D(A)$ there corresponds (by the Hahn-Banach theorem) one and only one $g \in X^*$ such that $\langle Ax, f \rangle = \langle x, g \rangle$ for all $x \in D(A)$. Now we define $A^*f = g$. The adjoint of a continuous operator is also continuous. If X and Y are Banach spaces, then A^* is a closed operator. Moreover, when A is closed we have $\overline{D(A^*)} = Y^*$ and $A^{**} = A$. If A is continuous, then $||A^*|| = ||A||$.

1.1.6. An operator $T \in L(X, Y)$ is said to be *completely continuous* or *compact* if for a neighborhood of zero U in X the image set $T(U)$ is relatively compact in Y (i.e. the closure $\overline{T(U)}$ is compact in Y). We denote the set of all compact operators by $\mathcal{K}(X, Y)$. In the case $X = Y$ the set $\mathcal{K}(X)$ forms a two-sided proper ideal in the ring $\mathcal{L}(X)$. If X is a Banach space, then the ideal $\mathcal{K}(X)$ is closed in the algebra $\mathcal{L}(X)$.

It is obvious that a compact operator mops every bounded set onto a relatively compact set. It is easy to see that in the case where X is a Banach space also the converse is true. Moreover, in the case of Banach spaces an operator $T \in \mathcal{L}(X, Y)$ is compact if and only if T^* is compact.

A linear operator is called *finite dimensional* (*of finite rank*) if its image space has a finite dimension. Every continuous finite dimensional operator is compact.

1.2. Projections and Normally Solvable Operators

In this section X and Y are arbitrary F-spaces.

1.2.1. An operator $P(X \to X)$, $D(P) = X$, is called a *projection* if $P^2 = P$. When $P \in \mathcal{L}(X)$ we call P a continuous projection. Obviously, if P is a projection the operator $Q = I - P$ (I the identity operator in X) is also a projection. This projection is called the *complementary projection* of P. In view of $\operatorname{im} P = \ker Q$ the image space of a continuous projection is always closed. For the complementary projection P and Q we obviously have the relations $PQ = QP = 0$.

In a Banach space the norm of a continuous projection $P \neq 0$ is not less than one. The norm of an orthogonal projection in a Hilbert space (i.e. a continuous projection P with $\operatorname{im} P \perp \ker P$) is always equal to unity.

1.2.2. Let M and N be subspaces of X. The *sum* $M + N$ of these subspaces is defined to be the set of all elements of the form $x + y$ with $x \in M$ and $y \in N$. Obviously, $M + N$ is a subspace of X. If in addition we have $M \cap N = \{0\}$ (0 the zero element of X), then $M + N$ is called a *direct sum* and is denoted by $M \oplus N$.

It is obvious that in X every projection P generates a direct sum $X = M \oplus N$ with

$$M = \operatorname{im} P\,, \qquad N = \ker P\,. \tag{2.1}$$

Conversely, every direct sum $X = M \oplus N$ defines a projection P with the properties (2.1). This projection is called the projection P of the space X onto the subspace M *parallel* to N.

X is called the *topological direct sum* of M and N if $X = M \oplus N$, and the projection of X onto M parallel to N is continuous (then the projection $Q = I - P$ of X onto N is continuous and M, N are closed). In the sequel, we denote the topological direct sum by $X = M \dotplus N$.

Observe that: *If the F-space X is a direct sum of its closed subspaces M and N, then X is the topological direct sum $X = M \dotplus N$.*

In fact, since M and N are closed subspaces of the F-space X, they are themselves F-spaces. Hence also the topological product $M \times N$ is an F-space, and by Banach's theorem on homomorphisms the continuous operator $(x_1, x_2) \to x_1 + x_2$ of $M \times N$ onto X has a continuous inverse. From this the continuity of the projection P follows immediately since by the definition of the topological product the mapping $(x_1, x_2) \to x_1$ of $M \times N$ onto M is also continuous.

In general, the direct sum $M \oplus N$ of two closed subspaces M and N of an F-space X (or even of a B-space) need not be again closed (compare e.g. K. Hoffman [1] Chap. 9). But we have:

The sum $M + N$ of a closed subspace M and a finite dimensional subspace N is always closed.

In fact, let φ be the canonical mapping of the space X onto the quotient spape X/M. Then $\varphi(N)$ is a finite dimensional and consequently a closed subspace of X/M. Since φ is continuous, the preimage $\varphi^{-1}(\varphi(N)) = M + N$ *closed* in X.

1.2.3. Let M be a subspace of X. Then every subspace $N \subset X$ with the property that X is the direct sum of M and N is called an *algebraic comple-*

ment of M in X. Such a complements exists for an arbitrary subspace M (see A. P. ROBERTSON and W. ROBERTSON [1]). Of course, it is not in generat unique.

Analogously, every subspace $N \subset X$ with $X = M \dotplus N$ is called a *topological complement* of M in X. Every topological complement N, if it exists, is isomorphic with the quotient space X/M (see A. P. ROBERTSON and W. ROBERTSON [1], V. 7). The dimension of the quotient space X/M is called the *codimension* of the subspace M and is denoted by codim M.

Obviously, only closed subspaces can possess topological complements. A Hilbert space has the property that in it every closed subspace has a topological complement (e.g. such a complement is the orthogonal complement). This property is also characteristic of Hilbert spaces[1]). But the following statement can be proved.

Lemma 2.1. *If the closed subspace M of the F-spaces X has a finite dimension or a finite codimension, then M has a topological complement in X.*

Proof. First of all, let $m = \dim M < \infty$ and $e_1, e_2, \ldots, e_m$ be a basis in M. By the Hahn-Banach theorem there exists a corresponding biorthogonal system of functionals $f_1, f_2, \ldots, f_m$ from X^* (i.e. $\langle e_i, f_j \rangle = \delta_{i,j}; i, j = 1, 2, \ldots, m$). We put $N = \bigcap_{i=1}^{m} \ker f_i$. Then N is a closed subspace of X, and we can easily see that $X = M \dotplus N$.

If, on the other hand, $n = \dim X/M$ is finite, and $\xi_1, \xi_2, \ldots, \xi_n$ is a basis in the quotient space X/M, we choose in the coset ξ_k ($k = 1, 2, \ldots, n$) an arbitrary element $x_k \in X$. Then for the linear subspace N spanned by the linearly independent elements $x_1, x_2, \ldots, x_n$ we have $X = M \dotplus N$.

Corollary 2.1. *Let $M \subset X$ be a closed subspace and $M^\perp = \{f \in X^* : \langle x, f \rangle = 0, x \in M\}$.*

The codimension of M is finite if and only if $M^\perp$ is finite dimensional, and in this case we have codim $M = \dim M^\perp$.

Proof. First of all, let $n = \dim M^\perp$ be finite and $f_1, f_2, \ldots, f_n$ a basis in $M^\perp$. We choose a system of elements $e_1, e_2, \ldots, e_n$ in X which is biorthogonal to the system of functionals f_i (see L. V. KANTOROVIČ and G. P. AKILOV [1], p. 131). By N we denote the subspace spanned by these elements. It is easy to see that $X = M \dotplus N$. Then as a consequence of the fact that the spaces X/M and N are isomorphic we obtain

$$\dim X/M = \dim N = \dim M^\perp .$$

1) J. LINDENSTRAUSS and L. TZAFRIRI [1] stated recently the remarkable result that in every Banach space which is not isomorphic to a Hilbert space there is a closed subspace which has no topological complement.

Conversely, if $\dim X/M$ is finite, then by Lemma 2.1 we obtain $X = M \dot{+} N$ with $\dim N = \dim X/M$. Since obviously $\dim M^{\perp} = \dim N$, from this the assertion follows immediately.

In the sequel we often use the following

Lemma 2.2. *If the F-space X be the topological direct sum of the closed subspace X_1 and the finite dimensional subspace X_0 and D is a dense subspace of X, then we have*

(1) $D_1 = D \cap X_1$ is dense in X_1, and

(2) X is representable as a topological direct sum $X = X_1 \dot{+} X_0'$ with $X_0' \subset D$.

Proof. Let $e_1, e_2, \ldots, e_m$ be a basis in X_0 and $f_1, f_2, \ldots, f_m$ a corresponding biorthogonal system of functionals from X^* for which the statements

$$\langle e_i, f_j \rangle = \delta_{ij} \quad (i, j = 1, \ldots, n), \qquad \langle x, f_i \rangle = 0 \quad (x \in X, \quad i = 1, \ldots, n)$$

hold. Then X_1 is the set of all elements from X which are annihilated by all functionals f_i $(i = 1, 2, \ldots, n)$. Now for every $i = 1, 2, \ldots, n$ we chose an element $\tilde{e}_i \in D$ in such a small neighborhood of e_i that the determinant $\det \{\langle \tilde{e}_i, f_j \rangle\} \neq 0$.

We take an arbitrary element $x \in X_1$ and show that there exists a sequence $\{\tilde{x}_k\} \subset D_1$ with $\tilde{x}_k \to x$. Once this is done, then the assertion (1) is proved.

In view of $\overline{D} = X$ there exists a sequence $\{x_k\} \subset D$ with $x_k \to x$. We make the statement

$$\tilde{x}_k = x_k + \sum_{j=1}^{m} a_{kj} \tilde{e}_j \qquad (k = 1, 2, \ldots),$$

where the numbers a_{kj} are determined by the condition $\tilde{x}_k \in D_1$. The last condition can be written in the form $\langle \tilde{x}_k, f_i \rangle = 0$ $(i = 1, 2, \ldots, m)$ or in an equivalent form as the system of equations

$$\sum_{j=1}^{m} a_{kj} \langle \tilde{e}_j, f_i \rangle + \langle x_k, f_i \rangle = 0$$

$$(i = 1, \ldots, m; \qquad k = 1, 2, \ldots).$$

Since the determinant of this system is non-zero, this system of equations has a unique solution $\{a_{k1}, a_{k2}, \ldots, a_{km}\}$ for arbitrary $k = 1, 2, \ldots$. From

$$\lim_{k \to \infty} \langle x_k, f_i \rangle = \langle x, f_i \rangle = 0 \qquad (i, 1, \ldots, m)$$

it now follows that $\lim_{k \to \infty} a_{kj} = 0$ $(j = 1, 2, \ldots, m)$; hence

$$\lim_{k \to \infty} x_k = \lim_{k \to \infty} \tilde{x}_k = x.$$

For the proof of assertion (2) we denote the subspace spanned by the linearly independent elements $\tilde{e}_1, \tilde{e}_2, \ldots, \tilde{e}_m$ by X_0'. Obviously, we have $X_0' \subset D$, and it is also easy to see that $X = X_1 \dot{+} X_0'$.

1.2.4. An operator $A(X \to Y)$ is sald to be *normally solvable* if it possesses the following property: The equation $Ax = y$ $(y \in Y)$ has at least one solution $x \in D(A)$ if (and only if)

$$\langle y, f\rangle = 0 \quad \text{for all} \quad f \in (\operatorname{im} A)^{\perp} \tag{2.2}$$

holds.

We recall that by the definition of tbe adjoint operator, $(\operatorname{im} A)^{\perp} = \ker A^*$ provided $D(A)$ is dense in X.

We now prove: *The operator A is normally solvable if and only if its image space* $\operatorname{im} A$ *is closed.*

Obviously, it is sufficient to show that the normal solvability of A follows from the closedness of $\operatorname{im} A$. Let $y_0 \in Y$ be an arbitrary fixed element which does not belong to $\operatorname{im} A$. The assertion well be proved if we can show that there exists a functional $f_0 \in (\operatorname{im} A)^{\perp}$ with $\langle y_0, f_0\rangle \neq 0$.

We put $\langle y + \lambda y_0, f_0\rangle = \lambda$ on the closed subspace $Y_0 = \operatorname{im} A + \{\lambda y_0\}$. Then f_0 is a continuous linear functional on Y_0 which by the Hahn-Banach theorem can be extended to the whole space Y. This extension is the functional sought.

Another criterion for normal solvability which will be used frequently in the sequel follows easily from the statement we have just proved.

1.2.5. Let $A(X \to Y)$ be a linear operator. By the Sections 1.2.2 and 1.2.3 always there is a projection P_A of X onto $\ker A$, and X is representable as a direct sum $X = X_1 \oplus \ker A$ with $X_1 = \ker P_A$.

We put $D_1 = D(A) \cap X_1$, and by A_1 we denote the operator with domain $D(A_1) = D_1$ which coincides with A on D_1. A_1 is called the *restriction of the operator* A to the subspace D_1 and is denoted by $A_1 = A|D_1$ or $A_1 = A_{D_1}$.

Obviously, $\ker A_1 = \{0\}$ and $\operatorname{im} A_1 = \operatorname{im} A$. Hence for A_1 there exists the inverse mapping A_1^{-1} of $\operatorname{im} A$ onto D_1. For the operator A_1^{-1} we have

$$AA_1^{-1}y = y \qquad (y \in \operatorname{im} A)\,, \tag{2.3}$$

$$A_1^{-1}Ax = x - P_A x \quad (x \in D(A))\,. \tag{2.4}$$

In the sequel we assume that there exists a *continuous projection* P_A of X onto $\ker A$, i.e. $\ker A$ has a topological complement in X. Under these assumptions we obtain the following

Lemma 2.3. *The closed operator A is normally solvable if and only if the operator A_1^{-1} is continuous.*

Proof. Since, by the continuity of the projection P_A, the subspace X_1 is closed, A_1 and hence also A_1^{-1} is a closed operator. If now A is normally solvable, then $\operatorname{im} A$ is closed and thus an F-space. By the closed graph theorem A_1^{-1} is continuous.

Conversely, let A_1^{-1} be continuous and $\{y_n\} \subset \operatorname{im} A$ a sequence which converges to the element y in Y. By the completeness of the space X we obtain $x_n = A_1^{-1} y_n \to x \in X$, and by the closedness of the operator A_1 we have $y = A_1 x \in \operatorname{im} A$. Hence im A is closed, i.e. A is normally solvable.

Remark. By Lemma 2.1 there is a continuous projection P_A of X onto ker A provided ker A is finite dimensional. In this case P_A is more-over a finite dimensional and thus a completely continuous operator.

1.2.6. Banach-Hausdorff Theorem. *Let X and Y be arbitrary F-spaces and $A(X \to Y)$ a closed operator with a dense domain $D(A)$. The following properties are equivalent:*

(a) im A *is closed.*
(b) A *is normally solvable, i.e.* $\operatorname{im} A = \{y \in Y : \langle y, f \rangle = 0$ *for all* $f \in \ker A^*\}$.
(c) A^* *is normally solvable, i.e.* $\operatorname{im} A^* = \{f \in X^* : \langle x, f \rangle = 0$ *for all* $x \in \ker A\}$.
(d) im A^* *is closed in* X^*.

The equivalence of the statements (a) and (b) has been proved in section 1.2.4. The assertion (c) $\Rightarrow$ (d) is obvious. Also it is easily seen that the condition (d) follows from (a).

In fact, let A be closed and $f_n = A^* g_n$ ($g_n \in Y^*$) be a sequence which converges (weakly) to a functional $f \in X^*$. Then for all $x \in D(A)$ we have

$$\langle Ax, g_n \rangle = \langle x, f_n \rangle \to \langle x, f \rangle .$$

Thus the restrictions of the functionals g_n to the F-space im A are weakly convergent to a certain limit g_0 which consequently is a continuous linear functional on im A. By the Hahn-Banach theorem there exists an extension g of g_0 to the whole space Y. Obviously, we have $\langle Ax, g \rangle = \langle Ax, g_0 \rangle = \langle x, f \rangle$, i.e. $g \in D(A^*)$ and $A^* g = f$. Thus the (weak and hence also the strong) closedness of im A^* in X^* follows.

The implications (a) $\Rightarrow$ (c) and (d) $\Rightarrow$ (a) follows from the next two lemmas.

Lemma 2.4. *Let X and Y be arbitrary F-spaces and $A(X \to Y)$ a closed operator. Then* im A *is closed if and only if the following holds: For an arbitrary sequence $\{y_n\} \subset \operatorname{im} A$, $y_n \to 0$ there exists a sequence $\{x_n\} \subset D(A)$ with $x_n \to 0$ and $Ax_n = y_n$.*

Proof. Let $\hat{X}$ be the quotient space $X/\ker A$ and $\hat{A}$ the linear operator which is defined on $D(\hat{A}) = D(A)/\ker A$ by the equation $\hat{A}\hat{x} = Ax \, (x \in D(A))$. Obviously, we have $\operatorname{im} \hat{A} = \operatorname{im} A$ and $\ker \hat{A} = \{0\}$. It is easily seen that $\hat{A}$ is a closed operator. By Lemma 2.3 im $\hat{A}$ is closed if and only if $\hat{A}^{-1}$ is continuous. From this the assertion of the lemma follows immediately.

Now we prove that (a) implies (c) i.e. $\operatorname{im} A^* = (\ker A)^\perp$. The inclusion $\operatorname{im} A^* \subset (\ker A)^\perp$ follows immediately from the definition of the adjoint operator. We show the converse inclusion. With each functional $g \in (\ker A)^\perp$ we associate a linear continuous functional f_0 defined on the set $\operatorname{im} A \subset Y$

by putting $f_0(y) = \langle x, g\rangle$ for arbitrary $y = Ax \in \operatorname{im} A$. It is obvious that f_0 is a unique function. Its continuity follows from Lemma 2.4. By the Hahn-Banach theorem there exists an extension $f \in Y^*$ of the functional f_0, and $\langle Ax, f\rangle = \langle Ax, f_0\rangle = \langle x, g\rangle$ for arbitrary $x \in D(A)$, i.e. $A^*f = g$. Hence $(\ker A)^\perp \subset \operatorname{im} A^*$.

At the proof of the assertion (d) $\Rightarrow$ (a) we restrict ourselves to the case of Banach spaces.

Lemma 2.5. *Let X and Y be Banach spaces and $A(X \to Y)$ a closed operator with a dense domain $D(A)$. If the set* $\operatorname{im} A^*$ *is closed in X^* then* $\operatorname{im} A$ *is closed.*

Proof. We put $Y_1 = \overline{\operatorname{im} A}$. It is sufficient to show that the image $A(K)$ of the unit ball $K = \{x \in D(A): \|x\| \leqq 1\}$ is dense in a ball of the form $S_n = \left\{y \in Y_1: \|y\| \leqq \frac{1}{n}\right\}$ $(n = 1, 2, 3, \ldots)$. If this condition is satisfied, then the relation $\operatorname{im} A = Y_1 = \overline{\operatorname{im} A}$ follows from the closedness of the operator A (see L. V. KANTOROVIČ and G. P. AKILOV [1], Lemma (1.XII)).

Assume that in no ball S_n $(n = 1, 2, 3, \ldots)$ the image $A(K)$ is dense. Then there exists a sequence of elements $y_n \in S_n$ with $y_n \notin \overline{A(K)}$. Since the set $\overline{A(K)}$ is closed and convex and $y \in \overline{A(K)}$ implies $-y \in \overline{A(K)}$, it is possible to find functionals $s_n \in Y_1^*$ such that

$$f_n(y_n) > 1\,, \qquad |f_n(Ax)| \leqq 1 \qquad (x \in K\,;\quad n = 1, 2, \ldots)$$

(see L. V. KANTOROVIČ and G. P. AKILOV [1], Satz 5 (2.XI)). By regarding A as an operator A_1 which maps $D(A)$ in to Y_1 we obtain from the last inequalities

$$\|A_1^* f_n\| < \|f_n\|\ \|y_n\| \qquad (n = 1, 2, \ldots)\,.$$

Since obviously $\ker A_1^* = \{0\}$, we obtain from Lemma 2.4 that A_1^* is continuously invertible. But this contradicts the preceding estimate because $\lim_{n\to\infty} y_n = 0$.

The reader can find a proof of the Lemma 2.5 for arbitrary F-spaces in the book of H. H. SCHAEFER [2] (Chapter IV, § 7).

1.3. Noether Operators

In this section by X, Y, Z we denote arbitrary F-spaces.

1.3.1. Let $A(X \to Y)$ be a linear operator. The quotient space $\operatorname{coker} A = Y/\overline{\operatorname{im} A}$ is called the *cokernel* of the operator A. The dimensions

$$\alpha(A) = \dim \ker A\,, \qquad \beta(A) = \dim \operatorname{coker} A$$

are called the *nullity* and the *deficiency* of the operator A, respectively.

By Corollary 2.1 we have: The deficiency $\beta(A)$ is finite if and only if the subspace $(\operatorname{im} A)^{\perp} \subset Y^*$ (called the *defect space* of the operator A) is finite dimensional, and in this case both dimensions coincide.

In the case of a dense domain $D(A)$ this means:

The numbers $\beta(A)$ and $\alpha(A^*)$ are both finite or both infinite, and when they are finite, we have

$$\beta(A) = \alpha(A^*) . \tag{3.1}$$

The ordered pair of numbers $(\alpha(A), \beta(A))$ is called the *d-characteristic* of the operator A. If at least one of the numbers $\alpha(A)$ or $\beta(A)$ is finite, then the difference $\operatorname{Ind} A = \alpha(A) - \beta(A)$ is called the *index* of the operator A.

The operator A is said to have a *finite d-characteristic* or a *finite index* if both of the numbers $\alpha(A)$ and $\beta(A)$ are finite.

A closed normally solvable operator A is called a *Noether operator* or *Φ-operator*[1]) if its d-characteristic is finite. It is called a *semi-Noether operator* if at least one of the numbers $\alpha(A)$ or $\beta(A)$ is finite. In the class of semi-Noether operators we shall in the sequel distinguish between *Φ_+-operators* $(\alpha(A) < \infty)$ and *Φ_--operators* $(\beta(A) < \infty)$.

The set of all Φ-operators (resp. $\Phi_\pm$-operators) $A(X \to Y)$ is denoted by $\Phi(X, Y)$ (resp. $\Phi_\pm(X, Y)$).

The class $\Phi(X) = \Phi(X, X)$ contains e.g. the operators of the form $I + T$ (T compact) for which moreover $\operatorname{Ind}(I + T) = 0$ holds (see e.g. L. V. Kantorovič and G. P. Akilov [1], D. Przeworska-Rolewicz and S. Rolewicz [1]).

1.3.2. For the further considerations the following criterion for Φ_+-operators is useful.

Lemma 3.1. *Let X and Y be two F-spaces with the generating system of semi-norms* (1.1) *resp.* $\{\|y\|'_n\}_1^\infty$.

The closed operator $A(X \to Y)$ is a Φ_+-operator if and only if there exist compact operators $T_j \in \mathcal{K}(X)$ $(j = 1, 2, \dots, l)$ with the following properties: For every n there exist a number $m = m_n$ $(n, m = 1, 2, \dots)$ and a constant $C_n > 0$ such that the following a priori estimate holds:

$$\|x\|_n \leq C_n(\|Ax\|'_m + \sum_{j=1}^{l} \|T_j x\|_m) . \qquad x \in D(A) . \tag{3.2}$$

If A is a Φ_+-operator, then (3.2) *is satisfied with $l = 1$.*

Proof. Necessity. Let A be a Φ_+-operator. By P_A we denote the continuous projection of X onto $\ker A$ introduced in 1.2.5 and by A_1 the restric-

[1]) The terms "Fredholm operator" or "F-operator" are also usual. We use in this book the terminology of the Φ- and $\Phi_\pm$-operators introduced by I. C. Gohberg and M. G. Krein [1]. Such operators with a non-zero index first appeared in a paper of F. Noether [1] in the investigation of singular integral operators.

tion of A to $D_1 = D(A) \cap \operatorname{im}(I - P_A)$. The operator $T_1 = P_A$ is compact, and A_1^{-1} is continuous by Lemma 2.3. Thus for every n there exist a number m and a constant $C'_n > 0$ such that

$$\|x_1\|_n \leqq C'_n \|A_1 x_1\|'_m$$

for all $x_1 \in D_1$. Here we can assume $m \geqq n$.

Every $x \in D(A)$ is representable in the form $x = x_1 + x_2$ with $x_1 \in D_1$ and $x_2 \in \ker A$. Since obviously $Ax = A_1 x_1$ and $T_1 x = x_2$, we obtain for all $x \in D(A)$ the estimate

$$\|x\|_n \leqq \|x_1\|_n + \|x_2\|_m \leqq C_n(\|Ax\|'_m + \|T_1 x\|_m)$$

with $C_n = \max(1, C'_n)$.

Sufficiency. Suppose the estimate (3.2) is satisfied for certain compact operators T_j. First we show that the closed subspace $X_0 = \ker A$ is finite dimensional. By (3.2) we have for all $x \in X_0$

$$\|x\|_n \leqq C_n \sum_{j=1}^{k} \|T_j x\|_m . \tag{3.3}$$

Because of the compactness of the operators T_j there exists a neighborhood of zero U in X such that all image sets $T_j(U)$ are relatively compact. It then follows from (3.3) that $U_0 = X_0 \cap U$ is a relatively compact neighborhood of zero in the F-space X_0. By a known criterion of Kolmogorov (see L. V. Kantorovič and G. P. Akilov [1]) it is therefore necessary that X_0 is finite dimensional.

It remains to establish the closedness of the image set $\operatorname{im} A$. Suppose $y_n = Ax_n \to y$. We have to show that $y \in \operatorname{im} A$. Obviously, we can choose $x_n \in D_1$. Let $U = \{x \in X : \|x\|_p < \varepsilon\}$ be the above-mentioned neighborhood of zero. First we show that the sequence of norms $\|x_n\|_p$ $(n = 1, 2, \dots)$ is bounded. Assume the contrary. By taking a subsequence, if necessary, we can obtain $\|x_n\|_p \to \infty$ for $n \to \infty$. Now we put

$$\tilde{x}_n = \frac{r}{\|x_n\|_p} x_n , \qquad \tilde{y}_n = \frac{r}{\|x_n\|_p} y_n ,$$

where r is a fixed number with $0 < r < \varepsilon$. Obviously, we have $\tilde{y}_n = A\tilde{x}_n$, $\tilde{y}_n \to 0$, $\|\tilde{x}\|_p = r$, and therefore $\tilde{x}_n \in U$. Hence there exists a subsequence $\{\tilde{x}_{n_k}\}$ such that the sequences $\{T_j \tilde{x}_{n_k}\}$ $(j = 1, \dots, l)$ converge. From (3.2) it follows now that $\{\tilde{x}_{n_k}\}$ is a fundamental sequence which converges to an element $x \in \operatorname{im}(I - P_A)$. Thus we have

$$\tilde{x}_{n_k} \to x , \qquad A\tilde{x}_{n_k} \to 0$$

from which $x \in D(A)$ and $Ax = 0$ follow because of the closedness of A. But since $D_1 \cap \ker A = \{0\}$, we obtain $x = 0$. On the other hand, we find $\|x\|_p = \lim_k \|\tilde{x}_{n_k}\|_p = r$. This contradiction shows that the sequence $\{\|x_n\|_p\}_{n=1}^{\infty}$ is bounded. Therefore we can assume that $x_n \in U$ for $n \geqq n_0$.

By applying the above arguments to the sequence $\{x_n\}$ instead of $\{\tilde{x}_n\}$ we obtain the existence of an element $x \in D(A)$ with $Ax = y$. This proves the lemma.

Remark. For the closed operator $A(X \to Y)$ to be a Φ_+-operator it is sufficient that the condition (3.2) hold for certain compact operators $T_j \in \mathcal{K}(X, X_j)$ (X_j, $j = 1, \ldots, l$, also F-spaces).

This follows immediately from the second part of the proof of Lemma 3.1.

Theorem 3.1. *Let $A(X \to Y)$ be a closed operator and $B \in \mathcal{L}(Y, Z)$. Then $BA \in \Phi_+(X, Z)$ implies $A \in \Phi_+(X, Y)$.*

Proof. Let $BA \in \Phi_+(X, Z)$. By Lemma 3.1 there is an operator $T \in \mathcal{K}(X)$ such that

$$||x||_n \leqq C_n(||BAx||''_m + ||Tx||_m)\,, \qquad x \in D(A)$$

holds for all $n = 1, 2, \ldots$ and certain numbers $m = m_n$ and $C_n > 0$. Here $\{||z||''_n\}_1^\infty$ denotes a generating system of semi-norms in Z. By the continuity of the operator B for every m there exist a number $p \geqq m$ and a constant $C'_m > 0$ with

$$||By||''_m \leqq C'_m ||y||'_p\,, \qquad y \in Y\,.$$

From this together with (3.4) we obtain the estimate

$$||x||_n \leqq C''_n(||Ax||'_p + ||Tx||_p)\,, \qquad x \in D(A)\,,$$

where C''_n is a constant depending only on n. By Lemma 3.1 we therefore have $A \in \Phi_+(X. Y)$.

Remark. We can see from the preceding proof that the following sharpening of Theorem 3.1 holds: If the operator BA has a closed extension $C \in \Phi_+(X, Z)$ of, then $A \in \Phi_+(X, Y)$.

Corollary 3.1. *$A \in \Phi_+(X, Y)$ and $T \in \mathcal{K}(X, Y)$ imply $A + T \in \Phi_+(X, Y)$.*

Proof. By Lemma 3.1 the estimate (3.2) holds for $l = 1$. Since $||Ax||'_m \leqq$ $\leqq ||(A + T)\,x||'_m + ||Tx||'_m$, we obtain

$$||x||_n \leqq C'_n[||(A + T)\,x||'_m + ||Tx||'_m + ||T_1x||_m]\,,$$

and the assertion follows from the remark to Lemma 3.1.

Theorem 3.2. *Let $A(X \to Y)$ be a closed operator and $B \in \mathcal{L}(X, Z)$ a continuous operator with a finite dimensional kernel. Then $BA \in \Phi(X, Z)$ implies $A \in \Phi(X, Y)$.*

Proof. By Theorem 3.1 we have $A \in \Phi_+(X, Y)$. Contrary to the assertion we assume $\beta(A) = +\infty$.

We decompose the finite dimensional subspace ker B in to a topological direct sum ker $B = N_1 \dotplus N_2$ with $N_1 = \ker B \cap \operatorname{im} A$ and put $Y_1 = \operatorname{im} A \dotplus$ $\dotplus N_2$. Since the subspace N_2 is finite dimensional, obviously we have $\dim Y_1^\perp = \infty$.

Hence for any natural number m we can find linearly independent functionals $f_1, f_2, \dots, f_m$ in $Y_1^{\perp}$. Let $y_1, y_2, \dots, y_m$ be a corresponding biorthogonal system of elements in Y. By Y_2 we denote the subspace spanned by these (linearly independent) elements. Obviously, $Y_1 \cap Y_2 = \{0\}$.

We consider the direct sum $M = \operatorname{im} A \oplus N_2 \oplus Y_2$. It is easy to see that $\operatorname{im} BA \cap B(Y_2) = \{0\}$. Since by hypothesis $\operatorname{im} BA$ is closed, $B(M) = \operatorname{im} BA \dotplus B(Y_2)$ is a closed subspace of Y (compare 1.2.2), and

$$\dim (X/B(M)) = \beta(BA) - m$$

holds. In view of the inclusion $B(M) \subset \operatorname{im} B$ we obtain

$$\dim (X/B(M)) \geqq \dim (X/\overline{\operatorname{im} B}) = \beta(B) ,$$

i.e. $\beta(BA) \geqq \beta(B) + m$. Since m is arbitrary, this implies $\beta(BA) = \infty$ which contradicts the hypothesis of the theorem.

Hence $\beta(A)$ is finite, and thus $A \in \Phi(X, Y)$.

Theorem 3.3. *Let $A \in \mathcal{L}(X, Y)$ be a continuous operator and $B(Y \to Z)$ a closed operator. If* $\operatorname{im} A \subset D(B)$, *then* $BA \in \Phi_-(X, Z)$ *implies* $B \in \Phi_-(X, Z)$.

Proof. We put $BA = C$. Because of the assumption of the theorem obviously we have

$$\operatorname{im} C \subset \operatorname{im} B . \tag{3.5}$$

From the last relation we immediately obtain $\beta(B) \leqq \beta(C) < \infty$.

Futhermore it follows from (3.5) and the finiteness of the codimension of $\operatorname{im} C$ that

$$\operatorname{im} B = \operatorname{im} C \oplus N$$

holds for a certain finite dimensional subspace N. Since $\operatorname{im} C$ is a closed subspace, we conclude from Section 1.2.2 that $\operatorname{im} B$ is closed. Hence B is normally solvable, i.e. $B \in \Phi_-(Y, Z)$.

Theorem 3.4. *For the closed operator $A(X \to Y)$ to be a Φ-operator it is necessary and sufficient that there exist two continuous operators $B_1, B_2 \in \mathcal{L}(Y, X)$ such that* $\operatorname{im} B_2 \subset D(A)$ *and*

$$B_1A = I - T_1 , \qquad AB_2 = I - T_2 , \tag{3.6}$$

where T_1 is a compact operator on X, and T_2 is a compact operator on Y.

B_1 and B_2 can be chosen such that $B_1 = B_2$ holds and T_1, T_2 are continuous projections of finite rank.

Proof. Sufficiency. If the operators B_1 and B_2 exist, then $A \in \Phi(X, Y)$ follows from Theorem 3.1 and Theorem 3.3.

Necessity. Conversely, let $A \in \Phi(X, Y)$. By P_A we denote a continuous projection of the space X onto the finite dimensional subspace $\ker A$ and by A_1 the operator introduced in Section 1.2.5. By Lemma 2.3, A_1^{-1} is a continuous operator defined on the closed subspace $\operatorname{im} A$. Since $\beta(A)$ is finite, im A has a topological complement Y_1 in Y (see Lemma 2.1). By K we denote the continuous projection of Y onto Y_1 parallel to im A. From dim $Y_1 = \beta(A)$ we see that K is a projection of finite rank.

We put $B = A_1^{-1}(I - K)$. Obviously, $B \in \mathcal{L}(Y, X)$ and im $B \subset D(A)$. The relations (2.3) and (2.4) then imply the equations (3.6) with $T_1 = P_A$, $T_2 = K$ and $B_1 = B_2 = B$.

Theorem 3.5. *Let $A \in \Phi(X, Y)$ be a Φ-operator with a dense domain. Then A^* is normally solvable, i.e.*

$$\text{im}\, A^* = \{f \in X^* \colon \langle x, f\rangle = 0 \quad \textit{for all} \quad x \in \ker A\}\,.$$

Moreover A^* has a finite index, and we have

$$\alpha(A^*) = \beta(A)\,, \qquad \beta(A^*) = \alpha(A)\,.$$

Proof. Let the functional $f \in X^*$ satisfy the condition

$$\langle x, f\rangle = 0 \quad \text{for all} \quad x \in \ker A\,. \tag{3.7}$$

We have to show that we can find a functional $g \in Y^*$ such that $A^*g = f$.

Let P_A be a continuous projection of X onto ker A and Q a continuous projection of Y onto the closed subspace im A (compare Lemma 2.1). Then $A_1^{-1}Q \in \mathcal{L}(Y, X)$. We put $g = (A_1^{-1}Q)^* f$. By condition (3.7) for arbitrary $x \in D(A)$, we obtain

$$\langle Ax, g\rangle = \langle A_1^{-1}QA(I - P_A)\, x, f\rangle = \langle x, f\rangle + \langle P_A x, f\rangle = \langle x, f\rangle\,,$$
$$\text{i.e. } A^*g = f\,.$$

It as follows that the quotient space $X^*/\text{im}\, A^*$ is isomorphic to ker A, i.e. $\beta(A^*) = \alpha(A)$. Since moreover the relation (3.1) holds, the theorem follows.

Lemma 3.2. *Let $A(X \to Y)$ be a closed normally solvable operator and X_0 a closed subspace of X. For $A(X_0)$ to be closed it is sufficient, and in the case of a continuous operator A also necessary, that the set $X_0 +$ ker A be closed.*

Proof. If A is continuous and if $A(X_0)$ is a closed set, then the complete pre-image $A^{-1}[A(X_0)] = X_0 + \ker A$ is closed.

Conversely, let $X_0 + \ker A$ be a closed set. By φ we denote the canonical mapping of X onto the F-space $\hat{X} = X/\ker A$ (quotient space of the F-space X with respect to the closed subspace ker A). Since φ is a homomorphism, the set

$$\varphi(X - [X_0 + \ker A]) = \hat{X} - \varphi(X_0)$$

is open. Hence $\varphi(X_0)$ is closed in $\hat{X}$.

The closed operator A in natural way generates a closed operator $\hat{A}(\hat{X} \to Y)$ with ker $\hat{A} = \{0\}$ and im $\hat{A}$ = im A defined by the relation

$$\hat{A}\varphi = A\,.$$

Since by hypothesis im A is closed and thus im A is an F-space, the operator $\hat{A}^{-1}$ is continuous by the closed graph theorem. Thus

$$A(X_0) = \hat{A}\varphi(X_0)$$

as a complete pre-image of the closed set $\varphi(X_0) \subset \hat{X}$ with respect to the continuous mapping $\hat{A}^{-1}$, is a closed set. This proves the lemma.

The following corollary is an immediate consequence of Lemma 3.2 and of the remarks from Section 1.2.2.

Corollary 3.2. *Let $A(X \to Y)$ be a closed normally solvable operator, $B \in \Phi_+(Y, Z)$ and $C \in \Phi_-(Z, X)$. Then the operators $BA(X \to Z)$ and $AC(Z \to Y)$ are normally solvable.*

Theorem 3.6. *$A \in \Phi(X, Y)$, $B \in \Phi(Y, Z)$ and $\overline{D(B)} = Y$ imply $BA \in \Phi(X, Z)$ and*

$$\text{Ind } BA = \text{Ind } A + \text{Ind } B . \tag{3.8}$$

Proof. By Corollary 3.2 the operator BA is normally solvable. Hence im BA is closed.

Let n_1 be the dimension of the finite dimensional subspace $Y_1 = \text{im } A \cap \cap \ker B$. Then obviously

$$\alpha(BA) = \alpha(A) + n_1 , \tag{3.9}$$

and the finite dimensional subspace ker B can be split up into the topological direct sum of the subspace Y_1 and a certain subspace Y_2 with $\dim Y_2 = = \alpha(B) - n_1$.

By Lemma 2.2 the space Y is representable in the form $Y = \text{im } A \dotplus Y_2 \dotplus \dotplus Y_3$, where $Y_3 \subset D(B)$ is a finite dimensional subspace. We put $n_3 = \dim Y_3$. Then by the isomorphism of the space $Y/\text{im } A$ with the space $Y_2 \dotplus Y_3$ we obtain the equation $\beta(A) = \dim Y_2 + \dim Y_3$ or

$$\alpha(B) - \beta(A) = n_1 - n_3 . \tag{3.10}$$

Now we use the representation

$$D(B) = D(B) \cap (\text{im } A \dotplus Y_2 \dotplus Y_3) = D_1 \oplus Y_2 \oplus Y_3 ,$$

where $D_1 = D(B) \cap \text{im } A$. By applying the operator B and because of $B(D_1) = \text{im } BA$ we obtain

$$\text{im } B = \text{im } BA \dotplus B(Y_3) . \tag{3.11}$$

Hence

$$\beta(BA) = \beta(B) + \dim B(Y_3) = \beta(B) + n_3 . \tag{3.12}$$

From the equations (3.9), (3.10) and (3.12) we now obtain

$$\begin{aligned} \text{Ind } BA &= \alpha(BA) - \beta(BA) = \alpha(A) + n_1 - \beta(B) - n_3 = \\ &= \alpha(A) - \beta(A) + \alpha(B) - \beta(B) = \text{Ind } A + \text{Ind } B . \end{aligned}$$

It remains to show that the product BA is a closed operator. For this reason we consider the restriction $\bar{B}$ of the operator B to the subspace D_1:

$$\bar{B}y = By\,, \qquad D(\bar{B}) = D_1\,.$$

It is easy to see that $\bar{B}$ is a closed operator. From the obvious relation im $\bar{B}$ = im BA the normal solvability of the operator $\bar{B}$ follows immediately.

Since ker $\bar{B}$ = ker $B \cap$ im A, there exists a finite dimensional subspace X_1 with

$$\ker A \subset X_1 \subset (DA)\,, \qquad A(X_1) = \ker \bar{B}\,.$$

Let P be a continuous projection of X onto X_1 and $Q = I - P$. Then

$$\operatorname{im} A = \operatorname{im} AQ \dotplus \ker \bar{B}\,.$$

Now we take an arbitrary sequence $\{x_n\} \subset D(BA)$ such that

$$x_n \to x\,, \qquad BAx_n \to z\,.$$

Let $\bar{B}_1$ be the restriction of the operator $\bar{B}$ to the subspace im AQ. In view of the continuity of the operator $\bar{B}_1^{-1}$ we then obtain

$$\bar{B}_1^{-1}(BAx_n) = AQx_n \to \bar{B}_1^{-1}z\,.$$

Since the operator A is closed we have

$$Qx \in D(A)\,, \qquad AQx = \bar{B}_1^{-1}z\,.$$

Hence

$$x \in D(A)\,, \qquad Ax = AQx + APx \in D(B)$$

and consequently

$$x \in D(BA)\,, \qquad BAx = z\,.$$

This proves that the operator BA is closed the assertion of the theorem follows.

From the proof of the Theorem 3.6 we immediately obtain the

Corollary 3.3. *$A \in \Phi_\pm(X, Y)$, $B \in \Phi_\pm(Y, Z)$ and $\overline{D(B)} = Y$ imply $BA \in \Phi_\pm(X, Z)$.*[1])

The following corollary completes the statements of Theorems 3.1.—3.3.

Corollary 3.4. *Let $A \in \mathcal{L}(X, Y)$ and $B \in \mathcal{L}(Y, Z)$. Then we have:*

(a) *$BA \in \Phi_+(X, Z)$ and $A \in \Phi(X, Y)$ imply $B \in \Phi_+(Y, Z)$.*

(b) *$BA \in \Phi_-(X, Z)$ and $B \in \Phi(Y, Z)$ imply $A \in \Phi_-(X, Y)$.*

---

[1]) Here the operator BA is not necessarily closed.

Proof. We prove assertion (a). By Theorem 3.2 and Theorem 3.4 there exists a continuous operator $R \in \Phi(Y, X)$ such that

$$RA = I - T_1, \qquad AR = I - T_2$$

where T_1 and T_2 are finite dimensional operators. By Corollary 3.3 we then have

$$BAR = B - BT_2 \in \Phi_+(Y, Z).$$

Since BT_2 is a continuous finite dimensional operator, in view of the remarks from Section 1.2.2, we conclude from the last formula that $B \in \Phi_+(Y, Z)$.

Assertion (b) is proved analogously.

From Theorem 3.2, Theorem 3.3 and Corollary 3.4 we obtain the following counterpart to Theorem 3.6.

Corollary 3.5. *Let $A \in \mathcal{L}(X, Y)$, $B \in \mathcal{L}(Y, Z)$ and $BA \in \Phi(X, Z)$. Then the operators A and B are both together Φ-operators or not.*

Theorem 3.7. *Let $A \in \Phi(X, Y)$ and $T \in \mathcal{K}(X, Y)$. Then $A + T \in \Phi(X, Y)$ and* Ind $(A + T) =$ Ind A.

Proof. By Theorem 3.4 there exists an operator $B \in \mathcal{L}(Y, X)$ with im $B \subset D(A)$ such that

$$BA = I - T_1, \qquad AB = I - T_2 \tag{3.13}$$

holds. Here T_1 and T_2 are compact operators in X and Y, respectively.

From (3.13) we obtain the equations

$$B(A + T) = I - T_3, \qquad (A + T)B = I - T_4,$$

where $T_3 = T_1 - BT$ and $T_4 = T_2 - TB$ are compact operators on X and Y, respectively. Moreover, since $A + T$ is a closed operator with domain $D(A + T) = D(A)$, Theorem 3.4 yields $A + T \in \Phi(X, Y)$.

We shall now prove the index equation. Let

$$||x||_1 \leqq ||x||_2 \leqq \dots \leqq ||x||_n \leqq \dots$$

and

$$||y||_1' \leqq ||y||_2' \leqq \dots \leqq ||y||_n' \leqq \dots$$

be generating system of semi-norms in X and Y, respectively. In the domain $D(A)$ we define a system of semi-norms by the formulae

$$|x|_n = ||x||_n + ||Ax||_n' \qquad (n = 1, 2, \dots). \tag{3.14}$$

The locally convex topology generated by the system (3.14) transforms $D(A)$ into a metrizable space (see L. V. Kantorovič and G. P. Akilov [1], Satz 2 (2.XI)) which we denote by D. Since A is closed it follows immediately that D is complete, and thus D is an F-space.

By A' we denote the operator of D into Y defined by equations

$$D(A') = D, \qquad A'x = Ax. \tag{3.15}$$

Analogously, let B' be the operator B regarded as a mapping of Y into D. From the estimates

$$||A'x||'_n = ||Ax||'_n \leqq |x|_n \qquad (n = 1, 2, \ldots)$$

and

$$|B'y|_n = ||By||_n + ||ABy||'_n \geqq C_n ||y||'_{m_n} ,$$

where C_n is a constant (dependent on n) and $m_n \geqq n$ a natural number, we conclude that the operators A' and B' are continuous. Since moreover the image spaces and the kernels of the operators A and A' coincide, A' is a Φ-operator, and we have

$$\operatorname{Ind} A = \operatorname{Ind} A' . \tag{3.16}$$

In view of relation

$$A'B' = I - T_2 \in \Phi(Y)$$

we obtain from Theorem 3.2 that B' is also a Φ-operator. Theorem 3.6 and formula (3.16) now imply

$$\operatorname{Ind} A = \operatorname{Ind} (I - T_2) - \operatorname{Ind} B' = - \operatorname{Ind} B' . \tag{3.17}$$

If we repeat the argument starting with definition (3.15) for the operator $A + T$ instead of A, we obtain in an analogous way equation

$$\operatorname{Ind} (A + T) = - \operatorname{Ind} B' . \tag{3.18}$$

The assertion now follows from (3.17) and (3.18).

The following theorem holds for Banach spaces. It implies immediately that the set of the continuous Φ-operators $A(X \to Y)$ is an open set in the space $\mathscr{L}(X, Y)$ normed by (1.3).

Theorem 3.8. *Let X and Y be Banach spaces. For every operator $A \in \Phi(X, Y)$ there exists a number $\varrho > 0$ such that $C \in \mathscr{L}(X, Y)$ and $||C|| < \varrho$ imply $A + C \in \Phi(X, Y)$ and*

$$\operatorname{Ind} (A + C) = \operatorname{Ind} A . \tag{3.19}$$

Proof. By Theorem 3.4 there exists an operator $B \in \mathscr{L}(Y, X)$ with $\operatorname{im} B \subset D(A)$ such that B satisfies the relations (3.13). We show that the number $\varrho = ||B||^{-1}$ has the properties specified in Theorem 3.8.

From $||C|| < \varrho$ we obtain first that

$$||BC|| \leqq ||B|| \, ||C|| < 1 .$$

Thus the continuous inverse operator $(I + BC)^{-1} \in \mathscr{L}(X)$ exists. Hence

$$B(A + C) = I - T_1 + BC = (I + BC)(I - T) ,$$

where $T = (I + BC)^{-1} T_1$ is compact on the space X. Consequently, the operator $B_1 = (I + BC)^{-1} B$ satisfies the condition

$$B_1(A + C) = I - T . \tag{3.20}$$

In an analogous way we can show that for the operator $B_2 = B(I + CB)^{-1}$ and a certain operator $T_0 \in \mathcal{K}(Y)$ the equation

$$(A + C)\, B_2 = I - T_0$$

holds. Hence it follows from Theorem 3.4 that $A + C \in \Phi(X, Y)$.

Using the same considerations as in the proof of Theorem 3.7 we conclude from the relation

$$(A + C)\, B = (I - T_0)\,(I + CB)$$

that the formula

$$\operatorname{Ind}(A + C) = \operatorname{Ind}(I - T_0) + \operatorname{Ind}(I + CB) - \operatorname{Ind} B' = -\operatorname{Ind} B'$$

holds. This formula together with (3.17) shows that equation (3.19) holds. This proves Theorem 3.8.

The properties proved in Theorems 3.7 and 3.8 can be carried over to Φ_+- and Φ_--operators in Banach spaces.

Theorem 3.7*. *Let* $A \in \Phi_+(X, Y)$ $(A \in \Phi_-(X, Y))$ *and* $T \in \mathcal{K}(X, Y)$. *Then* $A + T \in \Phi_+(X, Y)$ $(A + T \in \Phi_-(X, Y))$, *and the numbers*

$$\beta(A + T)\,, \qquad \beta(A) \qquad (\alpha(A + T)\,, \qquad \alpha(A))$$

are both together finite or not.

Proof. For Φ_+-operators the assertion of the theorem follows already from Corollary 3.1 and Theorem 3.7.

Now let $A \in \Phi_-(X, Y)$. In addition, we suppose that $D(A)$ is dense in X. By the Banach-Hausdorff theorem we have $A^* \in \Phi_+(Y^*, X^*)$, and from Corollary 3.1 we conclude that $A^* + T^* \in \Phi_+(Y^*, X^*)$.

From this it follows on the other hand that $A + T \in \Phi_-(X, Y)$.

Theorem 3.8*. *For every operator* $A \in \Phi_\pm(X, Y)$ *there exists a number* $\varrho > 0$ *such that* $C \in \mathcal{L}(X, Y)$ *and* $\|C\| < \varrho$ *imply* $A + C \in \Phi_\pm(X, Y)$ *and* $\alpha(A + C) \leqq \alpha(A)$, $\beta(A + C) \leqq \beta(A)$, $\operatorname{Ind}(A + C) = \operatorname{Ind} A$.

Proof. First let $A \in \Phi_+(X, Y)$. From the proof of Lemma 3.1 it follows that there exists a finite dimensional operator $T \in \mathcal{K}(X)$ with $\dim \operatorname{im} I = \alpha(A)$ and

$$\|x\|_X \leqq \gamma(\|Ax\|_Y + \|Tx\|_X)\,, \qquad x \in D(A)\,,$$

where $\gamma > 0$ is a constant. We put $\varrho = 1/\gamma$. Then for $\|C\| < \varrho$ the last inequality implies the estimate

$$\|x\|_X \leqq \frac{\gamma}{1 - \|C\|\,\gamma}\,(\|(A + C)\,x\|_Y + \|Tx\|_X)\,, \qquad x \in D(A)\,.$$

From this it follows that $A + C \in \Phi_+(X, Y)$ (compare Lemma 3.1) and that the finite dimensional spaces $X_0 = \ker(A + C)$ and $\operatorname{im}(T|X_0)$ are isomorphic. Hence $\alpha(A + C) \leqq \alpha(A)$.

Let $A \in \Phi_-(X, Y)$. Changing over to the adjoint operator as in the proof of the preceding theorem we obtain $A + C \in \Phi_-(X,Y)$ and $\beta(A + C) \leqq \beta(A)$ for $||C|| < \varrho$.

In view of Theorem 3.8 we have only to show that $\operatorname{Ind} A = \infty$ implies $\operatorname{Ind}(A + C) = \infty$ for $||C|| < \varrho$ and sufficiently small ϱ. We prove this assertion for the case that $X = Y = H$ is an Hilbert space and A is a continuous operator. Thus let $A \in \Phi_+(H)$. Then from the proof of the Theorem 3.4 it follows easily that there exists an operator $B \in \mathscr{L}(H)$ such that $BA - I$ is compact. We put $\varrho = 2^{-1}||B||^{-1}$ and $||C|| < \varrho$. As in the proof of the Theorem 3.8 we then obtain the formula (3.20) and thus $A + C \in \Phi_+(H)$ (compare Theorem 3.1). We assume that $\beta(A + C)$ is finite and show that then $\beta(A)$ is also finite. From $A + C \in \Phi(H)$ together with (3.20) we obtain the equality

$$AB_1 = (I - CB_1) + T_0 , \tag{3.21}$$

where T_0 is compact (compare the Corollary 4.3 proved later). Here $||CB_1|| < \frac{1}{2} ||(I + BC)^{-1}|| \leqq 1$. Hence the operator $I - CB_1$ is invertible. By Theorem 3.3 and (3.21) the number $\beta(A)$ is finite.

For operators in Banach spaces the last assertion of Theorem 3.8* is proved by estimating the gap of the defect spaces of the operator A and $A + C$ and then using a known criterion of Krein, Krasnosielski and Milman (for this compare I. C. Gohberg and M. G. Krein [1] or also D. Przeworska-Rolewicz and S. Rolewicz [1] as well as T. Kato [1]).

To conclude our discussion of Φ-operators we prove a simple theorem which will be useful in several applications.

Theorem 3.9. *Let X, Y and Z be F-spaces, where $Z \subset X$ and $Z \subset Y$ hold in the sense of continuous embeddings. Furthermore let $A(X \to Y)$ be an operator with the following properties:* (1) *$Ax \in Z$ for all $x \in Z \cap D(A)$ and* (2) *$Ax \in Z (x \in X)$ implies $x \in Z$.*

By A_Z we denote the restriction of the operator A to the space Z. Then we have

(a) *The normal solvability of A implies the normal solvability of A_Z.*

(b) *If Z is everywhere dense in Y, then $A \in \Phi(X, Y)$ implies $A_Z \in \Phi(Z)$ and*

$$\alpha(A) = \alpha(A_Z) , \qquad \beta(A) = \beta(A_Z) .$$

Proof. (a) Let A be normally solvable. We choose an arbitrary sequence $\{y_n\} \subset \operatorname{im} A_Z$ converging in Z to an element $y \in Z$. Then $y_n \to y$ holds also with respect to convergence in Y, and hence $y \in \operatorname{im} A$. Since by the hypothesis of the theorem

$$Z \cap \operatorname{im} A = \operatorname{im} A_Z \tag{3.22}$$

obiously holds, we have $y \in \operatorname{im} A_Z$. Thus A_Z is normally solvable.

(b) It is dear that if A is closed then A_Z is closed and we have $\ker A = \ker A_Z$.

Since $A \in \Phi(X, Y)$, the space Y can be representated as a topological direct sum $Y = \operatorname{im} A \dotplus N$, where N is a $\beta(A)$-dimensional subspace. By Lemma 2.2 we can choose N such that $N \subset Z$. Consequently by (3.22) we have

$$Z = Z \cap Y = \operatorname{im} A_z \dotplus N .$$

Hence

$$\operatorname{codim} \operatorname{im} A = \operatorname{codim} \operatorname{im} A_Z = \dim N ,$$

and the assertion is proved.

Remark. The concept of the Noether operator can also be introduced for operators an general locally convex spaces. In order to obtain analogous statements as in the case of the F-spaces in this case it is necessary to demand in addition to the conditions of Section 1.3.1 that in locally convex spaces a Φ-operator be a homomorphism (see Section 1.1.2). An analogous statement holds for semi-Noether operators. By Banach's theorem the $\Phi(\Phi_{\pm})$-operators on F-spaces introduced in Section 1.3.1 are necessarily homomorphisms.

The properties of Noether operators in F-spaces proved in Section 1.3.2 also hold for continuous operators on general locally convex spaces (see H. H. SCHAEFER [1], A. PIETSCH [1—2], D. PRZEWORSKA-ROLEWICZ and S. ROLEWICZ [1]; compare also the survey given in S. N. KRAČKOVSKI and A. S. DIKANSKI [1].) It would be of interest to carry over the results of this section to closed operators on arbitrary locally convex spaces (for this compare also U. KÖHLER and B. SILBERMANN [1—2]).

1.3.3. An important subclass of Noether operators is the class of the invertible respectively, of the one-sided invertible operators.

An operator $A(X \to Y)$ is said to be (continuously) *invertible* if there exists an operator $A^{-1} \in \mathcal{L}(Y, X)$ with $\operatorname{im} A^{-1} \subset D(A)$ such that the following relations hold:

$$A^{-1}Ax = x(x \in D(A)) , \qquad AA^{-1}y = y(y \in Y) .$$

Lemma 2.3 immediately implies the following assertion.

1°. *The closed operator* $A(X \to Y)$ *is invertible if and only if* $\ker A = \{0\}$ *and* $\operatorname{im} A = Y$.

The operator $A(X \to Y)$ is said to be *left* (resp. *right*) *invertible* if there exists an operator $A^{(-1)} \in \mathcal{L}(Y, X)$ with $\operatorname{im} A^{(-1)} \subset D(A)$ such that

$$A^{(-1)}Ax = x , \qquad x \in D(A) \qquad (\text{resp. } AA^{(-1)}y = y , \qquad y \in Y) .$$

The operator $A^{(-1)}$ is called a *left* (resp. *right*) *inverse* of A.

By similar considerations as in Section 1.2.5 the two following assertions are easily proved.

2°. *The closed operator* $A(X \to Y)$ *is left invertible if and only if* $\ker A = \{0\}$ *and there exists a continuous projection of* Y *onto* $\operatorname{im} A$.

3°. *The closed operator* $A(X \to Y)$ *is right invertible if and only if* $\operatorname{im} A = Y$ *and there exists a continuous projection of* X *onto* $\ker A$.

The following property is easily verified.

4°. *If $A(X \to Y)$ has a left inverse $A^{(-1)}$, then the equation*

$$Ax = y$$

is solvable if and only if $AA^{(-1)}y = y$. If the last equality holds, then $x = A^{(-1)}y$ is the unique solution of the equation (3.23).

If A has a right inverse $A^{(-1)}$, then $x = A^{(-1)}y$ is one of the solutions of the equation (3.23) *for any $y \in Y$. In this case we have*

$$\ker A = \operatorname{im}\,(I - A^{(-1)}A)\,.$$

5°. *Let X and Y be Banach spaces. If the operator $A \in \mathcal{L}(X, Y)$ is onesided invertible, then all operators $B \in \mathcal{L}(X, Y)$ with*

$$\|B - A\| < \|A^{(-1)}\|^{-1}$$

are invertible on the same side as A, and we have

$$\dim \ker B = \dim \ker A\,, \qquad \dim \operatorname{coker} B = \dim \operatorname{coker} A\,.$$

For the proof of the assertion 5° it suffices to remark that the operator B is representable in the form

$$B = [I - (A - B)\,A^{(-1)}]\,A$$

if A is left invertible and in the form

$$B = A[I - A^{(-1)}\,(A - B)]$$

if A is right invertible. Here the operators in the square brackets are invertible.

6°. *Let R be a Banach algebra with unit and $a(\lambda)$ a continuous mapping of the interval $\Delta = \{\lambda: 0 \leqq \lambda \leqq 1\}$ into the algebra R. If the element $a(\lambda)$ is at least one-sided invertible for each $\lambda \in \Delta$ and (twosided) invertible for at least one $\lambda_0 \in \Delta$, then $a(\lambda)$ is invertible for each $\lambda \in \Delta$.*

Proof. In the sorne way as in the proof of the assertion 5° the following statement can be proved: If an element $x \in R$ is only one-sided invertible, then all elements y in a sufficiently small neighborhood of x are also invertible on the same side as x.

We denote by $\Delta_1(\Delta_2)$ the set of all $\lambda \in \Delta$ for which the element $a(\lambda)$ is invertible (one-sided invertible). Then Δ_1, Δ_2 are disjoint open sets, where $\Delta = \Delta_1 \cup \Delta_2$. Since the interval Δ is a connected set and the set Δ_1 by hypothesis has at least one point, Δ_2 is the empty set.

1.4. Operators with Bounded Regularizer

In this section we shall prove some important properties for closed operators in F-spaces having a bounded regularizer.

Let $A(X \to Y)$ (X and Y arbitrary F-spaces) be an arbitrary linear operator. An operator $B(Y \to X)$ is called a *left* (*right*) *regularizer* of the operator

A if $D(B) \supset \operatorname{im} A(D(A) \supset \operatorname{im} B)$ and moreover

$$BA = I_X + T_X \qquad (AB = I_Y + T_Y)$$

holds, where I_X (I_Y) is the identity operator and $T_X(T_Y)$ is a compact operator on the space $X(Y)$. If such an operator B exists, then A is said to be *left (right) regularizable*. An operator B which is both a left and a right regularizer of the operator A is called a *two-sided regularizer*.

Obviously, we have: If B is a bounded (left or right) regularizer of the operator A, then B is also a regularizer of $A + T$ for any $T \in \mathcal{K}(X, Y)$. If B is a not necessarily bounded regularizer of the bounded operator $A \in \mathcal{L}(X, Y)$, then for any $T' \in \mathcal{K}(Y, X)$ the operator $B + T'$ is a regularizer of A.

Theorem 4.1. *If the closed operator A has a bounded left regularizer, then A is a Φ_+-operator.*

Proof. Let $B \in \mathcal{L}(X, Y)$ be a left regularizer of the operator A. Then the operator $C = I_X + T_X \in \mathcal{L}(X) \cap \Phi(X)$ is a continuous and consequently also a closed extension of the operator BA. Now we obtain the assertion immediately from the remark to Theorem 3.1.

Theorem 4.2. *If the closed operator A has a bounded right regularizer, then A is a Φ_--operator.*

Since $AB = I_Y + T_Y \in \mathcal{L}(Y) \cap \Phi(Y)$, this theorem is a direct consequence of Theorem 3.3.

The following theorem together with the Corollary 4.1 gives an equivalent formulation of Theorem 3.4.

Theorem 4.3. *A closed operator is a Φ-operator if and only if it possesses both a left and a right regularizer which is bounded.*

Corollary 4.1. *If the closed operator $A(X \to Y)$ is a Φ-operator, then it possesses a two-sided regularizer.*

Corollary 4.2. *Two arbitrary bounded regularizers of a Φ-operator differ from each other only by a compact operator.*

Proof. Let B_1 and B_2 be two bounded regularizers of the Φ-operator A. If B_1 is a left and B_2 a right regularizer, then

$$B_1AB_2 = (I_X + T_X) B_2 = B_1(I_Y + T_Y),$$

and thus $B_1 - B_2 = T_XB_2 - B_1T_Y$ is compact.

Now let B_1 and B_2 be left regularizers of A. By Theorem 4.3 there exists a bounded right regularizer B of the operator A, and by the first part of the proof $B_1 - B_2 = (B_1 - B) + (B - B_2)$ is compact.

The assertion in the case where both B_1 and B_2 are right regularizers is proved similarly.

Corollary 4.3. *Every bounded left (right) regularizer of a bounded Φ-operator is also a right (left) regularizer.*

Remark. In the case of a continuous operator A necessary and sufficient conditions for the existence of a bounded regularizer can be stated. We have

1°. *A possesses a bounded left regularizer if and only if A is a Φ_+-operator and there exists a continuous projection of Y onto* im A.

2°. *A possesses a bounded right regularizer if and only if A is a Φ_--operator and there exists a continuous projection of X onto* ker A.

In the case of a closed operator in F-spaces the conditions formulated in Theorems 1° and 2° are sufficient for the existence of the corresponding bounded regularizer. These assertions are easily obtained from the proof of the Theorems 5.1 and 5.2 of the next section by using Lemma 2.3. Finally, we remark that for Hilbert spaces there the projectious mentioned in Theorems 1° and 2° always exist. Consequently for Hilbert spaces the set of all $\Phi_+(\Phi_-)$-operators coincides with the set of all closed operators having a bounded left (right) regularizer.

1.5. Operators with Unbounded Regularizer

In this section we shall study the properties of operators having a not necessarily bounded regularizer. We shall see that the results of Section 1.4 can on principle be carried over to such operators, with the difference however that the renunciation of the continuity of the regularizer implies the loss of its normal solvability.

In the sequel X and Y are again two arbitrary F-spaces.

1.5.1. For unbounded regularizers we obtain the following simple version to Theorem 1° formulated at the end of Section 1.4.

Theorem 5.1. *The operator $A(X \to Y)$ possesses a left regularizer $B(Y \to X)$ if and only if its nullity $\alpha(A)$ is finite.*

Proof. If B is a left regularizer of A, then

$$\alpha(A) \leqq \alpha(BA) \leqq \alpha(I_X + T_X) < \infty .$$

Conversely, suppose $\alpha(A)$ is finite. Then there exists a continuous projection P_A of X onto the finite dimensional subspace ker A. Being a continuous finite dimensional operator, P_A is compact. Since $D(A_1^{-1}) =$ im A, it follows from (2.4) that the operator A_1^{-1} is a left regularizer of the operator A.

Theorem 5.2. *The operator $A(X \to Y)$ possesses a right regularizer $B(Y \to X)$ with a dense domain if and only if its deficiency $\beta(A)$ is finite.*

Proof. Let B be a right regularizer of A with $\overline{D(B)} = Y$. Then obviously $(\text{im } A)^\perp \subset (\text{im } AB)^\perp$ and hence $\beta(A) \leqq \beta(AB)$. Since the domain $D(AB) = D(B)$ is dense in Y, $\beta(AB)$ is equal to the deficiency of the operator $I_Y + T_Y$ considered on the whole space Y. Since this number is finite, the same is also true for $\beta(A)$.

Conversely, if $\beta(A)$ is finite, then $\overline{\operatorname{im} A}$ has a topological complement Y_1 in Y. By K we denote the continuous projection of Y onto Y_1 parallel to $\overline{\operatorname{im} A}$. We consider the operator $B = A_1^{-1}(I_Y - K)$ with domain $D(B) = \operatorname{im} A \oplus Y_1$. Obviously, $\overline{D(B)} = Y$ and $\operatorname{im} B \subset D(A)$. By (2.3) it follows that

$$ABy = y - Ky \qquad (y \in D(B)) .$$

Hence B is a right regularizer of A since the operator K is compact.

Corollary 5.1. *The operator $A(X \to Y)$ has a finite index if and only if it has both a left and a right regularizer $B(Y \to X)$ with a dense domain.*

For the proof it is sufficient to remark that when A has a finite index the operator $B = A_1^{-1}(I_Y - K)$ is also a left regularizer of A.

With regard to the applications to singular integral operators in subsequent chapters, we shall in this section deal mainly with operators possessing a left regularizer.

Theorem 5.3. *Let $A(X \to Y)$ be a closed operator with a finite nullity $\alpha(A)$. A is normally solvable if and only if all its left regularizers are bounded on the image space* $\operatorname{im} A$.

Proof. First let A be normally solvable and let B a left regularizer of A. Then A_1 is a bounded operator (compare Lemma 2.3), and by (2.3) the equality $BAx = x + Tx$ $(x \in D(A))$ gives

$$By = A_1^{-1}y + TA_1^{-1}y , \qquad y \in \operatorname{im} A .$$

Hence the operator B is bounded on $\operatorname{im} A$.

We prove the converse now. Let B be a left regularizer of A which is bounded on $\operatorname{im} A$. By B_0 we denote the operator defined on the subspace $\operatorname{im} A$ by $B_0 y = By$ $(y \in \operatorname{im} A)$. Since by hypothesis B_0 is bounded, it has a continuous extension $\overline{B_0}$ to the closure $Y_0 = \overline{\operatorname{im} A}$. Thus the closed operator $A(X \to Y_0)$ has a bounded left regularizer $\overline{B_0}$. By Theorem 4.1 the image set $\operatorname{im} A$ is closed in Y_0 and consequently also in Y. This shows that A is normally solvable.

Theorem 5.4. *If a closed operator with a finite index has an unbounded left regularizer, then it is not normally solvable.*

Proof. Let $A(X \to Y)$ be a closed operator with a finite index and B an unbounded left regularizer of A. Contrary to the assertion we assume that A is normally solvable.

Then $\operatorname{im} A$ is a closed subspace, and by Theorem 5.3 B is bounded on $\operatorname{im} A$. Since the deficiency $\beta(A)$ is finite, Y can be represented as a topological direct sum $Y = \operatorname{im} A \dotplus Y_1$ with $\dim Y_1 = \beta(A)$. Hence by $\operatorname{im} A \subset D(B)$ we have $D(B) = \operatorname{im} A \oplus N$, where $N = D(B) \cap Y_1$. Since every linear operator is bounded on a finite dimensional subspace, B must be bounded on N.

Thus B is a bounded operator contrary to hypothesis. This contradiction proves the theorem.

Corollary 5.2. *A closed operator with a finite index cannot have both a bounded and an unbounded left regularizer.*

Theorem 5.5. *Suppose the operator $A(X \to Y)$, $D(A) = X$, has a left regularizer with a dense domain and with a finite dimensional kernel. Then A has a finite index.*

Proof. By Theorem 5.1 $\alpha(A)$ is finite. The proof of Theorem 3.2 shows the finiteness of the deficiency $\beta(A)$ if it is modified in the following way: Here we put

$$N_1 = \ker B \cap \overline{\operatorname{im} A}\,, \qquad Y_1 = \overline{\operatorname{im} A} \dotplus N_2\,,$$

and instead of the biorthogonal system $y_1, y_2, \dots, y_m$ we choose a system of elements of $D(B)$ with $\det\{f_i(y_j)\} \neq 0$. By the denseness of $D(B)$ and the linear independence of the functionals $f_1, f_2, \dots, f_m$ there such elements exist. The rest of the argument is the same as in the proof of Theorem 3.2.

Corollary 5.3. *The operator $A(X \to Y)$, $D(A) = X$, has a finite index if and only if A possesses a left regularizer with a dense domain and with a finite dimensional kernel.*

Proof. By Theorem 5.5 we have only to show: If A has a finite index, then there exists a left regularizer with the specified properties. It is easily verified that the regularizer B constructed in the proof of Theorem 5.2 is such an operator.

1.5.2. As we have seen in Sections 1.4 and 1.5.1 the existence of a regularizer implies important properties of the operator A. On the other hand the knowledge of a regularizer permits one to carry over the functional equations

$$Ax = y \tag{5.1}$$

into an equation with a compact operator of the form

$$x + Tx = By \tag{5.2}$$

if B is a left regularizer: $BA = I + T$ or

$$ABz = y \tag{5.3}$$

if B is a right regularizer. Under these conditions we speak of the *regularization* of the equation (5.1). The equations (5.2) and (5.3) are called *Riesz-Schauder functional equations* or *equations of the second kind*. In many cases they can be solved more easily than the general functional equation (5.1).

It is obvious that for any $y \in D(B)$ every solution of the equation (5.1) is also a solution of the Riesz-Schauder equation (5.2). For $y \notin D(B)$ the equation (5.1) is not solvable since $D(B) \supset \operatorname{im} A$. If the converse also holds, i.e. if for any $y \in D(B)$ every solution $x \in X$ of the equation (5.2) is also a solution of the equation (5.1), and if in addition $\overline{D(B)} = Y$, then B is called an *equivalent left regularizer*.[1])

It is easy to see that a left regularizer B of the operator $A(X \to Y)$ is an equivalent regularizer if and only if it has the following properties: 1) $\ker B = \{0\}$, 2) for every solution $x \in X$ of the equation (5.2) we have $x \in D(A)$.

As for equation (5.3), we can first state the following: If (5.3) is solvable and if z is a solution, then

$$x = Bz \tag{5.4}$$

is a solution of equation (5.1). But in general equation (5.1) can have also other solutions besides the solution (5.4). If for any y ($\in \operatorname{im} A$) every solution of (5.1) has the form (5.4) (i.e. if $\operatorname{im} B = D(A)$) and if in addition $\overline{D(B)} = Y$, then B is called an *equivalent right regularizer*.

Thus in the case of an equivalent right regularizer we have: The equations (5.1) and (5.3) are both together either solvable or not solvable and in the case of solvability we obtain the solution of one of these equations from the solution of the other equation by the aid of the (for resolvable z) substitution (5.4).

In the following two theorems we state criteria for the existence of an equivalent regularizer.

Theorem 5.6. *An operator $A(X \to Y)$ with a dense domain $D(A)$ has an equivalent left regularizer if and only if A has a finite and non-negative index.*

Proof. Necessity. Let B be an equivalent left regularizer of the operator A. Then $\alpha(A) = \alpha(I + T)$ is a finite number. By Theorem 5.2 the deficiency $\beta(B)$ of the regularizer B is also finite. Using the relations

$$BAx = x + Tx \qquad (x \in D(A))$$

and $\overline{D(A)} = X$ we obtain $\overline{\operatorname{im} BA} = \operatorname{im} (I + T)$. This implies

$$\beta(BA) = \beta(I + T) = \alpha(A) .$$

By Lemma 2.2 Y kann be represented as a direct topological sum $Y = \overline{\operatorname{im} A} \dotplus Y_1$ with $Y_1 \subset D(B)$ and $\dim Y_1 = \beta(A)$. As the definition of the equivalent regularizer shows immediately, we can without loss of generality assume that $D(B) = \operatorname{im} A \oplus Y_1$. Hence $\operatorname{im} B = \operatorname{im} BA \dotplus B(Y_1)$ and con-

---

[1]) We remark that for a left regularizer B of A the relation $BAx = (I + T)\,x$ holds for all $x \in D(A)$. From the definition of the equivalent left regularizer it follows in particular that every solution $x \in X$ of the equation (5.2) also belongs to $D(A)$.

sequently

$$\beta(BA) = \beta(A) + \beta(B)$$

since by the equivalence of the equations (5.1) and (5.2) we have im $(I + T) \cap \cap B(Y_1) = \{0\}$. Thus we obtain

$$\operatorname{Ind} A = \alpha(A) - \beta(A) = \beta(B) \geqq 0 .$$

Sufficiency. Now let $A(X \to Y)$ be an operator with a finite non-negative index. Then the closed image space $\overline{\operatorname{im} A}$ has a topological complement Y_1 in Y, where $n = \dim Y_1 = \beta(A)$. In Y_1 we choose a basis $y_1, y_2, \ldots, y_n$ and in $\ker A$ a basis $x_1, x_2, \ldots, x_m$. The stated hypothesises imply that $m = \alpha(A) \geqq n$.

The operator A_1^{-1} (compare Section 1.2.5) is defined on the subspace im A. If $n = 0$ (i.e. $\overline{\operatorname{im} A} = Y$), then A_1^{-1} is already the desired equivalent regularizer. This follows immediately from formulae (2.3)—(2.4).

In the case $n \geqq 1$ we extend A_1^{-1} to a linear operator $B(Y \to X)$ defined on the set $D(B) = \operatorname{im} A \oplus Y_1$ which is dense in Y. For this it suffices to put

$$By_i = x_i \qquad (i = 1, 2, \ldots, n) .$$

Since $BA = A_1^{-1}A = I - P_A$, B is a left regularizer of the operator A. We show that every solution of the equation

$$x - P_A x = By \qquad (y \in D(B)) \tag{5.5}$$

satisfies equation (5.1). First, we remark that by im $B \subset D(A)$ every solution $x \in X$ of equation (5.5) belongs to $D(A)$.

Let $y = y' + y''$ ($y' \in \operatorname{im} A$, $y'' \in Y_1$) and x a solution of equation (5.5). Then

$$x - P_A x - A_1^{-1}y' = By'' .$$

where the right-handside of the last equation belongs to im P_A and the left-hand side to ker P_A. Since im $P_A \cap \ker P_A = \{0\}$, we obtain $By'' = 0$ and thus $y'' = 0$,

$$Ax = y' = y .$$

Hence B is an equivalent left regularizer of the operator A, and the theorem is proved.

We remark that the denseness of the set $D(A)$ assumed in Theorem 5.6 has been needed only for the proof of the necessity. In the case of a bounded regularizer, even for the necessity this assumption is not needed. Namely we have

Corollary 5.4. *A closed operator $A(X \to Y)$ has a bounded equivalent left regularizer if and only if A is a Φ-operator with a non-negative index.*

The sufficiency of the conditions mentioned in Corollary 5.4 follows immediately from the second part of the proof of Theorem 5.6 since in the case of a closed normally solvable operator A the regularizer B constructed in that proof is bounded and defined on the whole space Y.

We establish the necessity as follows: If A has a bounded equivalent left regularizer B, then by Theorem 4.1 A is a Φ_+-operator. On the other hand, in this case equation (5.1) is solvable if and only if

$$\langle y, B^*\varphi_j\rangle = 0\,, \qquad j = 1, \dots, n\,,$$

where $n = \alpha(A) = \beta(I + T)$ and $\varphi_1, \varphi_2, \dots, \varphi_n$ are the linearly independent solutions of the adjoint homogeneous equation $\varphi + T^*\varphi = 0$. Thus the functionals $B^*\varphi_j$ $(j = 1, \dots, n)$ belong to the defect space $(\operatorname{im} A)^\perp$, and they span this space completely. Hence

$$\beta(A) = \dim\,(\operatorname{im} A)^\perp \leqq n\,,$$

which was to be proved.

Theorem 5.7. *An operator $A(X \to Y)$ has an equivalent right regularizer if and only if it has a finite non-positive index.*

Proof. Necessity. Let B be an equivalent right regularizer of the operator A. By Theorem 5.2 the deficiency $\beta(A)$ is finite.

$\operatorname{im} B = D(A)$ immediately implies

$$\alpha(A) \leqq \alpha(AB)\,, \qquad \operatorname{im} A = \operatorname{im} AB\,,$$

and thus $\beta(AB) = \beta(A)$. Finally, from $AB = I + T$ and $\overline{D(B)} = Y$ it follows that $\overline{\operatorname{im} AB} = \operatorname{im}\,(I + T)$ from which we conclude

$$\alpha(AB) \leqq \alpha(I + T) = \beta(I + T) = \beta(AB)\,.$$

Combining the last inequalities we now obtain $\alpha(A) \leqq \beta(A)$, i.e. $\operatorname{Ind} A \leqq 0$.

Sufficiency. Let $A(X \to Y)$ be an operator with a finite non-positive index. We use the notations introduced in the proof of sufficiency in Theorem 5.6.

If $n = \beta(A) = 0$, then $\alpha(A) \leqq \beta(A)$ implies $\ker A = \{0\}$ and hence $\operatorname{im} A_1^{-1} = D(A)$. Since moreover the relation (2.3) holds, in this case $A^{-1} = A_1^{-1}$ is an equivalent right regularizer of the operator A.

Now we assume $n \geqq 1$ and extend A_1^{-1} to a linear operator B defined on the set $D(A) = \operatorname{im} A \oplus Y_1$ which is dense in Y by putting

$$By_i = \begin{cases} x_i & \text{for} \quad i = 1, \dots, m-1\,, \\ x_m & \text{for} \quad i = m, \dots, n\,. \end{cases}$$

We show that B is an equivalent right regularizer of the operator A.

By K we denote the continuous projection of Y onto Y_1 parallel to $\overline{\operatorname{im} A}$. Then

$$B = A_1^{-1}(I - K) + BK$$

from which, using $\operatorname{im}\,(BK) \subset \ker A$ the relation

$$AB = AA_1^{-1}(I - K) = I - K$$

follows. Since K as a finite dimensional continuous operator is also compact, B is a right regularizer of A.

Thus it remains to show that $\operatorname{im} B = D(A)$. Every element $x \in D(A)$ is uniquely representable as a sum $x = x' + x''$ with $x' \in \ker A$ and $x'' \in D(A) \cap \cap \ker P_A = D(A_1)$. Hence $x' = \sum_{i=1}^{m} \alpha_i x_i$. We put

$$y' = \sum_{i=1}^{m} \alpha_i y_i\,, \qquad y'' = A_1 x''\,, \qquad y = y' + y''\,.$$

Then it follows that $y' \in Y_1$, $y'' \in \operatorname{im} A$ and thus

$$By = \sum_{i=1}^{m} \alpha_i x_i + A_1^{-1} A_1 x'' = x\,.$$

Consequently $D(A) \subset \operatorname{im} B$. Since the converse inclusion obviously holds, this proves the theorem.

Corollary 5.5, *A closed operator A has a bounded equivalent right regularizer if and only if A is a Φ-operator with a non-positive index.*

The assertion is immediately obtained from the proof of Theorem 5.7 using Theorem 4.2 and Lemma 2.3.

Remark. It is easy to see that the left regularizers of the operator A constructed in the proofs of Theorems 5.1 and 5.6 and the Corollary 5.3 are closed operators if A itself is a closed operator.

1.5.3. As we have seen in the preceding two sections the operator equation (5.1) can always be reduced to a Riesz-Schauder functional equation if the operator A has a finite index. The equivalence of these equations depends on the sign of the number $\operatorname{Ind} A$.

For applications it often suffices to reduce the equation (5.1) to an equation with a Φ-operator. This method is frequently called *normalisation* of the equation (5.1). Here an operator $N(Y \to X)$ is called a *left (right) normalizer* of the operator $A(X \to Y)$ if $D(N) \supset \operatorname{im} A$ $\big(D(A) \supset \operatorname{im} N\big)$ and if NA (AN) is a Φ-operator.

Obviously, every regularizer is also a normalizer. Conversely, if N is a left (right) normalizer of A, then RN (resp. NR) is a left (right) regularizer of A, where R is an arbitrary regularizer of the Φ-operator NA (resp. AN).

If the operator A has a finite index, then there exists a left (right) normalizer of A with a dense domain $D(N)$ and $\operatorname{im} N = X$ such that the equations

$$Ax = y \quad \text{and} \quad NAx = Ny$$

$$(\text{respectivily } Ax = y \text{ and } ANz = y)$$

are equivalent for any $y \in Y$ independently of the sign of the index of A. This follows immediately from the following theorem.

Theorem 5.8. *Let $A \in \mathscr{L}(X, Y)$, where the spaces X and Y have the following property: Every closed subspace of finite codimension is isomorphic to the whole space.*

(a) *If (and only if) $\alpha(A)$ is finite, then A has the respresentation $A = BC$, where $C \in \mathscr{L}(X)$ is a Φ-operator and the kernel of the operator $B \in \mathscr{L}(X, Y)$ contains only the zero element.*

(b) *If (and only if) the operator A has a finite index, then A is representable in the form $A = C'D$, where $C' \in \mathscr{L}(Y)$ is a Φ-operator and for the operator $D \in \mathscr{L}(X, Y)$ the relations $\alpha(D) = \beta(D) = 0$ hold.*

Here the operators B and D are normally solvable if and only if A is normally solvable.

Proof. (a) In the case $\alpha(A) < \infty$ we conclude from (2.4) that $A = A_1(I - P_A)$. By hypothesis there exists an isomorphism F of $X_1 = \operatorname{im}(I - P_A)$ onto X. Now we can put

$$B = A_1 F^{-1}, \qquad C = F(I - P_A)$$

and obtain the equation $A = BC$.

(b) Suppose A has a finite index. First, let $\operatorname{Ind} A \geqq 0$. We consider the topological direct sums

$$Y = \overline{\operatorname{im} A} \dotplus Y_1. \qquad \overline{\operatorname{im} A} = R_0 \dotplus R_1,$$

where $\dim Y_1 = \beta(A)$ and $\dim R_1 = \operatorname{Ind} A$. Let Y_2 be the $\alpha(A)$-dimensional subspace $Y_1 \dotplus R_1$. By hypothesis there exists an operator $G \in \mathscr{L}(Y)$ with $\ker G = Y_1$, $\operatorname{im} G = R_0$. Let $K \in \mathscr{L}(X, Y)$ be an arbitrary operator with $\ker K = X_1$, $\operatorname{im} K = Y_2$ and $G^{(-1)}$ the inverse of the operator $G: \overline{\operatorname{im} A} \to R_0$. It is easily verified that $C' = G^{(-1)}P$ and $D = GA + K$ are the required operators, where P denotes the continuous projection of Y onto R_0 parallel to Y_2.

In the case $\operatorname{Ind} A < 0$ we consider the direct sums

$$Y_1 = Y_2 \dotplus Y_3, \qquad R_0 = \overline{\operatorname{im} A} \dotplus Y_3,$$

where $\dim Y_2 = \alpha(A)$, and we define C' and D by the same formulae as for $\operatorname{Ind} A \geqq 0$.

Thus statement (b) is proved. The last assertion of the theorem follows immediately from the construction of the operators B and D.

Remark. If $\operatorname{Ind} A$ is finite, then as in the proof of assertion (b) we can show that there is a representation of the form $A = BC$ with $\alpha(B) = \beta(B) = 0$. In this case $N = B^{-1}$ and $N = D^{-1}$ are the left and the right normalizer, respectively, of the operator A with the above-mentioned properties.

1.6. Comments and References

1.2. The results of the Sections 1.2.1 and 1.2.2 including Lemma 2.1 and Corollary 2.1 can be found in several books on topological vector spaces (e.g.

see A. P. Robertson and W. Robertson [1]). Lemma 2.2 was proved by I. C. Gohberg.

The concept of normal solvability was introduced by F. Hausdorff [1]. He also gave several equivalent definitions of this concept.

Theorem 1.2.6 was first proved by F. Hausdorff [1] and S. Banach [1] for continuous operators on B-spaces. The generalization to the case of closed operators in F-spaces likely goes back to F. E. Browder [1].

1.3. A complete representation of the theory of the $\Phi(\Phi_\pm)$-operators in Banach spaces was given in the survey articles of I. C. Gohberg and M. G. Krein [1] and I. C. Gohberg, A. S. Markus and I. A. Feldman [1]. In these articles also detailed references to the bibliography until 1960 can be found. An estimate of the state of development of the theory until the year 1969 is given by the article of S. N. Kračkovski and A. S. Dikanski [1]. Various aspects of the theory and the application of the Noether operators are treated in the following books: M. Breuer [1], R. G. Douglas [1], I. C. Gohberg and N. J. Krupnik [4], S. Goldberg [1], T. Kato [1], S. G. Krein [1], S. G. Mihlin [3], R. S. Palais [1], D. Przeworska-Rolewicz [1], D. Przeworska-Rolewicz and S. Rolewicz [1], M. Schechter [1]. Notwithstanding the conformities which the present chapter has with the mentioned books it differs essentially from these in the contents as well as in the mode of presentation.

The Theorems 3.1 to 3.3 and the Corollaries 3.3, 3.4 were stated for continuous operators on B-spaces by B. Yood [1] and on F-spaces by A. Pietsch [1—2]. Under the conditions mentioned here the first two theorems were proved by the author in [4] (the proof given there holds also for arbitrary closed operators on F-spaces). Lemma 3.1, used here to provide a simpler proof, was stated for B-spaces by U. Köhler in his diploma paper (1971) (for a special case it was already proved in 1965 by M. S. Agranovič). The criterion mentioned in Lemma 3.1 (and by this also simultaneously Theorem 3.1 and Corollary 3.1) were recently carried over to general locally convex (l. c.) vector spaces by U. Köhler and B. Silbermann [1].

Theorem 3.4 was proved for continuous operators on B-spaces by F. V. Atkinson [1] in B-spaces and on l. c. spaces by H. H. Schaefer [1].

For the proof of Theorem 3.5 we follow the author paper [4]. Lemma 3.2. is due to I. C. Gohberg, A. S. Markus and I. A. Feldman [1].

Theorem 3.6 was stated for the first time by F. Noether [1] for singular integral operators. This theorem was proved for bounded operators on B-spaces by F. V. Atkinson [1] and for arbitrary closed operators on B-spaces by I. C. Gohberg. It was carried over to continuous operators on l. c. spaces by H. Schaefer [1]. Corollary 3.5 also is due to this author.

Theorem 3.7 was discovered for the special case of singular integral operators by F. Noether [1]. Its abstract formulation was first given by S. G. Mihlin [1] for continuous operators on Hilbert space and by F. V. Atkinson [1] and B. Yood [1] for continuous operators on B-spaces. Theorem 3.7 was proved in the case of closed operators on B-spaces by B. Sz.-Nagy and I. C. Gohberg and for continuous operators on arbitrary l. c. spaces by H. H. Schaefer [1].

Theorem 3.8 was proved for bounded operators by J. Dieudonné and in the general case by M. G. Krein and M. A. Krasnosielski.

For closed operators in B-spaces Theorems 3.7* and 3.8* are due to I. C. Gohberg and M. G. Krein [1]. In the case of continuous Φ_+-operators Theorem 3.7* had been previously stated by G. Köthe [1] for general l.c. spaces. A. Pietsch [2] proved this theorem for continuous operators on F-spaces and J. N. Vladimirski [1—2] for closed operators on F-spaces (by another method).

Theorem 3.9 was stated by the author [6] (compare also S. PRÖSSDORF and L. v. WOLFERSDORF [1]).

Theorems 1°--5° can be found in the book I. C. GOHBERG and I. A. FELDMAN [1], Theorem 6° in the nomograph I. C. GOHBERG and M. G. KREIN [2].

1.4. Theorems 4.1 and 4.2 are due to S. G. MIHLIN [3]. Corollary 4.2 was stated by I. C. GOHBERG in his thesis (1955).

Theorems 1° and 2° were proved by B. YOOD [1] for B-spaces and by A. DEPRIT and A. PIETSCH [1] for general l.c. spaces.

1.5. The theorem on the equivalent left (bounded) regularization (see Corollary 5.4) was descovered by S. G. MIHLIN [3]. All the other results of the Sections 1.5.1. and 1.5.2 are due to the author [1], [3], [9].

The results from 1.5.3. were stated by M. I. HAIKIN [3].

CHAPTER 2

ABSTRACT SINGULAR EQUATIONS OF NORMAL TYPE

The classes of singular integral operators considered in this book can be interpreted as concrete realizations of an algebra of paired operators whose coefficients are continuous functions of a certain invertible operator. In this chapter the relevant operator algebra is constructed and its most important properties are investigated. Among other things, we discoss questions about the factorization of functions and their applications to the inversion of operators in this algebra.

The abstract approach to the theory of singular equations (of normal type) described here is due to I. C. Gohberg [2]. In this chapter we essentially use results from the book I. C. Gohberg and I. A. Feldman [1] and the papers of M. S. Budjanu and I. C. Gohberg [1—2].

2.1. An Algebra of Operator Functions

2.1.1. The Concept of the Symbol. Let M be a subalgebra of the algebra $\mathcal{L}(X)$ of all continuous linear operators on a Banach space X and let K be compact set. By $C(K)$ we denote the Banach algebra of all continuous complex valued functions on K with the norm (1), We assume that with every operator $A \in \mathfrak{M}$ there is in an unique way associated a function $\mathcal{A}(x) \in C(K)$ such that the following conditions are satisfied $(A, B \in \mathfrak{M})$:
complex valued functions on K with the norm

$$||f(x)||_C = \max_{x \in K} |f(x)| \qquad (f(x) \in C(K)) .$$

We assume that with every operator $A \in \mathfrak{M}$ there is in an unique way associated a function $\mathcal{A}(x) \in C(K)$ such that the following conditions are satisfied $(A, B \in \mathfrak{M})$:

σ.1. *$C = \alpha A + \beta B$ implies $\mathcal{C} = \alpha\mathcal{A} + \beta\mathcal{B}$.*

σ.2. *$C = AB$ implies $\mathcal{C} = \mathcal{A}\mathcal{B}$.*

σ.3. $||\mathcal{A}(x)||_C \leqq ||A||$

The function $\mathcal{A}(x)$ is called the *symbol* of the operator A. Thus the symbol produces a continuous homomorphism of the operator algebra $\mathfrak{M}$ into the function algebra $C(K)$.

If the algebra $\mathfrak{M}$ contains the two-sided ideal $\mathfrak{K} = \mathcal{K}(X)$ of compact operators on X, then besides the conditions σ.1.—σ.3. the following condition is to be satisfied:

σ.4. $T(x) \equiv 0$ *for any* $T \in \mathfrak{K}$.

In this case the conditions σ.1, σ.2 and σ.4 imply the inequality

$$||\mathcal{A}(x)||_C \leqq \inf_{T \in K} ||A + T||$$

for any $A \in \mathfrak{M}$ so that the symbol also indices a continuous homomorphism of the factor algebra $\mathfrak{M}/\mathfrak{K}$ into the algebra $C(K)$. We remark that conversely the last inequality also implies the condition σ.4.

For the concrete algebras $\mathfrak{M}$ which we shall consider later the homomorphism $\mathfrak{M}/\mathfrak{K} \to C(K)$ referred to above is even an isomorphism. In these cases K will be the unit circle, the topological product of the unit circle or the topological product of a finite closed system of curves. Moreover, the symbol will exhibit several other remarkable properties. An operator $A \in \mathfrak{M}$ usually is a Noether operator if and only if its symbol is unequal to zero everywhere on K. If this is the case, then the index of A can be explicitly calculated by means of the symbol.

2.1.2. Polynomials of Invertible Operators. In the sequel X cull be a Banach space and $U \in \mathcal{L}(X)$ an invertible operator which together with its inverse U^{-1} satisfies the following two conditions:

(I) *The spectral radii of the operators* U *and* U^{-1} *are both equal to one*: $r(U) = r(U^{-1}) = 1$.

(II) *There exists a projection* $P \in \mathcal{L}(X)$ *such that*

$$UP = PUP\,, \qquad UP \neq PU\,, \qquad PU^{-1} = PU^{-1}P\,. \tag{1.1}$$

Obviously, the first two conditions in (1.1) mean that the image space im P is an invariant subspace of the operator U which is mapped by U on to a proper subset of im P. The last condition in (1.1) which can also be written in the form

$$U^{-1}Q = QU^{-1}Q$$

with $Q = I - P$ means that im Q is an invariant subspace of the operator U^{-1}.

By V and $V^{(-1)}$ we denote the restrictions of the operators PUP and $PU^{-1}P$, respectively, to the closed subspace $X_+ = \text{im } P$. As is easily seen, the operator V is only left invertible, and $V^{(-1)}$ is a left inverse for V.

Namely, by (1.1) we have

$$PU^{-1}P^2UP = PU^{-1}UP = P\,, \qquad PUP^2U^{-1}P = UPU^{-1} \neq P\,.$$

We show that *the spectrum of both operators* U *and* U^{-1} *coincides with the unit circle* $|z| = 1$.

Since for every λ with $|\lambda| < 1$ the equality

$$U - \lambda I = U(I - \lambda U^{-1})$$

holds, where the operator $I - \lambda U^{-1}$ is invertible, the spectrum of U lies on the unit circle. The assumption that the spectrum is not the whole unit circle leads to contradiction. In fact, in this case the regular values of U form a connected set. Obsiously, for all points λ of this set the operator $V - \lambda I \in \mathcal{L}(X_+)$ is left invertible and consequently for all λ either only left invertible or two-sided invertible (compare 1.3, assertion 5°). But this is not the case since the operator V is only left invertible whereas the operator $V - \lambda I$ is invertible for sufficiently large $|\lambda|$. An analogous proof can be given for the operator U^{-1}.

In the sequel we denote by $\mathfrak{E}(U)$ the set of all polynomials with complex coefficients of the form

$$R = \sum_{j=-m}^{n} \alpha_j U^j .$$

i.e. the linear span of the operators U^j ($j = 0, \pm 1, \ldots$). Analogously, let $\mathfrak{E}^+(U)$ and $\mathfrak{E}^-(U)$ be the linear spans of the operator U^j for $j = 0, 1, \ldots$ and for $j = 0, -1, \ldots$, respectively.

From condition (II) it follows easily that the coefficients α_j are uniquely determined by the operator $R \in \mathfrak{E}(U)$. With the operator R we associate the polynomial $R(z) = \sum_{j=-m}^{n} \alpha_j z^j$ ($|z| = 1$). Then we have a one-to-one mapping of the set $\mathfrak{E}(U)$ onto the set of all polynomials $R(z)$. In the sequel we often use the notation $R = R(U)$.

Lemma 1.1. *For every operator $R \in \mathfrak{E}(U)$ the inequality*

$$\max_{|z|=1} |R(z)| \leqq ||R|| \tag{1.2}$$

holds.

Proof. Let z_0 be an arbitrary point on the unit circle. We put $\lambda = R(z_0)$ and consider the operator $R - \lambda I \in \mathfrak{E}(U)$. To this operator there corresponds the polynomial $R(z) - \lambda$ which has a zero at the point z_0. Consequently we have

$$R(z) - \lambda = R_1(z)\,(z - z_0)$$

for a certain polynomial $R_1(z)$. From this we obtain

$$R - \lambda I = R_1(U)\,(U - z_0 I) = (U - z_0 I)\,R_1(U) .$$

Since z_0 is a point of the spectrum of U, as proved befor, it follows from the last equation that the operator $R - \lambda I$ is not invertible.

Consequently the inequality $|\lambda| \leqq ||R||$ holds. Since z_0 is an arbitrary point with $|z_0| = 1$, (1.2) is proved.

Remark. It follows from the considerations of this subsection that the condition (I) is equivalent to the following condition (I′):

$$\text{(I}'\text{)} \qquad r(U) \leqq 1 , \qquad r(U^{-1}) \leqq 1 .$$

2.1.3. Continuous Functions of Invertible Operators. By $\Re(U)$, $\Re^+(U)$ and $\Re^-(U)$ we denote the closures (in the operator norm) of the linear spans $\mathfrak{E}(U)$, $\mathfrak{E}^+(U)$ and $\mathfrak{E}^-(U)$, respectively. It is easily seen that $\Re(U)$ is a commutative Banach algebra and $\Re^\pm(U)$ are its subalgebras. By (1.2) with each operator $A \in \Re(U)$ a certain continuous function $A(z)$ on the unit circle can be associated as follows: We choose an arbitrary sequence $\{R_n\} \subset \mathfrak{E}(U)$ which converges to A in the operator norm. By (1.2) the corresponding sequence of polynomials $R_n(z)$ converges uniformly on the unit circle. The limit function $A(z)$ ($|z| = 1$) of this sequence is also a continuous function. Obviously, the function $A(z)$ is independent of the special choice of the sequence $\{R_n\} \subset \mathfrak{E}(U)$. Consequently $A(z)$ is uniquely determined by the operator $A \in \Re(U)$.

Obviously, by (1.2) we have

$$\max_{|z|=1} |A(z)| \leqq ||A|| \qquad (A \in \Re(U)). \tag{1.3}$$

Thus for the algebra $\mathfrak{M} = \Re(U)$ conditions $\sigma.1-\sigma.3$ from 2.1.1 are satisfied. Hence the function $A(z)$ ($|z| = 1$) is the *symbol* of the operator $A \in \Re(U)$. We remark that the mapping $A \to A(z)$ is one-to-one and consequently an isomorphism of $\Re(U)$ into the algebra of all continuous functions on the unit circle if and only if $\Re(U)$ is an algebra without radical (compare 2.1.4).

It follows immediately from the estimate (1.3) that the symbol $A(z)$ ($|z| = 1$) of an operator $A \in \Re^+(U)$ ($\Re^-(U)$) admits an analytic continuation into the interior of the unit circle: $|z| < 1$ (into the exterior of the unit circle: $|z| > 1$).

In the sequel we denote by $\Re(z)$ the algebra of all symbols $A(z)$ ($|z| = 1$) with $A \in \Re(U)$. The following theorem gives some information about the structure of this algebra $\Re(z)$. Here and in the sequel by W we shall denote the so-called *Wiener algebra* of all functions $A(z)$ on the unit circle which can be expanded into an absolutely convergent Fourier series

$$A(z) = \sum_{j=-\infty}^{\infty} a_j z^j \qquad (|z| = 1).$$

As is well-known, with the norm

$$||A(z)||_W = \sum_{-\infty}^{\infty} |a_j|$$

W is a Banach algebra.

Theorem 1.1. (a) *If*

$$||U|| = ||U^{-1}|| = 1, \tag{1.4}$$

then $\Re(z)$ contains the algebra W, and for the operators $A \in \Re(U)$ with a symbol $A(z) \in W$ the estimate

$$||A|| \leqq ||A(z)||_W \tag{1.5}$$

holds.

(b) *If $X = H$ is a Hilbert space and U is an unitary operator on H, then $\mathfrak{R}(z)$ coincides with the algebra of all continuous functions on the unit circle. For any operator $A \in \mathfrak{R}(U)$ the equality*

$$||A|| = \max_{|z|=1} |A(z)| \tag{1.6}$$

holds.

Proof. (a) Suppose

$$A(z) = \sum_{j=-\infty}^{\infty} a_j z^j \qquad (|z| = 1)\,, \qquad \sum_{-\infty}^{\infty} |a_j| < \infty\,.$$

Then the series $\sum_{-\infty}^{+\infty} a_j U^j$ converges in the operator norm. Consequently its sum S belongs to $\mathfrak{R}(U)$, and we have $S(z) = A(z)$ ($|z| = 1$). The estimate (1.5) is obtained by passing to the limit in the inequality

$$\left\| \sum_{j=-n}^{n} a_j U^j \right\| \leqq \sum_{j=-n}^{n} |a_j|\,.$$

(b) Let H be a Hilbert space and U an unitary operator on H. Then the spectral decomposition of unitary operators at once implies the estimate

$$||R|| \leqq \max_{|z|=1} |R(z)|$$

for any operator $R \in \mathfrak{E}(U)$ with the symbol $R(z)$. Since along with (1.2) the reverse inequality also holds, we obtain equation

$$||R|| = \max_{|z|=1} |R(z)|\,. \tag{1.7}$$

Now let $A(z)$ be an arbitrary continuous function on the unit circle and $R_n(z)$ ($n = 1, 2, \ldots$) a sequence of polynomials converging uniformly to $A(z)$. By (1.7) the sequence $R_n = R_n(U) \in \mathfrak{E}(U)$ converges to an operator $A \in \mathfrak{R}(U)$ whose symbol is thus equal to $A(z)$. Passing to the limit for $n \to \infty$ in the relation (1.7) we obtain equation (1.6).

Hence the theorem is proved.

We remark that every isometric operator U which maps the Banach space X onto itself obviously satisfies conditions (I) and (1.4). For the shift operator U on the space of absolutely summable sequences we even have the equations $\mathfrak{R}(z) = W$ and $||A|| = ||A(z)||_W$ (compare Section 3.1.2).

Under the suppositions of Theorem 1.1, (b) $\mathfrak{R}(U)$ is isometrically isomorphic to the algebra of all continuous functions on the unit circle. It follows from the estimate (1.5) that under the conditions (1.4) the mapping $A \to A(z) \in W$ is one-to-one.

2.1.4. Inversion of Continuous Functions of Invertible Operators. For the sequel we shall need the following well-known results from the theory of Banach algebras (see I. M. Gelfand, D. A. Raikov and G. E. Šilov [1]):

With every element x of a commutative Banach algebra $\mathfrak{R}$ we can associate a complex valued function $x(M)$ ($M \in \mathfrak{M}$) on the set $\mathfrak{M}$ of the maximal ideals of this algebra. Here $\hat{M}(x) = x(M)$ defines a continuous linear and multiplicative functional on $\mathfrak{R}$ with $\hat{M}(e) = 1$ (e the unit of $\mathfrak{R}$). The range of the function $x(M$ on $\mathfrak{M}$ coincides with the spectrum of the element $x \in \mathfrak{R}$. The element x is invertible in $\mathfrak{R}$ if and only if $x(M) \neq 0$ (for all $M \in \mathfrak{M}$) or, equivalently, x belongs to no maximal ideal of the algebra $\mathfrak{R}$. The correspondence $x \to x(M)$ is one-to-one if and only if $\mathfrak{R}$ is an algebra without radical (i.e. the intersection of all maximal ideals consists only of the zero element or, equivalently, there exists no element $x \in \mathfrak{R}$ unequal to the zero element with $\lim_{n\to\infty} \sqrt[n]{\|x^n\|} = 0$).

The following lemma holds.

Lemma 1.2. *Let z be a arbitrary point on the unit circle and*

$$M_z = \{A \in \mathfrak{R}(U) : A(z) = 0\} .$$

Then M_z is a maximal ideal of the algebra $\mathfrak{R}(U)$.

Conversely, to every maximal ideal M of the algebra $\mathfrak{R}(U)$ there corresponds a point z ($|z| = 1$) such that $M = M_z$. Here the function $A(M_z)$ belonging to the element $A \in \mathfrak{R}(U)$ coincides with the symbol $A(z)$ on the set of the maximal ideals:

$$A(M_z) = A(z) \qquad (|z| = 1) .$$

Proof. We define a functional on the algebra $\mathfrak{R}(U)$ by the relation

$$f_z(A) = A(z) \qquad (A \in \mathfrak{R}(U)) .$$

The functional f_z is linear, multiplicative and by (1.3) also continuous. Then, as is well-known, $\ker f_z = M_z$ is a maximal ideal (see I. M. Gelfand, D. A. Raikov and G. A. Šilov [1], § 4).

Now let M be an arbitrary maximal ideal of $\mathfrak{R}(U)$. Since the spectrum of the operator U is the unit circle, we have

$$U(M) = z_0 \quad \text{with} \quad |z_0| = 1 .$$

From the multiplicativity of the functional $\hat{M}$ it follows that $U^{-1}(M) = z_0^{-1}$ and consequently $R(M) = R(z_0)$ for arbitrary $R \in \mathfrak{C}(U)$. By passing to the limit (see inequality (1.3)) we finally obtain

$$A(M) = A(z_0) \qquad \forall\, A \in \mathfrak{R}(U) .$$

It is well-known that $A \in \mathfrak{R}(U)$ belongs to M if and only if $A(M) = 0$. Hence $M = M_{z_0}$, and the assertion is proved.

We remark that the mapping $M_z \to z$ yields a homeomorphism of the compact space all maximal ideals of the algebra $\mathfrak{R}(U)$ onto the unit circle.

From Lemma 1.2 we obtain the following important theorem.

Theorem 1.2. *An operator $A \in \mathfrak{R}(U)$ is invertible in $\mathcal{L}(X)$ if and only if its symbol has no zeros:*

$$A(z) \neq 0 \qquad (|z| = 1) . \tag{1.8}$$

If this condition is satisfied, then $A^{-1} \in \mathfrak{R}(U)$.

Proof. If the condition (1.8) is satisfied, then it follows from Lemma 1.2 that A is invertible in $\Re(U)$.

Now we assume that the operator A is invertible in $\mathscr{L}(X)$ and that $A(z_0) = 0$ for a certain z_0 ($|z_0| = 1$). Then by assertion 5°, Section 1.3, every operator $R \in \mathscr{L}(X)$ which satisfies the inequality

$$||A - R|| < \varrho \qquad (\varrho = ||A^{-1}||^{-1}) \tag{1.9}$$

is invertible.

We choose an operator $R_0 \in \mathfrak{E}(U)$ such that $||A - R_0|| < \dfrac{\varrho}{2}$. Then by (1.3)

$$\max_{|z|=1} |A(z) - R_0(z)| < \frac{\varrho}{2}$$

holds, and hence $|R_0(z_0)| < \dfrac{\varrho}{2}$.

The operator $R = R_0 - R_0(z_0) I \in \mathfrak{E}(U)$ satisfies condition (1.9) and is therefore invertible. Since the symbol of the operator R vanishes at the point z_0, the like as operator $R - \lambda I$ in the proof of Lemma 1.1 the operator R can be represented in the form

$$R = R_1(U - z_0 I) \qquad \left(R_1 \in \mathfrak{E}(U) \right). \tag{1.10}$$

Then with R the operator $U - z_0 I$ must also be invertible. But this is not case. This contradiction proves the theorem.

2.1.5. Wiener-Hopf Operators. In the sequel we shall use the following expression: With every operator A from a certain set $\mathfrak{A} \subset \mathscr{L}(X)$ there shall be associated a real number $\mu(A)$. We say that the *invertibility of the operator A corresponds to the functional* μ provided the operator A is invertible, only left invertible or only right invertible if and only if the number $\mu(A)$ is zero, positive or negative, respectively.

In the sequel the functional μ will be the index of a certain continuous function. The *index* of a function $a(z)$ defined on the unit circle Γ ($|z| = 1$) with $a(z) \neq 0$ ($z \in \Gamma$) is the integer

$$\operatorname{ind} a(z) = \frac{1}{2\pi} [\arg a(z)]_\Gamma = \frac{1}{2\pi i} [\ln a(z)]_\Gamma .$$

Here the symbol [ ]$_\Gamma$ denotes the increment of the expression enclosed in brackets produced by a single circuit of the curve Γ in the positive direction (i.e. counter-clockwise). The index of a continuous function on an arbitrary closed Jordan curve or on an orientated system of curves which consists of finitely many closed Jordan courves without common points is defined analogously. In the case of a system of curves of this type the index is equal to the sum of the indices on the individual curves (where the orientation fixed on these curves is to be used).

In the sequel by $\hat{A} = PAP \mid \text{im } P$ we shall always denote the restriction of the operator PAP to the closed subspace $X_+ = \text{im } P$. $\hat{\mathfrak{R}}(U)$ is the collection of all operators $\hat{A}$ with $A \in \mathfrak{R}(U)$. We agree that the symbol $A(z)$ ($|z| = 1$) of the operator A is also called the *symbol* of the operator $\hat{A}$. The operators $\hat{A} \in \hat{\mathfrak{R}}(U)$ are called *singular operators of Wiener-Hopf type* or simply *Wiener-Hopf operators*.

Operators $\hat{A}$ and $\hat{A}'$ from $\hat{\mathfrak{R}}(U)$ whose symbols are equal to $A(z)$ and $A\left(\dfrac{1}{z}\right)$ ($|z| = 1$), respectively, are said to be *transposed* to each other.

Theorem 1.3. *The operator $\hat{A} \in \hat{\mathfrak{R}}(U)$ is at least one-sided invertible if and only if its symbol $A(z)$ has no zeros. If this condition is satisfied, then the invertibility of the operator $\hat{A}$ corresponds to the index of the symbol*

$$\varkappa = \text{ind } A(z) \,.$$

If condition (1.8) is not satisfied, then $\hat{A}$ is neither a Φ_+- nor a Φ_--operator in the space $X_+ = \text{im } P$.

Proof. 1. First let the condition (1.8) be satisfied. We choose an operator $R \in \mathfrak{E}(U)$ such that

$$R^{-1}A - I = C \,, \qquad \|C\| < \|P\|^{-1} \qquad (C \in \mathfrak{R}(U)) \,. \tag{1.11}$$

From (1.3) and (1.11) we obtain

$$\max_{|z|=1} |R^{-1}(z)\, A(z) - 1| \leqq \|C\| < 1 \,.$$

Hence

$$R(z) \neq 0 \qquad (|z| = 1) \,, \qquad \text{ind } R(z) = \text{ind } A(z) = \varkappa \,.$$

Thus the polynomial $R(z)$ can be represented in the form

$$R(z) = c \prod_{k=1}^{p} (1 - z_k^+ z^{-1})\, z^\varkappa \prod_{l=1}^{q} (z - z_l^-) \tag{1.12}$$

with $|z_k^+| < 1$ ($k = 1, \dots, p$) and $|z_l^-| > 1$ ($l = 1, \dots, q$).

Replacing z by U in the last formula we obtain the representation

$$R = R_- U^\varkappa R_+ \,,$$

for the operator R, where $R_\pm \in \mathfrak{E}^\pm(U)$ are invertible operators with $R_\pm^{-1} \in \mathfrak{R}^\pm(U)$.

The equations (1.1) immediately imply the relations

$$A_+ P = PA_+ P \,, \qquad PA_- = PA_- P \qquad (A_\pm \in \mathfrak{R}^\pm(U)) \tag{1.13}$$

or the equations

$$QA_+ = QA_+ Q \,, \qquad A_- Q = QA_- Q \qquad (Q = I - P)$$

which are equivalent to (1.13). Hence

$$PAP = PR_-(I + C)\, U^\varkappa R_+ P = \hat{R}_-[P(I + C)\, U^\varkappa P]\, \hat{R}_+ ,$$

where $\hat{R}_\pm = PR_\pm P$.

By (1.11) the operator $P + PCP$ is invertible in $\mathcal{L}(X_+)$, and the operators $\hat{R}_\pm$ possess the inverses $PR_\pm^{-1}P$. Consequently in the case $\varkappa = 0$ the operator $\hat{A} = PAP \mid X_+$ is invertible.

It follows from equations (1.13) that

$$P(I + C)\, U^\varkappa P = (P + PCP)\, PU^\varkappa P \qquad (\varkappa > 0) ,$$

$$P(I + C)\, U^\varkappa P = PU^\varkappa P(P + PCP) \qquad (\varkappa < 0) .$$

By relations (1.1) and (1.13) we can conclude, that in $\mathcal{L}(X_+)$ the operator $PU^\varkappa P$ is only left invertible for $\varkappa > 0$ and only right invertible for $\varkappa < 0$. Consequently the same is true for the operator $\hat{A}$, and thus the sufficiency of the condition (1.8) is proved.

2. It only remains to prove the last assertion of the theorem. This proof is carried out analogously to the proof of Theorem 1.2.

Let $A(z_0) = 0$ $(|z_0| = 1)$. Suppose that $\hat{A} \in \Phi_+(X_+)$. By Theorem 3.8*, Chapter 1, there exists a number $\varrho > 0$ such that $B \in \Phi_+(X_+)$ for all operators $B \in \mathcal{L}(X_+)$ with $\|\hat{A} - B\| < \varrho$.

Now we choose an operator $R_0 \in \mathfrak{S}(U)$ with

$$\|A - R_0\| < \frac{\varrho}{2} \|P\|^{-1} .$$

Then (1.3) implies $|R_0(z_0)| < \dfrac{\varrho}{2}$. We put

$$B = PRP \mid X_+ \quad \text{for} \quad R = R_0 - R_0(z_0)\, I \in \mathfrak{S}(U) .$$

It is easily seen that $\|\hat{A} - B\| < \varrho$ and thus $B \in \Phi_+(X_+)$.

On the other hand, from (1.0) and (1.11) we obtain the formula

$$B = PR_1P(V - z_0 I_+) ,$$

where I_+ denotes the identity operator in X_+. Then by Theorem 3.1, Chapter 1, we have $V - z_0 I_+ \in \Phi_+(X_+)$. We show that the relations

$$\operatorname{Ind}(V - zI_+) = 0 \quad (|z| > 1) , \qquad \operatorname{Ind}(V - zI_+) \neq 0 \quad (|z| < 1) \qquad (1.14)$$

hold.

In fact, from the conditions (I) and (II) of 2.1.2 it is easily seen that the spectral radii of the operators V and $V^{(-1)}$ $(\in \mathcal{L}(X_+))$ are not greater then one. Consequently the operator $V - zI_+$ is invertible in $\mathcal{L}(X_+)$ for any z with $|z| > 1$, and the first equation in (1.14) is proved. For any z with $|z| < 1$ the operator $I_+ - zV^{(-1)}$ is invertible, and hence together with V also the operator

$$V - zI_+ = (I_+ - zV^{(-1)})\,V$$

is only left invertible. Thus also the second relation in (1.14) holds.

However, together with the assertion $V - z_0 I_+ \in \Phi_+(X_+)$ the relations (1.14) contradict the statements of the Theorems 3.8 and 3.8* from Chapter 1. Hence $\hat{A}$ cannot be a Φ_+-operator.

Using the representation

$$B = (V^{(-1)} - z_0^{-1} I_+)\, P R_2 P \qquad (R_2 \in \mathfrak{C}(U))$$

one shows in a completely similar way that $\hat{A}$ cannot be a Φ_--operator.

Thus the theorem is proved.

If the condition (1.8) is satisfied, then in the sequel we shall say that the symbol does *not degenerate*. In this case the operators $\hat{A} \in \hat{\mathfrak{R}}(U)$ and $A \in \mathfrak{R}(U)$ are called *operators of normal type*.

Remark 1. The construction of the one-sided inverse of the operator $\hat{A} = PAP \,\mathrm{im}\, | P$ given in the preceding proof in general is not very effective since it uses the knowledge of the inverse operator of $P + PCP$. More effective formulae for the inverse operators will be obtained in Section 2.4.

Remark 2. The properties of the operator $\hat{A} = PAP \mid \mathrm{im}\, P$ formulated in Theorem 1.3 also hold for the operator $AP + Q \in \mathscr{L}(X)$.

In fact, obviously we have

$$AP + Q = (PAP + Q)\,(I + QAP)\,.$$

From $(QAP)^2 = 0$ it follows that $I + QAP$ is an invertible operator with

$$(I + QAP)^{-1} = I - QAP\,.$$

It is easily seen that $PAP + Q$ is a $\Phi_\pm$-operator or one-sided invertible in X if and only if the operator $\hat{A}$ has the corresponding properties in the space $X_+ = \mathrm{im}\, P$.

Moreover we have: There is a one-to-one correspondence between the solution sets of the equations

$$\hat{A}\varphi = Pf \tag{1.15}$$

and

$$(AP + Q)\,\psi = f\,. \tag{1.16}$$

Namely, if ψ is a solution of (1.16), then obviously $\varphi = P\psi$ is a solution of (1.15). Conversely, if φ is a solution of (1.15), then $\psi = (I - QA)\,P\varphi + Qf$ is a solution of the equation (1.16).

Because of the last remark the equations (1.15) and (1.16) can be identified. In the sequel equations of the form (1.15) resp. (1.16) shall be called *singular equations of Wiener-Hopf type* or simply *Wiener-Hopf equations*.

More general equations with operators of the form $AP + BQ$ $(A, B \in \mathfrak{R}(U))$ will be investigated in the next section.

2.2. Abstract Singular Equations

2.2.1. Paired Operators. Let $\mathfrak{M}$ be a closed subalgebra of the algebra $\mathcal{L}(X)$ of all linear bounded operators on a Banach space X, P a continous projection on X and $Q = I - P$.

Every operator of the form $AP + BQ$ or $PA + QB$ with $A \in \mathfrak{M}$ and $B \in \mathfrak{M}$ is called a *paired operator*. The paired operators $AP + BQ$ and $PB + QA$ are called *transposed* to each other. It can be shown that for certain special algebras $\mathfrak{M}$ there exists a simple relationship between the paired operators and the operators of the form PMP or QMQ ($M \in \mathfrak{M}$) (the last operators acting on the subspaces im P resp. im Q). In the sequel we shall assumme that the algebra $\mathfrak{M}$ has the following two properties:

a) *The operators invertible in* $\mathfrak{M}$ *form a dense set in* $\mathfrak{M}$.

b) *The operator* $M \in \mathfrak{M}$ *is invertible in* M *if one of the operators* $PMP \mid$ im P *or* $QMQ \mid$ im Q *is at least one-sided invertible.*

Later on, the algebra $\mathfrak{R}(U)$ constructed in the preceding section (as well as others) will play the part of the algebra $\mathfrak{M}$, where the invertible operator $U \in \mathcal{L}(X)$ satisfies conditions (I) and (II) from 2.1.2. By Theorems 1.2 and 1.3 the algebra $\mathfrak{R}(U)$ satisfies the condition b). We show that $\mathfrak{R}(U)$ also satisfies condition a). In fact, each operator $M \in \mathfrak{R}(U)$ can be approximated by operators $R \in \mathfrak{E}(U)$ to within any degree of accuracy. Now the operator $R \in \mathfrak{E}(U)$ can be represented as the product

$$R = U^l \prod_{k=1}^{m} (U - z_k I)\,, \tag{2.1}$$

where l is an integer, m a natural number and z_k are certain complex numbers. If $|z_k| \neq 1$ ($k = 1, \ldots, m$), then by Theorem 1.2 the operator R is invertible in $\mathfrak{R}(U)$. If, on the other hand, some of the numbers z_k lie on the unit circle, then we obtain the desired operator by replacing these numbers in (2.1) by such numbers which do not lie on the unit circle and whose distance to z_k is sufficiently small.

Furthermore, it follows from the Theorems 1.2 and 1.3 that the algebra $\mathfrak{M} = \mathfrak{R}(U)$ satisfies the following stronger condition:

b′) *The operator* $M \in \mathfrak{M}$ *is invertible in* $\mathfrak{M}$ *if one of the operators* $PMP \mid$ im P *or* $QMQ \mid$ im Q *is a* Φ_+- *or a* Φ_--*operator.*

For any algebra $\mathfrak{M}(\subset \mathcal{L}(X))$ satisfying the conditions a), b) we have the following theorem.

Theorem 2.1. *Let the algebra* $\mathfrak{M}$ *satisfy the conditions* a), b), *and let* A, $B \in \mathfrak{M}$. *For the operator* $AP + BQ$ *to be at least one-sided invertible it is necessary and sufficient that the operators* A *and* B *be invertible in* $\mathfrak{M}$ *and that the operator* $PB^{-1}AP \mid$ im P *be invertible on the same side as the operator* $AP + BQ$.

Theorem 2.1 remains true if in its formulation the operator $PB^{-1}AP \mid \operatorname{im} P$ is replaced by the operator $QA^{-1}BQ \mid \operatorname{im} Q$.

Proof. Let the operators A and B be invertible in $\mathfrak{M}$. Then the operator $AP + BQ$ can be represented in the following form:

$$AP + BQ = B(PB^{-1}AP + Q)\,(I + QB^{-1}AP)\,. \tag{2.2}$$

Here $I + QB^{-1}AP$ is an invertible operator, and obviously

$$(I + QB^{-1}AP)^{-1} = I - QB^{-1}AP\,.$$

Analogously we obtain the representation

$$AP + BQ = A(P + QA^{-1}BQ)\,(I + PA^{-1}BQ)\,, \tag{2.3}$$

where $I + PA^{-1}BQ$ is an invertible operator:

$$(I + PA^{-1}BQ)^{-1} = I - PA^{-1}BQ\,.$$

The equations (2.2) and (2.3) immediately imply the following statement: If one of the operators $AP + BQ$, $PB^{-1}AP \mid \operatorname{im} P$, $QA^{-1}BQ \mid \operatorname{im} Q$ is one-sided invertible, then the other two operators are also invertible on the same side.

Now we assume that the operator $AP + BP$ is at least one-sided invertible. The theorem will be proved if we can show that the operators A and B are invertible in $\mathfrak{M}$.

First we show: If one of the operators A, B is invertible, then the other is also invertible. Let e.g. A be invertible. The representation (2.3) shows that the operator $QA^{-1}BQ \mid \operatorname{im} Q$ is invertible on the same side as the operator $AP + BQ$. Then by the condition b) $A^{-1}B$ and hence also B is invertible in $\mathfrak{M}$.

The case in which neither A nor B is invertible is not possible under the above formulated assumptions. In fact, assume that A and B are not invertible. By condition a) the operator A can be approximated by invertible operators $A' \in \mathfrak{M}$ to within any degree of accuracy. If the norm $\|A - A'\|$ is sufficiently small, then the operator $A'P + BQ$ is invertible on the same side as the operator $AP + BQ$ (see Theorem 5°, Section 1.3). Then, by the first part of the proof, the operator B is invertible which contradicts the assumption.

This proves Theorem 2.1.

Equations (2.2) and (2.3) immediately imply:

Corollary 2.1. *If the operators A and B from $\mathfrak{M}$ are invertible, then the following formulae hold:*

$$\begin{aligned} \dim \ker (AP + BQ) &= \dim \ker (PB^{-1}AP \mid \operatorname{im} P) \\ &= \dim \ker (QA^{-1}BQ \mid \operatorname{im} Q) \end{aligned} \tag{2.4}$$

and

$$\dim \operatorname{coker} (AP + BQ) = \dim \operatorname{coker} (PB^{-1}AP \mid \operatorname{im} P$$
$$= \dim \operatorname{coker} (QA^{-1}BQ \mid \operatorname{im} Q)\,. \qquad (2.5)$$

Using the results of Section 1.3 the following assertion is proved completely analogously to Theorem 2.1.

Theorem 2.2. *Let the algebra* $\mathfrak{M}$ *satisfy the conditions* a) *and* b'). *For the operator* $AP + BQ(A, B \in \mathfrak{M})$ *to be a* $\Phi_+(\Phi_-)$*-operator it is necessary and sufficient that the operators* A *and* B *are invertible in* $\mathfrak{M}$ *and that the operator* $PB^{-1}AP \mid \operatorname{im} P$ *is a* $\Phi_+(\Phi_-)$*-operator.*

Theorem 2.2 remains true if in its formulation $PB^{-1}AP \mid \operatorname{im} P$ is replaced by $QA^{-1}BQ \mid \operatorname{im} Q$. There also the formulae (2.4) and (2.5) hold. Finally we remark that by Theorem 3.7*, Chapter 1, in the formulation of the Theorem 2.2 the operator $AP + BQ$ can be replaced by an operator of the form $AP + BQ + T$, where T is a compact operator on X.

Analogous theorems can be formulated for paired operators of the form $PA + QB$. The corresponding proofs differ from the proofs of the theorems 2.1 and 2.2 only in that the representations (2.2) and (2.3) are replaced by the following equalities:

$$PA + QB = (I + PAB^{-1}Q)\,(PAB^{-1}P + Q)\,B \qquad (2.2')$$

and

$$PA + QB = (I + QBA^{-1}P)\,(P + QBA^{-1}Q)\,A\,.$$

2.2.2. The Abstract Singular Operators. Now let the algebra $\mathfrak{M}(\subset \mathcal{L}(X))$ satisfy the following condition:

c) *The commutator* $[P, A] = PA - AP$ *is compact for any operator* $A \in \mathfrak{M}$.

First we remark: If the condition c) is satisfied, then for any $A \in \mathfrak{M}$ the operators hould

$$[Q, A]\,, \qquad PAQ\,, \qquad QAP$$

are compact since the relations

$$[Q, A] = [A, P]\,, \qquad PAQ = P[A, Q] = [P, A]\,Q$$

obviously.

An operator of the form

$$C = AP + BQ + T\,, \qquad (2.6)$$

where $A, B \in \mathfrak{M}$ and T is a compact operator in X is called an *abstract singular operator* or simply a *singular operator*. Sometimes the operators A and B are called *coefficients* of the operator (2.6).

The operator $C_0 = AP + BQ$ is called the *characteristic part* and T the *compact part* of the singular operator C.

By $\widetilde{\mathfrak{M}}$ we denote the collection of all singular operators of the form (2.6). It is easily seen that $\widetilde{\mathfrak{M}}$ is a (not necessarily closed) algebra.

In fact, obviously the addition of two elements of $\widetilde{\mathfrak{M}}$ and the multiplication of an element of $\widetilde{\mathfrak{M}}$ with a number again yield an element of $\widetilde{\mathfrak{M}}$. It remains to show that the product of two operators of $\widetilde{\mathfrak{M}}$ again lies in $\widetilde{\mathfrak{M}}$.

Let

$$C = A_1P + B_1Q + T_1\,, \qquad D = A_2P + B_2Q + T_2$$

be two operators from $\widetilde{\mathfrak{M}}$. Then

$$CD = A_1A_2P + B_1B_2Q + T \tag{2.7}$$

with

$$T = A_1[P, A_2]\,P + B_1[Q, B_2]\,Q + A_1(PB_2Q) + B_1(QA_2P) + CT_2 + \\ + T_1D - T_1T_2\,.$$

Since by condition c) the operator T is compact, (2.7) immediately implies $CD \in \widetilde{\mathfrak{M}}$.

The abstract singular operator (2.6) is called an *operator of normal type* if its coefficients A and B are invertible in $\mathfrak{M}$, i.e. if the inverse operators $A^{-1} \in \mathfrak{M}$ and $B^{-1} \in \mathfrak{M}$ exist.

From equality (2.7) it follows immediately that the product of two singular operators of normal type is again an operator of normal type.

Theorem 2.3. *Let the algebra $\mathfrak{M}$ satisfy the condition* c). *If the singular operator C is an operator of normal type, then the operator $R = A^{-1}P + B^{-1}Q$ is a two-sided regularizer of the operator C.*

Proof. For $A_1 = A^{-1}$, $B_1 = B^{-1}$, $A_2 = A$ and $B_2 = B$ the equality (2.7) shows that

$$RC = P + Q + T' = I + T'\,,$$

where T' is compact. If we interchange the roles of the operators A and A^{-1} (resp. B and B^{-1}) in the above substitutions, then we obtain

$$CR = I + T''$$

with a compact operator T''. Hence the assertion is proved.

Corollary 2.2. *Under the hypotheses of the Theorem 2.3 C is a Φ-operator.*

This follows immediately from Theorem 2.3 and Theorem 4.3. Chapter 1.

Using Theorem 3.7*, Chapter 1, we obtain the following theorem immediately from Theorem 2.2 and Corollary 2.2.

Theorem 2.4. *Let the algebra $\mathfrak{M}$ satisfy the above-mentioned conditions* a), b′) *and* c).

The abstract singular operator (2.6) *is a $\Phi_+(\Phi_-)$-operator if and only if it is an operator of normal type.*

The equation

$$C\varphi = AP\varphi + BQ\varphi + T\varphi = f \tag{2.6'}$$

will be called an *abstract singular equation* or simply a *singular equation*. If C is an operator of normal type then the equation (2.6′) is also said to be *of normal type*. In the case $T = 0$ the equation (2.6′) is called the *characteristic singular equation*.

In the sequel several representations of paired operators will play an important role. For stating such representations we denote the subalgebra of the algebra $\mathfrak{M}(\subset \mathcal{L}(X))$ which consists of all operators $A \in \mathfrak{M}$ with the property $AP = PAP (PA = PAP)$ by $\mathfrak{M}^+(\mathfrak{M}^-)$ (analogously to the notations introduced in 2.1.3.).

Let $R_+ \in \mathfrak{M}^+$ and $R_- \in \mathfrak{M}^-$ be given operators. For coefficients $A, B \in \mathfrak{M}$ of the form

$$\alpha)\ A = A_1R_+ , \qquad B = B_1R_- \qquad (A_1, B_1 \in \mathfrak{M})$$

the operator $AP + BQ$ can be represented in the following form:

$$AP + BQ = (A_1P + B_1Q)(R_+P + R_-Q) . \tag{2.8$_1$}$$

If in addition to condition α) condition

$$\beta)\ A - A_2R_- \in \mathfrak{M}^- , \qquad B - B_2R_+ \in \mathfrak{M}^+ \qquad (A_2, B_2 \in \mathfrak{M})$$

is satisfied, then the following representation holds:

$$PA + QB = (PA_1P + QB_1Q + PA_2Q + QB_2P)(R_+P + R_-Q) . \tag{2.8$_2$}$$

For coefficients of the form

$$\gamma)\ A = R_-A_1 , \qquad B = R_+B_1 \qquad (A_1, B_1 \in \mathfrak{M})$$

we have

$$PA + QB = (PR_- + QR_+)(PA_1 + QB_1) . \tag{2.8$_3$}$$

If in addition to condition γ) the condition

$$\delta)\ A - R_+A_2 \in \mathfrak{M}^+ , \qquad B - R_-B_2 \in \mathfrak{M}^- \qquad (A_2, B_2 \in \mathfrak{M})$$

is satisfied, then the following representation holds:

$$AP + BQ =$$
$$= (PR_- + QR_+)(PA_1P + QB_1Q + PB_2Q + QA_2P). \tag{2.8$_4$}$$

The validity of the formulae (2.8) can be easily checked.

If the algebra $\mathfrak{M}$ satisfies condition c), then obviously all expressions enclosed in parentheses in formulae (2.8) are operators from the algebra $\widetilde{\mathfrak{M}}$.

2.2.3. Singular Operators with Coefficients from $\mathfrak{R}(U)$. Now we consider the algebra $\widetilde{\mathfrak{M}}$ of the abstract singular operators introduced in the preceding section for the case $\mathfrak{M} = \mathfrak{R}(U)$. In the sequel we assume that the invertible operator $U \in \mathcal{L}(X)$ satisfies condition

$$\text{(III)}\ \dim \operatorname{coker} (U \mid \operatorname{im} P) < \infty$$

as well as conditions (I) and (II).

First we show that under these hypotheses the algebra $\mathfrak{M} = \mathfrak{R}(U)$ satisfies the condition c) mentioned in 2.2.2.

Lemma 2.1. *For any operator $A \in \mathfrak{R}(U)$ the commutator $[P, A] = PA - AP$ is compact.*

Proof. For arbitrary $k = 1, 2, 3, \dots$ the operator $P_k = P - U^k P U^{-k}$ is finite dimensional.

In fact, it is easy to see that P_k is a projection of X onto a certain complement of im $U^k P$ in the subspace im P. But by condition (III) this complement is finite dimensional.

Then together with P_k

$$[P, U^k] = P_k U^k; \qquad [P, U^{-k}] = -U^{-k} P_k$$

are also finite dimensional operators for all $k = 1, 2, \dots$. This shows that the commutator $[P, R]$ is a finite dimensional operator for any $R \in \mathfrak{C}(U)$.

Now let A be an arbitrary operator from $\mathfrak{R}(U)$ and $\{R_n\}_1^\infty$ a sequence of operators from $\mathfrak{C}(U)$ which converges uniformly to A. Then as the uniform limit of the sequence of the finite dimensional operators $[P, R_n]$, the commutator $[P, A]$ is a compact operator.

Thus the lemma is proved.

The algebra $\widetilde{\mathfrak{R}}(U)$ consists of all operators of the form

$$C = AP + BQ + T\,, \tag{2.6}$$

where $A, B \in \mathfrak{R}(U)$ and T is a compact operator in X. Let $A(z)$ and $B(z)$ be the symbols of the operators A and B.

By Theorem 1.2 the operator (2.6) is an operator of normal type if and only if the two conditions

$$A(z) \neq 0\,, \qquad B(z) \neq 0 \qquad (|z| = 1) \tag{2.9}$$

are satisfied.

Since $\mathfrak{R}(U)$ is a commutative algebra, as an immediate consequence of the relation (2.7) we now obtain:

The commutator $[C, D]$ of two singular operators $C, D \in \widetilde{\mathfrak{R}}(U)$ is compact.

With the operator (2.6) we associate the following function $C(z, \theta)$ of the two variables $z(|z| = 1)$ and $\theta = \pm 1$:

$$C(z, \theta) = A(z)\frac{1+\theta}{2} + B(z)\frac{1-\theta}{2} \qquad (|z| = 1, \theta = \pm 1)\,. \tag{2.10}$$

We show that the function $C(z, \theta)$ is the symbol of the operator $C \in \widetilde{\mathfrak{R}}(U)$ in the sense of the definition given in 2.1.1.[1])

Obviously, the conditions $\sigma.1$ and $\sigma.4$ are satisfied. The condition $\sigma.2$ is easily checked by taking into consideration the relation (2.7). The following theorem shows finally that condition $\sigma.3$ is satisfied.

---

[1]) Since $C(z, +1) = A(z)$, $C(z, -1) = B(z)$, the symbol (2.10) can also be identified with the function pair $(A(z), B(z))$.

Theorem 2.5. *For every operator of the form* (2.6) *the estimate*

$$\max |C(z, \theta)| \leqq ||C|| \tag{2.11}$$

holds, where the maximum is to form over all $z(|z| = 1)$ *and all* $\theta = \pm 1$.

Proof. We show that for all $z(|z| = 1)$ any $T \in \mathcal{K}(T)$ inequality

$$|A(z)| \leqq ||AP + BQ + T|| \tag{2.12}$$

holds. If this is not the case, then there exist a point z_0 ($|z_0| = 1$) and a compact operator $T \in \mathcal{K}(X)$ with

$$|A(z_0)| > ||C_0 + T|| , \qquad C_0 = AP + BQ .$$

We put $\lambda = 1/A(z_0)$ and consider the operator

$$D = I - \lambda(C_0 + T) .$$

By $||\lambda(C_0 + D)|| = |\lambda| \, ||C_0 + T|| < 1$ the operator D is invertible and thus a Φ-operator. Then by Theorem 3.7, Chapter 1, the operator

$$I - \lambda C_0 = (I - \lambda A) P + (I - \lambda B) Q$$

is a Φ-operator. But this contradicts the last assertion of the Theorem 2.4 since the symbol $1 - A(z)/A(z_0)$ of the operator $I - \lambda A \in \mathfrak{R}(U)$ vanishes at the point z_0. With this contradiction the relation (2.12) is proved. Analogously the inequality corresponding to (2.12) with $B(z)$ instead of $A(z)$ is obtained.

The estimate (2.11) now follows from (2.12) and the obvious relation

$$\max_{z, \theta} |C(z, \theta)| = \max \left\{\max_z |A(z)|, \max_z |B(z)|\right\} .$$

Corollary 2.3. *For the paired operator* $AP + BQ(A, B \in \mathfrak{R}(U))$ *to be compact it is necessary and if* $\mathfrak{R}(U)$ *is an algebra without radical also sufficient that its symbol is identically equal to zero.*

Corollary 2.4. *If* X *is a Hilbert space and* U *is an unitary operator on* X *which satisfies conditions* (II) *and* (III), *then* $\widetilde{\mathfrak{R}}(U)$ *is a closed subalgebra of the Banach algebra* $\mathcal{L}(X)$.

In fact, under the hypotheses of Corollary 2.4 the relation (2.1) and Theorem 1.1 show that the estimate

$$\max \{||A||, ||B||\} \leqq ||AP + BQ + T|| \tag{2.13}$$

holes for any $A, B \in \mathfrak{R}(U)$ and $T \in \mathcal{K}(X)$. From this the assertion follows easily.

Obviously, the relation (2.13) and thus also the statement of the Corollary 2.4 remains true if just equation (1.6) holds for all $A \in \mathfrak{R}(U)$.

It is obvious, that the conditions (2.9) are satisfied if and only if the symbol $C(z, \theta)$ does *not degenerate*, i.e. if

$$C(z, \theta) \neq 0 \qquad (|z| = 1, \theta \pm 1)$$

holds. The *index* of the non-degenerated symbol $C(z, \theta)$ is defined as the index of the quotient $A(z)/B(z)$:

$$\varkappa = \text{ind } C(z, \theta) = \text{ind } [A(z)/B(z)] =$$
$$= \text{ind } A(z) - \text{ind } B(z) .$$

In view of Theorems 1.2 and 1.3 we obtain the following theorems as a concrete realization of Theorems 2.1 and 2.2.

Theorem 2.6. *Let* $A, B \in \Re(U)$. *For the paired operator* $C = AP + BQ$ $(C = PA + QB)$ *to be at least one-sided invertible it is necessary and sufficient that its symbol does not degenerate.*

If the conditions (2.9) *are satisfied, then the invertibility of the operator* C *corresponds to the index of the symbol*

$$\varkappa = \text{ind } C(z, \theta) .$$

If at least one of the conditions (2.9) *is not satisfied, then the operator* C *is neither a* Φ_+- *nor a* Φ_--*operator in the space* X.

Remark. In the proof of the Theorem 2.6 the condition (III) has not been used.

Theorem 2.7. *The abstract singular operator* $C \in \tilde{\Re}(U)$ *is a* $\Phi_+(\Phi_-)$-*operator if and only if its symbol does not degenerate. If this condition is satisfied, then* C *is a* Φ-*operator and* $R = A^{-1}P + B^{-1}Q$ *is a twosided regularizer of the operator* C.

R is an equivalent left regularizer for $\varkappa = \text{ind } C(z, \theta) \leqq 0$ *and an equivalent right regularizer for* $\varkappa \geqq 0$.

The first assertion of the theorem is a direct consequence of the Theorems 2.3 and 2.4. The last assertion follows from the relations (compare Theorem 2.6)

$$\dim \ker R = 0 \quad (\varkappa \leqq 0), \quad \dim \text{coker } R = 0 \quad (\varkappa \geqq 0) .$$

2.3. Factorization of Functions

In the first part of the proof of the Theorem 1.3 the representation (1.12) of a polynomial $R(z)$ played the crucial role. Using this it was possible to compute the one-sided inverse of the operator $PRP(R \in \mathfrak{E}(U))$ effectively.

Analogous effective formulae can also be established for the one-sided inverses of the operator PAP and of the paired operators $AP + BQ$, $PA + QB$ for any $A, B \in \Re(U)$ if the functions of the algebra $\Re(z)$ permit a corresponding representation (factorization). Therefore in this section we study the factorization of functions which belong to certain Banach algebras of continuous functions.

2.3.1. Factorization in R-Algebras. Let Γ be a smooth closed (in general multiply connected) oriented plane curve which divides the complex plane in to two regions: a connected interior region D_+ and an exterior region D_- which contains the point at infinity $z = \infty$. Without loss of generality we assume that the origin of the coordinate system lies in D_+. By $G_\pm$ we denote the closures of the regions $D_\pm$, i.e. $G_\pm = D_\pm \cup \Gamma$.

Let $a(z)$ $(z \in \Gamma)$ be a continuous function. A representation of the form

$$a(z) = a_-(z)\, z^\varkappa a_+(z) \qquad (z \in \Gamma) \tag{3.1}$$

is called a *factorization* of $a(z)$. Here $\varkappa$ is an integer and the function $a_\pm(z)$ possesses a continuation which is analytic in the region $D_\pm$ and continuous in $G_\pm$ with

$$a_+(z) \neq 0 \ (z \in G_+)\,, \qquad a_-(z) \neq 0 \ (z \in G_-)\,.$$

The number $\varkappa$ is obviously determined by the equation

$$\varkappa = \operatorname{ind} a(z)\,.$$

By $C(\Gamma)$ we denote the Banach algebra of all continuous functions on Γ with the norm

$$\|a(z)\|_C = \max_{z \in \Gamma} |a(z)|$$

and by $R(\Gamma)$ the set (dense in $C(\Gamma)$) of all rational functions having no poles on Γ. Analogously $R^\pm(\Gamma)$ denotes the set of all rational functions with poles outside $G_\pm$ and $C^\pm(\Gamma)$ the closure of the set $R^\pm(\Gamma)$ in the norm of $C(\Gamma)$. Obviouly, $C^\pm(\Gamma)$ is a (closed) subalgebra of $C(\Gamma)$ which consists of all continuous functions on Γ possessing a continuation analytic in the region $D_\pm$ and continuous in $G_\pm$.

In the sequel by $\mathfrak{A}(\Gamma)$ we denote an arbitrary Banach algebra which consists of continuous functions on Γ and which has the following properties:

a) *$\mathfrak{A}(\Gamma)$ contains the set $R(\Gamma)$.*

b) *$a(z) \in \mathfrak{A}(\Gamma)$ and $a(z) \neq 0$ $(z \in \Gamma)$ imply $a^{-1}(z) \in \mathfrak{A}(\Gamma)$.*

We put $\mathfrak{A}^\pm(\Gamma) = \mathfrak{A}(\Gamma) \cap C^\pm(\Gamma)$. By the relation[1])

$$\|a(z)\|_C \leqq \|a(z)\|_{\mathfrak{A}} \tag{3.2}$$

which holds for all $a(z) \in \mathfrak{A}(\Gamma)$ we obtain that $\mathfrak{A}^+(\Gamma)$ and $\mathfrak{A}^-(\Gamma)$ are (closed) subalgebras of the Banach algebra $\mathfrak{A}(\Gamma)$. By $\mathring{\mathfrak{A}}^-(\Gamma)$ we denote the subalgebra of $\mathfrak{A}^-(\Gamma)$ which consists of all functions $a(z) \in \mathfrak{A}^-(\Gamma)$ with $a(\infty) = 0$.

The compact space of all maximal ideals of the algebra $\mathfrak{A}(\Gamma)$ ($\mathfrak{A}^\pm(\Gamma)$) is homeomorphic to the curve Γ (resp. to the point set $G_\pm$). Namely, we have

---

[1]) Compare I. M. Gelfand, D. A. Raikov and G. E. Šilov [1], § 9.

Lemma 3.1. *Every maximal ideal of the algebra* $\mathfrak{A}(\Gamma)$ ($\mathfrak{A}^{\pm}(\Gamma)$) *is of the form*

$$M_{z_0} = \{a(z) \in \mathfrak{A}(\Gamma) : a(z_0) = 0\} ,$$

where z_0 *runs through the point set* Γ (resp. $G_{\pm}$).

Proof. First we prove the assertion for the algebra $\mathfrak{A}(\Gamma)$. As at the proof of the Lemma 1.2 one shows that $M_{z_0} (z_0 \in \Gamma)$ is a maximal ideal.

Now let M be an arbitrary maximal ideal of the algebra $\mathfrak{A}(\Gamma)$. We assume that $M \neq M_z$ for all $z \in \Gamma$. It is easily seen (compare I. M. Gelfand, D. A. Raikov and G. E. Šilov [1], p. 10—11) that then there exists a system of functions $a_i(z) \in M$ ($i = 1, 2, \ldots, n$) with

$$\sum_{i=1}^{n} |a_i(z)|^2 > 0 \qquad (z \in \Gamma) .$$

We choose functions $r_i(z) \in R(\Gamma)$ ($i = 1, 2, \ldots, n$) such that the inequality

$$\max_{z \in \Gamma} |\overline{a}_i(z) - r_i(z)| < \varepsilon \qquad (i = 1, 2, \ldots, n)$$

holds. Then the function

$$f(z) = \sum_{i=1}^{n} r_i(z)\, a_i(z) \qquad (z \in \Gamma)$$

belongs to M, and for sufficiently small ε we have $|f(z)| > 0$ ($z \in \Gamma$). Consequently by condition b) the function $f(z)$ is invertible in $\mathfrak{A}(\Gamma)$. But this contradicts the fact that $f(z)$ longs to the maximal ideal M.

The assertion for the algebra $\mathfrak{A}^{\pm}(\Gamma)$ can be proved analogously by using the maximum principle for analytic functions.

Corollary 3.1. *Let the algebra* $\mathfrak{A}(\Gamma)$ *satisfy the conditions* a) *and* b). *Then* $a(z) \in \mathfrak{A}^{\pm}(\Gamma)$ *and* $a(z) \neq 0$ ($z \in G_{\pm}$) *imply* $a^{-1}(z) \in \mathfrak{A}^{\pm}(\Gamma)$.

The algebra $\mathfrak{A}(\Gamma)$ is called a *decomposing algebra* if $\mathfrak{A}(\Gamma)$ is the topological direct sum of $\mathfrak{A}^{+}(\Gamma)$ and $\mathring{\mathfrak{A}}^{-}(\Gamma)$:

$$\mathfrak{A}(\Gamma) = \mathfrak{A}^{+}(\Gamma) \dotplus \mathring{\mathfrak{A}}^{-}(\Gamma) .$$

By 1.2.2 $\mathfrak{A}(\Gamma)$ is a decomposing algebra if and only if each function $a(z) \in \mathfrak{A}(\Gamma)$ is representable in the form $a(z) = a_{+}(z) + a_{-}(z)$ with $a_{\pm}(z) \in \mathfrak{A}^{\pm}(\Gamma)$.

An example of a decomposing algebra is the Wiener algebra W (see 2.1.2). In this case W^{+} and $\mathring{W}^{-}$ consist of all functions of the form

$$\sum_{0}^{\infty} a_n z^n \qquad (|z| = 1)$$

and

$$\sum_{-1}^{-\infty} a_n z^n \qquad (|z| = 1) ,$$

respectively.

An algebra $\mathfrak{A}(\Gamma)$ of continuous functions on Γ is called an *R-algebra* if it satisfies condition a) and if moreover the set $R(\Gamma)$ is dense in $\mathfrak{A}(\Gamma)$.

An R-algebra necessarily satisfies condition b). This follows immediately from the fact that the statement of the Lemma 3.1 is holds for any R-algebra.

We prove the last assertion in the same way as Lemma 1.2: Let M be an arbitrary maximal ideal of the algebra $\mathfrak{A}(\Gamma)$ and $\lambda \notin \Gamma$ an arbitrary point. Since the spectrum of the element $(z - \lambda)^{-1} \in \mathfrak{A}(\Gamma)$ coincides with the range of the functions $(z - \lambda)^{-1}$ $(z \in \Gamma)$, there exists exactly one point $z_0 \in \Gamma$ for which

$$(z - \lambda)^{-1} (M) = (z_0 - \lambda)^{-1}$$

holds. Since the functional $\hat{M}$ is additive and multiplicative, we obtain easily that the point z_0 is independent of the choise of the point $\lambda \notin \Gamma$. Hence it follows that each rational function $a(z) \in R(\Gamma)$ satisfies the condition $a(z)(M) = a(z_0)$. By the density of $R(\Gamma)$ in $\mathfrak{A}(\Gamma)$ and inequality (3.2) we see that the last equation is true for any function $a(z) \in \mathfrak{A}(\Gamma)$. But this implies $M = M_{z_0}$, and the assertion is proved.

Obviously, the Wiener algebra W is a decomposing R-algebra.

For R-algebras we have the following extremely important theorem holds.

Theorem 3.1. *Let $\mathfrak{A}(\Gamma)$ be an R-algebra. For any function $a(z) \in \mathfrak{A}(\Gamma)$ with $a(z) \neq 0$ $(z \in \Gamma)$ to admit a factorization* (3.1) *with $a_\pm(z) \in \mathfrak{A}^\pm(\Gamma)$ it is necessary and sufficient that $\mathfrak{A}(\Gamma)$ be a decomposing algebra.*

Before the proof of this theorem we consider a general lemma which also later will find an application.

Lemma 3.2. *Let R be a Banach algebra with unit e, $R^\pm$ two subalgebras of R with $R = R^+ \dotplus R^-$, P the continuous projection of R onto R^+ parallel to R^- and $Q = I - P$ the complementary projection.*

If the element $a \in R$ satisfies the condition

$$||a|| < \min (||P||^{-1}, ||Q||^{-1}) , \tag{3.3}$$

then the element $e + a$ admits the following factorization:

$$e + a = (e + a_-) (e + a_+) , \tag{3.4}$$

where $a_\pm \in R^\pm$ and $(e + a_\pm)^{-1} - e \in R^\pm$.

Proof. By condition (3.3) the equation

$$x + Pax = e \tag{3.5}$$

has a unique solution $x \in R$ which obviously has the form $x = e + x_+$ $(x_+ \in R^+)$. (3.5) implies the relation

$$(e + a) (e + x_+) = e + a_- \qquad (a_- \in R^-) . \tag{3.6}$$

Considering the equation

$$x + Qxa = e$$

in a similar way we obtain

$$(e + x_-)\,(e + a) = e + a_+ \qquad (x_- \in R^-,\, a_+ \in R^+)$$

or

$$(e + a_+)\,(e + a)^{-1} = e + x_- \,. \tag{3.7}$$

Multiplication of the left-hand and right-hand sides of the equations (3.6) and (3.7) gives

$$a_+ + x_+ + a_+x_+ = x_- + a_- + x_-a_- \,.$$

Since R^+ and R^- have only the zero element in common, both sides of the last equation are equal to zero. This implies

$$(e + a_+)\,(e + x_+) = (e + x_-)\,(e + a_-) = e \,. \tag{3.8}$$

If R is a commutative algebra, then the assertion of the lemma already follows from (3.6) and (3.8).

In the case of a non-commutative algebra R it first follows from (3.8) only that the elements $e + a_\pm$ are one-sided invertible. For the proof of the invertibility of the elements $e + a_\pm$ in this case we remark that all previous considerations remain valid if the element a is replaced by λa $(0 \leqq \lambda \leqq 1)$. Then the elements $a_\pm$ are to be replaced by certain analytic functions $a_\pm(\lambda)$ $(0 \leqq \lambda \leqq 1)$, where the element $e + a_+(\lambda)$ $(e + a_-(\lambda))$ is right (left) invertible for every λ. Since the element $e + a_\pm(0) = e$ is invertible, the assertion 6°, Section 1.3, shows that also the element $e + a_\pm = e + a_\pm(1)$ is also invertible.

This proves Lemma 3.2.

P r o o f o f t h e T h e o r e m 3.1. Let $\mathfrak{A}(\Gamma)$ be a decomposing R-algebra and $a(z) \in \mathfrak{A}(\Gamma)$ a function which vanishes nowhere on Γ. Let $r(z)$ be a function from $R(\Gamma)$ such that the norm (in $\mathfrak{A}(\Gamma)$) of the function $b(z) = a(z)\, r^{-1}(z) - 1$ satisfies the condition

$$\|b\|_{\mathfrak{A}} < \min\,(\|P\|^{-1}, \|Q\|^{-1}) \,, \tag{3.9}$$

where P is the projection of $\mathfrak{A}(\Gamma)$ onto $\mathfrak{A}^+(\Gamma)$ parallel to $\mathring{\mathfrak{A}}^-(\Gamma)$ and $Q = I - P$.

Applying Lemma 3.2 to the algebras

$$R = \mathfrak{A}(\Gamma)\,, \qquad R^+ = \mathfrak{A}^+(\Gamma)\,, \qquad R^- = \mathring{\mathfrak{A}}^-(\Gamma)$$

we come to the factorization of the function $1 + b(z)$, i.e. to a relation of the form

$$a(z)\, r^{-1}(z) = b_-(z)\, b_+(z) \,, \tag{3.10}$$

where

$$b_\pm(z) \in \mathfrak{A}^\pm(\Gamma)\,, \qquad b_\pm(z) \neq 0 \qquad (z \in G_\pm)\,.$$

Since by relation (3.2) inequality

$$\max_{z\in\Gamma} |a(z)\, r^{-1}(z) - 1| \leqq \|b\|_{\mathfrak{A}} < 1$$

holds, we have

$$r(z) \neq 0 \; (z \in \Gamma) \quad \text{and} \quad \operatorname{ind} r(z) = \operatorname{ind} a(z) = \varkappa \,.$$

As in the proof of Theorem 1.3 (compare (1.12)) the function $r(z) \in R(\Gamma)$ can be factorized:

$$r(z) = r_-(z)\, z^{\varkappa} r_+(z)\,, \tag{3.11}$$

where

$$r_\pm(z) \in R^\pm(\Gamma)\,, \qquad r_\pm(z) \neq 0 \qquad (z \in G_\pm)\,.$$

From equations (3.10) and (3.11) we immediately obtain the factorization of the function $a(z)$.

Now we prove the necessity of the condition formulated in Theorem 3.1. Let $a(z)$ be an arbitrary function from $\mathfrak{A}(\Gamma)$ and $b(z) = e^{a(z)}$. Since

$$b(z) \neq 0 \;(z \in \Gamma)\,, \qquad \operatorname{ind} b(z) = 0\,,$$

the function $b(z) \in \mathfrak{A}(\Gamma)$ admits a factorization

$$b(z) = b_-(z)\, b_+(z) \qquad (b_\pm(z) \in \mathfrak{A}^\pm(\Gamma)\,, b_-(\infty) = 1)\,.$$

Since $\operatorname{ind} b(z) = 0$, it follows by a theorem of G. E. Šilov [1] that

$$a_\pm(z) = \ln b_\pm(z) \in \mathfrak{A}^\pm(\Gamma)\,, \qquad a_-(z) \in \mathring{\mathfrak{A}}^-(\Gamma)\,.$$

Thus

$$a(z) = a_+(z) + a_-(z)\,,$$

i.e. $\mathfrak{A}(\Gamma)$ is a decomposing algebra.

This proves the Theorem 3.1.

Remark. Obviously, $C(\Gamma)$ is an R-algebra. But it can be shown that $C(\Gamma)$ is not a decomposing algebra (see I. C. Gohberg and I. A. Feldman [1] p. 51).

Thus by Theorem 3.1 a continuous function $a(z)$ in general does not admit a factorization of the form (3.1) with $a^\pm(z) \in C^\pm(\Gamma)$.

2.3.2. Factorization in Algebras with Two Norms. From the proof of the Theorem 3.1 it follows easily that it is possible to factorize functions from more general algebras as the R-algebras, namely certain *algebras with two norms*.

Let $\mathfrak{A}(\Gamma)$ and $\widetilde{\mathfrak{A}}(\Gamma)$ be two Banach algebras of continuous functions on Γ which in addition to conditions a) and b) from 2.3.1 satisfy the following conditions:

α) *$\mathfrak{A}(\Gamma) \subset \widetilde{\mathfrak{A}}(\Gamma)$ in the sense of a continuous embedding (i.e. $\mathfrak{A}(\Gamma)$ is a subset of $\widetilde{\mathfrak{A}}(\Gamma)$ and the embedding operator $E: \mathfrak{A} \to \widetilde{\mathfrak{A}}$ is bounded).*

β) *$R(\Gamma)$ is dense in $\mathfrak{A}(\Gamma)$ in the norm of $\widetilde{\mathfrak{A}}(\Gamma)$.*

γ) *$\mathfrak{A}(\Gamma)$ and $\widetilde{\mathfrak{A}}(\Gamma)$ are decomposing algebras.*

δ) *There exists a constant $\varrho > 0$ such that for all functions $a(z) \in \mathfrak{A}(\Gamma)$ and $\varphi(z) \in \widetilde{\mathfrak{A}}(\Gamma)$ with $\|1 - a(z)\|_{\widetilde{\mathfrak{A}}} < \varrho$ from $\widetilde{P} a \widetilde{P} \varphi \in \mathfrak{A}(\Gamma)$ it implies that $a\widetilde{P}\varphi \in \mathfrak{A}(\Gamma)$, where $\widetilde{P}$ denotes the projection of $\widetilde{\mathfrak{A}}(\Gamma)$ onto $\widetilde{\mathfrak{A}}^+(\Gamma)$.*

Obviously, conditions γ) and δ) are satisfied if $\widetilde{\mathfrak{A}}(\Gamma)$ is a decomposing algebra and moreover $(\widetilde{P}a - a\widetilde{P})\, \widetilde{P}\varphi \in \mathfrak{A}(\Gamma)$ holds for all functions $a(z) \in$

$\in \mathfrak{A}(\Gamma)$ and $\varphi(z) \in \widetilde{\mathfrak{A}}(\Gamma)$. This follows immediately from the equation

$$a\widetilde{P}\varphi = (a\widetilde{P} - \widetilde{P}a)\,\widetilde{P}\varphi + \widetilde{P}a\widetilde{P}\varphi\,.$$

If $\mathfrak{A}(\Gamma)$ is a decomposing R-algebra, then obviously the conditions $\alpha)-\delta)$ are satisfied with $\widetilde{\mathfrak{A}}(\Gamma) = \mathfrak{A}(\Gamma)$. Later we shall see that the algebra of the Hölder continuous functions on Γ also satisfies the spezified conditions but is not an R-algebra.

Lemma 3.3. *Let the conditions $\alpha)$, $\gamma)$ and $\delta)$ be satisfied, and let $\widetilde{Q} = I - \widetilde{P}$. Then every function $a(z) \in \mathfrak{A}(\Gamma)$ which satisfies the condition*

$$\|1 - a(z)\|_{\widetilde{\mathfrak{A}}} < \min\,(\|\widetilde{P}\|^{-1}, \|\widetilde{Q}\|^{-1}, \varrho) \tag{3.12}$$

admits a factorization of the form

$$a(z) = a_-(z)\,a_+(z) \qquad (z \in \Gamma) \tag{3.13}$$

with $a_\pm(z)$ and $a_\pm^{-1}(z)$ from $\mathfrak{A}^\pm(\Gamma)$.

Proof. Let $a(z) \in \mathfrak{A}(\Gamma)$ be an arbitrary function which satisfies the condition (3.12). By Lemma 3.2 the function $a(z)$ admits a factorization of the form (3.13), where $a_\pm(z)$ and $a_\pm^{-1}(z)$ are functions from $\widetilde{\mathfrak{A}}^\pm(\Gamma)$. We show that these functions necessarily belong to $\mathfrak{A}^\pm(\Gamma)$.

Putting $\varphi(z) = a_+^{-1}(z) - a_-(z)$ $\big(\in \widetilde{\mathfrak{A}}(\Gamma)\big)$ and observing that the function $a_-(z) - c$ $\big(c = a_-(\infty)\big)$ belongs to $\mathring{\widetilde{\mathfrak{A}}}{}^-(\Gamma)$ we obtain the relations

$$a_+^{-1} = \widetilde{P}\varphi + c\,, \qquad a_- = -\,\widetilde{Q}\varphi + c\,. \tag{3.14}$$

The formulae (3.13) and (3.14) yield the equation

$$a\widetilde{P}\varphi + \widetilde{Q}\varphi = f\,,$$

where $f(z) = [1 - a(z)]\,c \in \mathfrak{A}(\Gamma)$. Thus

$$\widetilde{P}a\widetilde{P}\varphi = \widetilde{P}f \in \mathfrak{A}(\Gamma)\,,$$

from which by the condition $\delta)$ it follows that the function $a\widetilde{P}\varphi$ and thus also $\widetilde{P}\varphi$ belongs to $\mathfrak{A}(\Gamma)$. Consequently $a_+^{-1}(z) \in \mathfrak{A}(\Gamma)$ and

$$a_\pm(z) \in \mathfrak{A}^\pm(\Gamma)\,, \qquad a_\pm^{-1}(z) \in \mathfrak{A}^\pm(\Gamma)\,.$$

Hence the lemma is proved.

Theorem 3.2. *Let the Banach algebra $\mathfrak{A}(\Gamma)$ satisfy the conditions $\alpha)-\delta)$. Then every function $a(z) \in \mathfrak{A}(\Gamma)$ which vanishes nowhere on Γ admits a factorization of the form* (3.1) *with $a_\pm(z)$ and $a_\pm^{-1}(z)$* from $\mathfrak{A}^\pm(\Gamma)$.

Proof. By the condition $\beta)$ there exists a function $r(z) \in R(\Gamma)$ such that the function $b(z) = a(z)\,r^{-1}(z) - 1 \in \mathfrak{A}(\Gamma)$ satisfies the inequality

$$\|b\|_{\widetilde{\mathfrak{A}}} < \min\,(\|\widetilde{P}\|^{-1}, \|Q\|^{-1}, \varrho)\,.$$

Then by Lemma 3.3 the function $a(z)\,r^{-1}(z)$ admits a factorization of the form (3.10). The rest of the argument is the same as in the proof of sufficiency in the Theorem 3.1.

2.3.3. Factorization of Functions on the Real Line. If we choose the closed real line $\Big($i.e. more exactly the homeomorphic image of the unit circle with respect to the stereographic projection

$$\lambda = -\,i\,\frac{z+i}{z-i}\,, \qquad z = i\,\frac{\lambda - i}{\lambda + i}\Big)$$

as the curve Γ, then we obtain the concept of the factorization of a function on the real line. Here the upper half-plane $\operatorname{Im}\lambda > 0$ and the lower half-plane $\operatorname{Im}\lambda < 0$ play the roles of the regions D_+ and D_-, respectively.

Thus let C be the space of all continuous functions on the real line $(-\infty, +\infty)$, normed in the usual way, for which the (finite) limits $f(+\infty)$ and $f(-\infty)$ exist and are equal. Let $a(z) \in C$. The representation of $a(z)$ in the form

$$a(\lambda) = a_-(\lambda)\left(\frac{\lambda - i}{\lambda + i}\right)^{\varkappa} a_+(\lambda) \tag{3.15}$$

is called a *factorization* of $a(z)$. Here $\varkappa$ is an integer and the function $a_+(\lambda)$ $(a_-(\lambda))$ has a continuation analytic in the open half-plane $\operatorname{Im}\lambda > 0$ $(\operatorname{Im}\lambda < 0)$ and continuous in the closed half-plane $\operatorname{Im}\lambda \geqq 0$ $(\operatorname{Im}\lambda \leqq 0)$, and furthermore we have

$$a_+(\lambda) \neq 0 \ (\operatorname{Im}\lambda \geqq 0)\,, \qquad a_-(\lambda) \neq 0 \ (\operatorname{Im}\lambda \leqq 0)\,.$$

The number $\varkappa$ in (3.15) is determined by

$$\varkappa = \operatorname{ind} a(\lambda) = \frac{1}{2\pi}\,[\arg a(\lambda)]_{\lambda=-\infty}^{\infty}\,.$$

A Banach algebra $\mathfrak{C} \subset C$ is called an *R-algebra* if it contains the set of all rational functions with poles outside the real line and if this set is dense in $\mathfrak{C}$. It is easy to see that the linear span of the functions

$$\left(\frac{\lambda - i}{\lambda + i}\right)^{k} \tag{3.16}$$

$(k = 0, \pm 1, \ldots)$ is dense in the R-algebra $\mathfrak{C}$.

By $\mathfrak{C}^+$, $\mathfrak{C}^-$ and $\mathring{\mathfrak{C}}^-$ we denote the closure of the linear span of the functions (3.16) in the norm of $\mathfrak{C}$ for $k = 0, 1, 2, \ldots$; $k = 0, -1, -2, \ldots$ and $k = -1, -2, \ldots$, respectively.

The R-algebra $\mathfrak{C}$ is said to be *decomposing* if $\mathfrak{C} = \mathfrak{C}^+ \dotplus \mathring{\mathfrak{C}}^-$ holds. Using the stereographic projection we obtain the following theorem immediately from Theorem 3.1.

Theorem 3.3. *Let $\mathfrak{C} \subset C$ be an R-algebra. For every function $a(\lambda) \in \mathfrak{C}$ with*

$$a(\lambda) \neq 0 \qquad (-\infty \leqq \lambda \leqq +\infty)$$

to admit a factorization (3.15) *with $a_{\pm}(\lambda)$ and $a_{\pm}^{-1}(\lambda)$ from $\mathfrak{C}^{\pm}$ it is necessary and sufficient that $\mathfrak{C}$ be a decomposing algebra.*

It is also possible to formulate a corresponding version of the Theorem 3.2.

2.3.4. A Special Function Algebra. Let Γ be the curve considered in 2.3.1 and $\mathfrak{A} = \mathfrak{A}(\Gamma)$ a certain Banach algebra of continuous functions on Γ which satisfies the condition a) of 2.3.1. In the sequel we consider a subalgebra of the algebra $\mathfrak{A}$ will be is of great importance in the theory of singular equations of non-normal type.

By $\mathfrak{A}(\alpha, m)$, where α is an arbitrary point of Γ and m is an arbitrary natural number, we denote the algebra of all functions $a(z) \in \mathfrak{A}$ which admit a representation of the form

$$a(z) = \sum_{j=0}^{m-1} a_j(z - \alpha)^j + \tilde{a}(z)\,(z - \alpha)^m \qquad (z \in \Gamma) \tag{3.17}$$

for certain complex numbers $a_j (j = 0, 1, \ldots, m - 1)$ and a function $\tilde{a}(z) \in \mathfrak{A}$. Obviously, the representation (3.17) of the function $a(z) \in \mathfrak{A}(\alpha, m)$ is unique. With the norm

$$||a(z)||_{\mathfrak{A}(\alpha, m)} = \sum_{j=0}^{m-1} |a_j| + ||\tilde{a}(z)||_{\mathfrak{A}}$$

$\mathfrak{A}(\alpha, m)$ is a Banach algebra.

By $f^{\{n\}}(\alpha)$ $(n = 1, 2, \ldots)$ we denote the n-th Taylor derivative of the function $f(z) \in C(\Gamma)$ at the point $\alpha \in \Gamma$, i.e. the limit (if it exists)

$$f^{\{n\}}(\alpha) = n! \lim_{\substack{z \to \alpha \\ z \in \Gamma}} \frac{f(z) - \sum\limits_{k=0}^{n-1} \frac{1}{k!} f^{\{k\}}(\alpha)\,(z - \alpha)^k}{(z - \alpha)^n}$$

with $f^{\{0\}}(\alpha) = f(\alpha)$. Obviously, for the function $a(z) \in \mathfrak{A}(\alpha, m)$ of the form (3.17) the relations:

$$a_j = \frac{I}{j!} a^{\{j\}}(\alpha) \qquad (j = 0, 1, \ldots . n - 1)\,, \qquad \tilde{a}(\alpha) = \frac{1}{m!} a^{\{m\}}(\alpha)$$

hold.

Let $\boldsymbol{\alpha} = \{\alpha_1, \alpha_2, \ldots, \alpha_r\}$ be a system of points $\alpha_j \in \Gamma$ $(j = 1, 2, \ldots, r)$ with $\alpha_j \neq \alpha_i$ for $j \neq i$ and $\boldsymbol{m} = \{m_1, m_2, \ldots, m_r\}$ a system of natural numbers. We shall use the following notations

$$\mathfrak{A}(\boldsymbol{\alpha}, \boldsymbol{m}) = \bigcap_{j=1}^{r} \mathfrak{A}(\alpha_j, m_j)\,,$$

$$||a(z)||_{\mathfrak{A}(\boldsymbol{\alpha}, \boldsymbol{m})} = \sum_{j=1}^{r} ||a(z)||_{\mathfrak{A}(\alpha_j, m_j)} \qquad (a(z) \in \mathfrak{A}(\boldsymbol{\alpha}, \boldsymbol{m}))\,.$$

Obviously, $\mathfrak{A}(\boldsymbol{\alpha}, \boldsymbol{m})$ is a Banach algebra with norm $||a(z)||_{\mathfrak{A}(\boldsymbol{\alpha}, \boldsymbol{m})}$.

We state some properties of the algebra $\mathfrak{A}(\boldsymbol{\alpha}, \boldsymbol{m})$.

1°. *If the algebra* $\mathfrak{A}$ *satisfies condition* b) (*see* 2.3.1), *then also the algebra* $\mathfrak{A}(\boldsymbol{\alpha}, \boldsymbol{m})$ *satisfies this condition.*

2°. *If* $\mathfrak{A}$ *is an R-algebra, then* $\mathfrak{A}(\boldsymbol{\alpha}, \boldsymbol{m})$ *is also an R-algebra.*

3°. *If* $\mathfrak{A}$ *is a decomposing algebra, then* $\mathfrak{A}(\boldsymbol{\alpha}, \boldsymbol{m})$ *is also a decomposing algebra.*

The properties 1° and 2° can easily be checked. For the proof of the property 3° it is sufficient to show that the projection P of $\mathfrak{A}$ onto $\mathfrak{A}^+$ is a continuous projection in $\mathfrak{A}(\alpha, m)$ ($\alpha \in \Gamma$). First we remark that for an arbitrary polynomial $q(z)$ and an arbitrary function $\varphi(z) \in \mathfrak{A}$ the expression $f(z) = (Pq - qP)\,\varphi$ is a polynomial.

In fact, for an arbitrary natural number k the function

$$g(z) = (Pz^k - z^kP)\,\varphi = (P - z^kPz^{-k})\,z^k\varphi$$

is a function from $\mathfrak{A}^+$. It is easily seen that each function $g(z) \in \mathfrak{A}^+$ is representable in the form

$$g(z) = \sum_{j=0}^{k-1} g_jz^j + z^kh(z)$$

with certain complex numbers g_j and $h(z) \in \mathfrak{A}^+$ (compare also Lemma 2.1, Chapter 7). The last two equations imply

$$(Pz^k - z^kP)\,\varphi = \sum_{j=0}^{k-1} g_jz^j\,.$$

Thus we obtain

$$(Pq - qP)\,\varphi = \sum_{j=0}^{l} f_jz^j \tag{3.18}$$

and

$$\sum_{j=0}^{l} |f_j| \leqq c_1\,||f(z)||_{\mathfrak{A}} \leqq c_2\,||\varphi(z)||_{\mathfrak{A}}\,.$$

Now let $a(z) \in \mathfrak{A}(\alpha, m)$. Using the representation (3.17) we obtain

$$Pa = \sum_{j=0}^{m-1} a_j(z - \alpha)^j + [P(z - \alpha)^m - (z - \alpha)^m\,P]\,\tilde{a} + (z - \alpha)^m\,P\tilde{a}\,.$$

Hence $Pa \in \mathfrak{A}(\alpha, m)$ and

$$||Pa||_{\mathfrak{A}(\alpha, m)} \leqq c_3\left(\sum_{j=0}^{m-1} |a_j| + ||\tilde{a}||_{\mathfrak{A}}\right) = c_3\,||a||_{\mathfrak{A}(\alpha, m)}\,.$$

This proves assertion 3°.

It is easily seen that properties 1°—3° and the relation (3.18) imply the following property.

4°. *If the algebras* $\tilde{\mathfrak{A}}$ *and* $\mathfrak{A}$ *satisfy the conditions* α) *to* δ) *of* 2.3.2. *then the same holds for the algebras* $\mathfrak{A}(\boldsymbol{\alpha}, \boldsymbol{m})$ *and* $\tilde{\mathfrak{A}}(\boldsymbol{\alpha}, \boldsymbol{m})$.

5°. *$a(z) \in \mathfrak{A}(\boldsymbol{\alpha}, \boldsymbol{m})$ and $a(\alpha_j) = 0$ imply $a(z)\,(z - \alpha_j)^{-1} \in \mathfrak{A}(\boldsymbol{\alpha}, \boldsymbol{m}')$ with $\boldsymbol{m}' = \{m_1, \ldots, m_j - 1, \ldots, m_r\}$.*

Proof. Obviously, it is sufficient to assume $r = 2$ and $j = 1$. Then the function $a(z)$ is representable in the form

$$a(z) = f(z)\,(z - \alpha_1) = b_0 + g(z)\,(z - \alpha_2)$$

with

$$f(z) = \sum_{j=1}^{m_1-1} a_j (z - \alpha_1)^{j-1} + \tilde{a}(z)\,(z - \alpha_1)^{m_1-1},$$

$$g(z) = \sum_{j=1}^{m_2-1} \tilde{b}_j (z - \alpha_2)^{j-1} + \tilde{b}(z)\,(z - \alpha_2)^{m_2-1},$$

where $\tilde{a}(z), \tilde{b}(z) \in \mathfrak{A}$. Obviously, we have $f(z) \in \mathfrak{A}(\alpha_1, m_1 - 1)$ and $g(z) \in \mathfrak{A}(\alpha_2, m_2 - 1)$. It is easily seen that the equation

$$f(z) = g(\alpha_1) + \frac{f(z) - g(z)}{\alpha_1 - \alpha_2}\,(z - \alpha_2) \qquad (z \in \Gamma) \tag{3.19}$$

holds. From this it follows immediately that $f(z) \in \mathfrak{A}(\alpha_2, 1)$. Finally, by successive application of the formula (3.19) we obtain $f(z) \in \mathfrak{A}(\alpha_2, m_2 - 1)$. Hence it follows again from (3.19) that $f(z) \in \mathfrak{A}(\alpha_2, m_2)$. Consequently $a(z)\,(z - \alpha_1)^{-1} = f(z) \in \mathfrak{A}(\boldsymbol{\alpha}, \boldsymbol{m}')$.

The property 5° implies two further properties of the algebra $\mathfrak{A}(\boldsymbol{\alpha}, \boldsymbol{m})$.

6°. *Let $a(z) \in \mathfrak{A}(\boldsymbol{\alpha}, \boldsymbol{m})$. Then there exists a polynomial $p(z)$ of degree $\sum_1^r m_j - 1$ such that*

$$[a(z) - p(z)] \prod_{j=1}^{r} (z - \alpha_j)^{-m_j} \in \mathfrak{A}.$$

The (abviously unique) polynomial $p(z)$ is the *Hermitian interpolation polynomial* of the function $a(z)$ with the knots α_j.

$\boldsymbol{\alpha} = \{\alpha_1, \alpha_2, \ldots, \alpha_n\}$ is called a *system of zeros of multiplicity* $\boldsymbol{m} = \{m_1, m_2, \ldots, m_r\}$ for the function $a(z) \in \mathfrak{A}(\boldsymbol{\alpha}, \boldsymbol{m})$ if $a(z)$ can be represented in the form

$$a(z) = \prod_{j=1}^{r} (z - \alpha_j)^{m_j}\, b(z)$$

with $b(z) \in \mathfrak{A}$ and $b(\alpha_j) \neq 0$ $(j = 1, 2, \ldots, r)$.

7°. *$\boldsymbol{\alpha}$ is a system of zeros of the multiplicity $\boldsymbol{m}$ for the function $a(z) \in \mathfrak{A}(\boldsymbol{\alpha}, \boldsymbol{m})$ if and only if the following conditions are satisfied:*

$$a^{\{k\}}(\alpha_j) = 0, \qquad k = 0, 1, \ldots, m_j - 1, \qquad j = 1, 2, \ldots, r$$

and

$$a^{\{m_j\}}(\alpha_j) \neq 0, \qquad j = 1, 2, \ldots, r.$$

Corresponding algebras of functions on the real line can be introduced similarly. If $\mathfrak{C}$ is an R-algebra of continuous functions on the real line and α

a point on the real line, then by $\mathfrak{C}(\alpha, m)$ we denote the algebra of all functions $a(\lambda) \in \mathfrak{C}$ which are representable in the form

$$a(\lambda) = \sum_{j=0}^{m-1} a_j \left(\frac{\lambda - \alpha}{\lambda + i}\right)^j + \tilde{a}(\lambda) \left(\frac{\lambda - \alpha}{\lambda + i}\right)^m \quad (-\infty < \lambda < \infty)$$

with $\tilde{a}(\lambda) \in \mathfrak{C}$. The algebra $\mathfrak{C}(\boldsymbol{\alpha}, \boldsymbol{m})$ is defined in the same way as the algebra $\mathfrak{A}(\boldsymbol{\alpha}, \boldsymbol{m})$. $\mathfrak{C}(\boldsymbol{\alpha}, \boldsymbol{m})$ has properties similar to those of $\mathfrak{A}(\boldsymbol{\alpha}, \boldsymbol{m})$.

2.4. Application of the Factorization to the Solution of Singular Equations

Again let $U \in \mathscr{L}(X)$ be an invertible operator which together with its inverse U^{-1} and the projection $P \in \mathscr{L}(X)$ satisfies the conditions (I) and (II) from 2.1.2.

2.4.1. Wiener-Hopf Equations. It follows easily from the definition of the algebras $\mathfrak{R}(U)$ and $\mathfrak{R}(z)$ (see Section 2.1) that $\mathfrak{R}(z)$ is an R-algebra of continuous functions on the unit circle. By $\mathfrak{R}^+(z)$ and $\mathfrak{R}^-(z)$ we denote the subalgebras of $\mathfrak{R}(z)$ which are obtained when A runs through the subalgebras $\mathfrak{R}^+(U)$ and $\mathfrak{R}^-(U)$, respectively.

If $\mathfrak{R}(z)$ is a decomposing algebra and if the correspondence between $\mathfrak{R}(U)$ and $\mathfrak{R}(z)$ is one-to-one (i.e. if $\mathfrak{R}(U)$ is an algebra without radical), then the inverses of the operator

$$\hat{A} = PAP \mid \operatorname{im} P \in \hat{\mathfrak{R}}(U) \tag{4.1}$$

can be effectively constructed.

In fact, let $A \in \mathfrak{R}(U)$ be an arbitrary operator whose symbol $A(z)$ satisfies the condition

$$A(z) \neq 0 \qquad (|z| = 1)\,. \tag{4.2}$$

By Theorem 3.1 the function $A(z)$ permits a factorization of the form

$$A(z) = A_-(z)\, z^\varkappa A_+(z) \tag{4.3}$$

with $\varkappa = \operatorname{ind} A(z)$ and $A_\pm(z) \in \mathfrak{R}^\pm(z)$, $A_\pm^{-1}(z) \in \mathfrak{R}^\pm(z)$.

The factorization of the symbol (4.3) induces a factorization of the operator A:

$$A = A_- U^\varkappa A_+ (A_\pm \in \mathfrak{R}^\pm(U)\,, \qquad A_\pm^{-1} \in \mathfrak{R}^\pm(U))\,. \tag{4.4}$$

Here $A_\pm$ is the operator which belongs to the symbol $A_\pm(z)$. Using the relations (1.13) it is easily checked that the operator

$$\hat{A}^{(-1)} = A_+^{-1} P U^{-\varkappa} P A_-^{-1} \mid \operatorname{im} P \tag{4.5}$$

is the operator inverse, left inverse or right inverse to (4.1) if the number $\varkappa$ is equal to zero, positive or negative, respectively.

Finally, we remark that for the validity of the last assertion obviously it is enough to have a representation of the operator A in the form (4.4).

The following theorem complements the statements of Theorem 1.3. Again we use the notations introduced in 2.1.2:

$$V = \hat{U} = PUP \mid \operatorname{im} P\,, \qquad V^{(-1)} = PU^{-1}P \mid \operatorname{im} P\,.$$

Since $V^{(-1)}$ is a left inverse to V, obviously the relation

$$\dim \ker V^{(-1)} = \dim \operatorname{coker} V = \beta(\hat{U})$$

holds.

Theorem 4.1. *Let the operator $A \in \Re(U)$ satisfy the condition* (4.2), *and let $\varkappa = \operatorname{ind} A(z)$. Then*

$$\dim \ker \hat{A} = \max\,(-\varkappa, 0)\,\beta(\hat{U})\,, \tag{4.6}$$

$$\dim \operatorname{coker} \hat{A} = \max\,(\varkappa, 0)\,\beta(\hat{U})\,.$$

Moreover, if the operator A is representable in the form (4.4) *and $\varkappa < 0$, then there exists a closed subspace $L \subset \operatorname{im} P$ of the dimension $\beta(\hat{U})$ such that*

$$\ker \hat{A} = L \dotplus VL \dotplus \dots \dotplus V^{-\varkappa-1}L\,. \tag{4.7}$$

As L the subspace $\hat{A}_{+}^{-1}(\ker V^{(-1)})$ can be chosen.

Proof. By Theorem 1.3 we have

$$\dim \ker \hat{A} = 0 \quad (\varkappa \geqq 0)\,, \qquad \dim \operatorname{coker} \hat{A} = 0 \quad (\varkappa \leqq 0)\,.$$

First of all, we consider the simplest case $A_n = U^{-n}$ $(n = 1, 2, \dots)$, and we put $L_n = \ker \hat{A}_n = \ker V^{(-n)}$ with $V^{(-n)} = [V^{(-1)}]^n$. We show that the subspace L_n can be represented as the direct sum

$$L_n = L_1 \dotplus VL_1 \dotplus \cdots \dotplus V^{n-1}L_1\,. \tag{4.8}$$

For this it is sufficient to show that the equality

$$L_{k+1} = L_k \dotplus V^k L_1 \qquad (k = 1, 2, \dots) \tag{4.9}$$

holds.

The subspaces L_k and V^kL_1 have only the zero element as a common element since by relations (1.1)

$$V^k x = y \qquad (x \in L_1,\, y \in L_k)$$

implies $x = V^{(-k)}y = 0$. Now we choose an arbitrary element $z \in L_{k+1}$ and put $x = V^{(-k)}z$, $y = z - V^k x$. Then

$$z = y + V^k x\,,$$

where $x \in L_1$ and $y \in L_k$. This implies (4.9) and thus also (4.8).

Now let $A \in \Re(U)$ be an arbitrary operator which satisfies the condition (4.2). In the proof of Theorem 1.3 was shown that the operator $\hat{A}$ can be represented in the form

$$\hat{A} = \hat{R}_-\hat{D}V^\varkappa\hat{R}_+ \quad (\varkappa \geqq 0), \qquad \hat{A} = \hat{R}_-V^{(\varkappa)}\hat{D}\hat{R}_+ \quad (\varkappa < 0). \tag{4.10}$$

where the operators

$$\hat{R}_\pm = PR_\pm P \mid \operatorname{im} P, \qquad \hat{D} = (P + PCP) \mid \operatorname{im} P$$

are invertible. It follows immediately from the representations (4.10) that

$$\dim \operatorname{coker} \hat{A} = \dim \ker V^{(-\varkappa)} \qquad (\varkappa > 0) \tag{4.11}$$

and

$$\dim \ker \hat{A} = \dim \ker V^{(\varkappa)} \qquad (\varkappa < 0). \tag{4.12}$$

Together with (4.8) equations (4.11) and (4.12) yield the formulae (4.6).

If the operator A possesses the representation (4.4) and if $\varkappa < 0$. then $\hat{A} = \hat{A}_-V^{(\varkappa)}\hat{A}_+$ and it follows from (4.8) that

$$\ker \hat{A} = BL_1 \dotplus BVL_1 \dotplus \dots \dotplus BV^{-\varkappa-1}L_1 \tag{4.13}$$

with $B = \hat{A}_+^{-1}$. Since the operators $\hat{A}_+^{-1}$ and V commute, from (4.13) we obtain (4.7) with $L = \hat{A}_+^{-1}L_1$. Thus Theorem 4.1 is proved.

2.4.2. Abstract Singular Equations. Now we consider paired operators of the form $AP + BQ$ resp. $PA + QB$, where A and B are operators from $\Re(U)$ whose symbols $A(z)$ and $B(z)$ satisfy the conditions

$$A(z) \neq 0, \qquad B(z) \neq 0 \qquad (|z| = 1). \tag{4.14}$$

By $\varkappa$ we denote the index of the symbol of the operator:

$$\varkappa = \operatorname{ind} A(z) - \operatorname{ind} B(z).$$

The following theorem follows immediately from Theorem 4.1 and Corollary 2.1.

Theorem 4.2. *If the conditions* (4.14) *are satisfied, then the formulae*

$$\dim \ker C = \max(-\varkappa, 0)\,\beta(\hat{U}),$$

$$\dim \operatorname{coker} C = \max(\varkappa, 0)\,\beta(\hat{U})$$

hold for the paired operator $C = AP + BQ$ ($C = PA + QB$).

By Theorem 4.2 and Theorem 3.7, Chapter 1 we obtain

Corollary 4.1. *If the condition* (4.14) *are satisfied, then the relations*

$$\dim \ker (C + T) \geqq \dim \ker C, \qquad \dim \operatorname{coker} (C + T) \geqq \dim \operatorname{coker} C$$

hold for any compact operator T *in the space* X.

For paired operators the corresponding inverses can also be effectively constructed provided the operator $B^{-1}A$ admits a factorization of the form

$$B^{-1}A = C_-U^\varkappa C_+(C_\pm \in \Re^\pm(U),\, C_\pm^{-1} \in \Re^\pm(U)). \tag{4.15}$$

In fact, we first observe that by relation (1.13) the following representation holds:

$$A_1A_+P + A_2A_-Q = (A_1P + A_2Q)(A_+P + A_-Q)\,. \tag{4.16}$$

Here $A_\pm \in \mathfrak{R}^\pm(U)$ and $A_1, A_2 \in \mathfrak{R}(U)$ are arbitrary operators.

Similarly, we obtain

$$PA_1A_- + QA_2A_+ = (PA_- + QA_+)(PA_1 + QA_2)\,. \tag{4.17}$$

(We recall that $\mathfrak{R}(U)$ is a commutative algebra.)

Using the factorization (4.15) and formula (4.16) we can decompose the operator $AP + BQ$ into a product of three operators:

$$AP + BQ = BC_-(U^\varkappa P + Q)(C_+P + C_-^{-1}Q)\,. \tag{4.18$_1$}$$

Here the two outer operators on the right-hand side in (4.18$_1$) are invertible while the middle operator is inverted by $U^{-\varkappa}P + Q$ for $\varkappa \geqq 0$ from the left and for $\varkappa \leqq 0$ from the right. Hence

$$(AP + BQ)^{(-1)} = (C_+^{-1}P + C_-Q)(U^{-\varkappa}P + Q)\,C_-^{-1}B^{-1} \tag{4.19$_1$}$$

is a left inverse operator for $\varkappa \geqq 0$ and a right inverse operator for $\varkappa \leqq 0$ to $AP + BQ$.

Similarly, for the paired operator $PA + QB$ we obtain the formulae

$$PA + QB = (PC_- + QC_+^{-1})(PU^\varkappa + Q)\,C_+B \tag{4.18$_2$}$$

and

$$(PA + QB)^{(-1)} = B^{-1}C_+^{-1}(PU^{-\varkappa} + Q)(PC_-^{-1} + QC_+)\,. \tag{4.19$_2$}$$

The statements of the preceding subsection show that the factorization (4.15) can be easily obtained if the function $A(z)/B(z)$ admits a factorization and if $\mathfrak{R}(U)$ is an algebra without radical.

Remark. If the operator $B^{-1}A$ admits the factorization (4.15), then for $\varkappa < 0$ the formula

$$\ker(AP + BQ) = L \dotplus UL \dotplus \dots \dotplus U^{-\varkappa-1}L \tag{4.20}$$

holds, where

$$L = (P + U^{\varkappa+1}Q)(C_+^{-1}P + C_-Q)\,[\ker(P + UQ)]\,.$$

In fact, it is easily seen that the right-hand side of (4.20) is a direct sum and moreover a subset of $\ker(AP + BQ)$. From this the equation (4.20) follows by Theorem 4.2.

An analogous assertion holds for the operator $PA + QB$.

2.5. Comments and References

2.1. The concept of the symbol of a singular integral operator goes back to S. G. Mihlin [1] who also stated the properties σ.1, σ.2 and σ.4. By I. C. Gohberg it was first noticed that the symbol introduced by S. G. Mihlin also has the property σ.3.

All results of Subsections 2.1.2—2.1.5 are due to I. C. GOHBERG [2]. In our representation we essentially follow the book I. C. GOHBERG and I. A. FELDMAN [1], Kapitel 1, §§ 1—3.

2.2. Subsection 2.2.1 is taken from I. C. GOHBERG and I. A. FELDMAN [1], Kapitel 5, § 1.

The definitions and results given in 2.2.2 (with the exception of Theorem 2.4 and formulae (2.8)) can be found in the papers Z. I. HALILOV [1], Ju. I. ČERSKII [3] and in the monograph D. PRZEWORSKA-ROLEWICZ and S. ROLEWICZ [1], Chapter II.

2.2.3. The Lemma 2.1 is proved in the book I. C. GOHBERG and I. A. FELDMAN [1]. Formula (2.10) represents the definition of the symbol introduced for singular integral operators by S. G. MIHLIN. Theorem 2.7 was proved by S. G. MIHLIN [1]. The other results of this chapter are due to I. C. GOHBERG [1—2].

2.3.1. and 2.3.3. Here we essentially follow the articles M. S. BUDJANU and I. C. GOHBERG [1—2] and the book I. C. GOHBERG and I. A. FELDMAN [1], Kapitel 1, § 5.

2.3.2. M. S. BUDJANU and I. C. GOHBERG [2] studied the factorization of functions and matrix functions (compare Section 7.2) in algebras with two norms which satisfy conditions α) to γ) from the Subsection 2.3.2 and the following condition:

δ') For arbitrary functions $a(z)$ and $b(z)$ from $\mathfrak{A}(\Gamma)$ the estimate

$$||ab||_{\mathfrak{A}} \leqq c(||a||_{\mathfrak{A}} \, ||b||_{\tilde{\mathfrak{A}}} + ||a||_{\tilde{\mathfrak{A}}} \, ||b||_{\mathfrak{A}})$$

holds with a constant $c > 0$ independent of $a(z)$ and $b(z)$.

It can be proved that under the hypotheses α) to γ) condition δ') implies condition δ) (see S. PRÖSSDORF and G. UNGER [1]). Sometimes condition δ') can be checked more easily then condition δ). In the sequel conditions γ) and δ) will be essentially weakened (compare Section 7.2).

2.3.4. The results of this subsection are new (compare S. PRÖSSDORF [16]).

2.4. The results of this section are due to I. C. GOHBERG [2].

CHAPTER 3

SPECIAL SINGULAR EQUATIONS OF NORMAL TYPE

In this chapter the theory of discrete Wiener-Hopf equations, Wiener-Hopf integral equations, the corresponding paired equations and singular integral equations with continuous coefficients on closed curve systems is developed in the case of a non-degenerate symbol. All the essential properties of the Wiener-Hopf equations follow directly from the general theorems of the preceding chapter. The same holds for the singular integral equations in the space $L^p(1 < p < \infty)$ on the unit circle. By modifying the corresponding arguments from Chapter 2 these results can also be obtained for singular integral equations on general curve systems. For this some additional investigations of the properties of singular integrals are readed. Finally we obtain the theory of singular integral equations for the spaces of Hölder continuous functions from the theory for the spaces L^p by applying a theorem on the regularity of the solution which will be proved at the end of this chapter.

3.1. Discrete Wiener-Hopf Equations

3.1.1. By l^2 we denote the Hilbert space of the sequences of complex numbers

$$\xi = \{\xi_n\}_{-\infty}^{\infty}, \qquad ||\xi||^2 = \sum_{-\infty}^{\infty} |\xi_n|^2 < \infty$$

and by l_+^2 the Hilbert space of all sequences of complex numbers of the form

$$\xi_+ = \{\xi_n\}_0^{\infty}, \qquad ||\xi_+||^2 = \sum_{0}^{\infty} |\xi_n|^2 < \infty .$$

Let U be the operator defined on l^2 by

$$U\{\xi_n\} = \{\xi_{n-1}\} . \tag{1.1}$$

Obviously, U is invertible, and the corresponding inverse operator is given by the equation

$$U^{-1}\{\xi_n\} = \{\xi_{n+1}\} .$$

Since U is an unitary operator, it satisfies condition (I) of Section 2.1. By P we denote the projection defined on l^2 by equations

$$P\{\xi_n\} = \{\eta_n\}\,, \qquad \eta_n = \begin{cases} \xi_n\,, & n = 0, 1, \ldots \\ 0\,, & n = -1, -2, \ldots\,. \end{cases} \tag{1.2}$$

Then the operators U, U^{-1} and P obviously satisfy also the condition (II) from Section 2.1.

By Theorem 1.1 of Chapter 2 $\mathfrak{R}(U)$ is a Banach algebra without radical which is isomorphic with the Banach algebra of all continuous functions on the unit circle.

Now let $a(z)$ be an arbitrary continuous function on the unit circle and a_n $(n = 0, \pm 1, \ldots)$ its Fourier coefficients: $a(z) = \sum_{n=-\infty}^{+\infty} a_n z^n$. By $A \in \mathfrak{R}(U)$ we denote the operator whose symbol is equal to a(z). A is uniquely determined by the function $a(z)$. We show that the operator A is represented in the space l^2 by the so-called *Toeplitz matrix*

$$\{a_{j-k}\}_{j,k=-\infty}^{\infty}\,, \tag{1.3}$$

i.e. A acts according to the rule:

$$A\{\xi_j\} = \{\eta_j\}\,, \qquad \eta_j = \sum_{k=-\infty}^{\infty} a_{j-k}\xi_k \qquad (j = 0, \pm 1, \ldots)\,.$$

To this end we choose a sequence of polynomials of the form $R_n(z) = \sum_{j=-n}^{n} b_j^{(n)} z^j$ $(n = 1, 2, \ldots)$ which uniformly converges to the function $a(z)$ on the unit circle. We put $b_j^{(n)} = 0$ for $|j| > n$. The operator $R_n \in \mathfrak{E}(U)$ $(n = 1, 2, \ldots)$ is obviously represented in l^2 by the matrix

$$\{b_{j-k}^{(n)}\}_{j,k=-\infty}^{\infty}\,.$$

Theorem 1.1 of Chapter 2 shows that

$$\|A - R_n\| = \max_{|z|=1} |a(z) - R_n(z)| \underset{n\to\infty}{\to} 0\,.$$

On the other hand, the sequence β_n $\{b_j^{(n)}\}_{j=-\infty}^{\infty}$ converges to the sequence $\alpha = \{a_j\}_{-\infty}^{\infty}$ in the l^2-norm. Thus by applying the Schwarz inequality we obtain

$$\left|\sum_{k=-\infty}^{\infty} a_{j-k}\xi_k - \sum_{k=-\infty}^{\infty} b_{j-k}^{(n)}\xi_k\right| \leq \|\alpha - \beta_n\|\,\|\xi\| \underset{n\to\infty}{\to} 0 \tag{1.3$_1$}$$

for all $j = 0, \pm 1, \ldots$ Since norm convergence in l^2 implies coordinatewise convergence, the assertion is proved.

Now we embed the space l_+^2 into the space l^2 by identification of the element $\xi_+ = \{\xi_n\}_0^{\infty} \in l_+^2$ with the element $\xi = \{\eta_n\}_{-\infty}^{\infty} \in l^2$, where η_n is defined by (1.2). Such an identification of the space l_+^2 with the subspace

$\operatorname{im} P \subset l^2$ is possible since the mapping $\xi_+ \to \xi$ preserves the linear structure and the norm of l^2_+.

Then $V = PUP \mid \operatorname{im} P$ and $V^{(-1)} = PU^{-1}P \mid \operatorname{im} P$ are operators which are defined on l^2_+ by the equations

$$V\{\xi_n\}_0^\infty = \{0, \xi_0, \xi_1, \dots\}, \qquad V^{(-1)}\{\xi_n\}_0^\infty = \{\xi_{n+1}\}_0^\infty.$$

Here obviously

$$\dim \operatorname{coker} (U \mid \operatorname{im} P) = \dim \ker V^{(-1)} = 1. \tag{1.4}$$

Further $\hat{A} = PAP \mid \operatorname{im} P$ $(A \in \mathfrak{R}(U))$ is the operator which is represented in the space l^2_+ by the matrix

$$\{a_{j-k}\}_{j,k=0}^\infty. \tag{1.5}$$

According to Section 2.1 the function $a(z)$ is the *symbol* of the operator $\hat{A}$: $A(z) = a(z)$.

Now by Theorems 1.3 and 1.4 of Chapter 2 we obtain the following result for discrete Wiener-Hopf equations.

Theorem 1.1. *Let $a(z)$ be an arbitrary continuous function on the unit circle $|z| = 1$ and a_n $(n = 0, \pm 1, \dots)$ its Fourier coefficients. For the operator $\hat{A}$ defined on l^2_+ by the matrix* (1.5) *to be at least one-sided invertible it is necessary and sufficient that the function $a(z)$ $(|z| = 1)$ nowhere vanishes.*

If this condition is satisfied, then the invertibility of the operator $\hat{A}$ corresponds to the number

$$\varkappa = \operatorname{ind} a(z),$$

and

$$\dim \ker \hat{A} = \max(-\varkappa, 0), \qquad \dim \operatorname{coker} \hat{A} = \max(\varkappa, 0).$$

If $a(z)$ vanishes somewhere on the unit circle, then $\hat{A}$ is neither a Φ_+- nor a Φ_--operator.

Remark. The necessity of the condition mentioned in the preceding theorem can be stated more precisely as follows (see J. Leiterer [1]): If the symbol $a(z)$ $(|z| = 1)$ vanishes at least in one point and if it is not identically equal to zero, then $\hat{A}$ is not a normally solvable operator.

If the function $a(z)$ admits a factorization:

$$a(z) = a_-(z)\, z^\varkappa a_+(z) \tag{1.6}$$

(e.g. if $a(z)$ can be expanded into an absolutely convergent Fourier series), then the operator $\hat{A}$ possesses the representation (4.4), Chapter 2. In this case Theorem 4.1, Chapter 2, implies the existence of a vector $\xi = \{\xi_n\}_0^\infty \in l^2_+$ for $\varkappa < 0$ such that the vectors

$$\xi^{(m)} = \{\underbrace{0, \dots, 0}_{m-1}, \xi_0, \xi_1, \xi_2, \dots\} \qquad (m = 1, 2, \dots, -\varkappa)$$

represent all linearly independent solutions of the homogeneous system

$$\sum_{k=0}^{\infty} a_{j-k}\xi_k = 0 \qquad (j = 0, 1, \ldots). \tag{1.7}$$

The vector

$$\xi = \{b_0, b_1, b_2, \ldots\}$$

can be chosen for ξ, where b_n ($n = 0, 1, \ldots$) are the Fourier coefficients of the function $a_+^{-1}(z)$.

Now we show how the construction of the one-sided inverse of the operator $\hat{A}$ described in Section 2.4 can be performed for the discrete Wiener-Hopf equation. Again we assume that the function $a(z)$ possesses the factorization (1.6). Then the one-sided inverse operator $\hat{A}^{(-1)}$ is given by the formula (4.5) of Chapter 2.

In order to obtain the matrix of the operator $\hat{A}^{(-1)}$ we write the function $a^{-1}(z)$ in the form

$$a^{-1}(z) = b_-(z)\, b_+(z)\,,$$

where

$$b_+(z) = \begin{cases} a_+^{-1}(z)\,, & \varkappa \geqq 0 \\ a_+^{-1}(z)\, z^{-\varkappa}\,, & \varkappa < 0 \end{cases}; \qquad b_-(z) = \begin{cases} a_-^{-1}(z)\, z^{-\varkappa}\,, & \varkappa \geqq 0 \\ a_-^{-1}(z)\,, & \varkappa < 0\,. \end{cases}$$

By $b_n^{(1)}$ and $b_n^{(2)}$ ($n = 0, 1, \ldots$) we denote the Fourier coefficients of the functions $b_+(z)$ and $b_-\left(\frac{1}{z}\right)$, respectively. In view of the results of this section we easily obtain using formula (4.5), Chapter 2, that the operator $\hat{A}^{(-1)}$ is represented by the matrix

$$\{b_{jk}\}_{j,\,k=0}^{\infty}$$

with

$$b_{jk} = \sum_{l=0}^{\min(j,k)} b_{j-l}^{(1)} b_{k-l}^{(2)}\,.$$

3.1.2. In the sequel we denote by l^p ($1 \leqq p < +\infty$) the Banach space of all sequences of (complex) numbers $\xi = \{\xi_n\}_{-\infty}^{\infty}$ with a finite norm

$$\|\xi\| = \left[\sum_{-\infty}^{\infty} |\xi_n|^p\right]^{1/p}$$

and by m the space of all bounded sequences of numbers $\xi = \{\xi_n\}_{-\infty}^{\infty}$ which by under the norm

$$\|\xi\| = \sup_n |\xi_n| \tag{1.8}$$

also becomes a Banach space. Let c be the subspace of all convergent sequences $\xi \in m$ and c^0 the subspace of all null sequences. The norm in c and c^0

is defined in the same way as the norm in m by the equation (1.8). Then c and c^0 are also Banach spaces[1]).

Throughout the rest of this section E denotes any one of the above-mentioned sequence spaces, E_+ the corresponding space of sequences $\{\xi_n\}_0^\infty$ and P the projection defined in E by (1.2).

Obviously, the operator U defined in the space E by the equation (1.1) is an isometry, and im $U = E$. As we remarked in Section 2.1 such an operator satisfies the conditions (I) and (II) mentioned there.

In view of these remarks we obtain easily from the statements of Section 2.1 that all results of the Subsection 3.1.1 hold in each of the sequence spaces E_+, provided we assume that the function $a(z)$ belongs to the corresponding function algebra $\mathfrak{R}(z)$. For this, by Theorem 1.1, Chapter 2, it is sufficient that $a(z) \in W$, and in this case the operator $A \in \mathfrak{R}(U)$ belonging to the symbol $a(z) \in W$ is again represented in any space E by the Toeplitz matrix (1.3).

In fact, putting

$$R_n(z) = \sum_{j=-n}^{n} b_j^{(n)} z^j , \qquad b_j^{(n)} = \begin{cases} a_j & \text{for } |j| \leqq n , \\ 0 & \text{for } |j| > n \end{cases}$$

for all $n = 1, 2, \ldots$ and using the inequality (1.5), Chapter 2, we obtain

$$\|A - R_n\| \leqq \|a(z) - R_n(z)\|_W \underset{n\to\infty}{\to} 0 .$$

In this case, the left-hand side in (1.3_1) can be estimated by the quantity $\|\alpha - \beta_n\|_{l^1} \|\xi\|_E$ which also tends to zero for $n \to \infty$.

For the space $E = l^1$ the algebra $\mathfrak{R}(z)$ coincides with the algebra W of all functions which can be expanded into an absolutely convergent Fourier series on the unit circle. Here

$$\|A\| = \sum_{-\infty}^{\infty} |a_n| = \|a(z)\|_W$$

holds for the operator $A \in \mathfrak{R}(U)$ with the symbol $a(z)$, where a_n are the Fourier coefficients of the function $a(z)$. This assertion follows immediately from the easily verified relation

$$\|R\| = \sum |\alpha_n| ,$$

where $R \in \mathfrak{E}(U)$ and α_n are the Fourier coefficients of the polynomial $R(z)$.

Theorem 1.2 *Let the function $a(z) \in W$ be different from zero for all z ($|z| = 1$). Then the homogeneous system* (1.7) *has the same solutions in all spaces E_+.*

Proof. By Theorem 3.1, Chapter 2, the function $a(z) \in W$ admits a factorization of the form (1.6), where $a_\mp^{-1}(z) \in W^+$. Hence in the case $\varkappa < 0$ the

[1]) The reader can find a detailed treatment of these spaces e.g. in the book L. V. Kantorovič and G. P. Akilov [1] (p. 55ff.).

vectors

$$\xi^{(m)} = \{\underbrace{0, \ldots, 0}_{m-1}, b_0, b_1, b_2, \ldots\} \qquad (m = 1, 2, \ldots, -\varkappa),$$

where b_n are the Fourier coefficients of the function $a_+^{-1}(z)$, belong to all spaces E_+ simultaneously. As already proved these vectors are solutions of the system (1.7). Since by Theorem 1.1 the number of the linearly independent solutions of the system (1.7) in the space E_+ is equal to $-\varkappa$, thus $\xi^{(m)}$ $(m = 1, 2, \ldots, -\varkappa)$ are all solutions of (1.7), and the assertion is proved.

Theorem 1.3. *Let the function $a(z) \in W$ be different from zero for all z $(|z| = 1)$. If $\varkappa \leqq 0$, then the inhomogeneous system of equations*

$$\sum_{k=0}^{\infty} a_{j-k}\xi_k = \eta_j \qquad (j = 0, 1, \ldots) \tag{1.9}$$

is solvable in the space E_+ for any sequence $\{\eta_j\}_0^\infty \in E_+$, and the homogeneous system (1.7) has exactly $-\varkappa$ linearly independent solutions.

In the case $\varkappa > 0$ the homogeneous system transposed to (1.9)

$$\sum_{k=0}^{\infty} a_{k-j}\psi_k = 0 \qquad (j = 0, 1, \ldots) \tag{1.10}$$

has exactly $\varkappa$ linearly independent solutions $\psi^{(l)} = \{\psi_k^{(l)}\}_{k=0}^\infty$ $(l = 1, 2, \ldots . \varkappa)$, which simultaneously belong to all spaces E_+. The inhomogeneous system (1.9) has a solution $\xi = \{\xi_k\}_0^\infty \in E_+$ if and only if the following conditions are satisfied:

$$\sum_{k=0}^{\infty} \eta_k\psi_k^{(l)} = 0 \qquad (l = 1, 2, \ldots, \varkappa). \tag{1.11}$$

Proof. Let $\hat{A}$ be the operator defined on the space E_+ by the left-hand side of system (1.9). By $\hat{A}'$ we denote the operator transposed to $\hat{A}$ which is defined on E_+ by the left-hand side of system (1.10). The operator $\hat{A}'$ has the symbol $a(\bar{z})$, and

$$\operatorname{ind} a(\bar{z}) = -\varkappa .$$

Thus Theorem 1.3 with the exception of the last assertion on the solvability of the system (1.9) in the case $\varkappa > 0$ follows from the Theorems 1.1 and 1.2. Now we prove the last assertion of Theorem 1.3.

It is well-known that every linear functional f on the space l_+^p $(1 \leqq p < \infty)$ is determined by a sequence $\{f_k\}_0^\infty \in l_+^{p'}$ by means of the formula

$$\langle \xi, f\rangle = \sum_{k=0}^{\infty} \xi_k f_k \quad \text{with} \quad \xi = \{\xi_k\}_0^\infty \in l_+^p, \tag{1.12}$$

where $p' = \dfrac{p}{p-1}$ for $p > 1$, $p' = \infty$ for $p = 1$ and $l_+^\infty = m_+$.

It is easy to see that for arbitrary sequences $\xi = \{\xi_k\}_0^\infty \in l_+^p$ and $f = \{f_k\}_0^\infty \in$ $\in l_+^{p'}$ the commutation formula

$$\langle \hat{A}\xi, f\rangle = \sum_{j=0}^{\infty}\left(\sum_{k=0}^{\infty} a_{j-k}\xi_k\right) f_j = \sum_{k=0}^{\infty}\left(\sum_{j=0}^{\infty} a_{j-k}f_j\right)\xi_k \tag{1.13}$$

holds. Using relation (1.12) we conclude from (1.13) that $\hat{A}'$ (in $l_+^{p'}$) is the adjoint operator to $\hat{A}$ (in l_+^p, $1 \leqq p < \infty$). Since by Theorem 1.1 the operator $\hat{A}$ is normally solvable. the equation (1.9) is solvable if and only if the condition

$$\langle \eta, \psi \rangle = \sum_{k=0}^{\infty} \eta_k \psi_k = 0$$

is satisfied for any solution $\psi = \{\psi_k\}_0^\infty$ of the equation $\hat{A}^* \psi = 0$, i.e. the condition (1.11) is satisfied.

Now let E_+ be one of the spaces m, c or c^0. Then for any sequence $\{f_k\}_0^\infty \in \in l_+^1$ the formula (1.12) defines a linear functional f on the space E_+, and the formula (1.13) also holds for every $\xi \in E_+$. Hence each of the sequences $\psi^{(l)}$ $(l = 1, 2, \ldots, \varkappa)$ is a solution of the adjoint homogeneous equation $\hat{A}\psi^* = 0$. Since by Theorem 1.1 the number of the linearly independent solutions of the last equation is exactly equal to $\varkappa$, the sequences $\psi^{(l)}$ $(l = = 1, 2, \ldots, \varkappa)$ are just all these solutions. The rest of the argument is the same as in the case of the space l_+^p. Hence the Theorem 1.3 is completely proved.

3.2. Wiener-Hopf Integral Equations

In this section we consider Wiener-Hopf integral equations for a whole family of spaces of complex valued functions defined on the real line $(-\infty, +\infty)$ or on the positive half-line $(0, +\infty)$.

3.2.1. First we introduce some notations. By $L^p (1 \leqq p < \infty)$ we denote the space of all measurable complex valued functions $f(t)$ on $(-\infty, +\infty)$ for which $|f(t)|^p$ is integrable, where the norm is defined by

$$||f|| = \left[\int_{-\infty}^{\infty} |f(t)|^p \, dt \right]^{1/p}.$$

By M we denote the space of all measurable and almost everywhere bounded functions on the real line, where the norm is defined by

$$||f|| = \sup_{-\infty < t < \infty} \text{ess} \, |f(t)| \, .$$

M^c is the subspace of all continuous functions and M^u the subspace of all uniformly continuous functions from M.

Furthermore, let C be the space of all continuous functions on $(-\infty, +\infty)$ for which the (finite) limits

$$f(+\infty) = \lim_{t \to +\infty} f(t) \, , \qquad f(-\infty) = \lim_{t \to -\infty} f(t)$$

exist and are equal. C° is the subspace of all functions $f \in C$ with $f(+\infty) = f(-\infty) = 0$. In the spaces C and C^0 the norm is defined as follows:

$$||f|| = \max_{-\infty \leqq t \leqq +\infty} |f(t)| .$$

All these spaces mentioned before are complete, i.e. they are Banach spaces. It is well-known that $L^{p'}\left(\frac{1}{p}+\frac{1}{p'}=1\right)$ is the conjugate space of $L^p(1 < p < \infty)$ and $L^\infty = M$ is the conjugate space of L^1.

The space $L = L^1$ becomes a Banach algebra (without an unit) if the multiplication $f_1 * f_2$ of two functions $f_1, f_2 \in L$ is defined as *convolution*

$$(f_1 * f_2)(t) = \int_{-\infty}^{\infty} f_1(t-s)\, f_2(s)\, ds = \int_{-\infty}^{\infty} f_1(s)\, f_2(t-s)\, ds .$$

By $L^+(L^-)$ we denote the subalgebra of all functions from L, which are indentically equal to zero for $t < 0$ $(t > 0)$.

For any function $f \in L$ the *Fourier transform*

$$(Ff)(\lambda) = \int_{-\infty}^{\infty} e^{i\lambda t} f(t)\, dt \qquad (-\infty < \lambda < \infty)$$

is a function from C^0. Furthermore, the important convolution theorem holds: The Fourier transform of the convolution $f_1 * f_2$ of two functions f_1, $f_2 \in L$ is equal to the product of the Fourier transforms of these functions:

$$F(f_1 * f_2) = Ff_1 \cdot Ff_2 .$$

From this it follows that the set of all Fourier transforms $Ff (f \in L)$ form a certain Banach algebra $\mathring{\mathfrak{L}}$ (the norm of Ff is defined by $||f||_L$).

By $\tilde{L}$ we denote the Banach algebra obtained from L by formally adjoining of a unit δ (the Dirac δ-function).

Thus $\tilde{L}$ consists of all functions of the form $\mathfrak{f} = c\delta + f$, where c is an arbitrary constant and f is q function from L. We put $\delta * \delta = \delta$, $\delta * f = f * \delta = f$ and consequently

$$(c_1\delta + f_1) * (c_2\delta + f_2) = c_1c_2\delta + c_1f_2 + c_2f_1 + f_1 * f_2 .$$

The norm in $\tilde{L}$ is defined by the equation

$$||\mathfrak{f}|| = |c| + ||f||_L = |c| + \int_{-\infty}^{\infty} |f(t)|\, dt .$$

$\tilde{L}^+(\tilde{L}^-)$ is the subalgebra of all functions of the form $c\delta + f$ with $f \in L^+(L^-)$.

The Fourier transform of a function $\mathfrak{f} = c\delta + f \in \tilde{L}$ is defined by

$$\mathfrak{F}(\lambda) = c + \int_{-\infty}^{\infty} e^{i\lambda t} f(t)\, dt \qquad (-\infty < \lambda < \infty) .$$

The preceding remarks show that the set of Fourier transforms $\mathfrak{F}(\lambda)$ of all functions $\mathfrak{f} \in \tilde{L}$ forms a Banach algebra with the unit 1 in the norm

$||\mathfrak{F}|| = ||\mathfrak{f}||$. In the sequel we shall denote this algebra by $\mathfrak{L}$. Thus the Fourier transform $\mathfrak{f} \to \mathfrak{F}(\lambda)$ yields an isomorphism of the algebra $\tilde{L}$ onto the algebra $\mathfrak{L}$.

By $\mathfrak{L}^+(\mathfrak{L}^-)$ we denote the subalgebra of the algebra $\mathfrak{L}$ which consists of the Fourier transforms $\mathfrak{F}(\lambda)$ of all functions $\mathfrak{f} \in \tilde{L}^+(\tilde{L}^-)$. The subalgebra $\mathring{\mathfrak{L}}^-$ is formed by all functions $\mathfrak{F}(\lambda) \in \mathfrak{L}^-$ with $c = \mathfrak{F}(\pm\infty) = 0$. Obviously, $\mathfrak{L}$ is the direct sum $\mathfrak{L} = \mathfrak{L}^+ \dot{+} \mathring{\mathfrak{L}}^-$.

The functions $\dfrac{\lambda - i}{\lambda + i}$ and $\dfrac{\lambda + i}{\lambda - i}$ $(-\infty < \lambda < \infty)$ are the Fourier transforms of the functions

$$\mathfrak{f}_1 = \delta - 2e^{-t}(t > 0)\,, \qquad \mathfrak{f}_2 = \delta - 2e^{t} \qquad (t < 0)\,, \tag{2.1}$$

respectively, which belong to $\tilde{L}^+$ resp. $\tilde{L}^-$. Consequently $\dfrac{\lambda - i}{\lambda + i} \in \mathfrak{L}^+$ and $\dfrac{\lambda + i}{\lambda - i} \in \mathfrak{L}^-$.

From this with the above mentioned properties of the Fourier transformation we conclude easily that the linerar span of the functions

$$\left(\frac{\lambda - i}{\lambda + i}\right)^n \tag{2.2}$$

$(n = 0, \pm 1, \ldots)$ consists of the Fourier transforms of all functions of the form

$$c\delta - r(t)\,, \qquad r(t) = \begin{cases} e^{-t} p_1(t)\,, & 0 < t < \infty\,, \\ e^{t}\ p_2(t)\,, & -\infty < t < 0\,, \end{cases} \tag{2.3}$$

where c is an arbitrary constant and $p_1(t)$, $p_2(t)$ are arbitrary polynomials in t.

Since, as is well-known, the set of all functions $r(t)$ of the form (2.3) is dense in L, the algebra $\mathfrak{L}$ is the closure of the linear span of the functions (2.2) and thus is an R-algebra (compare Section 2.3.3). Moreover, we may conclude that the closures of the linear spans of the functions (2.2) for the integers $n \geqq 0$, $n \leqq 0$ or $n \leqq -1$ coincide with the subalgebras $\mathfrak{L}^+$, $\mathfrak{L}^-$ or $\mathring{\mathfrak{L}}^-$, respectively. From this it follows that $\mathfrak{L} = \mathfrak{L}^+ \dot{+} \mathring{\mathfrak{L}}^-$ is also a decomposing algebra. At the same time we have shown that the functions from $\mathfrak{L}^+(\mathfrak{L}^-)$ possess a continuation which is analytic in the open half-plane $\operatorname{Im} \lambda > 0$ $(\operatorname{Im} \lambda < 0)$ and continuous in the closed half-plane $\operatorname{Im} \lambda \geqq 0$ $(\operatorname{Im} \lambda \leqq 0)$.

Applying the results of the Section 2.3 to the algebra $\mathfrak{L}$ we obtain the following important theorem.

N. Wiener's Theorem. *$\mathfrak{F}(\lambda) \in \mathfrak{L}$ and $\mathfrak{F}(\lambda) \neq 0$ $(-\infty \leqq \lambda \leqq \infty)$ imply $[\mathfrak{F}(\lambda)]^{-1} \in \mathfrak{L}$.*

3.2.2. In the sequel by E we denote one of the spaces

$$L^p(1 \leqq p < \infty)\,, \qquad C^o \subset C \subset M^u \subset M^c \subset M \tag{2.4}$$

and by E_+ the corresponding space of functions on the positive half-line $[0, +\infty)$.

In this section we shall be concerned with the properties of integral equations of the form

$$(\hat{A}\varphi)\,(t) = c\varphi(t) - \int_0^\infty k(t-s)\,\varphi(s)\,ds = f(t) \qquad (0 \leqq t < \infty)\,, \tag{2.5}$$

where c is a constant and $k(t) \in L$.

As is well-known (see E. C. TITCHMARSH [1] or A. ZYGMUND [1]), the operator[1])

$$(K\varphi)\,(t) = \int_{-\infty}^{\infty} k(t-s)\,\varphi(s)\,ds = (k * \varphi)(t)\,\big(k(t) \in L)\big) \tag{2.6}$$

maps each of the functions spaces E into itself, where for the norm of this operator the following estimate holds:

$$||K||_E \leqq ||k||_L\,. \tag{2.7}$$

Hence the operator A defined in E by the equation

$$(A\varphi)\,(t) = c\varphi(t) - \int_{-\infty}^{\infty} k(t-s)\,\varphi(s)\,ds \qquad (-\infty < t < \infty) \tag{2.8}$$

is bounded.

We remark that, according to the notation introduced above the operator A can be considered as the operator obtained by convolution with the function $\mathfrak{k} = c\delta - k(t) \in \tilde{L}$:

$$(A\varphi)(t) = (\mathfrak{k} * \varphi)(t) \qquad (\mathfrak{k} \in \tilde{L}, \varphi \in E)\,. \tag{2.8'}$$

(2.7) implies the estimate

$$||A||_E \leqq ||\mathfrak{k}||_{\tilde{L}}\,. \tag{2.7'}$$

Furthermore, by the associativity of the convolution the relation

$$\mathfrak{h} * (\mathfrak{k} * \varphi) = (\mathfrak{h} * \mathfrak{k}) * \varphi \tag{2.9}$$

holds for arbitrary functions $\mathfrak{h}, \mathfrak{k} \in \tilde{L}$ and $\varphi \in E$.

For the operator A defined by (2.8′) the following relations are obvious:

$$AP = PAP(\mathfrak{k} \in \tilde{L}^+)\,, \qquad PA = PAP(\mathfrak{k} \in \tilde{L}^-)\,. \tag{2.10}$$

---

1) If $\varphi \in L^p$ $(1 \leqq p < \infty)$, then the integral (2.6) is to understand as the limit of the sequence $\int_{-n}^{m} k(t-s)\,\varphi(s)\,ds$ $(n, m \to \infty)$ with respect to the norm convergence in L^p. In all the other cases this integral exists in the common sense.

Here P denotes the projection of the space E acting according the following rule:

$$(P\varphi)(t) = \begin{cases} \varphi(t), & 0 < t < \infty, \\ 0, & -\infty < t < 0. \end{cases} \qquad (2.11)$$

Now we consider operators U and U^{-1} defined on the space E by the following equations:

$$(U\varphi)(t) = \varphi(t) - 2 \int\limits_{-\infty}^{t} e^{s-t}\, \varphi(s)\, ds, \qquad (-\infty < t < \infty) \qquad (2.12)$$

$$(U^{-1}\varphi)(t) = \varphi(t) - 2 \int\limits_{t}^{\infty} e^{t-s}\, \varphi(s)\, ds.$$

Obviously, U and U^{-1} are the operators obtained by convolution with the functions $\mathfrak{f}_1 \in \tilde{L}^+$ and $\mathfrak{f}_2 \in \tilde{L}^-$ defined by (2.1).

By $\mathfrak{f}_1 * \mathfrak{f}_2 = \delta$ the equation (2.9) immediately shows that U^{-1} is the inverse of U. Furthermore, it follows from (2.10) that the operators U, U^{-1} and P satisfy condition (II) of Section 2.1.

Now we show that the spectrum of the operators U and U^{-1} lies on the unit circle. This immediately implies condition (I), Section 2.1. Let $|\mu| \neq 1$. We write the operator $U - \mu I$ as a convolution operator:

$$(U - \mu I)\, \varphi = (\mathfrak{f}_1 - \mu\delta) * \varphi.$$

Since the Fourier transform $\mathfrak{F}(\lambda) = \dfrac{\lambda - i}{\lambda + i} - \mu$ of the function $\mathfrak{f} = \mathfrak{f}_1 - \mu\delta \in$ $\in \tilde{L}$ is unequal to zero for all $\lambda(-\infty \leq \lambda \leq \infty)$, N. Wiener's theorem implies the existence of a function $\mathfrak{g} \in \tilde{L}$ with $\mathfrak{g} * \mathfrak{f} = \delta$. By the relation (2.9) the convolution operator with the function $\mathfrak{g}$ is the inverse of $U - \mu I$. Hence the spectrum of U can only lie on the unit circle. Repeating the last argument with $\mathfrak{f}_2$ instead of $\mathfrak{f}_1$ we obtain the assertion for the operator U^{-1}.

It follows immediately from the statements of Subsection 3.2.1 that the linear span $\mathfrak{E}(U)$ of the operators U^n $(n = 0, \pm 1, \ldots)$ consists of all operators of the form

$$(R\varphi)(t) = (c\delta - r) * \varphi = c\varphi(t) - \int\limits_{-\infty}^{\infty} r(t - s)\, \varphi(s)\, ds, \qquad (2.13)$$

where the function $r(t)$ has the form (2.3). Here for the polynomial $R(z)$ $(|z| = 1)$ associated with the operator $R \in \mathfrak{E}(U)$ (the symbol of the operator R) the relation

$$R\left(\frac{\lambda - i}{\lambda + i}\right) = c - \int\limits_{-\infty}^{\infty} e^{i\lambda t}\, r(t)\, dt \qquad (2.14)$$

holds. Since the functions of the form $c\delta - r(t)$ form a set dense in $\tilde{L}$, as we have already remarked above, by inequality (2.7') we conclude that the algebra $\mathfrak{R}(U)$ contains all operators A of the form (2.8) with $k(t) \in L$. Thus for the symbol $A(z)$ ($|z| = 1$) of the operator $A \in \mathfrak{R}(U)$ defined by the equation (2.8) we obtain the following equation

$$A(z) = \mathcal{A}\left(i\frac{1+z}{1-z}\right), \qquad \mathcal{A}(\lambda) = c - \int_{-\infty}^{\infty} e^{i\lambda t}\, k(t)\, dt \qquad (2.15)$$
$$(-\infty \leqq \lambda \leqq \infty)\,.$$

In fact, let $A(z)$ be the function defined by (2.15). Then by (2.14) we obtain the estimate

$$\max_{|z|=1} |A(z) - R(z)| \leqq ||k - r||_L\,.$$

If now $\{r_n(t)\}_0^\infty$ is a sequence of functions of the form (2.3) for which $||k - r_n||_L \to 0$, then the last inequality and the estimate (2.7′) yield the relations

$$R_n(z) \rightrightarrows A(z)\ (|z| = 1)\,, \qquad R_n \to A\,.$$

From this it follows immediately that $A(z)$ is the symbol of A.

The function $\mathcal{A}(\lambda)$ $(-\infty \leqq \lambda \leqq \infty)$ of the real variable λ which by (2.15) is obtained from the symbol $A(z)$ by means of a bilinear transformation of the unit circle onto the closed real line will also be called the *symbol* of the operator A.

In order to be able to apply the results of the Section 2.1 to integral equations of the form (2.5) in analogy to the preceding section we embed the space E_+ of functions on the half-line into the space E by identification of the function $f(t) \in E_+$ with the function

$$\tilde{f}(t) = \begin{cases} f(t)\,, & t > 0 \\ 0\,, & t < 0\,. \end{cases}$$

Then $V = PUP \mid \operatorname{im} P$ and $V^{(-1)} = PU^{-1}P \mid \operatorname{im} P$ are operators defined on E_+ by equations

$$\begin{aligned} (V\varphi)(t) &= \varphi(t) - 2\int_0^t e^{s-t}\,\varphi(s)\,ds\,, \\ (V^{(-1)}\varphi)(t) &= \varphi(t) - 2\int_t^\infty e^{t-s}\,\varphi(s)\,ds\,. \end{aligned} \qquad (0 < t < \infty)$$

The operator $\hat{A} = PAP \mid \operatorname{im} P$ can be identified with the operator defined on the space E_+ by equation (2.5). The function $\mathcal{A}(\lambda)$ is also called the *symbol* of the operator $\hat{A}$.

It is easy to see that

$$\dim \operatorname{coker}\,(U \mid \operatorname{im} P) = \dim \ker V^{(-1)} = 1\,. \qquad (2.16)$$

In fact, differentiating both sides of the equation

$$\varphi(t) - 2\int_t^\infty e^{t-s}\,\varphi(s)\,ds = 0 \qquad (2.17)$$

by t and then subtracting the obtained equation from (2.17) yields $\varphi(t) = e^{-t}$ $(t > 0)$ as the unique solution of the equation (2.17).

Applying Theorems 1.3 and 4.1 from Chapter 2 we now obtain the following result for Wiener-Hopf integral equations on the half-line.

Theorem 2.1. *Let $k(t) \in L$. For the operator $\hat{A}$ defined on the space E_+ by equation* (2.5) *to be at least one-sided invertible it is necessary and sufficient that its symbol vanishes nowhere*:

$$\mathcal{A}(\lambda) = c - \int_{-\infty}^{\infty} e^{i\lambda t} k(t)\, dt \neq 0 \qquad (-\infty \leqq \lambda \leqq \infty)\,. \tag{2.18}$$

If condition (2.18) *is satisfied, then the invertibility of the operator $\hat{A}$ corresponds to the number*

$$\varkappa = \operatorname{ind} \mathcal{A}(\lambda) = \frac{1}{2\pi} [\arg \mathcal{A}(\lambda)]_{\lambda=-\infty}^{\infty}\,,$$

and

$$\dim \ker \hat{A} = \max(-\varkappa, 0)\,, \qquad \dim \operatorname{coker} \hat{A} = \max(\varkappa, 0)\,.$$

If $\mathcal{A}(\lambda) = 0$ for some λ $(-\infty \leqq \lambda \leqq \infty)$, then $\hat{A}$ is neither a Φ_+- nor a Φ_--operator on the space E_+.

Remark. As in the case of discrete Wiener-Hopf equations the necessity of condition (2.18) can be stated more precisely as follows (see J. Leiterer [1]): If the symbol $\mathcal{A}(\lambda)$ vanishes for at least one $\lambda(-\infty \leqq \lambda \leqq \infty)$ and if it is not identically equal to zero, then A is not a normally solvable operator.

In view of the properties formulated just now the operator $\hat{A}$ especially then is in particular neither not a Φ_+- nor a Φ_--operator (and not even normally solvable) on the space E_+ if

$$c = \mathcal{A}(\pm\infty) = 0\,.$$

In this case (2.5) is called an *equation of the first kind*. We shall investigate such equations in Chapter 5.

In the remainder of this section we consider only *equations of the second kind* for which we can always assume that $c = 1$. In this case the condition (2.18) has the following form:

$$\mathcal{A}(\lambda) = 1 - \int_{-\infty}^{\infty} e^{i\lambda t} k(t)\, dt \neq 0 \qquad (-\infty < \lambda < \infty)\,. \tag{2.18'}$$

We assume that the last condition is satisfied. Then the one-sided inverses of the operator

$$(\hat{A}\varphi)(t) = \varphi(t) - \int_0^{\infty} k(t-s)\, \varphi(s)\, ds \qquad (0 < t < \infty)$$

can be effectively constructed.

In fact, by Theorem 3.3, Chapter 2, the symbol $\mathcal{A}(\lambda)$ ($\in \mathfrak{L}$) admits a factorization of the form

$$\mathcal{A}(\lambda) = \mathcal{A}_-(\lambda)\left(\frac{\lambda - i}{\lambda + i}\right)^{\varkappa} \mathcal{A}_+(\lambda) \tag{2.19}$$

with $\mathcal{A}_\pm(\lambda) \in \mathfrak{L}^\pm$ and $\mathcal{A}_\pm^{-1}(\lambda) \in \mathfrak{L}^\pm$. Analogously to Section 3.1 we represent the function $\mathcal{A}^{-1}(\lambda)$ in the form

$$\mathcal{A}^{-1}(\lambda) = \mathcal{B}_-(\lambda)\,\mathcal{B}_+(\lambda)\,,$$

where

$$\mathcal{B}_+(\lambda) = \begin{cases} \mathcal{A}_+^{-1}(\lambda)\,, & \varkappa \geqq 0 \\ \mathcal{A}_+^{-1}(\lambda)\left(\dfrac{\lambda - i}{\lambda + i}\right)^{-\varkappa}, & \varkappa < 0\,, \end{cases}$$

$$\mathcal{B}_-(\lambda) = \begin{cases} \mathcal{A}_-^{-1}(\lambda)\left(\dfrac{\lambda - i}{\lambda + i}\right)^{-\varkappa}, & \varkappa \geqq 0 \\ \mathcal{A}_-^{-1}(\lambda)\,, & \varkappa < 0\,. \end{cases}$$

Obviously, we have $\mathcal{B}_\pm(\lambda) \in \mathfrak{L}^\pm$ so that these functions admit the representations

$$\mathcal{B}_+(\lambda) = 1 + \int_0^\infty \gamma_1(t)\, e^{i\lambda t}\, dt\,, \qquad \mathcal{B}_-(\lambda) = 1 + \int_{-\infty}^0 \gamma_2(t)\, e^{i\lambda t}\, dt$$

with $\gamma_1(t) \in L^+$, $\gamma_2(t) \in L^-$.

Applying formula (4.5) from Chapter 2 we obtain the following relation for the corresponding invers operator $\hat{A}^{(-1)}$ after some simple calculations:

$$(\hat{A}^{(-1)}\varphi)(t) = \varphi(t) + \int_0^\infty \gamma(t, s)\,\varphi(s)\, ds \qquad (0 \leqq t < \infty)\,,$$

where the kernel $\gamma(t, s)$ is determined by formula

$$\gamma(t, s) = \gamma_1(t - s) + \gamma_2(t - s) + \int_0^{\min(t, s)} \gamma_1(t - r)\,\gamma_2(r - s)\, dr\,.$$

Theorem 2.2. *If the condition (2.18′) is satisfied, then the homogeneous equation $\hat{A}\varphi = 0$ has the same solutions in all spaces E_+.*

Proof. Since for $\varkappa \geqq 0$ the equation $\hat{A}\varphi = 0$ has only the trivial solution, it is only necessary to consider the case $\varkappa < 0$.

Let A be the operator with symbol $\mathcal{A}(\lambda)$ defined on the space E by the equation (2.8) (for $c = 1$). Obviously, the symbol of the operator

$$B = PU^{-\varkappa}AP$$

considered in E_+ has index zero. Hence by Theorem 2.1 there exists an inverse operator B^{-1} which maps the space E_+ continuously onto itself.

Using the relations (1.13), Chapter 2, we obtain $PU^{\varkappa}B = PAP = \hat{A}$ and thus

$$\hat{A}B^{-1}PU^{-\varkappa}Q = 0 .$$

This implies $B^{-1}PU^{-\varkappa}$ (im Q) $\subset$ ker $\hat{A}$.

It is easy to check that the functions

$$e^{-t},\ e^{-t}t,\ \ldots,\ e^{-t}t^{-\varkappa-1} \qquad (t > 0)$$

form a basis for the subspace $PU^{-\varkappa}$(im Q). Thus this subspace has the dimension $-\varkappa$ and simultaneoulsy belongs to all spaces E_+. Therefore by dim ker $\hat{A}$ $= -\varkappa$ we have

$$B^{-1}PU^{-\varkappa}(\operatorname{im} Q) = \ker \hat{A} .$$

Since B^{-1} maps each of the spaces E_+ into itself, ker $\hat{A}$ also belongs to all the spaces E_+.

Theorem 2.3. *Let the symbol* $\mathcal{A}(\lambda)$ *of the operator* $\hat{A}$ *satisfy the condition* (2.18′) *and let* $\nu = -$ ind $\mathcal{A}(\lambda) > 0$. *Then the following statements hold*:

(a) *The Fourier transform of the general solution of the equation* $\hat{A}\varphi = 0$ *is given by the formula*

$$\Phi(\lambda) = \mathcal{A}_+^{-1}(\lambda)\,\frac{P_{\nu-1}(\lambda)}{(\lambda + i)^{\nu}} . \tag{2.20}$$

Here $P_{\nu-1}(\lambda)$ *is an arbitrary polynomial in* λ *of a degree* $\leqq \nu - 1$, *and the function* $\mathcal{A}_+(\lambda)$ *is determined by the factorization* (2.19).

(b) *There exists a basis* $\{\varphi_j(t)\}_{j=0}^{\nu-1}$ *of the null space* ker $\hat{A}$ *with the following properties*:

1. *The functions* $\varphi_j(t)$ $(j = 0, 1, \ldots, \nu - 1)$ *are absolutely continuous in every finite interval and tend to zero for* $t \to +\infty$.

2. $\varphi_{j+1}(t) = \dfrac{d\varphi_j(t)}{dt}, \qquad \varphi_j(0) = 0 \ (j = 0, \ldots, \nu - 2)$.

3. $\varphi_{\nu-1}(0) \neq 0$.

Proof. (a) In the proof of the Theorem 2.2 it was shown that the general solution of the equation $\hat{A}\varphi = 0$ has the form

$$\varphi = B^{-1}[e^{-t}\,R_{\nu-1}(t)] ,$$

where $R_{\nu-1}(t)$ is an arbitrary polynomial in t of degree $\leqq \nu - 1$.

By formula (4.5), Chapter 2, we have

$$B^{-1} = A_+^{-1}PA_-^{-1} ,$$

where $A_\pm^{-1} \in \Re^{\pm}(U)$ is the operator belonging to the symbol $\mathcal{A}_\pm^{-1}(\lambda)$. We put $\psi(t) = e^{-t}R_{\nu-1}(t)$ $(t > 0)$. Then the Fourier transform of ψ is

$$(F\psi)(\lambda) = \frac{\tilde{R}_{\nu-1}(\lambda)}{(\lambda + i)^{\nu}} .$$

Using the relations

$$(FA_{\pm}^{-1}f)(\lambda) = \mathcal{A}_{\pm}^{-1}(\lambda)\,(Ff)(\lambda) \qquad (f \in L)$$

we obtain

$$\Phi(\lambda) = \mathcal{A}_{+}^{-1}(\lambda)\,FPA_{-}^{-1}\psi$$

and

$$g(\lambda) = FPA_{-}^{-1}\psi = \mathcal{A}_{-}^{-1}(\lambda)\frac{\tilde{R}_{\nu-1}(\lambda)}{(\lambda+i)^{\nu}} - FQA_{-}^{-1}\psi\,,$$

where $Q = I - P$. The last equation implies

$$g(\lambda) - \frac{P_{\nu-1}(\lambda)}{(\lambda+i)^{\nu}} \in \mathring{\mathfrak{L}}^{-}\,.$$

But on the other hand, this function obviously belongs to $\mathfrak{L}^{+}$. Since $\mathfrak{L}^{+} \cap \mathring{\mathfrak{L}}^{-} = \{0\}$ it follows then that $g(\lambda) = \dfrac{P_{\nu-1}(\lambda)}{(\lambda+1)^{\nu}}$. This proves the formulae (2.20).

(b) We put

$$\Phi_0(\lambda) = \frac{\mathcal{A}_{+}^{-1}(\lambda)}{(\lambda+i)^{\nu}}\,, \qquad \Phi_j(\lambda) = (-i\lambda)^j\,\Phi_0(\lambda) \qquad (j = 1, \dots, \nu-1)\,.$$

By assertion (a) proved just now we have

$$\Phi_j(\lambda) = \int_0^{\infty} e^{i\lambda t}\,\varphi_j(t)\,dt \qquad (j = 0, 1, \dots, \nu-1)\,,$$

where the functions $\varphi_j(t)$ are solutions of the equation $\hat{A}\varphi = 0$ belonging to all spaces E_{+}. Thus

$$\int_0^{\infty} e^{i\lambda t}\,\varphi_{j+1}(t)\,dt = -\,i\lambda\int_0^{\infty} e^{i\lambda t}\,\varphi_j(t)\,dt \qquad (j = 0, \dots, \nu-2)\,,$$

from which by partial integration we obtain

$$\varphi_j(t) = \int_0^{t} \varphi_{j+1}(s)\,ds \qquad (j = 0, \dots, \nu-2)\,.$$

This proves the conditions 1 and 2 for the functions $\varphi_j(t)$, $j = 0, \dots, \nu-2$.

It is easily seen that the function $\varphi_{\nu-1}(t)$ has the representation

$$\varphi_{\nu-1}(t) = ce^{-t} + \int_0^{t} e^{s-t}\,f(s)\,ds \qquad (t > 0)$$

with $c = \text{const} \neq 0$ and $f \in L$. From this condition 3 and condition 1 for $j = \nu - 1$ follow immediately.

Thus the theorem is proved.

Theorem 2.4. *Let the condition* (2.18′) *be satisfied.*

In the case $\varkappa \leqq 0$ the inhomogeneous equation

$$\varphi(t) - \int_0^\infty k(t-s)\,\varphi(s)\,ds = f(t) \qquad (0 \leqq t < \infty) \tag{2.21}$$

is solvable in the space E_+ for any function $f \in E_+$, and the homogeneous equation

$$\varphi(t) - \int_0^\infty k(t-s)\,\varphi(s)\,ds = 0$$

has exactly $-\varkappa$ linearly independent solutions.

For $\varkappa > 0$ the homogeneous equation transposed to (2.21)

$$\psi(t) - \int_0^\infty k(s-t)\,\psi(s)\,ds = 0 \tag{2.22}$$

has exactly $\varkappa$ linearly independent solutions $\psi_j(t)$ $(j = 1, 2, \dots, \varkappa)$ belonging to all spaces E_+. The inhomogeneous equation (2.21) has a solution $\varphi \in E_+$ (which then is necessarily unique) if and only if the following conditions hold:

$$\int_0^\infty f(t)\,\psi_j(t)\,dt = 0 \qquad (j = 1, 2, \dots, \varkappa)\,.$$

Proof. Theorems 2.1 and 2.2 show that it is only necessary to prove the last assertion of the theorem about the solvability of the equation (2.21) in the case $\varkappa > 0$.

The proof of this assertion is analogous to the proof of the Theorem 1.3. The formula (1.12) is to interchange with the integral relation

$$\langle \varphi, \tilde{f} \rangle = \int_0^\infty \varphi(t)\,f(t)\,dt \qquad (f \in L_+^{p'},\ \varphi \in L_+^p)\,,$$

and the commutation formula (1.13) with

$$\langle \hat{A}\varphi, \tilde{f} \rangle = \int_0^\infty (\hat{A}\varphi)(t)\,f(t)\,dt = \int_0^\infty \varphi(t)\,(\hat{A}'f)(t)\,dt\,, \tag{2.23}$$

where $\hat{A}'$ is the operator defined on the space E_+ by the left-hand side of the equation (2.22). The rest of the argument is the same as in the proof of Theorem 1.3.

3.3. Paired Wiener-Hopf Equations

In the preceding two sections we have seen that the special invertible operators of the form (1.1) and (2.12) considered there satisfy the conditions (I) and (II) of 2.1 and the condition (III) of 2.2.3. Consequently all the results of Section 2.2 can be applied to paired operators of the form $AP + BQ$ and $PA + QB$, where A and B are discrete or integral operators of Wiener-

Hopf type. Here we shall formulate the most important cases of these applications.

3.3.1. Paired Systems. Let E be one of the sequence spaces l^p $(1 \leqq p \leqq \infty)$, c or c^0 introduced in Section 3.1 and U, P the operators defined in E by the relations (1.1) and (1.2).

We recall that the algebra $\mathfrak{R}(z)$ contains the algebra W of all functions on the unit circle whose Fourier series is absolutely convergent. In the case $E = l^1$ we have $\mathfrak{R}(z) = W$, and for the space $E = l^2$ the equality $\mathfrak{R}(z) = C\{|z| = 1\}$ holds.

Let $a(z)$ and $b(z)$ be two functions from $\mathfrak{R}(z)$ which in the case $E \neq l^2$ shall belong to the algebra W. By a_n, b_n $(n = 0, \pm 1, \ldots)$ we denote the Fourier coefficients of these functions and by $A, B \in \mathfrak{R}(U)$ the operators with the symbols $a(z)$, $b(z)$, respectively. By 3.1 the operators A and B are represented in E by the Toeplitz matrices $\{a_{j-k}\}_{j,k=-\infty}^{\infty}$ and $\{b_{j-k}\}_{j,k=-\infty}^{\infty}$, respectively. Hence the equation $(AP + BQ)\,\xi = \eta$ $(\xi, \eta \in E)$ can be written as the system

$$\sum_{k=0}^{\infty} a_{j-k}\xi_k + \sum_{k=-\infty}^{-1} b_{j-k}\xi_k = \eta_j \qquad (j = 0, \pm 1, \ldots)\,. \tag{3.1}$$

The equation $(PA + QB)\,\xi = \eta$ yields the system

$$\begin{cases} \displaystyle\sum_{k=-\infty}^{\infty} a_{j-k}\xi_k = \eta_j \qquad (j = 0, 1, \ldots)\,, \\ \displaystyle\sum_{k=-\infty}^{\infty} b_{j-k}\xi_k = \eta_j \qquad (j = -1, -2, \ldots)\,. \end{cases} \tag{3.2}$$

Applying the Theorems 2.5 and 4.2 from Chapter 2 and using the relation (1.4) we immediately obtain the following theorem.

Theorem 3.1. *For the operator $C = AP + BQ$ defined on the space E by the system* (3.1) *to be at least one-sided invertible it is necessary and sufficient that the following two conditions hold:*

$$a(z) \neq 0\,, \qquad b(z) \neq 0 \qquad (|z| = 1)\,. \tag{3.3}$$

If the conditions (3.3) *are satisfied, then the invertibility of the operator C corresponds to the index of the symbol*

$$\varkappa = \frac{1}{2\pi}\,[\arg a(e^{i\varphi})/b(e^{i\varphi})]_{\varphi=0}^{2\pi}\,, \tag{3.4}$$

and

$$\dim\ker C = \max(-\varkappa, 0)\,, \qquad \dim\operatorname{coker} C = \max(\varkappa, 0)\,. \tag{3.5}$$

If at least one of the conditions (3.5) *does not hold, then C is neither a Φ_+- nor a Φ_--operator on the space E.*

Theorem 3.1 is also true for the operator $PA + QB$ defined by the system (3.2). Furthermore, by the remarks made before it is sufficient to assume that the functions $a(z)$ and $b(z)$ belong to the algebra $\mathfrak{R}(z)$.

Using formulae (2.2) and (2.2′) of Chapter 2 we immediately obtain the following theorem from Theorem 1.2.

Theorem 3.2. *If the functions $a(z)$ and $b(z)$ from W satisfy the conditions* (3.3), *then the homogeneous systems belonging to* (3.1) *and* (3.2) *have in all spaces E the same solution.*

The following theorem holds analogously to the Theorem 1.3.

Theorem 3.3. *Let the functions $a(z)$ and $b(z)$ from W satisfy conditions* (3.3), *and let $\varkappa$ be the number determined by* (3.4).

If $\varkappa \leqq 0$, then the inhomogeneous system (3.1) *is solvable for any sequence $\{\eta_j\}_{-\infty}^{\infty} \in E$, and the homogeneous system belonging to* (3.1) *has exactly $-\varkappa$ linearly independent solutions.*

In the case $\varkappa > 0$ the homogeneous transposed system

$$\begin{cases} \sum\limits_{k=-\infty}^{\infty} a_{k-j}\xi_k = 0 & (j = 0, 1, \dots) \\ \sum\limits_{k=-\infty}^{\infty} b_{k-j}\xi_k = 0 & (j = -1, -2, \dots) \end{cases}$$

has exactly $\varkappa$ linearly independent solutions $\psi^{(l)} = \{\psi_k^{(l)}\}_{k=-\infty}^{\infty}$ $(l = 1, 2, \dots, \varkappa)$ belonging to all spaces E. The inhomogeneous system (3.1) *is solvable if and only if*

$$\sum_{k=-\infty}^{\infty} \eta_k \psi_k^{(l)} = 0 \qquad (l = 1, 2, \dots, \varkappa) .$$

The first part of the theorem follows immediately from Theorem 3.1. The second part can be proved analogously to Theorem 1.3.

A corresponding theorem can also be formulated for the paired system (3.2). Furthermore, by the aid of the formulae (4.19_1) and (4.19_2) from Chapter 2 the corresponding inverse operators $(AP + BQ)^{(-1)}$ and $(PA + QB)^{(-1)}$ can easily be constructed by factorization of the function $a(z)/b(z) \in W$.

By Theorem 2.7, Chapter 2, the operators $AP + BQ$ and $PA + QB$ have a two-sided regularizer in each of the spaces E if (and only if) the conditions (3.3) are satisfied. Such a regularizer is the operator defined in E by the system

$$\sum_{k=0}^{\infty} a_{j-k}^{(-1)}\xi_k + \sum_{k=-\infty}^{-1} b_{j-k}^{(-1)}\xi_k = \eta_j \qquad (j = 0, \pm 1, \dots) ,$$

where $a_n^{(-1)}$ and $b_n^{(-1)}$ $(n = 0, \pm 1, \dots)$ are the Fourier coefficients of the functions $1/a(z)$ and $1/b(z)$, respectively.

3.3.2. Paired Integral Equations. Now let E be any one of the function spaces (2.4). U and P the operators defined on E by (2.12) and (2.11), respectively.

It was proved in Section 3.2 that in this case the algebra $\mathfrak{R}(U)$ contains all integral operators of the form

$$(A\varphi)(t) = c_1\varphi(t) - \int_{-\infty}^{\infty} k_1(t-s)\,\varphi(s)\,ds$$

with $k_1(t) \in L$ and $c_1 = \text{const}$. If the operator $B \in \mathfrak{R}(U)$ has the form

$$(B\varphi)(t) = c_2\varphi(t) - \int_{-\infty}^{\infty} k_2(t-s)\,\varphi(s)\,ds$$

($k_2(t) \in L$, $c_2 = \text{const}$), then

$$(AP + BQ)\,\varphi = c\varphi(t) - \int_{0}^{\infty} k_1(t-s)\,\varphi(s)\,ds - \int_{-\infty}^{0} k_2(t-s)\,\varphi(s)\,ds \quad (3.6)$$

and

$$(PA + QB)\,\varphi = \begin{cases} c_1\varphi(t) - \int_{-\infty}^{\infty} k_1(t-s)\,\varphi(s)\,ds\,, & 0 < t < \infty\,, \\ c_2\varphi(t) - \int_{-\infty}^{\infty} k_2(t-s)\,\varphi(s)\,ds\,, & -\infty < t < 0\,, \end{cases} \quad (3.7)$$

where

$$c = \begin{cases} c_1\,, & t > 0 \\ c_2\,, & t < 0\,. \end{cases}$$

Usually, the equation $(PA + QB)\,\varphi = f$ is called a *paired integral equation* and the equation $(AP + BQ)\,\varphi = f$ an *equation with two kernels.*

According to the statements made in 2.2.3 and 3.2.2 the *symbol* of the operators defined by the equations (3.6) and (3.7) is the function

$$\mathscr{C}(\lambda, \theta) = \mathscr{A}(\lambda)\frac{1+\theta}{2} + \mathscr{B}(\lambda)\frac{1-\theta}{2}\;(-\infty \leqq \lambda \leqq \infty,\ \theta = \pm 1)$$

with

$$\mathscr{A}(\lambda) = c_1 - \int_{-\infty}^{\infty} e^{i\lambda t} k_1(t)\,dt\,, \qquad \mathscr{B}(\lambda) = c_2 - \int_{-\infty}^{\infty} e^{i\lambda t} k_2(t)\,dt\,.$$

Theorems 2.5 and 4.2. Chapter 2, and the results of the Section 3.2 immediately yield the following theorem:

Theorem 3.4. *For the operator $C = AP + BQ$ defined on the space E by equation* (3.6) *with $k_j(t) \in L$ ($j = 1, 2$) to be at least one-sided invertible it is necessary and sufficient that its symbol $\mathscr{C}(\lambda, \theta)$ does not degenerate:*

$$\mathscr{A}(\lambda) \neq 0\,, \qquad \mathscr{B}(\lambda) \neq 0 \qquad (-\infty \leqq \lambda \leqq \infty)\,. \quad (3.8)$$

If conditions (3.8) *are satisfied, then the invertibility of the operator C corresponds to the index of the symbol*

$$\varkappa = \frac{1}{2\pi} [\arg \mathcal{A}(\lambda)/\mathcal{B}(\lambda)]_{\lambda=-\infty}^{\infty} ,$$

and formulae (3.5) *hold.*

If one of the conditions (3.8) *is not satisfied, then the operator C is neither a Φ_+- nor a Φ_--operator on the space E.*

Theorem 3.4 also holds for the operator $PA + QB$ defined by equation (3.7)

Similarily as in Subsection 3.3.1 for paired systems Theorems 2.2 and 2.4 can be formulated for the paired integral operators defined by (3.6) and (3.7).

If the conditions (3.8) are satisfied, then by Theorem 2.7, Chapter 2, the operators (3.6) and (3.7) have a two-sided regularizer R in each of the spaces E which can be given by the equation

$$R\varphi = c^{-1}\varphi(t) - \int_0^\infty h_1(t-s)\,\varphi(s)\,ds - \int_{-\infty}^0 h_2(t-s)\,\varphi(s)\,ds .$$

Here the functions $h_1(t), h_2(t) \in L$ are defined by the relations (compare N. Wiener's theorem)

$$\mathcal{A}^{-1}(\lambda) = c_1^{-1} - \int_{-\infty}^{\infty} e^{i\lambda t}\,h_1(t)\,dt , \qquad \mathcal{B}^{-1}(\lambda) = c_2^{-1} - \int_{-\infty}^{\infty} e^{i\lambda t}\,h_2(t)\,dt .$$

3.4. Singular Integral Equations

3.4.1. Properties of the Cauchy Singular Integral in $H^\mu(\Gamma)$. Throughout section Γ shall be a *curve system*[1]) in the complex plane which consists of finitely many smooth closed curves without common points, and which divides the plane into two regions: a connected bounded region D_+ (interior region) and a region D_- (exterior region) which contains the point at infinity $z = \infty$. Without loss of generality we shall always assume in the sequel that the origin of the coordinate system lies in D_+. By $G_\pm = D_\pm \cup \Gamma$ we denote the closures of the regions $D_\pm$. As the positive orientation on Γ we will choose that orientation which, by following it, ensures that the region D_+ remains to the left.

As usual, $C(\Gamma)$ will be the Banach algebra of all continuous functions φ on Γ with the norm

$$\|\varphi\|_C = \max_{t\in\Gamma} |\varphi(t)| .$$

1) For the properties of smooth curve systems and of classes of Hölder continuous functions defined on such curves we refer to the book of N. I. Muskhelišvili [1], (Kapitel I).

$M(\Gamma)$ will be the Banach space of all measurable and almost everywhere bounded functions on Γ. By $H^\mu(\Gamma)$ $(0 < \mu \leqq 1)$ we denote the collection of all (complex valued) functions $\varphi(t)$ which are defined on Γ and satisfy a *Hölder condition* there (or, in other words, are *H-continuous*) with exponent μ, i.e.

$$|\varphi(t) - \varphi(t')| \leqq A\,|t - t'|^\mu$$

for all points $t, t' \in \Gamma$, where A is a certain constant (dependent on φ). By $H^{\mu,\lambda}(\Gamma \times \Gamma)$ we denote the set of all functions $\varphi(t, \tau)$ which are defined for all pairs (t, τ) and (t', τ') of points from Γ and which satisfy the condition

$$|\varphi(t, \tau) - \varphi(t', \tau')| \leqq A[|t - t'|^\mu + |\tau - \tau'|^\lambda] .$$

For $\lambda = \mu$ we simply write $H^\mu(\Gamma \times \Gamma)$.

In the sequel we shall always assume that the functions $\varphi(t) \in H^\mu(\Gamma)$ are equipped with the following norm:

$$||\varphi||_{H^\mu} = ||\varphi||_C + h_\mu(\varphi) , \qquad h_\mu(\varphi) = \sup_{t,t' \in \Gamma} \frac{|\varphi(t) - \varphi(t')|}{|t - t'|^\mu} .$$

With the norm defined in this way $H^\mu(\Gamma)$ is a Banach algebra. In particular, for arbitrary functions $\varphi(t)$, $\psi(t) \in H^\mu(\Gamma)$ we have the inequality

$$||\varphi\psi||_{H^\mu} \leqq ||\varphi||_{H^\mu}\, ||\psi||_{H^\mu} . \tag{4.1}$$

By $L^p(\Gamma)$ $(1 < p < \infty)$ we denote the Banach space of all measurable functions $\varphi(t)$ on Γ for which $|\varphi(t)|^p$ is integrable. The norm is defined by the formula

$$||\varphi||_{L^p} = \Big[\int_\Gamma |\varphi(t)|^p\, |dt|\Big]^{1/p} .$$

As is well-known (see I. I. Privalov [1] or N. I. Mushelišvili [1]), the value of the Cauchy integral

$$\Phi(z) = \frac{1}{2\pi i} \int_\Gamma \frac{\varphi(\tau)}{\tau - z}\, d\tau \tag{4.2}$$

exists for any function $\varphi(t) \in H^\mu(\Gamma)$ $(0 < \mu \leqq 1)$ as the Cauchy principal value at all points $z \in \Gamma$ (for $z \notin \Gamma$ the integral exists in the common sense). In other words, the limit

$$\int_\Gamma \frac{\varphi(\tau)}{\tau - t}\, d\tau = \lim_{\varepsilon \to 0} \int_{\Gamma - \Gamma_\varepsilon} \frac{\varphi(\tau)}{\tau - t}\, d\tau \qquad (t \in \Gamma) ,$$

exists where $\Gamma_\varepsilon = \{z : |z - t| < \varepsilon\} \cap \Gamma$. Furthermore we have

$$\int_\Gamma \frac{\varphi(\tau)}{\tau - t}\, d\tau = \int_\Gamma \frac{\varphi(\tau) - \varphi(t)}{\tau - t}\, d\tau + \pi i \varphi(t) , \tag{4.3}$$

where obviously the integral on the right-hand side of (4.3) is absolutely convergent. In particular for $\varphi(t) \equiv 1$ we obtain

$$\int_\Gamma \frac{d\tau}{\tau - t} = \pi i .$$

The integral

$$(S\varphi)(t) = \frac{1}{\pi i} \int_\Gamma \frac{\varphi(\tau)}{\tau - t} d\tau \qquad (t \in \Gamma) \tag{4.4}$$

is called a *Cauchy singular integral* or simply a *singular integral.* Usually the function φ is called the *density* of the singular integral and the expression $\frac{d\tau}{\tau - t}$ the *Cauchy kernel.* The operator S defined by the integral (4.4) is called a *singular operator* (with Cauchy kernel).

We mention the most important properties of the operator S.[1])

1°. *S is a bounded linear operator on the space $H^\mu(\Gamma)$ for any μ with $0 < \mu < 1$* (Plemelj-Privalov theorem).

2°. *Let $\varphi(t) \in H^\mu(\Gamma)$ $(0 < \mu \leqq 1)$. Then at every point $t \in \Gamma$ there exist the boundary values of the Cauchy integral* (4.2)

$$\Phi_+(t) = \lim_{\substack{z \to t \\ z \in D_+}} \Phi(z) , \qquad \Phi_-(t) = \lim_{\substack{z \to t \\ z \in D_-}} \Phi(z) , \tag{4.5}$$

and the following formulae of Sochozki-Plemelj *hold:*

$$\Phi_+(t) = \tfrac{1}{2} [\varphi(t) + S\varphi] , \qquad \Phi_-(t) = \tfrac{1}{2} [- \varphi(t) + S\varphi] . \tag{4.6}$$

3°. *If $\varphi(t, \tau) \in H^\mu(\Gamma \times \Gamma)$ $(0 < \mu \leqq 1)$, then the following commutation formula of* Poincaré-Bertrand *holds:*

$$\frac{1}{(\pi i)^2} \int_\Gamma \frac{d\tau}{\tau - t} \int_\Gamma \frac{\varphi(\varrho, \tau)}{\varrho - \tau} d\varrho = \frac{1}{(\pi i)^2} \int_\Gamma d\varrho \int_\Gamma \frac{\varphi(\varrho, \tau)\, d\tau}{(\tau - t)(\varrho - \tau)} + \varphi(t, t) . \tag{4.7}$$

From this the extremely important equality

$$S^2 = I \tag{4.8}$$

follows immediately, where I is the identity operator.

In the sequel the operators

$$P = \tfrac{1}{2} (I + S) , \qquad Q = \tfrac{1}{2} (I - S) \tag{4.9}$$

play a fundamental role. Obviously, $P + Q = I$, and from (4.8) we immediately obtain

$$P^2 = P , \qquad Q^2 = Q , \qquad PQ = QP = 0 . \tag{4.10}$$

---

1) For the proofs of the Theorems 1°—3° we refer again to the book of N. I. Muskhelišvili [1] (Kapitel I).

Hence P and Q are complementary projections on the Banach space $H^\mu(\Gamma)$ $(0 < \mu < 1)$.

4°. *The projection* $P(Q)$ *projects the space* $H^\mu(\Gamma)$ $(0 < \mu < 1)$ *onto the subspace* $H^\mu_+(\Gamma)\,(H^\mu_-(\Gamma))$ *of all functions from* $H^\mu(\Gamma)$ *which have a continuation analytic in the region* D_+ *(analytic in the region* D_- *and vanishing at the point at infinity).*

This theorem is a direct consequence of Theorem 2° and Cauchy's integral formulae.

Lemma 4.1. *Let* $K(t, \tau) \in H^{\lambda,\mu}(\Gamma \times \Gamma)$ $(0 < \mu < \lambda \leqq 1)$ *and let* T *be the integral operator*

$$(T\varphi)(t) = \int_\Gamma T(t, \tau)\, \varphi(\tau)\, d\tau \tag{4.11}$$

with the kernel

$$T(t, \tau) = \frac{K(t, \tau) - K(t, t)}{\tau - t}. \tag{4.12}$$

Then T *is a bounded operator of the space* $H^\nu(\Gamma)$ $(0 < \nu \leqq 1)$ *into the space* $H^\mu(\Gamma) : T \in \mathscr{L}[H^\nu(\Gamma), H^\mu(\Gamma)]$.

Proof. Let $\varphi(t) \in H^\nu(\Gamma)$ be an arbitrary function. We put $\psi(t) = (T\varphi)(t)$ $(t \in \Gamma)$. Then first we have

$$\|\psi\|_C \leqq c'\Big[\int_\Gamma |\tau - t|^{\mu-1}\, |d\tau|\Big]\, \|\varphi\|_C = c''\, \|\varphi\|_C\,, \tag{4.13}$$

where here and in the sequel the letter c denotes certain constants.

Now we have to estimate the quantity $h_\mu(\psi)$. Obviously, for this it is sufficient to assume that Γ consists of one smooth closed curve.

Let t_1 and t_2 be two different points on Γ whose distance is sufficiently small. We consider the circle with the radius $\delta = 2|t_1 - t_2|$ and center t_1, By Γ_δ we denote the part of Γ inside this circle and by a and b the end points of Γ_δ.

We have

$$\psi(t_1) - \psi(t_2) = \int_{\Gamma_\delta} T(t_1, \tau)\, \varphi(\tau)\, d\tau - \int_{\Gamma_\delta} T(t_2, \tau)\, \varphi(\tau)\, d\tau +$$

$$+ (t_1 - t_2) \int_{\Gamma - \Gamma_\delta} \frac{K(t_2, \tau) - K(t_2, t_2)}{(\tau - t_1)(\tau - t_2)}\, \varphi(\tau)\, d\tau + \int_{\Gamma - \Gamma_\delta} \frac{K(t_1, \tau) - K(t_2, \tau)}{\tau - t_1}\, \varphi(\tau)\, d\tau +$$

$$+ \int_{\Gamma - \Gamma_\delta} \frac{K(t_2, t_2) - K(t_1, t_1)}{\tau - t_1}\, \varphi(\tau)\, d\tau = \sum_{k=1}^{5} I_k.$$

The integrals I_1 and I_2 are easily estimated. Putting $r = |\tau - t_1|$ and using the relation

$$|d\tau| = |ds| \leqq \gamma |dr| \quad (s \text{ arc length}, \gamma = \text{const})$$

which is valid for smooth curves we at once obtain the estimate

$$|I_1| \leqq c_1 \,||\varphi||_C \int_0^\delta r^{\mu-1}\, dr = 2c_1 |t_1 - t_2|^\mu \,||\varphi||_C\,.$$

For the integral I_2 we obtain an analogous result.

For I_3 we have

$$|I_3| \leqq c_3 \,||\varphi||_C\, |t_1 - t_2| \int_{\Gamma-\Gamma_\delta} |\tau - t_1|^{\mu-2} \left|\frac{\tau - t_1}{\tau - t_2}\right|^{1-\mu} |d\tau|$$

$$\leqq c_4 \,||\varphi||_C\, |t_1 - t_2| \int_\delta^R r^{\mu-2}\, dr$$

with $R = \max\limits_{\tau \in \Gamma - \Gamma_\delta} |\tau - t_1|$ since obviously

$$\frac{|\tau - t_1|}{|\tau - t_2|} \leqq \frac{|\tau - t_2| + |t_2 - t_1|}{|\tau - t_2|} \leqq 1 + \frac{|t_2 - t_1|}{\delta} = \frac{3}{2}\,.$$

Calculating the last integral we obtain

$$|I_3| \leqq c_5 \,||\varphi||_C\, |t_1 - t_2|^\mu\,.$$

For the integral I_4 we find the estimate

$$|I_4| \leqq c_6 \,||\varphi||_C\, |t_1 - t_2|^\lambda \int_\delta^R \frac{dr}{r}\,.$$

Furthermore

$$\int_\delta^R \frac{dr}{r} \leqq c_7 \,|\ln |t_1 - t_2|| \leqq c_8\, |t_1 - t_2|^{-\varepsilon}\,,$$

where ε is an arbitrary small positive number. Hence

$$|I_4| \leqq c_9 \,||\varphi||_C\, |t_1 - t_2|^\mu\,.$$

We write the integral I_5 in the form

$$I_5 = [K(t_2, t_2) - K(t_1, t_1)] \left\{ \int_{\Gamma-\Gamma_\delta} \frac{\varphi(\tau) - \varphi(t_1)}{\tau - t_1}\, d\tau + \varphi(t_1) \int_{\Gamma-\Gamma_\delta} \frac{d\tau}{\tau - t_1} \right\}.$$

The second integral on the right-hand side can easily be calculated:

$$\int_{\Gamma-\Gamma_\delta} \frac{d\tau}{\tau - t_1} = \ln \frac{a - t_1}{b - t_1}\,.$$

Thus this integral is bounded by a constant independent of $t_1 \in \Gamma$. For the first integral we obtain the estimate

$$\left| \int\limits_{\Gamma-\Gamma_\delta} \frac{\varphi(\tau) - \varphi(t_1)}{\tau - t_1} \, d\tau \right| \leqq c_{10} h_\nu(\varphi) \int\limits_\delta^R r^{\nu-1} \, dr \leqq c_{11} h_\nu(\varphi) \, .$$

Hence

$$|I_5| \leqq c_{12} \, ||\varphi||_{H^\nu} \, |t_1 - t_2|^\mu \, .$$

Now the estimates for the integrals I_k $(k = 1, \ldots, 5)$ together with (4.13) yield the desired relation

$$||\psi||_{H^\mu} \leqq c_{13} \, ||\varphi||_{H^\nu} \, .$$

Remark. For $K(t, \tau) = f(\tau)$, $\varphi(\tau) \equiv 1$ and $\mu = \nu$ the theorem of Plemelj-Privalov is obtained from the proof of the lemma by using relation (4.3).

Furthermore, the preceding proof immediately yields the following corollary.

Corollary 4.1. *Let* $K(t, \tau) \in H^\mu(\Gamma \times \Gamma)$ $(0 < \mu \leqq 1)$ *and* $0 < \nu < \mu$. *Then* $T \in \mathscr{L}[M(\Gamma), H^\nu(\Gamma)]$.

Corollary 4.2. *Let* $a(t) \in H^\mu(\Gamma)$ $(0 < \mu < 1)$ *and let* T_a *be the commutator*[1])

$$T_a \varphi = (Sa - aS)\, \varphi = \frac{1}{\pi i} \int\limits_\Gamma \frac{a(\tau) - a(t)}{\tau - t} \, \varphi(\tau) \, d\tau \, . \tag{4.14}$$

Then $T_a \in \mathscr{L}[H^\nu(\Gamma), H^\mu(\Gamma)]$ $(0 < \nu \leqq 1)$.

This assertion follows from Lemma 4.1 for $K(t, \tau) = a(\tau)$.

Corollary 4.3. *Under the hypotheses of Lemma* 4.1 T *is a compact operator from* $H^\nu(\Gamma)$ $(0 < \nu \leqq 1)$ *into the space* $H^\mu(\Gamma)$.

Proof. Obviously, it is sufficient to show that the embedding operator of the space $H^\mu(\Gamma)$ into the space $H^\nu(\Gamma)$ is compact for $0 < \nu < \mu$, i.e. that every bounded set $E \subset H^\mu(\Gamma)$ is relatively compact as a subset of $H^\nu(\Gamma)$.

Let $E \subset H^\mu(\Gamma)$ be a bounded set. For all functions $f \in E$ we have $||f||_{H^\mu} \leqq C$ with a constant C independent of f. Hence

$$|f(t) - f(\tau)| \leqq C \, |t - \tau|^\mu \qquad (t, \tau \in \Gamma) \, , \tag{4.15}$$

i.e. the functions $f \in E$ are uniformly bounded and equicontinuous. Thus, by Arzelá's theorem, in every infinite subset of E there exists a sequence $\{f_n\}_1^\infty$ which converges uniformly on Γ to a limit function f_0. (4.15) shows that $f_0 \in H^\mu(\Gamma)$ and $||f_0||_{H^\mu} \leqq C$.

---

[1]) In the whole book we use the following notation: If $a(t)$ is a certain function, then by a (sometimes also by $a(t)$ or aI) we denote the operator of the multiplication by this function (in a function space mentioned in each case).

We put

$$F_n(t, \tau) = \frac{f_n(t) - f_n(\tau)}{|t - \tau|^\nu} \qquad (n = 0, 1, 2, \dots ; t, \tau \in \Gamma)$$

with $F_n(t, t) = 0$. It is well-known that then $F_n \in H^\lambda(\Gamma \times \Gamma)$ for $\lambda = \mu - \nu$ and

$$|F_n(t, \tau) - F_n(t', \tau')| \leqq ||f_n||_{H^\mu} [|t - t'|^\lambda + |\tau - \tau'|^\lambda]$$

(see N. I. Mushelišvili [1], pp. 13—14). Consequently by (4.15) the functions $F_n(t, \tau)$ $(n - 1, 2, \dots)$ are uniformly bounded and equicontinuous on $\Gamma \times \Gamma$. From this we conclude again by Arzelá's theorem that the sequence $\{F_n\}$ contains a subsequence $\{F_{n_k}\}$ which is uniformly convergent on $\Gamma \times \Gamma$. But since

$$F_{n_k}(t, \tau) \underset{k \to \infty}{\to} F_0(t, \tau) \qquad (t, \tau \in \Gamma)$$

in the sense of pointwise convergence, the sequence $\{F_{n_k}\}$ converges uniformly to F_0. Thus

$$||f_0 - f_{n_k}||_{H^\nu} = ||f_0 - f_{n_k}|| + h_\nu(f_0 - f_{n_k}) \underset{k \to \infty}{\to} 0 ,$$

and the assertion is proved.

Corollary 4.4. *Let $a(t) \in H^\mu(\Gamma)$ $(0 < \mu < 1)$. The commutator $T_a = Sa - aS$ is a compact operator from $H^\nu(\Gamma)$ $(0 < \nu \leqq 1)$ into the space $H^\mu(\Gamma)$.*

Corollary 4.5. *If*

$$T(t, \tau) = \frac{k(t, \tau)}{|t - \tau|^\lambda} \qquad (0 \leqq \lambda < 1) , \tag{4.16}$$

where $k(t, \tau) \in H^\nu(\Gamma \times \Gamma)$ $(0 < \nu \leqq 1)$, then the integral operator (4.11) is compact on the space $H^\mu(\Gamma)$ for any $\mu < \min (1 - \lambda, \nu)$.

Proof. We put $\lambda' = \min (1 - \lambda, \nu) > 0$ and

$$K(t, \tau) = (\tau - t) |\tau - t|^{-\lambda} k(t, \tau) .$$

Then $K(t, \tau) \in H^{\lambda'}(\Gamma \times \Gamma)$ (see N. I. Mushelišvili [1], p. 15), and moreover the representation (4.12) holds. Thus the assertion follows from Corollary 4.3.

Remark 1. If $T(t, \tau) \in H^\mu(\Gamma \times \Gamma)$, then obviously $T \in \mathscr{L}[M(\Gamma), H^\mu(\Gamma)]$. Hence T is a compact operator in the space $H^\mu(\Gamma)$.

Remark 2. Let $K(t, \tau) \in H^\mu(\Gamma \times \Gamma)$. Then the kernel (4.12) is representable in the form (4.16), where $k(t, \tau) \in H^{\mu+\lambda-1}(\Gamma \times \Gamma)$ and $1 - \mu < \lambda < 1$. λ can be chosen arbitrarily in the relevant interval (see N. I. Mushelišvili [1], §§ 5—6).

Kernels of the form (4.16) are said to be *weakly singular*.

Corollaries 4.3 and 4.4 can be generalized. To this end we introduce the (linear) set $H_*^{\lambda,\mu}(\Gamma \times \Gamma)$ $(0 < \lambda, \mu \leqq 1)$ of all complex valued functions $\varphi(t, \tau)$ which for all points $t, t', \tau \in \Gamma$ satisfy the following condition:

$$|\Delta_\varphi(t, t', \tau)| \equiv |\varphi(t, \tau) - \varphi(t', \tau) - \varphi(t, t) + \varphi(t', t)| \leqq A|t - t'|^\lambda \cdot |\tau - t'|^\mu .$$

First we state some simple properties of the class $H_*^{\lambda,\mu}(\Gamma \times \Gamma)$ which can easily be checked.

1) $a(t) \cdot b(\tau) \in H_*^{\lambda,\mu}(\Gamma \times \Gamma)$ for arbitrary functions $a(t) \in H^\lambda(\Gamma)$ and $b(t) \in H^\mu(\Gamma)$.

2) $H^{\lambda,\mu}(\Gamma \times \Gamma) \subset H_*^{\nu,\varepsilon}(\Gamma \times \Gamma)$ with $\nu = \lambda(1 - \delta)$ and $\varepsilon = \mu\delta$, where δ is a arbitrary number from the interval $0 < \delta < 1$.

3) If $K(t, \tau) = (t - \tau)\, T(t, \tau)$ and $T(t, \tau) \in H^{\lambda,\mu}(\Gamma \times \Gamma)$, then $K(t, \tau) \in H^{\lambda,\mu}(\Gamma \times \Gamma) \cap H_*^{\lambda,\mu}(\Gamma \times \Gamma)$.

4) For any two functions $a(t)$, $b(t)$ bounded on Γ and $\varphi(t, \tau) = a(t) + b(\tau)$ there holds $\Delta_\varphi = 0$ and thus $\varphi(t, \tau) \in H_*^{\lambda,\mu}(\Gamma \times \Gamma)$.

Furthermore the following lemma holds.

Lemma 4.2. *Let* $K(t,\tau) \in H^\mu(\Gamma \times \Gamma) \cap H_*^{\mu,\varepsilon}(\Gamma \times \Gamma)$, *where* $0 < \mu < 1$ *and* $\varepsilon > 0$. *Then the integral operator* (4.11) *with the kernel* (4.12) *is a compact operator from* $H^\nu(\Gamma)$ $(0 < \nu \leqq 1)$ *into the space* $H^\mu(\Gamma)$.

Proof. We see from the proofs of Lemma 4.1 and Corollary 4.3, that it is sufficient to obtain the estimate $|I_4| \leqq c|t_1 - t_2|^\mu$ for the integral I_4. We transform this integral as follows:

$$I_4 = \int\limits_{\Gamma - \Gamma_\delta} \frac{K(t_1, \tau) - K(t_2, \tau)}{\tau - t_1} [\varphi(\tau) - \varphi(t_1)]\, d\tau$$

$$+ \varphi(t_1) \left\{ \int\limits_{\Gamma - \Gamma_\delta} \frac{\Delta_k(t_1, t_2, \tau)}{\tau - t_1}\, d\tau + [K(t_1, t_1) - K(t_2, t_1)] \int\limits_{\Gamma - \Gamma_\delta} \frac{d\tau}{\tau - t_1} \right\}.$$

The integrands of the first two integrals can be estimated by the quantities $|t_1 - t_2|^\mu\, |\tau - t_1|^{\nu - 1}\, ||\varphi||_{H^\nu}$ and $|t_1 - t_2|^\mu\, |\tau - t_1|^{\varepsilon - 1}$, respectively. Hence for I_4 we obtain similar estimates as for the integrals I_3 and I_5 in the proof of Lemma 4.1. This completes the proof of Lemma 4.2.

By properties 1) and 2) of the class $H_*^{\lambda,\mu}(\Gamma \times \Gamma)$ Corollaries 4.3 and 4.4 are special cases of Lemma 4.2.

Corollary 4.6. *The algebra* $\mathfrak{A}(\Gamma) = H^\mu(\Gamma)$ $(0 < \mu < 1)$ *satisfies all the conditions mentioned in Section* 2.3.2.

In fact, first it is obvious that $H^\mu(\Gamma)$ satisfies conditions a) and b) of 2.3.1. Now we show that with $\widetilde{\mathfrak{A}}(\Gamma) = H^\nu(\Gamma)$ for any ν, $0 < \nu < \mu$, conditions α)—δ) of 2.3.2 are satisfied.

Condition α), i.e. the boundedness of the embedding operator $H^\mu(\Gamma) \to H^\nu(\Gamma)$ is obvious. We establish the validity of the condition β) as follows. First of all, any function from $H^\mu(\Gamma)$ can be approximated with respect to the norm of $H^\nu(\Gamma)$ by piecewise linear functions and these functions by continuously differentiable functions. Finally any continuously differentiable function can be approximated by rational functions. By Theorem 4° $H^\nu(\Gamma)$ is a decomposing algebra: $H^\nu(\Gamma) = H_+^\nu(\Gamma) \dotplus H_-^\nu(\Gamma)$, and by $Pa - aP = Sa - aS$ condition δ) is immediately obtained from Corollary 4.2.

We remark that $H^\mu(\Gamma)$ is not an R-algebra in the sense of 2.3 since, as is well-known, this space is not separable.

From Theorem 3.2, Chapter 2, and Corollary 4.4 we immediately obtain the following theorem.

Theorem 4.1. *Every function* $a(t) \in H^\mu(\Gamma)$ $(0 < \mu < 1)$ *which vanishes nowhere on* Γ *admits a factorization of the form*

$$a(t) = a_-(t)\, t^\varkappa a_+(t) \qquad (t \in \Gamma) \tag{4.17}$$

with $a_\pm(t) \in H^\mu_\pm(\Gamma)$ *and*

$$a_+(z) \neq 0 \quad (z \in G_+)\,, \qquad a_-(z) \neq 0 \quad (z \in G_-)\,,$$

$$\varkappa = \frac{1}{2\pi i}\,[\ln a(t)]_\Gamma\,.$$

The factorization (4.17) can be explicitely stated. Namely, by the formula of Sochozki-Plemelj it is easy to see that the functions

$$a_\pm(z) = e^{\pm b(z)} (z \in D_\pm)\,, \qquad b(z) = \frac{1}{2\pi i} \int_\Gamma \frac{\ln\,[\tau^{-\varkappa} a(\tau)]\, d\tau}{\tau - z}$$

satisfy condition (4.17).

3.4.2. The Singular Integral Equation with Cauchy Kernel in $H^\mu(\Gamma)$. Now we consider the singular integral operator defined by the equation

$$(A\varphi)\,(t) = a(t)\,\varphi(t) + b(t)\,(S\varphi)\,(t) + \int_\Gamma T(t, \tau)\,\varphi(\tau)\,d\tau\,, \tag{4.18}$$

where the kernel $T(t, \tau)$ has the form

$$T(t, \tau) = \frac{K(t, \tau) - K(t, t)}{\tau - t}\,.$$

Here $a(t), b(t) \in H^\mu(\Gamma)$ $(0 < \mu < 1)$ and $K(t, \tau) \in H^\lambda(\Gamma \times \Gamma)$ $(\mu < \lambda \leqq 1)$ are given functions, and S is the singular operator (4.4).[1])

By Corollary 4.3 the integral operator T with the kernel $T(t, \tau)$ is a compact operator on the space $H^\mu(\Gamma)$, and thus the operator A is a bounded linear operator on this space (compare (4.1) and Theorem 1°).

By means of the projections P and Q from (4.9) the operator A can be written in the form

$$A = c(t)\,P + d(t)\,Q + T$$

with

$$c(t) = a(t) + b(t)\,, \qquad d(t) = a(t) - b(t)\,. \tag{4.19}$$

Consequently A is an operator of the form (2.6), Chapter 2, where here the algebra of all operators of the multiplication by functions from $H^\mu(\Gamma)$ plays the role of the algebra $\mathfrak{M} \subset \mathscr{L}(H^\mu(\Gamma))$. By Corollary 4.4 this algebra satisfies condition c) of 2.2.2.

[1]) For the sequel it suffices that $K(t, \tau)$ satisfies the following weaker condition: $K(t, \tau) \in H^\mu(\Gamma \times \Gamma) \cap H^{\mu,\varepsilon}_*(\Gamma \times \Gamma) \cap H^{\varepsilon,\mu}_*(\Gamma \times \Gamma)$, $\varepsilon > 0$ (compare Lemma 4.2).

The operator

$$A_0 = a(t)\, I + b(t)\, S = c(t)\, P + d(t)\, Q \tag{4.20}$$

is the *characteristic part* of the operator A.

By the definition given in 2.2.2 the singular operator A is called an *operator of normal type* if the two functions $c(t)$ and $d(t)$ of (4.19) vanish nowhere on Γ.[1])

Applying the Theorem 2.3, Chapter 2, we first obtain the following result.

Theorem 4.2. *Let the singular operator* (4.18) *be of normal type:*

$$a^2(t) - b^2(t) \neq 0 \qquad (t \in \Gamma)\,. \tag{4.21}$$

Then A is a Φ-operator in the space $H^\mu(\Gamma)$, and the operator $R = c^{-1}P + d^{-1}Q$ is a two-sided regularizer of the operator A.

Using the commutation formula (4.7) it is easy to see that for arbitrary functions $\varphi(t)$ and $\psi(t)$ from $H^\mu(\Gamma)$ the following relation holds:

$$\int_\Gamma (A\varphi)\, \psi\, dt = \int_\Gamma \varphi A'\psi\, dt\,. \tag{4.22}$$

Here A' is the *transpose operator of A* which is defined by the formula

$$\begin{aligned}(A'\psi)\,(t) &= a(t)\,\psi(t) - S(b\psi) + \int_\Gamma T(\tau, t)\,\psi(\tau)\, d\tau \\ &= (Pd + Qc + T')\,\psi\,.\end{aligned} \tag{4.23}$$

We remark that the transpose operator also can be represented in the form (4.18). For, obviously

$$(A'\psi)\,(t) = a(t)\,\psi(t) - b(t)\,(S\psi)\,(t) + \int_\Gamma T_1(t, \tau)\,\psi(\tau)\, d\tau$$

with

$$T_1(t, \tau) = T(\tau, t) - \frac{b(\tau) - b(t)}{\tau - t}\,.$$

The properties of the operator A formulated in Theorem 4.2 now can be stated more precisely.

Theorem 4.3. *Let the condition* (4.21) *be satisfied. Then the following statements hold:*

(a) *For the equation*

$$A\varphi = f \qquad (f \in H^\mu(\Gamma)) \tag{4.24}$$

to have a solution $\varphi \in H^\mu(\Gamma)$ it is necessary and sufficient that

$$\int_\Gamma f(t)\,\psi(t)\, dt = 0 \tag{4.25}$$

for any solution $\psi \in H^\mu(\Gamma)$ of the transposed homogeneous equation $A'\psi = 0$.

---

[1]) As is well-known, an element $a(t)$ of the algebra $C(\Gamma)$ (resp. $H^\mu(\Gamma)$) is invertible (in this algebra) if and only if $a(t) \neq 0$ $(t \in \Gamma)$ (see I. M. Gelfand, D. A. Raikov and G. E. Šilov [1], § 2).

(b) *The numbers* $\alpha(A)$ *and* $\alpha(A')$ *of the linearly independent solutions of the homogeneous equations* $A\varphi = 0$ *and* $A'\psi = 0$ *are finite.*

(c) *The difference* $\alpha(A) - \alpha(A')$ *depends only on the characteristic part of the operator* A *and can be calculated by the following formula*

$$\alpha(A) - \alpha(A') = \frac{1}{2\pi i}\left[\ln \frac{d(t)}{c(t)}\right]_\Gamma . \tag{4.26}$$

The assertions (a)—(c) are generally called the *Noether theorems* for the singular integral equation (4.24).

Assertion (b) is a direct consequence of Theorem 4.2. The necessity of the conditions (4.25) follows immediately from (4.23). We obtain the sufficiency of these conditions and the asertion (c) in the following from the corresponding properties of the singular operator in the space $L^p(\Gamma)$.

Remark. By Theorem 4.2 the singular operator A (under the condition 4.21)) is normally solvable in the space $H^\mu(\Gamma)$, i.e. the equation (4.24) is solvable in $H^\mu(\Gamma)$ if and only if the condition

$$\langle f, \chi \rangle = 0$$

is satisfied for all functionals $\chi \in [H^\mu(\Gamma)]^*$ which are solutions of the adjoint homogeneous equation $A^*\chi = 0$. On the other hand, each function $\psi(t) \in H^\mu(\Gamma)$ obviously generates a functional $\chi_\psi \in [H^\mu(\Gamma)]^*$ by the formula

$$\langle f, \chi_\psi \rangle = \int_\Gamma f(t)\, \psi(t)\, dt \qquad (f \in H^\mu(\Gamma)) , \tag{4.27}$$

and by (4.22) we have the relation

$$A^*\chi_\psi = A'\psi .$$

for the functionals of the form (4.27).

Thus assertion (a) of Theorem 4.3 means that the adjoint homogeneous equation $A^*\psi = 0$ (considered in the dual space of $H^\mu(\Gamma)$) has only functionals of the form (4.27) with $A'\psi = 0$ as solutions. In particular $\alpha(A^*) = \alpha(A')$.

In fact, suppose the contrary. Let $\psi_1, \dots, \psi_n \in H^\mu(\Gamma)$ be all linearly independent solutions of the equation $A'\psi = 0$ and $\chi \in [H^\mu(\Gamma)]^*$ a solution of the equation $A^*\chi = 0$ which is linearly independent of the functionals $\chi_{\psi_1}, \dots, \chi_{\psi_n}$. Then, as is well-known, there exists a function $f(t) \in H^\mu(\Gamma)$ (see L. V. Kantorovič and G. P. Akilov [1]) such that

$$\langle f, \chi_{\psi_j} \rangle = 0 \qquad (j = 1, \dots, n) , \qquad \langle f, \chi \rangle = 1 . \tag{4.28}$$

It follows from the first n conditions in (4.28) by assertion (a) of Theorem 4.3 that equation (4.24) is solvable in $H^\mu(\Gamma)$. But then we have necessarily $\langle f, \chi \rangle = 0$, and this contradicts the last condition in (4.28).

3.4.3. The Hilbert Singular Integral and the Integral Equation with Hilbert Kernel.

The so-called *Hilbert singular integral*

$$(H\varphi)(s) = \frac{1}{2\pi}\int_0^{2\pi} \cot \frac{\sigma - s}{2}\, \varphi(\sigma)\, d\sigma , \qquad 0 \leqq s \leqq 2\pi \tag{4.29}$$

stands in a close connection with the Cauchy singular integral. The integral (4.29) is again to be understood in the sencse of the Cauchy principal value. As it is easily seen, the integral (4.29) with arbitrary s exists for every function $\varphi(\sigma) \in H^\mu[0, 2\pi]$ $(0 < \mu \leqq 1)$ with $\varphi(0) = \varphi(2\pi)$, and we have

$$\frac{1}{2\pi}\int_0^{2\pi} \cot\frac{\sigma - s}{2}\,\varphi(\sigma)\,d\sigma = \frac{1}{2\pi}\int_0^{2\pi} \cot\frac{\sigma - s}{2}\,[\varphi(\sigma) - \varphi(s)]\,d\sigma\,.$$

In order to establish the connection between the Cauchy kernel $\dfrac{d\tau}{\tau - t}$ and the Hilbert kernel $\dfrac{1}{2}\cot\dfrac{\sigma - s}{2}\,d\sigma$ we assume for simplicity that Γ consists of only one closed smooth curve which satisfies a Ljapunov condition. We can assume, without loss of generality. that the length of the curve is equal to 2π. Then Γ possesses a parametric representation of the form

$$t = t(s) = x(s) + iy(s)\,, \qquad 0 \leqq s \leqq 2\pi$$

with s as arc length. Here the derivative $t'(s)$ is an H-continuous function and $t(0) = t(2\pi)$. By σ we denote the parameter value which corresponds to the point $\tau \in \Gamma : \tau = t(\sigma)$.

Then obviously

$$\frac{d\tau}{\tau - t} - \frac{1}{2}\cot\frac{\sigma - s}{2}\,d\sigma = \frac{k(s, \sigma)}{\sigma - s}\,d\sigma\,, \tag{4.30}$$

where

$$k(s, \sigma) = \frac{t'(\sigma)}{\dfrac{t(\sigma) - t(s)}{\sigma - s}} - \frac{\sigma - s}{2}\cot\frac{\sigma - s}{2}\,.$$

It is easy to see that the function $k(s, \sigma)$ satisfies an H-condition, and we have $k(s, s) = \lim\limits_{\sigma \to s} k(s, \sigma) = 0$.

If we now put $\varphi(\tau) = \varphi[t(\sigma)] = \tilde{\varphi}(\sigma)$ and

$$h(s, \sigma) = \frac{1}{\pi i}\,\frac{k(s, \sigma) - k(s, s)}{\sigma - s}\,, \tag{4.31}$$

then we obtain

$$(S\varphi)\,(t) = \frac{1}{i}\,(H\tilde{\varphi})\,(s) + \int_0^{2\pi} h(s, \sigma)\,\tilde{\varphi}(\sigma)\,d\sigma\,. \tag{4.32}$$

Here the estimate

$$|h(s, \sigma)| \leqq \frac{C}{|\sigma - s|^\lambda}\,, \qquad 0 \leqq \lambda < 1\,, \tag{4.33}$$

holds for the kernel $h(s, \sigma)$.

For Γ we may choose in particular the unit circle $\Gamma_0 = \{t : |t| = 1\}$. In this case by

$$t(s) = e^{is}\,, \qquad d\tau = i\tau\, d\sigma$$

the simple relation

$$\frac{d\tau}{\tau - t} - \frac{1}{2}\frac{d\tau}{\tau} = \frac{1}{2}\cot\frac{\sigma - s}{2}\, d\sigma$$

holds, and the formula (4.32) assumes the form

$$\frac{1}{\pi i}\int\limits_{\Gamma_0} \frac{\varphi(\tau)}{\tau - t}\, d\tau - \frac{1}{2\pi i}\int\limits_{\Gamma_0} \frac{\varphi(\tau)}{\tau}\, d\tau = \frac{1}{i}\,(H\tilde{\varphi})\,(s)\,. \tag{4.34}$$

Now we consider the singular integral equation with a Hilbert kernel of the form

$$(B\tilde{\varphi})\,(s) = \tilde{a}(s)\,\tilde{\varphi}(s) + \frac{\tilde{b}(s)}{2\pi}\int\limits_0^{2\pi} \cot\frac{\sigma - s}{2}\,\tilde{\varphi}(\sigma)\, d\sigma + \int\limits_0^{2\pi} h(s, \sigma)\,\tilde{\varphi}(\sigma)\, d\sigma = \tilde{g}(s)\,,$$

where $\tilde{a}(s)$, $\tilde{b}(s)$ and $\tilde{g}(s)$ are certain H-continuous and 2π-periodic functions and $h(s, \sigma)$ is a (in both variables) 2π-periodic function of the form (4.31) with a certain H-continuous function $k(s, \sigma)$. Obviously, the operator B is representable in the form (4.18):

$$(B\tilde{\varphi})\,(s) = a(t)\,\varphi(t) + ib(t)\,(S_0\varphi)\,(t) + \int\limits_{\Gamma_0} T(t, \tau)\,\varphi(\tau)\, d\tau\,. \tag{4.36}$$

Here we have put $a(t) = \tilde{a}(s)$, $b(t) = \tilde{b}(s)$ with $t = e^{is}$, and S_0 is the Cauchy singular integral operator from (4.34) with the unit circle as domain of integration.

Applying Theorem 4.3 to the operator defined by the right-hand side of the formula (4.36) we easily obtain the Noether Theorems for the integral equations of the form (4.35)[1]. We leave the formulations of the corresponding theorems to the reader.

3.4.4. The Singular Integral in $L^p(\Gamma)$. In the sequel we shall always assume that the curve system Γ satisfies a Ljapunov condition.

As is well-known, for every integrable function φ there exists the integral (4.29) exists almost everywhere in the interval $(0, 2\pi)$, and the operator H defined by this integral is a bounded operator in the space $L^p(0, 2\pi)$ $(1 < p < \infty)$ (see A. Zygmund [1]. Chapter VII, and N. Dunford and J. T. Schwartz [1], XI. 7). From this by using the relation (4.32) immediately we obtain the following theorem.

---

[1]) Especially for these equations F. Noether [1] stated the theorems called later after him.

1°. *The Cauchy singular integral* (4.4) *exists for every integrable function* φ on Γ *at almost every points* $t \in \Gamma$, *and the operator* S *defined by this integral is a bounded operator on the space* $L^p(\Gamma)$ $(1 < p < \infty)$.

Theorem 2° of 3.4.1 can be generalized as follows (see I. I. PRIVALOV [2], p. 136):

2°. *For every integrable function* φ *the Cauchy integral* (4.2) *has the boundary values* (4.5) *almost everywhere on* Γ *if the point* z *tends to the point* $t \in \Gamma$ *on an arbitrary non-tangential path. The boundary values can be represented by the formulae* (4.6).

3°. *The following commutation formulae hold:*

a) *Let* $\varphi \in L^p(\Gamma)$ $(1 < p < \infty)$ *and let* $T(t, \tau)$ *be the kernel defined by* (4.12) *with* $K(t, \tau) \in H^\mu(\Gamma \times \Gamma)$ $(0 < \mu \leqq 1)$. *Then for almost all* $t_0 \in \Gamma$ *the relations*

$$\int_\Gamma \frac{dt}{t - t_0} \int_\Gamma T(t, \tau)\, \varphi(\tau)\, d\tau = \int_\Gamma \varphi(\tau)\, d\tau \int_\Gamma \frac{T(t, \tau)}{t - t_0}\, dt \tag{4.37$_1$}$$

and

$$\int_\Gamma T(t_0, t)\, dt \int_\Gamma \frac{\varphi(\tau)\, d\tau}{\tau - t} = \int_\Gamma \varphi(\tau)\, d\tau \int_\Gamma \frac{T(t_0, t)}{\tau - t}\, dt \tag{4.37$_2$}$$

hold.

b) *If* $\varphi \in L^p(\Gamma)$ $(1 < p < \infty)$ *and* $\psi \in L^q(\Gamma)$ *with* $q > p' = \dfrac{p}{p-1}$, *then the equality*

$$S(\psi S\varphi) + S(\varphi S\psi) = \varphi\psi + S\varphi S\psi \tag{4.38}$$

holds almost everywhere on Γ. *In particular* $(\psi = 1)$ *the formula* (4.8) *holds, where* I *is the identity operator in the space* $L^p(\Gamma)$.

For the proof of formulae (4.37) and (4.38) we first remark that in the case of H-continuous functions φ and ψ they are a direct consequence of (4.7). Furthermore since the operator S is continuous with respect to the norm of $L^p(\Gamma)$ and $H^\mu(\Gamma)$ is dense in $L^p(\Gamma)$, it is easily seen that these formulae also hold in the general case.

Theorem 2° immediately implies

4°. *The operator* $P(Q)$ *defined in* $L^p(\Gamma)$ $(1 < p < \infty)$ *by* (4.9) *projects the space* $L^p(\Gamma)$ *onto the subspace* $L^p_+(\Gamma)$ $(L^p_-(\Gamma))$ *of all functions* $\varphi \in L^p(\Gamma)$ *for which the Cauchy integral* (4.2) *has the boundary values*

$$\Phi_+(t) = \varphi(t)\,, \qquad (\Phi_-(t) = -\varphi(t)) \tag{4.39}$$

almost everywhere on Γ.

We remark that by formulae (4.6) the relations (4.39) are equivalent to the following conditions:

$$S\varphi = \varphi \qquad (S\varphi = -\varphi)\,.$$

In the sequel as in 2.3, by $C^\pm(\Gamma)$, we denote the subalgebra of $C(\Gamma)$ consisting of all continuous functions on Γ which have a continuation analytic in the region $D_\pm$ and continuous in $G_\pm$. Obviously for arbitrary p, $1 < p < \infty$, we have

$$C^+(\Gamma) \subset L_+^p(\Gamma)\,, \qquad C^-(\Gamma) \subset L_-^p(\Gamma) \dotplus \{c\} \qquad (c = \text{const.})\,.$$

5°. $\varphi \in L_\pm^p(\Gamma)$ $(1 < p < \infty)$ *and* $a \in C^\pm(\Gamma)$ *imply* $a\varphi \in L_\pm^p(\Gamma)$.

Proof. Let $\varphi \in L_\pm^p(\Gamma)$ and $a \in C^+(\Gamma)$. Then the relations $S\varphi = \varphi$, $Sa = a$ hold almost everywhere on Γ. Hence the formula (4.38) yields

$$2S(a\varphi) = S(aS\varphi) + S(\varphi Sa) = 2a\varphi\,.$$

Thus $S(a\varphi) = a\varphi$, i.e. $a\varphi \in L_+^p(\Gamma)$.

In the case $\varphi \in L_-^p(\Gamma)$, $a \in C^-(\Gamma)$ we have $S\varphi = -\varphi$ and $Sa = -a + c$ $(c = \text{const})$, and from (4.38) we conclude that $S(a\varphi) = -a\varphi$, i.e. $a\varphi \in L_-^p(\Gamma)$.

Lemma 4.3. *Let* $a(t) \in C(\Gamma)$. *Then the commutator* $T_a = Sa - aS$ *is compact on the space* $L^p(\Gamma)$ $(1 < p < \infty)$.

Proof. If $a(t) \in H^\mu(\Gamma)$ $(0 < \mu \leqq 1)$, then the commutator T_a is an operator with a weakly singular kernel and thus compact on $L^p(\Gamma)$ (see e.g. L. V. Kantorovič and G. P. Akilov [1]. p. 296).

Now let $a(t) \in C(\Gamma)$. Then there exists a sequence of functions $a_n(t) \in H^\mu(\Gamma)$ which converges to $a(t)$ in the norm of the space $C(\Gamma)$. We put

$$T_{a_n}\varphi = (Sa_n - a_nS)\,\varphi \qquad (\varphi \in L^p(\Gamma))\,.$$

Then

$$\begin{aligned}\|T_a\varphi - T_{a_n}\varphi\|_{L^p} &\leqq \|S[(a - a_n)\,\varphi]\|_{L^p} + \|(a - a_n)\,S\varphi\|_{L^p} \leqq \\ &\leqq 2\,\|S\|_{L^p}\,\|\varphi\|_{L^p}\,\|a - a_n\|_C\end{aligned}$$

and consequently

$$\|T_a - T_{a_n}\|_{L^p} \leqq 2\,\|S\|_{L^p}\,\|a - a_n\|_C \underset{n\to\infty}{\rightarrow} 0\,.$$

Since T_a is the limit of a uniformly convergent sequence of compact operators, T_a is compact.

Thus the lemma is proved.

3.4.5. The Case of the Unit Circle. Let Γ be the unit circle $(|z| = 1)$ and $p = 2$. In this case the operators S, P and Q posses an extremely simple representation in the space $L^2(\Gamma)$.

In fact, let $\varphi(t) \in L^2(\Gamma)$ be an arbitrary function and let

$$\varphi(t) = \sum_{k=-\infty}^{\infty} c_k t^k \qquad (|t| = 1)$$

be its Fourier series. Since this series converges with respect to the norm of the space $L^2(\Gamma)$ and P is continuous, using Theorem 4° of 3.4.4 we obtain

$$P\varphi = \lim_{n\to\infty} P\left(\sum_{k=-n}^{n} c_k t^k\right) = \lim_{n\to\infty} \sum_{k=0}^{n} c_k t^k = \sum_{k=0}^{\infty} c_k t^k\,.$$

Analogously it follows that

$$Q\varphi = \sum_{k=-\infty}^{-1} c_k t^k .$$

Hence

$$S\varphi = (P - Q)\,\varphi = \sum_{k=0}^{\infty} c_k t^k - \sum_{k=-\infty}^{-1} c_k t^k .$$

It is easy to see that the sets im P and im $Q = \ker P$ are orthogonal in the sense of the scalar product

$$[f, g]_{L^2} = \int_\Gamma f(t)\,\overline{g(t)}\,ds \qquad (f, g \in L^2(\Gamma)), \tag{4.40}$$

i.e. in the present case P and Q are orthogonal projections in the Hilbert space $L^2(\Gamma)$ (compare 1.2) and are thus also self-adjoint.

Furthermore by the Parseval equation we obtain

$$\|S\varphi\|_{L^2}^2 = 2\pi \sum_{k=-\infty}^{\infty} |c_k|^2 = \|\varphi\|_{L^2}^2 ,$$

i.e. S in an isometric and hence unitary operator since by (4.8) obviously im $S = L^2(\Gamma)$. In particular $\|S\|_{L^2} = 1$.

Remark 1. The following relations hold generally for the norm of the sinuglar operator S in the space L^p on the unit circle (see I. C. Gohberg and N. J. Krupnik [1]):

$$\|S\|_{L^p} = \cot \frac{\pi}{2p} \quad \text{for} \quad p = 2^n \qquad (n = 1, 2, \dots)$$

and

$$\|S\|_{L^p} \geqq \cot \frac{\pi}{2p} \quad \text{for any} \quad p > 2 .$$

For $1 < p < 2$ the norm can be determined by the formula

$$\|S\|_{L^p} = \|S\|_{L^{p'}}$$

with $p' = \dfrac{p}{p-1} > 2$.

Remark 2. The corresponding representation for the Hilbert operator follows at once by (4.34) from the above representation of the operator S: If $\psi(s) = \sum_{k=-\infty}^{\infty} c_k e^{iks} \in L^2(0, 2\pi)$ then

$$H\psi = i \sum_{k=1}^{\infty} c_k e^{iks} - i \sum_{k=-\infty}^{-1} c_k e^{iks} .$$

3.4.6. The Singular Integral Equation in $L^p(\Gamma)$. In the sequel let Γ be a general Ljapunov curve system of the form described in 3.4.1 and S the singular operator defined in the space $L^p(\Gamma)$ $(1 < p < \infty)$ by (4.4).

We consider in $L^p(\Gamma)$ the singular integral equation

$$A\varphi = a(t)\,\varphi(t) + b(t)\,(S\varphi)\,(t) + T\varphi = f(t) . \tag{4.41}$$

Here $f(t) \in L^p(\Gamma)$ is a given function, T an arbitrary compact operator on $L^p(\Gamma)$, and $a(t)$, $b(t)$ are arbitrary continuous functions on Γ. Obviously, A is a bounded operator on the space $L^p(\Gamma)$.

It will be useful to represent the operator A again by means of the formulae (4.9) and (4.19) in the form

$$A = c(t)\, P + d(t)\, Q + T\,. \tag{4.42}$$

Then it can be seen at once that A is an operator of the form (2.6), Chapter 2. Here now the collection of all operators obtained by multiplication by a continuous function in the space $L^p(\Gamma)$ plays the role of the algebra $\mathfrak{M} \subset \mathscr{L}[L^p(\Gamma)]$. In other words, an operator $W \in \mathfrak{M}$ is defined by an equation of the form

$$(W\varphi)\,(t) = w(t)\,\varphi(t) \qquad (t \in \Gamma;\, \varphi \in L(\Gamma))\,, \tag{4.43}$$

where $w(t) \in C(\Gamma)$ is a continuous function uniquely determined by the operator W.

According to the definition introduced in Section 2.2 the function

$$A(t, \theta) = a(t) + \theta b(t) \qquad (t \in \Gamma;\; \theta = \pm 1) \tag{4.44}$$

is called the *symbol* of the operator A defined by (4.42). Thus in particular the function $w(t)$ in (4.43) is the symbol of the multiplication operator W. The operator A is of *normal type* if its symbol does not degenerate[1]). Obviously, this is the case if and only if the condition (4.21) or, what is the same, the following two conditions are satisfied:

$$a(t) + b(t) \neq 0\,, \qquad a(t) - b(t) \neq 0 \qquad (t \in \Gamma)\,, \tag{4.45}$$

It turns out that all essential properties of the equation (4.41) can be deduced from the results of the Section 2.2. To this end we have to show that the algebra $\mathfrak{M}$ satisfies all the conditions of 2.2.

Condition c) of 2.2 is satisfied according to Lemma 4.3. In order to check conditions a) and b) (resp. b′) of 2.2 we first prove a simple lemma.

Lemma 4.4. *The spectrum of the operator* (4.43) *coincides with all points of the curve*

$$\lambda = w(t) \qquad (t \in \Gamma)\,.$$

Proof. Obviously, it is sufficient to show that the operator W is not invertible in $\mathscr{L}[L^p(\Gamma)]$ if the function $w(t)$ has a zero on Γ. Thus let $w(t_0) = 0$ $(t_0 \in \Gamma)$. By Γ_0 we denote the closed subcurve of Γ which contains the point t_0 and by s_0 the curve abscissa of the point $t_0 \in \Gamma_0$. Obviously, without loss of generality we can assume that $s_0 > 0$.

We choose a sufficiently small number ε, and by Γ_ε we denote the curve segment of Γ_0 with the length ε and the initial point t_0 lying on that side of t_0

[1]) Compare the footnote on page 106.

which is given by the decreasing curve abscissa s. Then we put

$$\varphi_\varepsilon(t) = \begin{cases} \varepsilon^{-1/p} & \text{for} \quad t \in R_\varepsilon\,, \\ 0 & \text{for} \quad t \in \Gamma - \Gamma_\varepsilon\,. \end{cases}$$

Obviously, $\|\varphi_\varepsilon\|_{L^p} = 1$ and

$$\|W\varphi_\varepsilon\|_{L^p} = \left[\frac{1}{\varepsilon}\int_{\Gamma_\varepsilon} |w(t)|^p\, ds\right]^{1/p} \leqq \max_{t\in\Gamma_\varepsilon} |w(t)|\,.$$

Since $w(t)$ is continuous and $w(t_0) = 0$, it holds that $\max_{t\in\Gamma_\varepsilon} |w(t)| \to 0$ and thus also $\|W\varphi_\varepsilon\|_{L^p} \to 0$ for $\varepsilon \to 0$. Hence the operator W is not invertible in $\mathcal{L}[L^p(\Gamma)]$.

Corollary 4.7. *Let W be the operator defined on $L^p(\Gamma)$ by* (4.43). *Then*

$$\|W\|_{L^p} = \max_{t\in\Gamma} |w(t)|\,. \tag{4.46}$$

Proof. By Lemma 4.4 we obtain for the spectral radius $r(W) = \max_{t\in\Gamma} |w(t)|$. This immediately implies

$$\|W\|_{L^p} \geqq \max_{t\in\Gamma} |w(t)|$$

The reverse inequality is obvious.

Corollary 4.8. *The operator $W \in \mathfrak{M}$ is invertible in $\mathcal{L}[L^p(\Gamma)]$ if and only if its symbol $w(t)$ has no zeros on Γ. If this is the case, then $W^{-1} \in \mathfrak{M}$.*

We conclude from Corollary 4.8 that the algebra $\mathfrak{M}$ satisfies condition a) of 2.2 since it is easy to see that each function $w(t) \in C(\Gamma)$ can be approximated in the norm of the space $C(\Gamma)$ to within any degree of accuracy by such continuous functions (e.g. rational functions) as vanish nowhere on Γ.

If Γ is the unit circle then it can be seen at once that condition b') of 2.2 is also satisfied since in this case obviously $\mathfrak{M} = \mathfrak{R}(U)$, where U is the operator of the multiplication by the independent variable $t \in \Gamma$:

$$(U\varphi)\,(t) = t\varphi(t) \qquad (t \in \Gamma)\,. \tag{4.47}$$

By Lemma 4.4 the operator U satisfies the conditions (I) and (II) of 2.1.

Although in the case of an arbitrary curve system Γ the algebra $\mathfrak{R}(U)$ is only a proper subalgebra of $\mathfrak{M}$ (since the linear span of the functions $\{t_n\}_{-\infty}^{\infty}$ is no longer dense in $C(\Gamma)$), it is nevertheless possible to state Theorem 1.3 of Chapter 2 for the algebra $\mathfrak{M}$.

Lemma 4.5. *The operator $\hat{W} = PWP$ considered in the space $X_+ = \operatorname{im} P$ is at least one-sided invertible if and only if the symbol $w(t)$ $(t \in \Gamma)$ has no zeros. If this condition is satisfied, then the invertibility of $\hat{W}$ corresponds to the number* ind $w(t)$.

If $w(t)$ assumes the value zero on Γ, then $\widehat{W}$ is neither a Φ_+- nor a Φ_--operator.

The proof is a simple modification of the proof of Theorem 1.3, Chapter 2. In fact, the function $w(t) \in C(\Gamma)$ can be approximated to within any degree of accuracy by functions $R(t) \in R(\Gamma)$ (rational functions having no poles on Γ). It is easy to see that the function $R(t)$ can be represented in the form (compare (3.11), Chapter 2):

$$R(t) = R_-(t)\, t^{\varkappa} R_+(t) \qquad (R_\pm(t) \in R^\pm(\Gamma)) .$$

Finally, since by the assertion 5° from 3.4.4 the relations

$$a_+P = Pa_+P\,, \qquad Pa_- = Pa_-P \qquad (a_\pm \in C^\pm(\Gamma)) \tag{4.48}$$

also hold, all considerations from point 1 of the proof of the Theorem 1.3. Chapter 2 onward can be repeated (with the operator U defined by (4.47)). Thus we obtain the first part of Lemma 4.5.

The last assertion of the lemma can be proved by the same considerations as in point 2 of the proof of Theorem 1.3, Chapter 2. By (4.48) it is easy to see that the relations (1.14) of Chapter 2 hold. In the case in question these equations assume the form

$$\operatorname{Ind} V_z = 0 \qquad (z \in D_-)\,, \qquad \operatorname{Ind} V_z \neq 0 \qquad (z \in D_+)$$

with $V_z = P(t - z)\, P|X_+$, and we have: For $z \in D_-$ the operator $V_z^{(-1)} = P(t - z)^{-1}\, P|X_+$ is an inverse and for $z \in D_+$ only a left inverse of V_z.

Hence Lemma 4.5 is proved.

From the preceding lemma and Corollary 4.8 we conclude that the algebra $\mathfrak{M}$ also satisfies condition b′) of Section 2.2. Consequently all the results 2.2 can be applied to integral equations of the form (4.41).

We formulate the most important of these results.

Theorem 4.4. *The conditions* (4.45) *are necessary and sufficient for the operator $A_0 = a(t)\, I + b(t)\, S$ to be at least one-sided invertible in the space $L^p(\Gamma)$ $(1 < p < \infty)$.*

If the conditions (4.45) *are satisfied, then the invertibility of the operator A_0 corresponds to the index of the symbol*

$$\varkappa = \frac{1}{2\pi}\left[\arg \frac{a(t) + b(t)}{a(t) - b(t)}\right]_\Gamma, \tag{4.49}$$

and

$$\dim \ker A_0 = \max\,(-\varkappa, 0)\,, \qquad \dim \operatorname{coker} A_0 = \max\,(\varkappa, 0)\,.$$

The same assertions also hold for the operator $\overline{A}_0 = aI + Sb$.

Theorem 4.5. *The conditions* (4.45) *are necessary and sufficient for the operator* (4.42) *to be a Φ_+- or a Φ_--operator on the space $L^p(\Gamma)$.*

If the conditions (4.45) are satisfied, then the operator A is a Φ-operator in $L^p(\Gamma)$ with the index

$$\operatorname{Ind} A = \frac{1}{2\pi}\left[\arg \frac{a(t) - b(t)}{a(t) + b(t)}\right]_\Gamma,$$

and the operator $R = c^{-1}P + d^{-1}Q$ is a two-sided regularizer of the operator A.

Remark. As far as the necessity of conditions (4.45) is concerned the statement of the Theorem 4.5 can be further sharpened (see J. LEITERER [2]): If the operator A is normally solvable then each of the functions $c(t)$ and $d(t)$ is either identically equal to zero or nowhere equal to zero on every closed subcurve of Γ.

Theorem 4.6. *Let $a(t)$, $b(t) \in C(\Gamma)$ and T an arbitrary compact operator on $L^p(\Gamma)$ $(1 < p < \infty)$. Then the following estimate holds:*

$$\max_{t \in \Gamma,\, \theta = \pm 1} |a(t) + \theta b(t)| \leqq ||aI + bS + T||_{L^p}. \tag{4.51}$$

Remark. Since, as it is easily seen, the inequality

$$\max \{||a||_C, ||b||_C\} \leqq \max_{t,\,\theta} |a(t) + \theta b(t)| \tag{4.52}$$

holds, by (4.51) and (4.46) the factor norm $|A_0| = \inf\limits_T ||A_0 + T||_{L^p}$ is equivalent to the quantity defined by the left-hand side in (4.52).

Corollary 4.9. *The singular integral operators of the form (4.42) with arbitrary continuous coefficients $c(t)$ and $d(t)$ form a closed subalgebra of the Banach algebra $\mathscr{L}[L^p(\Gamma)]$.*

3.4.7. Solution of the Characteristic Equation. Now we consider the characteristic equation

$$A_0\varphi = a(t)\,\varphi(t) + b(t)\,(S\varphi)\,(t) = f(t) \tag{4.53}$$

$(f, \varphi \in L^p(\Gamma))$ under the more special hypothesis $a(t)$, $b(t) \in H^\mu(\Gamma)$ $(0 < \mu < 1)$. Furthermore, we assume that the conditions (4.45) are satisfied. In this case using the results from 2.4.2 we obtain a simple formula for the one-sided inverses of the operator A_0.

In fact, by (4.17) the function $c(t)/d(t)$ permits a factorization of the form

$$\frac{c(t)}{d(t)} = c_-(t)\, t^\varkappa c_+(t) \qquad (t \in \Gamma)$$

with $c_\pm(t) \in H^\mu_\pm(\Gamma)$, where the number $\varkappa$ is given by (4.49). Consequently (4.19_1), Chapter 2, yields the representation

$$A_0^{(-1)} = (c_+^{-1}P + c_-Q)\,(t^{-\varkappa}P + Q)\,c_-^{-1}d^{-1}$$

for the one-sided inverses of the operator A_0. Thus in view of Theorem 4.4 we obtain the following results depending on $\varkappa$.

1. $\varkappa = 0$: The equation (4.53) has a unique solution $\varphi = A_0^{(-1)}f$ for any right-hand side f.

2. $\varkappa < 0$: The equation (4.53) has the solution $\varphi = A_0^{(-1)} f$ for any f. The homogeneous equation $A_0 \varphi = 0$ possesses $m = -\varkappa$ linearly independent solutions[1])

$$\varphi_k(t) = t^k (c_+^{-1} - c_- t^{\varkappa}) \qquad (k = 0, 1, \dots, m - 1) .$$

3. $\varkappa > 0$: Equation (4.53) is solvable if and only if the right-hand side f satisfies the $\varkappa$ conditions

$$\langle f, \psi_j \rangle = 0 \qquad (j = 0, 1, \dots, \varkappa - 1) \tag{4.54}$$

in which the ψ_j are the linearly independent solutions of the adjoint homogeneous equation $A_0^* \psi = 0$. If these conditions are satisfied, then $\varphi = A_0^{(-1)} f$ is the unique solution of the equation (4.53).

We can also transform the conditions (4.54) into a somewhat different form. First of all, we remark that the relation (4.22) is also valid for arbitrary functions $\varphi \in L^p(\Gamma)$ and $\psi \in L^{p'}(\Gamma)$ $\left(p > 1, \dfrac{1}{p} + \dfrac{1}{p'} = 1\right)$. This follows easily from the continuity of the operators A and A' in $L^p(\Gamma)$ and from the denseness of $H^\mu(\Gamma)$ in $L^p(\Gamma)$.

On the other hand, as is well-known, the dual space of $L^p(\Gamma)$ can be identified with the space $L^{p'}(\Gamma)$. In the sequel we shall use this identification in the sense that the element $\psi(t) \in L^{p'}(\Gamma)$ is identified with the functional $\psi \in [L^p(\Gamma)]^*$ defined by

$$\langle f, \psi \rangle = \int_\Gamma f(t)\, \psi(t)\, dt \qquad (f \in L^p(\Gamma)) . \tag{4.55}$$

Then using the formula (4.22) we obtain for the adjoint operator A_0^* the relation

$$A_0^* \psi = A_0' \psi = a(t)\, \psi(t) - S(b\psi) . \tag{4.56}$$

This gives the following form for the conditions (4.54):

$$\int_\Gamma f(t)\, \psi_j(t)\, dt = 0 \qquad (j = 0, 1, \dots, \varkappa - 1) ,$$

where now $\psi_j(t) \in L^{p'}(\Gamma)$ are the linearly independent solutions of the homogeneous transposed equation $A_0' \psi = 0$. It is easily checked that

$$\psi_j(t) = \frac{c_+(t)\, t^j}{c(t)} \qquad (j = 0, 1, \dots, \varkappa - 1) .$$

Analogous results are obtained for the equation

$$a(t)\, \varphi(t) + S(b\varphi) = f(t) .$$

3.4.8. The Riemann-Hilbert Boundary Value Problem. Using the formulae of Sochozki-Plemelj we can write the equation (4.53) in the form

$$c(t)\, \Phi_+(t) = d(t)\, \Phi_-(t) + f(t) \qquad (t \in \Gamma) , \tag{4.57}$$

where $\Phi_\pm(t) \in L^p(\Gamma)$ are the boundary values of the Cauchy integral (4.2) with the density function $\varphi(t) \in L^p(\Gamma)$. Thus the singular integral equation (4.53) is equivalent to the following boundary value problem of the theory

[1]) It is easy to see that $A_0 \varphi_k = 0$ $(k = 0, 1, \dots, m - 1)$.

of complex functions (usually called the *Riemann-Hilbert boundary value problem* or *coupling problem*):

To be found is a function $\Phi(z)$ in the form of the Cauchy integral (4.2) with a density function $\varphi(t) \in L^p(\Gamma)$ whose boundary values $\Phi_\pm(t)$ satisfy condition (4.57) almost everywhere on Γ.

Thus the solution of the equation (4.53) given in the preceding subsection is also the solution of the boundary value problem in question.

3.4.9. On the Continuity of the Solution. Now we return to the singular integral equation

$$(A\varphi)(t) = a(t)\,\varphi(t) + b(t)\,(S\varphi)(t) + \int_\Gamma T(t,\tau)\,\varphi(\tau)\,d\tau = f(t) \quad (4.58)$$

considered in 3.4.2 in which $a(t)$, $b(t)$ and $f(t)$ are functions from $H^\mu(\Gamma)$ $(0 < \mu < 1)$ and the kernel $T(t, \tau)$ satisfies the conditions formulated in 3.4.2. Furthermore we assume that the conditions (4.45) are satisfied. We shall show that Theorem 4.3 is a consequence of Theorem 4.5.

Theorem 4.7. *Every solution* $\varphi \in L^p(\Gamma)$ $(1 < p < \infty)$ *of the equation* (4.58) *belongs to* $H^\mu(\Gamma)$.

Proof. We assume that equation (4.58) is solvable in $L^p(\Gamma)$ and that $\varphi \in L^p(\Gamma)$ is one of its solutions. By Theorem 4.5 we have

$$\varphi - V\varphi = Rf \quad (4.59)$$

with $R = c^{-1}P + d^{-1}Q$ and

$$V = \frac{1}{2}\left(\frac{1}{d} - \frac{1}{c}\right)(T_cP + T_dQ) - RT\,.$$

Here $g = Rf \in H^\mu(\Gamma)$. Using the commutation formulae (4.37) we see easily that V is an integral operator with a weakly singular kernel $V(t, \tau)$. Hence for sufficiently great n the n-th iterated kernel $V_n(t, \tau)$ is bounded (see S. G. Mihlin [2], p. 85).

From (4.59) we obtain

$$\varphi - V^n\varphi = \sum_{k=0}^{n-1} V^k g\,,$$

where the right-hand side belongs to $H^\mu(\Gamma)$. Since by the Hölder inequality

$$|V^n\varphi| = \left|\int_\Gamma V_n(t,\tau)\,\varphi(\tau)\,d\tau\right| \leq \|\varphi\|_{L^p}\left\{\int_\Gamma |V_n(t,\tau)|^{p'}\,|d\tau|\right\}^{1/p'},$$

we have $\varphi \in M(\Gamma)$. Corollary 4.1 shows that then $V\varphi \in H^\nu(\Gamma)$ $(0 < \nu < \mu)$ from which it follows by (4.58) that $\varphi \in H^\nu(\Gamma)$. Applying the Lemma 4.1 we now obtain $V\varphi \in H^\mu(\Gamma)$, and finally (4.58) implies $\varphi \in H^\mu(\Gamma)$.

This completes the proof of the theorem.

By the remark made at the end of the Subsection 3.4.7 the operator A' defined on $L^{p'}(\Gamma)$ by equation (4.23) is the adjoint operator of A (in $L^p(\Gamma)$), and by the theorem proved just now every solution $\psi \in L^{p'}(\Gamma)$ of the homogeneous equation $A'\psi = 0$ belongs to $H^{\mu}(\Gamma)$.

Using these remarks we easily obtain the Theorem 4.3 from the Theorem 4.5.

Remark. In the case $p = 2$ besides the operator A' we have the operator adjoint to A in the dense of the scalar product (4.40) which we will denote by $A[*]$ in order to distinguish it form the operator A'. Using the formula (4.22) we can easily see that the following relation holds between the operators $A[*]$ and A' :

$$A[*] = JA'J\,, \qquad J = J_1J_2$$

with $J_2\varphi = \overline{\varphi(t)}$, $J_1\varphi = e^{-\theta(t)}\,\varphi(t)$, where $\theta(t)$ is the angle which the tangent to Γ in the point t makes with the positive direction of the x-axis.

3.5. Comments and References

3.1. These results were first stated by M. G. Krein [1] for the case in which the symbol can be expanded into an absolutely convergent Fourier series. The generalizations of the results of Krein given in 3.1.1 are taken from the monograph I. C. Gohberg and I. A. Feldman [1].

3.2. All the results of this section are due to M. G. Krein [1]. The proofs of Theorems 2.2 and 2.3 are based on an idea of G. N. Čebotarev [4].

3.3. The results of this section were discovered by I. C. Gohberg and M. G. Krein [4] and by I. C. Gohberg [2].

3.4. The boundedness of the singular operator (4.4) was proved by J. Plemelj (1908) and I. I. Privalov (1916) for the space $H^{\mu}(\Gamma)$, by N. N. Lusin (1915) for the space $L^2(\Gamma)$ and by M. Riesz (1927) for the space $L^p(\Gamma)$ $(1 < p < \infty)$.

The compactness of the commutator $Sa - aS$ in the space $L^p(\Gamma)$ (Lemma 4.3) was found by S. G. Mihlin [1]. For the case of the space $H^{\mu}(\Gamma)$ and $a(t) \in H^{\mu}(\Gamma)$ $(0 < \mu < 1)$ this fact was first stated by I. A. Feldman (compare Budjanu and Gohberg [2]). The compactness of the more general integral operator (4.11) to (4.12) in the space $H^{\mu}(\Gamma)$ was proved by W. Pogorzelski [1] under the restricting hypotheses $K(t, \tau) \in H^{\lambda}(\Gamma \times \Gamma)$ and $\mu < 1/2\,\lambda$. Likely, the Lemma 4.2 is probably new.

Theorem 4.1 was first found by F. D. Gahov (see [1]) and B. V. Hvedelidze [1]. Essentially earlier F. Noether [1] proved the Theorems 4.2 and 4.3.

The second part of Theorem 4.5 was first stated by S. G. Mihlin [1], [4] for the space $L^p(\Gamma)$. B. V. Hvedelidze [1] carried over these results to the space $L^p(\Gamma)$ for arbitrary p, $1 < p < \infty$. The first part of Theorem 4.5 and Theorem 4.6 are due to I. C. Gohberg [1—2]. Theorem 4.4 was proved by B. V. Hvedelidze [1—2], I. B. Simonenko [1] and I. C. Gohberg [2]. Under essentially stronger hypotheses on the kernel $T(t, \tau)$ B. V. Hvedelidze proved Theorem 4.7.

The connection between the singular integral equation (4.53) and the Riemann-Hilbert boundary value problem (4.57) was first established by T. Carleman [1] who simultaneously worked out a method for solving the problem (4.53) in a special case. Using the Carleman method F. D. Gahov [1] then gave a complete solution of the Riemann-Hilbert problem and thus of the equation (4.53).

CHAPTER 4

ABSTRACT SINGULAR EQUATIONS OF NON-NORMAL TYPE

This chapter provides the theoretical foundations for subsequent chapters in which several special singular equations of non-normal type are investigated. We begin with the description of a general method by which it is possible to reduce certain classes of non-normally solvable equations to normally solvable equations. This method is based on a special representation of the operator of the equation and consists of the following: The space in which the right-hand side of the equation is given is restricted in a suitable way or the space in which the solution is sought is correspondingly extended. The method is applied to Wiener-Hopf equations and to abstract singular equations whose symbol possesses finitely many zeros of integral or fractional order. The end of this chapter deals with another method of reducing non-normally solvable to normally solvable equations. This method assumes a knowledge of the closed left regularizer. Both methods are confronted.

4.1. The Method of Special Factorization

4.1.1. Let X be a Banach space and $A \in \mathscr{L}(X)$ a bounded linear operator in X. We assume that the operator A is not normally solvable but admits a factorization of the form

$$A = BCD\,, \tag{1.1}$$

where $C \in \mathscr{L}(X)$ is a normally solvable operator and the operators B and D ($\in \mathscr{L}(X)$) satisfy the following conditions:

(1) dim ker $B = 0$.

(2) *The operator D possesses a continuous extension $\widetilde{D}$ which is a one-to-one mapping of an extended Banach space $\widetilde{X}$ ($\supset X$) onto X.*

We introduce the space $\overline{X} = \operatorname{im} B$ (image space of the operator B), where the norm of an element $y = Bf \in \overline{X}$ is defined by

$$||y||_{\overline{X}} = ||f||_X\,. \tag{1.2}$$

Obviously, $\overline{X}$ is a Banach space, and for any $y \in \overline{X}$ the estimate

$$||y||_X \leqq ||B||_X\,||y||_{\overline{X}}$$

holds. i.e. the space $\overline{X}$ is continuously embedded into the space X. From Banach's theorem on homomorphisms (see Section 1.1) we conclude easily that the norms $\|\cdot\|_X$ and $\|\cdot\|_{\overline{X}}$ defined on $\overline{X}$ are equivalent if and only if the operator $B \in \mathcal{L}(X)$ is normally solvable.

In the sequel by $\overline{B}$ we denote the operator B considered as a mapping of X into $\overline{X}$:

$$\overline{B}x = Bx \qquad (x \in X) .$$

Because of (1.2) the operator $\overline{B}$ yields an isometric isomorphism of X onto $\overline{X}$. In particular, $\overline{B}$ is invertible. By condition (2) $\widetilde{D} \in \mathcal{L}(\widetilde{X}, X)$ is also invertible.

Using representation (1.1) and condition (2) we can extend the domain of the operator A to $\widetilde{X}$. We regard this extension of the operator A as an operator from $\widetilde{X}$ into $\overline{X}$ and denote it by $\widetilde{A}$:

$$\widetilde{A} = \overline{B}C\widetilde{D} . \tag{1.3}$$

Obviously, the operator $\widetilde{A} \in \mathcal{L}(\widetilde{X}, \overline{X})$ thus obtained is normally solvable.

Furthermore, (1.3) implies the following statement: If C is a Φ-operator ($\Phi_{\pm}$-operator or invertible from some side, respectively), then $\widetilde{A}$ is also such an operator and conversely. We have

$$\begin{aligned} \dim \ker \widetilde{A} &= \dim \ker C , \\ \dim \operatorname{coker} \widetilde{A} &= \dim \operatorname{coker} C . \end{aligned} \tag{1.4}$$

Thus for the equation

$$\widetilde{A}x = y \tag{1.5}$$

we obtain the following solvability statements. For (1.5) to possess a solution $x \in \widetilde{X}$ it is necessary that $y \in \overline{X}$. Let $y \in \overline{X}$, then (1.5) is solvable in $\widetilde{X}$ if and only if $\langle B^{-1}y, f\rangle = 0$ for any solution $f \in X^*$ of the equation $C^*f = 0$. If this is the case, then the general solution of equation (1.5) has the form $x = \widetilde{D}^{-1}z$, where z is the general solution of equation

$$Cz = B^{-1}y .$$

It is easy to see that the condition (2) is satisfied if the following condition (2′) holds.

(2′) *There exist a linear operator $D^{(-1)}$ defined on X with a range $\widetilde{X} = D^{(-1)}(X) \supset X$ and a linear extension $\widetilde{D}$ of the operator D defined on $\widetilde{X}$ such that the following two relations hold:*

$$D^{(-1)}Dx = x , \qquad \widetilde{D}D^{(-1)}x = x \qquad \forall\, x \in X . \tag{1.6}$$

The operator $D^{(-1)}$ in the condition (2') is a certain "formal" inverse of D. Now in $\tilde{X}$ we define a norm by the formula

$$||f||_{\tilde{X}} = ||\tilde{D}f||_X \qquad (f \in \tilde{X}).$$

Using the second relation (1.6) we see easily that $\tilde{X}$ is a Banach space which the operator $\tilde{D}$ maps isometrically onto X. The operator $D^{(-1)}$ is the inverse of $\tilde{D}$. Finally, the continuous embedding $X \subset \tilde{X}$ follows from the first relation (1.6).

In the applications which will follow in the next chapters the condition (2') can always be simply checked. We remark that condition (2') is satisfied if the operator $D \in \mathcal{L}(X)$ has the following properties:

$$\dim \ker D = \dim \operatorname{coker} D = 0. \tag{1.7}$$

In fact, in this case $\tilde{X}$ can be chosen as the completion of the space X in the norm $|x| = ||Dx||_X$. Then X is continuously and densely embedded into $\tilde{X}$. If $\tilde{x} \cong \{x_n\}_1^\infty$ $(x_n \in X)$ is an arbitrary element of the completion $\tilde{X}$, then $||Dx_n - Dx_m||_X \to 0$ holds. Hence there exists an element $x \in X$ with the property $||Dx_n - x||_X \to 0$. We put

$$\tilde{D}\tilde{x} = x. \tag{1.8}$$

It is easy to see that the operator $\tilde{D}$ is uniquely defined by formula (1.8) and that $\tilde{D}$ is a linear extension of the operator D. Moreover, since $\tilde{D}$ is a one-to-one mapping of the space $\tilde{X}$ onto X, $D^{(-1)} = \tilde{D}^{-1}$ also exists.

It is obvious that the constructions carried out above are also possible in the case where the operators for the representation (1.1) act between two Banach spaces.

4.1.2. It follows from Theorem 5.8, Chapter 1, that every operator A with a finite index can be represented in the form (1.1) in such a way that B satisfies condition (1), D the conditions (1.7) and C is a Φ-operator. The factors in (1.1) are not uniquely determined. In fact we can always realize the two limiting cases $B = I$ and $D = I$ to which correspond the space pairs $(\tilde{X}, X)$ and $(X, \bar{X})$, respectively (for this compare Section 4.2). In these cases the equation (1.5) can be interpreted as follows:

1°. $B = I$: The right-hand side y of the equation is given in the space X and a generalized solution $x \in \tilde{X}$ is sought.

2°. $D = I$: $y \in \bar{X}$ is given and an ordinary solution $x \in X$ is sought.

In applications (see Chapter 5 and 6) we have to pay attention to the following problem: The operators B and D of the representation (1.1) are to be chosen in such a way that the spaces $\tilde{X}$ and $\bar{X}$ admit a simple analytic description.

4.1.3. In applications (compare Chapter 5) we sometimes use representations of the form

$$A = BC\,, \tag{1.9}$$

where $C \in \mathcal{L}(X, Y)$ is a Φ-operator and the operator $B \in \mathcal{L}(Y)$ satisfies the condition

$$0 < \dim \ker B < \infty\,.$$

This case can easily be reduced to the case $\dim \ker B = 0$ considered in 4.1.1.

To this end we introduce the factor space $\boldsymbol{Y} = Y/\ker B$. By F we denote the canonical mapping of Y onto $\boldsymbol{Y}$ which associates the class $\hat{y} = y + \ker B$ with the element $y \in Y$. Then the operator B naturally generates an operator $\boldsymbol{B} \in \mathcal{L}(\boldsymbol{Y}, Y)$ with $\dim \ker \boldsymbol{B} = 0$ which is defined by the equation

$$\boldsymbol{B}\hat{y} = By$$

Obviously, $\boldsymbol{B}F = B$. Thus by (1.9) we obtain the representation

$$A = \boldsymbol{B}\boldsymbol{C}\,,$$

for the operator A, where $\boldsymbol{C} = FC \in \mathcal{L}(X, \boldsymbol{Y})$ is a Φ-operator with

$$\operatorname{Ind} \boldsymbol{C} = \operatorname{Ind} C + \dim \ker B \tag{1.10}$$

and the operator $\boldsymbol{B} \in \mathcal{L}(\boldsymbol{Y}, Y)$ satisfies the condition

$$\dim \ker \boldsymbol{B} = 0\,.$$

Consequently, the scheme described in 4.1.1 is applicable to the operator A, i.e. the operator $A\colon X \to \overline{Y}$ is a Φ-operator with the index (1.10), where the norm of an element $z \in \overline{Y} = \operatorname{im} B$ is defined by

$$\|z\|_{\overline{Y}} = \|\boldsymbol{B}^{-1}z\|_{\boldsymbol{Y}} = \inf_{By=z} \|y\|_Y\,.$$

Remark. The results of Section 1.3 imply: An operator of the form (1.9) is a $\Phi_\pm$-operator from X into $\overline{Y}$ if and only if $C \in \mathcal{L}(X, Y)$ is such an operator.

In the following two statements we assume that the operator A possesses the representation (1.9), where B and C have the above-mentioned properties.

First we remark that for the solvability of the equation $Ax = y$ the relation $y \in \operatorname{im} B$ is obviously necessary.

Lemma 1.1. *Each solution x of the equation*

$$Ax = y \qquad (y \in \operatorname{im} B) \tag{1.11}$$

is a solution of the equation $Cx = z$ for a certain z with $Bz = y$ and conversely.

Proof. Let $Ax = y$. Putting $Cx = z$ we obtain $Bz = BCx = y$.

Conversely, $Cx = z$ with $Bz = y$ implies $y = BCx = Ax$.

Lemma 1.2. *The equation* (1.11) *is solvable if and only if*

$$\langle z, f\rangle = 0 \quad \textit{for all } f \in (\ker C^*) \cap (\ker B)^{\perp}, \tag{1.12}$$

where z is an arbitrary element with $Bz = y$.

Proof. Let x be a solution of the equation (1.11). Then $Cx = z_0$ for a certain z_0 with $Bz_0 = y$. From this it follows that $\langle z_0, f\rangle = 0$ for all $f \in \ker C^*$, and thus also the condition (1.12) holds.

Conversely, let now the condition (1.12) be satisfied. Then $z \in \operatorname{im} C + \ker B$, i.e. for a certain $h \in \ker B$ the equation $z - h = Cx$ holds. Since $B(z - h) = y$, by Lemma 1.1 we obtain $Ax = y$.

Corollary. *For the operator A of the form* (1.9) *the following formula holds:*

$$\dim \operatorname{coker} A = \dim [(\ker C^*) \cap (\ker B)^{\perp}] = \operatorname{codim} (\operatorname{im} C + \ker B).$$

Remark. The considerations of this subsection can also be carried over to the case that $\dim \ker B = \infty$ if, in addition, it is assumed that the sum $\ker B + \operatorname{im} C$ is closed. In this case A is normally solvable resp. a Φ_--operator if and only if C is such an operator.

This follows immediately from the representations $A = \boldsymbol{BC}$, $\boldsymbol{C} = FC$ and from Lemma 3.2, Chapter 1.

4.1.4. A further statement which we shall use later in applications is the following:

Let the operator $D \in \mathcal{L}(X)$ satisfy the condition (2) mentioned in 4.1.1, and let $Z \subset X$ be a Banach space (continuously embedded into X) on which the restriction

$$D_Z = D|Z : Z \to Z$$

is a bounded operator.

Then it is possible to construct a Banach space $\widetilde{Z}$ with $Z \subset \widetilde{Z} \subset \widetilde{X}$ and an extension $\widetilde{D}_Z$ of the operator D_Z to the space $\widetilde{Z}$ such that $\widetilde{D}_Z$ is an invertible continuous linear operator from $\widetilde{Z}$ onto Z. For this it is sufficient to put

$$\widetilde{Z} = \widetilde{D}^{-1}(Z), \qquad \widetilde{D}_Z x = \widetilde{D}x \qquad (x \in \widetilde{Z})$$

and to define a new norm on $\widetilde{Z}$ by

$$||x||_0 = ||\widetilde{D}x||_Z .$$

4.2. Wiener-Hopf Equations

We now apply the results of the preceding section to Wiener-Hopf operators of the form PAP with $A \in \mathfrak{R}(U)$, where the invertible operator $U \in \mathcal{L}(X)$ satisfies the conditions (I) and (II) from 2.1.

4.2.1. Let $A \in \mathfrak{R}(U)$ be an operator of non-normal type, i.e. its symbol $A(z)$ ($|z| = 1$) has zeros on the unit circle.

We assume that the operator A has the form

$$A = A_- C A_+ \quad \text{mit} \quad A_\pm \in \mathfrak{R}^\pm(U)\,, \qquad C \in \mathfrak{R}(U)\,, \tag{2.1}$$

where C is an operator of normal type. Then, using the relations (1.13) from Chapter 2, for the operator $\hat{A} = PAP \mid \text{im } P$ we obtain the representation

$$\hat{A} = \hat{B}\hat{C}\hat{D} \tag{2.2}$$

with the operators

$$\hat{B} = PA_-P\,, \qquad \hat{C} = PCP\,, \qquad \hat{D} = PA_+P \tag{2.3}$$

considered in the space $X_+ = \text{im } P$.

If the operators $\hat{B}$ and $\hat{D}$ in the representation (2.2) satisfy the conditions (1) and (2) (resp. (2')) of 4.1, then it is possible to construct the spaces $\overline{X}_+ \subset X_+ \subset \widetilde{X}_+$ according to the scheme given in 4.1 in such way that the operator $\hat{A}: \widetilde{X}_+ \to \overline{X}_+$ is a Φ_+-operator.

4.2.2. In the sequel we assume that $\mathfrak{R}(U)$ is an algebra without a radical.

Then the representation (2.1) holds and thus also (2.2) if the symbol $A(z)$ of the operator A permits a factorization of the form

$$A(z) = \varrho_-(z)\, C(z)\, \varrho_+(z) \tag{2.4}$$

with

$$\varrho_\pm(z) \in \mathfrak{R}^\pm(z)\,, \qquad C(z) \in \mathfrak{R}(z)\,, \qquad C(z) \neq 0 \qquad (|z| = 1)\,.$$

In this case the operators $\hat{B}$, $\hat{C}$ and $\hat{D}$ in the representation (2.2) are the operators from the algebra $\mathfrak{R}(U)$ which belong to the symbols $\varrho_-(z)$, $C(z)$ and $\varrho_+(z)$, respectively (compare 2.1).

We remark that by the uniqueness theorem for analytic functions (see I. I. Privalov [2], p. 212) the symbols $\varrho_+(z)$ and $\varrho_-(z)$ can vanish at most on a point set of the Lebesgue measure zero on the unit circle.

4.2.3. Now we consider the special case that the symbol $A(z)$ of the operator $A \in \mathfrak{R}(U)$ has finitely many zeros of integral orders.

Let $\alpha_1, \alpha_2, \dots, \alpha_r$ be distinct points of the unit circle and $m_1, m_2, \dots, m_r$ positive integers. Let the symbol $A(z)$ have the form

$$A(z) = \prod_{j=1}^{r} (z - \alpha_j)^{m_j}\, B(z) \tag{2.5}$$

with $B(z) \in \mathfrak{R}(z)$ and $B(z) \neq 0$ $(|z| = 1)$.

We choose arbitrary non-negative integers m_j' and m_j'' such that

$$m_j' + m_j'' = m_j \qquad (j = 1, 2, \dots, r)\,.$$

Putting

$$\varrho_+(z) = \prod_{j=1}^{r} (z - \alpha_j)^{m_j'}\,, \qquad \varrho_-(z) = \prod_{j=1}^{r} (z^{-1} - \alpha_j^{-1})^{m_j''} \qquad (|z| = 1) \tag{2.6}$$

and

$$C(z) = \prod_{j=1}^{r} (-\alpha_j z)^{m_j''} B(z)$$

we obtain the representation (2.4) for the symbol $A(z)$.

In the sequel we shall use the following notations:

$$B_\alpha = U^{-1} - \alpha^{-1} I \,, \qquad D_\alpha = U - \alpha I \,. \tag{2.7$_1$}$$

The operators $B_\alpha \in \mathfrak{R}^-(U)$, $D_\alpha \in \mathfrak{R}^+(U)$ have the symbols $z^{-1} - \alpha^{-1}$, $z - \alpha$, respectively. For the operators $\hat{B}$ and $\hat{D}$ from (2.3) the relations (1.13), Chapter 2, imply the representations

$$\hat{B} = \prod_{j=1}^{r} \hat{B}_{\alpha j}^{m_j''} \,, \qquad \hat{D} = \prod_{j=1}^{r} \hat{D}_{\alpha j}^{m_j'} \,. \tag{2.7$_2$}$$

Theorem 2.1. *Let* $m_j'' > 0$ *and*

$$\ker \hat{B}_{\alpha j} \cap \operatorname{im} \hat{B}_{\alpha j} = \{0\} \qquad (j = 1, 2, \ldots, r) \,. \tag{2.8}$$

Then the kernel of the operator $\hat{B}$ *is the direct sum*

$$\ker \hat{B} = \ker \hat{B}_{\alpha_1} \dotplus \ldots \dotplus \ker \hat{B}_{\alpha_r} \,.$$

Proof. Obviously, it is sufficient to prove the assertion for $r = 2$. First, by (2.8) we have $\ker \hat{B}_{\alpha j}^{m_j''} = \ker \hat{B}_{\alpha j}$. Furthermore, $\alpha_1 \neq \alpha_2$ immediately implies $\ker \hat{B}_{\alpha_1} \cap \ker \hat{B}_{\alpha_2} = \{0\}$.

Since for arbitrary $\varphi \in \ker \hat{B}_{\alpha_1}$ the equation

$$\hat{B}_{\alpha_2} \varphi = PU^{-1} \varphi - \alpha_2^{-1} \varphi = (\alpha_1^{-1} - \alpha_2^{-1}) \varphi$$

holds, we obtain $\hat{B}_{\alpha_2} (\ker \hat{B}_{\alpha_1}) = \ker \hat{B}_{\alpha_1}$.

Now let $\hat{B} f = 0$ $(f \in X_+)$. Then $\hat{B}_{\alpha_2}^{m_2''} f \in \ker \hat{B}_{\alpha_1}$. Consequently, by the relation proved before we obtain

$$\hat{B}_{\alpha_2}^{m_2''} f = \hat{B}_{\alpha_2}^{m_2''} f_1 \,, \qquad f_1 \in \ker \hat{B}_{\alpha_1} \,.$$

From this we conclude that $f_2 = f - f_1 \in \ker \hat{B}_{\alpha_2}$ and $f = f_1 + f_2$. The assertion is proved.

Remark. If the operator U satisfies the condition

$$\sup_n \|PU^{-n}\| < \infty \qquad (n = 0, 1, 2, \ldots) \,,$$

then the relation (2.8) holds for any point α $(|\alpha| = 1)$.

In fact, if for $f \in X_+$ and $g \in \ker \hat{B}_\alpha$ the equation

$$\hat{B}_\alpha f = PU^{-1} f - \alpha^{-1} f = g$$

holds, then for $n = 1, 2, \ldots$ we easily obtain the relations

$$\alpha^{n-1} PU^{-n} f - \alpha^{-1} f = ng \,, \qquad \|PU^{-n} f\| \geq n \|g\| - \|f\|$$

which by the hypothesis imply $g = 0$.

In the sequel we suppose that $\ker \hat{B}$ is finite dimensional. Then by 4.1.3 we can construct the Banach space $\overline{X}_+(\varrho_-) = \operatorname{im} \hat{B}$, where the norm of an element $\varphi \in \overline{X}_+(\varrho_-)$ is defined by

$$||\varphi||_{\overline{X}_+(\varrho_-)} = \inf_{\hat{B}\psi=\varphi} ||\psi||_{X_+} .$$

The spaces $\overline{X}_+^{(j)} = \operatorname{im} B_{\alpha_j}^{m_j''}$ $(j = 1, 2, \dots, r)$ are defined analogously.

If $||\varphi||^{(j)}$ denotes the norm in the space $\overline{X}_+^{(j)}$, then the intersection $\bigcap_{j=1}^{r} \overline{X}_+^{(j)}$ is a Banach space with the norm $\sum_{j=1}^{r} ||\varphi||^{(j)}$. The following theorem shows that the analytic description of the image space $\operatorname{im} \hat{B}$ can be reduced to the description of the space $\operatorname{im} \hat{B}_\alpha^m$ in *one* point α.

Theorem 2.2. *We have*

$$\overline{X}_+(\varrho_-) = \bigcap_{j=1}^{r} \overline{X}_+^{(j)} .$$

Before the proof of the theorem we give the following lemma.

Lemma 2.1. *Let X be a Banach space, let A and B be two commuting operators from $\mathcal{L}(X)$ and $C = AB$.*

If there exist two operators R_1 and R_2 $(\in \mathcal{L}(X))$ such that the equations

$$AR_1 + BR_2 = I , \qquad AR_2 = R_2A , \qquad BR_1 = R_1B$$

hold, then

$$\operatorname{im} C = \operatorname{im} A \cap \operatorname{im} B , \tag{2.9}$$

and on $\operatorname{im} C$ *the norms*

$$||f||_{\operatorname{im} C} = \inf_{Cx=f} ||x||_X \qquad (f \in \operatorname{im} C)$$

and $||f||' = ||f||_{\operatorname{im} A} + ||f||_{\operatorname{im} B}$ *are equivalent.*

Proof. Obviously, by $C = AB = BA$ we have

$$\operatorname{im} C \subset \operatorname{im} A \cap \operatorname{im} B .$$

Let $f \in \operatorname{im} A \cap \operatorname{im} B$, i.e.

$$f = Ax = By \qquad (x, y \in X) .$$

Then for $z = R_1 y + R_2 x$ we obtain

$$Cz = AR_1By + BR_2Ax = (AR_1 + BR_2) f = f ,$$

i.e. $f \in \operatorname{im} C$. Thus (2.9) is proved.

The estimates

$$||f||_{\operatorname{im} C} \leqq ||z|| \leqq ||R_1|| \, ||y|| + ||R_2|| \, ||x||$$

imply

$$||f||_{\operatorname{im} C} \leqq M_1 ||f||' .$$

Using the definition of the norms $||f||_{\mathrm{im}\,C}$ and $||f||'$ we easily see that an estimate in the reverse direction of the form

$$||f||' \leqq M_2 ||f||_{\mathrm{im}\,C}$$

also holds. This proves the lemma.

Now we shall give the proof of Theorem 2.2. Obviously, it is sufficient to consider the case $r = 2$.

It is easily seen that there exist two polynomials $P_j(z)$ $(j = 1, 2)$ with the property grad $P_j(z) \leqq m_j''$ such that the identity

$$(1 - z\alpha_1^{-1})^{m_1''} P_2(z) + (1 - z\alpha_2^{-1})^{m_2''} P_1(z) = z^{m_1''+m_2''}$$

is satisfied. By $S_j \in \mathfrak{E}(U)$ we denote the operator with symbol $P_j(z)/z^{m_j''}$ $(j = 1, 2)$. It is easy to check that the operators

$$A = \hat{B}_{\alpha_1}^{m_1''}\,, \qquad B = \hat{B}_{\alpha_2}^{m_2''}\,, \qquad R_1 = \hat{S}_2\,, \qquad R_2 = \hat{S}_1$$

satisfy the hypotheses of Lemma 2.1 in the space $X_+ = \mathrm{im}\,P$. This proves the assertion of the theorem.

In the sequel we assume that the operator $\hat{D}$ defined on the space $X_+ = \mathrm{im}\,P$ by the second relation (2.7_2) satisfies condition (2′) of 4.1. Then we can introduce the space $\widetilde{X}_+(\varrho_+) = \hat{D}^{(-1)}(X_+)$ in which the norm is defined by

$$||f||_{\widetilde{X}_+} = ||\hat{D}f||_{X_+}\,.$$

The spaces $\widetilde{X}_+(\varrho_+)$ and $\overline{X}_+(\varrho_-)$ are called the *spaces generated by the zeros of the symbol.*

From the results of the Chapter 2 and the Section 4.1 we obtain the following theorem.

Theorem 2.3. *For the operator*

$$\hat{A}: \widetilde{X}_+(\varrho_+) \to \overline{X}_+(\varrho_-)$$

to be a Φ_+- or a Φ_--operator it is necessary and sufficient that the function $B(z)$ in the representation (2.5) *(or, what is the same, the function $C(z)$ in* (2.4), *vanishes nowhere on the unit circle.*

If this condition is satisfied, then $\hat{A}$ is a Φ-operator from $\widetilde{X}_+(\varrho_-)$ into $\overline{X}_+(\varrho_-)$, and for its index the following formula holds:

$$\mathrm{Ind}\,\hat{A} = \dim\ker\hat{B} - \varkappa\beta(\hat{U})\,, \tag{2.10}$$

where

$$\varkappa = \mathrm{ind}\,C(z) = \mathrm{ind}\,B(z) + \sum_{j=1}^{r} m_j''\,.$$

Furthermore, Lemma 1.2 shows that under the hypotheses of the Theorem 2.3 the equation

$$\hat{A}x = y \qquad (y \in \overline{X}_+(\varrho_-)) \tag{2.11}$$

has a solution $x \in \widetilde{X}_+(\varrho_+)$ if and only if $\langle z, f \rangle = 0$ for all $f \in (\ker \hat{C}^*) \cap \cap (\ker \hat{B})^\perp$, where $z \in X_+$ is an arbitrary element with $\hat{B}z = y$.

Remark 1. If $\dim \ker \hat{B} = 0$ and $C(z)$ vanishes nowhere on the unit circle then the invertibility of the operator $\hat{A}$ corresponds to the number $\varkappa$, and

$$\dim \ker \hat{A} = \max(-\varkappa, 0)\, \beta(\hat{U}), \qquad \dim \operatorname{coker} \hat{A} = \max(\varkappa, 0)\, \beta(\hat{U}).$$

This is a direct consequence of the results of the Section 4.1 and the Theorems 1.3 and 4.1, Chapter 2.

Remark 2. Since the non-negative integers m_j' and m_j'' satisfy only the condition $m_j' + m_j'' = m_j$ $(j = 1, \dots, r)$ and besides this can be arbitrarily chosen, we can especially consider the following two limiting cases (depending on the properties of the right-hand side of equation (2.11) and the demanded properties of the solution):

1°. $m_j' = m_j, \quad m_j'' = 0 \quad (j = 1, \dots, r)$.

2°. $m_j' = 0, \quad m_j'' = m_j \quad (j = 1, \dots, r)$.

In the first case we have $\widetilde{X}_+(\varrho_-) = X_+$ and in the second case $\widetilde{X}_+(\varrho_+) = X_+$.

4.2.4. The considerations of Subsection 4.2.3 can also be carried over to the case where the symbol $A(z)$ has finitely many zeros of non-integral orders.

Let $\alpha_1, \alpha_2, \dots, \alpha_r$ be distinct points of the unit circle. $m_1', m_1', \dots, m_r'$ and $m_1'', m_2'', \dots, m_r''$ arbitrary non-negative real numbers. We assume that the symbol $A(z)$ of the operator $A \in \mathfrak{R}(U)$ is representable in the form (2.4), where $C(z) \in \mathfrak{R}(z)$ and the functions $\varrho_\pm(z)$ are given by (2.6)[1].

It is easy to see that the functions $\varrho_+(z)$ and $\varrho_-(z)$ are absolutely continuous on the unit circle and satisfy a Hölder condition. Thus these functions can be expanded into an absolutely convergent Fourier series (see A. Zygmund [1], Chapter VI, Theorem (3.6)). Furthermore, it is easily seen that $\varrho_+(z)$ $(\varrho_-(z))$ can be continued analytically into the region $|z| < 1$ $(|z| > 1)$. Hence $\varrho_\pm(z) \in W^\pm$ and thus also $\varrho_\pm(z) \in \mathfrak{R}^\pm(z)$. By $\hat{D}$ and $\hat{B}$ we denote the Wiener-Hopf operators with the symbols $\varrho_+(z)$ and $\varrho_-(z)$, respectively.

The considerations of Subsection 4.2.3 show that in this case Theorem 2.3 remains true if the formula (2.10) is interchanged with

$$\operatorname{Ind} \hat{A} = \dim \ker \hat{B} - \beta(\hat{U}) \operatorname{ind} C(z). \tag{2.10'}$$

Now we prove that under the hypotheses of this subsection (i.e. in the case of arbitrary real numbers $m_j'' \geqq 0$) Theorem 2.2 also remains true. The proof is a corresponding modification of the proof given in the case of integral orders m_j''.

We show the existence of two functions $f_0(z)$ and $g_0(z)$ from W^- for which

$$(z^{-1} - \alpha_1^{-1})^{m_1''} f_0(z) + (z^{-1} - \alpha_2^{-1})^{m_2''} g_0(z) = 1 \qquad (|z| \geqq 1) \tag{2.12}$$

[1]) We remark that in the case of non-integers m_j $(j = 1, 2, \dots, r)$ the representation (2.4) does not follow from the representation (2.5) since the functions z^{m_j} $(|z| = 1)$ are no longer continuous.

holds. To this end we consider the collection J of all functions of the form

$$(z^{-1} - \alpha_1^{-1})^{m_1''} f(z) + (z^{-1} - \alpha_2^{-1})^{m_2''} g(z) \tag{2.13}$$

with $f, g \in W^-$. The assertion (2.12) is proved if we can show that $J = W^-$. We assume that $J \neq W^-$. It is easy to see that J is an ideal of the algebra W^-. Since J is a proper ideal, it is contained in a maximal ideal of W^- (see I. M. Gelfand, D. A. Raikov and G. E. Šilov [1]). Hence by Lemma 3.1, Chapter 2, there exists a point $z_0(|z_0| \geqq 1)$ in which the functions (2.13) attain the value zero for any $f, g \in W^-$. Since $\alpha_1 \neq \alpha_2$, we can suppose that $z_0 \neq \alpha_1$. Now putting $f(z) \equiv 1$ and $g(z) \equiv 0$ we obtain $(z_0^{-1} - \alpha_1^{-1})^{m_1''} = 0$. This contradiction proves the assertion (2.12).

Now by S_1, S_2, A_1 and A_2 we denote the operators from $\Re^-(U)$ with symbols $f_0(z)$, $g_0(z)$, $(z^{-1} - \alpha_1^{-1})^{m_1''}$ and $(z^{-1} - \alpha_2^{-1})^{m_2''}$, respectively. It is easy to see that the operators

$$A = PA_1, \qquad B = PA_2, \qquad R_1 = PS_1, \qquad R_2 = PS_2$$

then satisfy the hypotheses of Lemma 2.1 in the space $X_+ = \operatorname{im} P$. From this the assertion of the theorem follows immediately.

4.3. Abstract Singular Equations

4.3.1. For operators from the algebra $\widetilde{\Re}(U)$, i.e. operators of the form

$$A = A_1P + A_2Q + T \qquad (A_1, A_2 \in \Re(U)),$$

where T is compact on X, Theorem 5.8, Chapter 1, can be sharpened in that way that in the present case the operators B, C and D in the representation (1.1) can be taken from the algebra $\widetilde{\Re}(U)$.

We suppose that the invertible operator $U \in \mathcal{L}(X)$ satisfies conditions (I) and (II) of 2.1 and in addition the condition

$$\beta(\hat{U}) = \beta(UP \mid \operatorname{im} P) = 1\,.$$

Under these hypotheses the following theorem holds.

Theorem 3.1. *Let* $A \in \widetilde{\Re}(U)$.

(a) *If* dim ker A *is finite, then* A *possesses the representation* $A = BC$.
(b) *If* A *has a finite index, then* A *is representable in the form* $A = CD$.

Here B, C, D *are operators from the algebra* $\widetilde{\Re}(U)$. C *is a* Φ*-operator, and for* B, D *the following relations hold*:

$$\dim\ker B = 0\,, \qquad \dim\ker D = \dim\operatorname{coker} D = 0\,.$$

Proof. We show that under the hypotheses assumed above the operators B, C and D constructed in the proof of Theorem 5.8, Chapter 1, belong to

$\widetilde{\mathfrak{R}}(U)$. The algebra $\widetilde{\mathfrak{R}}(U)$ contains the identity operator I and all compact operators and thus also all finite dimensional continuous operators. As we can easily be seen from the proof of Theorem 5.8. Chapter 1, it is sufficient to show that for two arbitrary closed subspaces X_1 and X_2 of the space X of finite codimension there exists an operator in $\widetilde{\mathfrak{R}}(U)$ which maps X_1 onto X_2 isomorphically.

If codim $X_1 =$ codim X_2, then there exists a closed subspace X_0 with $X = X_1 \dotplus X_0 = X_2 \dotplus X_0$. By P_j $(j = 1, 2)$ we denote the (finite dimensional) projection of X onto X_0 parallel to X_j. Then $F = I + P_1 - P_2 \in \widetilde{\mathfrak{R}}(U)$ is the desired isomorphism $(F^{-1} = I - P_1 + P_2)$.

To complete the proof it is sufficient to show that for any natural number m there exist a closed subspace X_m of codimension m and an operator from $\widetilde{\mathfrak{R}}(U)$ which maps X onto X_m isomorphically. We put

$$W = UP + Q\,, \qquad W^{(-1)} = U^{-1}P + Q\,.$$

Then $W^{(-1)}W = I$, and by Theorem 4.2. Chapter 2, $X_m = \operatorname{im} W^m$ is the desired subspace and $W^m \in \widetilde{\mathfrak{R}}(U)$ the desired isomorphism.

This proves the theorem.

4.3.2. In the sequel we consider in particular paired operators of the form

$$A_1P + A_2Q \qquad (A_1, A_2 \in \mathfrak{R}(U))$$

(without the compact part) and the corresponding transpose operators $PA_2 + QA_1$.

In the following we always suppose that the invertible operator $U \in \mathcal{L}(X)$ satisfies the conditions (I) and (II) of 2.1 and that $\mathfrak{R}(U)$ is an algebra without radical.

We assume that the operators A_1 and A_2 have the following form

$$A_1 = A_+A_0F\,, \qquad A_2 = A_-B_0F\,, \tag{3.1}$$

where A_0 and B_0 $(\in \mathfrak{R}(U))$ are operators of normal type, $A_\pm \in R^\pm(U)$ and $F \in \mathfrak{R}(U)$. Then we can easily state a representation of the form (1.1) for the above-mentioned paired operators.

Namely, using formulae (4.16), (4.17) of Chapter 2 we obtain the equations

$$A_1P + A_2Q = F(A_0P + B_0Q)\,(A_+P + A_-Q) \tag{3.2}$$

and

$$PA_2 + QA_1 = (PA_- + QA_+)\,(PB_0 + QA_0)\,F\,. \tag{3.3}$$

We remark that for the null space (image space) of the operator $B = PA_- + QA_+$ on the one hand and for the null spaces (image spaces) of the operators $PA_- = PA_-P$, $QA_+ = QA_+Q$ on the other hand we obviously have the relations

$$\ker B = \ker PA_- \dotplus \ker QA_+\,, \qquad \operatorname{im} B = \operatorname{im} PA_- \oplus \operatorname{im} QA_+\,. \tag{3.4}$$

4.3.3. By $A(z)$ and $B(z)$ we denote the symbols of the operators A_1 and A_2 ($\in \Re(U)$), respectively. We consider the case in which the functions $A(z)$ and $B(z)$ ($|z| = 1$) have finitely many zeros of integral orders.

Let $\alpha_1, \alpha_2, \dots, \alpha_r, \beta_1, \beta_2, \dots, \beta_s$ and $\gamma_1, \gamma_2, \dots, \gamma_q$ be certain (not necessarily different) points of the unit circle and $m_1, m_2, \dots, m_r, n_1, n_2, \dots, n_s$ and $l_1, l_2, \dots, l_q$ non-negative integers. We assume that the functions $A(z)$ and $B(z)$ ($|z| = 1$) have the form

$$A(z) = \varrho_+(z)\,\varrho(z)\,A_0(z)\,,$$
$$B(z) = \varrho_*(z)\,\varrho(z)\,B_1(z)\,, \tag{3.5}$$

where $A_0(z)$ and $B_1(z)$ are functions from $\Re(z)$ which vanish nowhere on the unit circle, and

$$\varrho_+(z) = \prod_{j=1}^{r} (z - a_j)^{m_j}\,, \qquad \varrho(z) = \prod_{j=1}^{q} (z - \gamma_j)^{l_j}\,,$$
$$\varrho_*(z) = \prod_{j=1}^{s} (z - \beta_j)^{n_j}\,.$$

Remark. Obviously, without loss of generality we can always assume that $\varrho(z) \equiv 1$ in the representations (3.5) for the functions $A(z)$ and $B(z)$. But by a suitable choise of the points γ_j ($j = 1, \dots, q$) we can arrange, of necessary, that the points α_k and β_l ($k = 1, \dots, r$; $l = 1, \dots, s$) are distinct. In the last case the points γ_k are just the common zeros of the functions $A(z)$ and $B(z)$.

For the following it is necessary to represent the functions $B(z)$ in the form

$$B(z) = \varrho_-(z)\,\varrho(z)\,B_0(z)\,, \tag{3.6}$$

where

$$\varrho_-(z) = \prod_{j=1}^{s} (z^{-1} - \beta_j^{-1})^{n_j}\,, \qquad B_0(z) = \prod_{j=1}^{s} (-\beta_j z)^{n_j}\,B_1(z)\,;$$

Now we put (see (2.7_1))

$$A_+ = \prod_{j=1}^{r} D_{\alpha_j}^{m_j}\,, \qquad A_- = \prod_{j=1}^{s} B_{\beta_j}^{n_j}\,, \qquad F = \prod_{j=1}^{q} D_{\gamma_j}^{l_j}\,,$$
$$B = PA_- + QA_+\,, \qquad D = A_+P + A_-Q$$

and denote with A_0 and B_0 the operators from the algebra $\Re(U)$ whose symbols are $A_0(z)$ and $B_0(z)$ by respectively. Then the operators A_1 and A_2 can be represented in the form (3.1). Hence the representations (3.2) and (3.3) also hold for the paired operators.

4.3.4. We maintain the hypotheses of the preceding subsection and suppose now, in addition, that the null space of the operators B and F are finite dimensional: $\dim \ker B < \infty$, $\dim \ker F < \infty$.

By $\overline{X}(\varrho_+, \varrho_-)$ we denote the image space im B with the norm

$$\|\varphi\|_{\overline{X}(\varrho_+,\varrho_-)} = \inf_{B\psi=\varphi} \|\psi\|_X \qquad (\varphi \in \operatorname{im} B)$$

and by $\overline{X}(\varrho)$ the image space im F with the norm

$$||\varphi||_{\overline{X}(\varrho)} = \inf_{F\psi=\varphi} ||\psi||_X \qquad (\varphi \in \text{im } F) .$$

Furthermore, we assume that the operators D and F satisfy the condition (2′) from 4.1.1. Then we can introduce the spaces $\widetilde{X}(\varrho_+, \varrho_-) = D^{(-1)}(X)$ and $\widetilde{X}(\varrho) = F^{(-1)}(X)$, where the norms are defined by

$$||f||_{\widetilde{X}(\varrho_+,\varrho_-)} = ||\widetilde{D}f||_X , \qquad ||f||_{\widetilde{X}(\varrho)} = ||\widetilde{F}f||_X .$$

The two following theorems are a direct consequence of the results of the Section 4.1 and of the Theorems 2.6 and 4.2 of Chapter 2.

Theorem 3.2. *For the operator*

$$A = A_1P + A_2Q: \widetilde{X}(\varrho_+, \varrho_-) \to \overline{X}(\varrho)$$

to be a Φ_+-operator or a Φ_--operator it is necessary and sufficient that the functions $A_0(z)$ and $B_0(z)$ in (3.5) *resp.* (3.6) *vanish nowhere on the unit circle*

If these conditions are satisfied, then A is a Φ-operator, and for its index the formula

$$\text{Ind } A = \dim \ker F - \varkappa\beta(\hat{U})$$

holds with $\varkappa = \text{ind } A_0(z) - \text{ind } B_0(z)$.

Theorem 3.3. *For the operator*

$$A' = PA_2 + QA_1: \widetilde{X}(\varrho) \to \overline{X}(\varrho_+, \varrho_-)$$

to be a Φ_+-operator or a Φ_--operator it is necessary and sufficient that the functions $A_0(z)$ and $B_0(z)$ nowhere on the unit circle vanish.

If these conditions are satisfied, then A' is a Φ-operator, and for its index the formula

$$\text{Ind } A' = \dim \ker B + \varkappa\beta(\hat{U})$$

holds.

Remark. If in Theorem 3.2 (Theorem 3.3) $\dim \ker F = 0$ ($\dim \ker B = 0$) holds and if the symbols $A_0(z)$ and $B_0(z)$ do not degenerate, then the invertibility of the operator $A(A')$ corresponds to the number $\varkappa(-\varkappa)$, where

$$\dim \ker A = \max(-\varkappa, 0)\, \beta(\hat{U})\ (= \dim \text{coker } A') ,$$

$$\dim \text{coker } A = \max(\varkappa, 0)\, \beta(\hat{U})\ (= \dim \ker A') .$$

The spaces $\overline{X}(\varrho_+, \varrho_-)$, $\overline{X}(\varrho)$, $\widetilde{X}(\varrho_+, \varrho_-)$ and $\widetilde{X}(\varrho)$ are called the *spaces generated by the zeros of the symbol.*

We point out that by the remark from 4.3.3 we can assume without loss of generality that $F = I$ and thus $\overline{X}(\varrho) = \widetilde{X}(\varrho) = X$.

The following theorem gives some information on the structure of the space $\overline{X}(\varrho_+, \varrho_-)$. We put

$$\varrho_+^{(j)}(z) = (z - \alpha_j)^{m_j} \quad (j = 1, \dots, r), \quad \varrho_-^{(k)}(z) = (z^{-1} - \beta_k^{-1})^{n_k} \quad (k = 1, \dots, s).$$

Theorem 3.4. *The following formula holds:*

$$\overline{X}(\varrho_+, \varrho_-) = \overline{X}(\varrho_+, 1) \cap \overline{X}(1, \varrho_-) = \left[\bigcap_{j=1}^{r} \overline{X}(\varrho_+^{(j)}, 1)\right] \cap \left[\bigcap_{j=1}^{s} \overline{X}(1, \varrho)_-^{(k)}\right].$$

The proof is analogous to the proof of Theorem 2.2.

4.3.5. We recall that in the Subsection 4.2.3 we could realize the two limiting cases $B = I$ and $D = I$ of the representation (1.1) of the Wiener-Hopf operator by a corresponding factorization of the symbol $(\varrho_-(z) \equiv 1$ and $\varrho_+(z) \equiv 1$, respectively). An analogous statement also holds for paired operators under corresponding hypotheses on the symbol.

Under the conditions of the preceding two subsections we put $\varrho(z) \equiv 1$ and thus $F = I$.

We choose arbitrary non-negative integers m_j', m_j'' $(j = 1, \dots, r)$ and n_k', n_k'' $(k = 1, \dots, s)$ such that for all j and k

$$m_j' + m_j'' = m_j . \qquad n_k' + n_k'' = n_k .$$

Analogously to 4.3.4 we can introduce the spaces $\overline{X}(\varrho_+'', \varrho_-'')$ and $\widetilde{X}(\varrho_+', \varrho_-')$. Here

$$\varrho_+'(z) = \prod_{j=1}^{r} (z - \alpha_j)^{m_j'}, \qquad \varrho_-'(z) = \prod_{k=1}^{s} (z^{-1} - \beta_k^{-1})^{n_k'},$$

$$\varrho_+''(z) = \prod_{k=1}^{s} (z - \beta_k)^{n_k''}, \qquad \varrho_-''(z) = \prod_{j=1}^{r} (z^{-1} - \alpha_j^{-1})^{m_j''}. \tag{3.7}$$

Then for the symbols $A(z)$ and $B(z)$ we obtain the representations

$$A(z) = \varrho_+'(z)\, \varrho_-''(z)\, A_0''(z), \qquad B(z) = \varrho_-'(z)\, \varrho_+''(z)\, B_0''(z) \tag{3.8}$$

with

$$A_0''(z) = \prod_{j=1}^{r} (-\alpha_j z)^{m_j''} A_0(z), \qquad B_0''(z) = \prod_{k=1}^{s} (-\beta_k z)^{-n_k''} B_0(z).$$

By A_+', A_+'', A_0'' we denote the opreators from $\mathfrak{R}(U)$ with symbols ϱ_+', ϱ_+'', $A_0''(z)$, respectively. A_-', A_-'' and B_0'' have a corresponding meaning. At last we put

$$B'' = PA_-'' + QA_+'', \qquad D' = A_+'P + A_-'Q.$$

Obviously,

$$A_1 = A_+'A_-''A_0''. \qquad B_1 = A_-'A_+''B_0''$$

and consequently

$$A = A_1P + A_2Q = (A_-''A_0''P + A_+''B_0''Q)\, D'.$$

Now we assume that the commutator

$$T = [A''_- A''_0, P] + [A''_+ B''_0, Q] \tag{3.9}$$

is a compact operator from X into the space $\overline{X}(\varrho''_+, \varrho''_-)$. (In the next section (comp. Corollary 4.1) we shall seee that this is always the case if the functions $A(z)$ and $B(z)$ are differentiable to a sufficiently high order in a certain neighborhood of their zeros.)

Then we obtain

$$A = B''(PA''_0 + QB''_0)\, D' + TD' ,$$

and from the results of the Section 4.1 the next theorem follows immediately.

Theorem 3.5. *Let the commutator* (3.9) *be a compact operator from* X *into* $\overline{X}(\varrho''_+, \varrho''_-)$.

For the operator $A = A_1 P + A_2 Q$ *to be a* Φ_+*-operator or a* Φ_-*-operator from* $\widetilde{X}(\varrho'_+, \varrho'_-)$ *into* $\overline{X}(\varrho''_+, \varrho''_-)$ *it is necessary and sufficient that the functions* $A_0(z)$ *and* $B_0(z)$ *from* (3.5) *and* (3.6), *respectively, vanish nowhere on the unit circle.*

If these conditions are satisfied, then A *is a* Φ*-operator, and for its index the following formula holds:*

$$\operatorname{Ind} A = \dim \ker B'' - \varkappa'' \beta(\hat{U}) \tag{3.10}$$

with

$$\varkappa'' = \operatorname{ind} A''_0(z) - \operatorname{ind} B''_0(z) = \sum_{j=1}^{r} m''_j + \sum_{k=1}^{s} n''_k + \varkappa ,$$

where the number $\varkappa$ *is determined as in Theorem* 3.2.

Analogously we obtain

$$PA_1 + QA_2 = B''(A''_0 P + B''_0 Q)\, D' + B'' \widetilde{T}$$

with

$$\widetilde{T} = [P, A'_+ A''_0] + [Q, A'_- B''_0] . \tag{3.11}$$

This implies the following theorem.

Theorem 3.6. *Let the commutator* (3.11) *be a compact operator from* $X(\varrho'_+, \varrho'_-)$ *into* X.

For the operator $PA_1 + QA_2$ *to be a* Φ_+*-operator or a* Φ_-*-operator from* $\widetilde{X}(\varrho'_+, \varrho'_-)$ *into* $\overline{X}(\varrho''_+, \varrho''_-)$ *it is necessary and sufficient that the functions* $A_0(z)$ *and* $B_0(z)$ *from* (3.5) *and* (3.6), *respectively, vanish nowhere on the unit circle.*

If these conditions are satisfied, then $PA_1 + QA_2$ *is a* Φ*-operator, and its index can be determined by the formula* (3.10).

4.3.6. If we use the factorization of functions (cf. Section 2.3), then for the paired operator

$$A = A_1 P + A_2 Q$$

we obtain a representation of the form $A = BC$ without the compact part (3.9). This is of particular importance for the effective solution of the equation

$$A\varphi = f \qquad (\varphi \in X, f \in \operatorname{im} B)\,.$$

Thus we assume that the functions $A_0(z)$ and $B_0(z)$ from (3.5) and (3.6), respectively, vanish nowhere on the unit circle. In the representations (3.8) of the symbols $A(z)$ and $B(z)$ we choose

$$m_j'' = m_j\,, \qquad n_k'' = n_k \qquad (j = 1, \dots, r;\ k = 1, \dots, s)\,,$$

i.e. $\varrho'_+(z) = \varrho'_-(z) \equiv 1$. In the sequel we suppose that the function $B_0(z)/A_0(z)$ admits a factorization. Let

$$B_0''(z)/A_0''(z) = C_-(z) z^{\varkappa_0} C_+(z)\,,$$

where $C_\pm(z) \in \Re^\pm(z)$, $C_\pm^{-1}(z) \in \Re^\pm(z)$. By $C_\pm \in \Re(U)$ we denote the operator with the symbol $C_\pm(z)$. Then for the operator A we obtain the representation

$$A = A_0'' C_+ (A_-'' P + A_+'' U^{\varkappa_0} Q)\, C \tag{3.12}$$

with $C = C_+^{-1} P + C_- Q$.

We assume that the condition

$$\dim \ker (A_+'' P + A_-'' Q) = 0 \tag{3.13}$$

is satisfied. Let $\varkappa_0 > 0$. Then by L_k $(k = 0, 1, \dots, \varkappa_0 - 1)$ we denote the $\beta(\hat{U})$-dimensional subspace

$$L_k = (A_+'' U_k P + A_-'' U^{k-\varkappa_0+1} Q)\, [\ker (P + UQ)]\,.$$

It is easily seen that $L_j \cap L_k = \{0\}$ $(j \neq k)$. Let X_0 be the direct sum

$$X_0 = L_0 \dotplus L_1 \dotplus \dots \dotplus L_{\varkappa_0 - 1} \qquad (\varkappa_0 > 0)$$

(for $\varkappa_0 \leqq 0$ we put $X_0 = \{0\}$).

It is easy to see that X is the direct sum

$$X = \operatorname{im} (P + U^{-\varkappa_0} Q) \dotplus X_0\,.$$

Moreover we have $X_0 \subset \ker (A_-'' P + A_+'' U^{\varkappa_0} Q)$. By Q_0 we denote the projection of X onto X_0 parallel to $\operatorname{im} (P + U^{-\varkappa_0} Q)$.

Lemma 3.1. *The following representation holds:*

$$A_-'' P + A_+'' U^{\varkappa_0} Q = (A_-'' P + A_+'' Q)\,(P + U^{\varkappa_0} Q)\,(I - Q_0)\,. \tag{3.14}$$

Proof. Since $Q_0 = 0$ für $\varkappa_0 \leqq 0$, in this case the assertion is obvious.

Let $\varkappa_0 > 0$. Each element $f \in X$ can be represented as a sum $f = f_1 + f_2$ with $f_1 \in \operatorname{im} (P + U^{-\varkappa_0} Q)$ and $f_2 \in X_0$. Then by $U^{\varkappa_0} Q f_1 \in \operatorname{im} Q$ we obtain

$$\begin{aligned}(A_-'' P + A_+'' Q)\,(P + U^{\varkappa_0} Q)\,(I - Q_0)\, f &= (A_-'' P + A_+'' U^{\varkappa_0} Q)\, f_1 \\ &= (A_-'' P + A_+'' U^{\varkappa_0} Q)\, f\,.\end{aligned}$$

This proves the assertion.

Now (3.12) and (3.14) imply the following representation for the operator A:

$$A = BA_0 \tag{3.15}$$

with

$$B = A_0''C_+(A_-''P + A_+''Q)\,, \qquad A_0 = (P + U^{\varkappa_0}Q)\,(I - Q_0)\,C\,.$$

Lemma 3.2. *The operator*

$$A_0^{(-1)} = C^{-1}(P + U^{-\varkappa_0}Q)$$

is a right inverse of A_0 for $\varkappa_0 \geqq 0$ and a left inverse of A_0 for $\varkappa_0 \leqq 0$.

Proof. Let $\varkappa_0 \geqq 0$. $(I - Q_0)\,(P + U^{-\varkappa_0}Q) = P + U^{-\varkappa_0}Q$ implies

$$A_0A_0^{(-1)} = (P + U^{\varkappa_0}Q)\,(I + Q_0)\,(P + U^{-\varkappa_0}Q) = I\,.$$

For $\varkappa_0 \leqq 0$ analogously we obtain

$$A_0^{(-1)}A_0 = C^{-1}(P + U^{-\varkappa_0}Q)\,(P + U^{\varkappa_0}Q)\,C = I\,.$$

Theorem 3.7. *In addition to* (3.13) *let the condition*

$$\dim\ker\,(A_-''P + A_+''Q) = 0$$

be satisfied. Then the invertibility of the operator $A\colon X \to \operatorname{im} B$ corresponds to the number

$$-\varkappa_0 = \operatorname{ind} B_0''(z) - \operatorname{ind} A_0''(z)\,,$$

and the following equations hold:

$$\dim\ker A = \max\,\{\varkappa_0, 0\}\,\beta(\hat{U})\,, \qquad \dim\operatorname{coker} A = \max\,\{-\varkappa_0, 0\}\,\beta(\hat{U})\,.$$

The corresponding one-sided inverse has the form

$$A^{(-1)} = C^{-1}(P + U^{-\varkappa_0}Q)\,B^{-1}\,.$$

The kernel of A consists of all elements of the form $C^{-1}\psi$ ($\psi \in X_0$).

4.3.7. The remarks made in 4.2.4 are also true for paired operators.

Namely, it is easy to see from the considerations of the preceding subsections that Theorems 3.2, 3.3, 3.5 and 3.6 are also hold in the case of finitely many zeros of non-integral orders of the symbol if we suppose that the functions $A(z)$ and $B(z)$ are representable in the form (3.5) and (3.6) resp. (3.8) and that all functions in these representations belong to the algebra $\Re(z)$. The considerations from 4.3.6 can also be applied to this case.

4.4. An Algebra of Singular Operators

In this section we suppose that the invertible operator $U \in \mathcal{L}(X)$ satisfies conditions (I) and (II) of 2.1.2 and (III) from 2.2.3 and that $\Re(U)$ is an algebra without radical.

Let $\boldsymbol{\alpha} = \{\alpha_1, \alpha_2, \dots, \alpha_r\}$ and $\boldsymbol{\beta} = \{\beta_1, \beta_2, \dots, \beta_s)$ be systems of points of the unit circle and $\boldsymbol{m} = \{m_1, m_2, \dots, m_r\}$ as well as $\boldsymbol{n} = \{n_1, n_2, \dots, n_s\}$ be systems of natural numbers. For the symbol algebra $\mathfrak{R} = \mathfrak{R}(z)$ we form the algebras $\mathfrak{R}(\boldsymbol{\alpha}, \boldsymbol{m})$, $\mathfrak{R}(\boldsymbol{\beta}, \boldsymbol{n})$ (comp. 2.3.4) and

$$\mathfrak{R}(\boldsymbol{\alpha}, \boldsymbol{m}; \boldsymbol{\beta}, \boldsymbol{n}) = \mathfrak{R}(\boldsymbol{\alpha}, \boldsymbol{m}) \cap \mathfrak{R}(\boldsymbol{\beta}, \boldsymbol{n}) .$$

We consider the set of all singular operators of the form

$$A = A_1 P + A_2 Q + T , \tag{4.1}$$

where T is a compact operator on X and A_1, A_2 are operators from $\mathfrak{R}(U)$ with the symbols $A_1(z)$, $A_2(z)$ from the algebra $\mathfrak{R}(\boldsymbol{\alpha}, \boldsymbol{m}; \boldsymbol{\beta}, \boldsymbol{n})$. From formula (2.7), Chapter 2, it follows immediately that the operators of the form (4.1) form an algebra.

4.4.1. Similarly as in 4.3 we introduce the operator B by the formula $B = PA_- + QA_+$ with

$$A_- = \prod_{j=1}^{r} (U^{-1} - \alpha_j^{-1} I)^{m_j} , \qquad A_+ = \prod_{k=1}^{s} (U - \beta_k I)^{n_k} .$$

Here we suppose that the operator B satisfies the following condition:

(1) $\dim \ker B = 0$.

By $\overline{X} = \overline{X}(\varrho_+, \varrho_-)$ we denote the image space $\operatorname{im} B$ which we equip with the norm (1.2) ($\varrho_\pm$ are the symbols of the operator $A_\pm$).

In the sequel we investigate the properties of the singular operator (4.1) in the pair of spaces (X, $\overline{X}$ resp. in the space $\overline{X}$).

Lemma 4.1. *Let $A_0(z) \in \mathfrak{R}(\boldsymbol{\alpha}, \boldsymbol{m}; \boldsymbol{\beta}, \boldsymbol{n})$ and $A_0 \in \mathfrak{R}(U)$ be the operator with symbol $A_0(z)$. Then the commutator $[P, A_0]$ is a compact operator from X into the space $\overline{X}$.*

Proof. First, we remark that the operator $[P, U^l] \in \mathcal{L}(X)$ $(l = 1, 2, \ldots)$ is finite dimensional (cf. the proof of Lemma 2.1, Chapter 2). Using the Theorem 3.4 we obtain easily from equation

$$[P, U^l] = -\alpha P(U^{-1} - \alpha^{-1} I)\, PU^l Q + \alpha PU^{l-1} Q \qquad (|\alpha| = 1)$$

that $\operatorname{im} [P, U^l] \subset \operatorname{im} B$ and thus $[P, U^l] \in \mathcal{K}(X, \overline{X})$.

By Theorem 6° of 2.3.4 the operator A_0 admits the representations

$$A_0 = R_1 + GA_- = R_2 + HA_+ \tag{4.2}$$

with $G, H \in \mathfrak{R}(U)$ and certain operator polynomials (in U) R_1 and R_2. Hence

$$[P, A_0] = [P, R_1] + PA_-[P, G] - QA_-GP .$$

The commutator $[P, R_1]$ is a compact operator from X into $\overline{X}$. The same holds for the operator

$$PA_-[P, G] = BP[P, G]$$

since $[P, G] \in \mathcal{K}(X)$ by Lemma 2.1, Chapter 2. Finally, using the relation (4.2) we obtain

$$QA_-GP = Q(R_2 - R_1)\, P + BQHP$$

and thus $QA_-GP \in \mathcal{K}(X, \overline{X})$. Hence the lemma is proved.

Corollary 4.1. *If $A_0(z) \in \mathfrak{R}(\boldsymbol{\beta}, \boldsymbol{n})$ and $B_0(z) \in \mathfrak{R}(\boldsymbol{\alpha}, \boldsymbol{m})$, then the hypothesis of the Theorem 3.5 is satisfied (i.e. the commutator (3.9) is a compact operator from X into $\overline{X}$).*

Corollary 4.2. *The singular operators of the form (4.1) with $T \in \mathcal{K}(X, \overline{X})$ and $A_j(z) \in \mathfrak{R}(\boldsymbol{\alpha}, \boldsymbol{m}; \boldsymbol{\beta}, \boldsymbol{n})$ $(j = 1, 2)$ form an algebra.*

Lemma 4.2. *Let $A_0(z) \in \mathfrak{R}(\boldsymbol{\alpha}, \boldsymbol{m}; \boldsymbol{\beta}, \boldsymbol{n})$ and $A_0 \in \mathfrak{R}(U)$ the operator with symbol $A_0(z)$. Then $A_0 \in \mathcal{L}(\overline{X})$.*

By Lemma 4.1 the assertion follows immediately from the relation

$$A_0B = BA_0 + [A_0, P]\,(A_- - A_+)\,.$$

As a consequence of Lemma 4.1 and Lemma 4.2 we obtain

Lemma 4.3. *Let*

$$A = A_1P + A_2Q\,, \qquad A' = A_1'P + A_2'Q$$

with $A_j(z), A_j'(z) \in \mathfrak{R}(\boldsymbol{\alpha}, \boldsymbol{m}; \boldsymbol{\beta}, \boldsymbol{n})$ $(j = 1, 2)$. Then $[A, A'] \in \mathcal{K}(X, \overline{X})$.

Theorem 4.1. *Let $A_j(z) \in \mathfrak{R}(\boldsymbol{\alpha}, \boldsymbol{m}; \boldsymbol{\beta}, \boldsymbol{n})$ $(j = 1, 2)$. For the operator $A = A_1P + A_2Q$ to be a $\Phi(\Phi_\pm)$-operator on the space $\overline{X}$ it is necessary and sufficient that the condition*

$$A_1(z) \neq 0\,, \qquad A_2(z) \neq 0 \qquad (|z| = 1)$$

be satisfied.

Proof. By Lemma 4.3 we have $AB = BA + T$ with $T \in \mathcal{K}(X, \overline{X})$. Hence A is a $\Phi(\Phi_\pm)$-operator in $\overline{X}$ if and only if it is such an operator in X. Now the assertion is a direct consequence the Theorem 2.7, Chapter 2.

Theorem 4.2. *Let $A_1(z)$, $A_2(z)$ be functions from the algebra $\mathfrak{R}(\boldsymbol{\alpha}, \boldsymbol{m}; \boldsymbol{\beta}, \boldsymbol{n})$ and $A = A_1P + A_2Q$ $(A = PA_1 + QA_2)$. For the operator A to be a Φ_+- or a Φ_--operator which acts from X into the space $\overline{X}$ it is necessary and sufficient that the functions $A_1(z)$ and $A_2(z)$ have the representations*

$$A_1(z) = \prod_{j=1}^{r} (z - \alpha_j)^{m_j}\, A_0(z)\,, \qquad A_2(z) = \prod_{k=1}^{s} (z - \beta_k)^{n_k}\, B_0(z)\,, \tag{4.3}$$

where $A_0(z)$ and $B_0(z)$ are functions from the algebra $\mathfrak{R}(z)$ which vanish nowhere on the unit circle.

If the conditions (4.3) *are satisfied, then A is a continuous Φ-operator from X into $\overline{X}$ with the index*

$$\operatorname{Ind} A = \left[\operatorname{ind}\left(B_0(z)/A_0(z)\right) - \sum_{j=1}^{r} m_j\right] \beta(\hat{U}) .$$

Proof. If the functions $A_1(z)$ and $A_2(z)$ possess the representations (4.3), then the operators A_1 and A_2 can be represented in the form

$$A_1 = A_- A_0' , \qquad A_2 = A_+ B_0$$

with A_0', $B_0 \in \mathfrak{R}(U)$, where the operator A_0' has the symbol $A_0(z) \prod_{j=1}^{r} (-\alpha_j z)^{m_j}$. Using Lemma 4.1 we obtain

$$A = B(PA_0' + QB_0) + T \tag{4.4}$$

with $T \in \mathcal{K}(X, \overline{X})$. Hence $A \in \mathcal{L}(X, \overline{X})$. If the functions $A_0(z)$ and $B_0(z)$ vanish nowhere on the unit circle, then by Theorem 2.7, Chapter 2, we have $PA_0' + QB_0 \in \Phi(X)$ and thus $A \in \Phi(X, \overline{X})$.

Now we shall show the necessity of conditions (4.3). Let A be a Φ_+-operator or a Φ_--operator which acts from X into the space $\overline{X}$. By $\alpha(\beta)$ we denote an arbitrary one of the points $a_j(\beta_k)$ $(j = 1, \ldots, r;\ k = 1, \ldots, s)$, and we put $m = m_j$, $n = n_k$. The functions $A_1(z)$ and $A_2(z)$ possess the representations

$$A_l(z) = A_l'(z) + A_l''(z) \qquad (l = 1, 2)$$

with

$$A_1'(z) = \sum_{j=0}^{m-1} a_j(z^{-1} - \alpha^{-1})^j , \qquad A_1''(z) = G(z)\,(z^{-1} - \alpha^{-1})^m ,$$

$$A_2'(z) = \sum_{j=0}^{n-1} b_j(z - \beta)^j , \qquad A_2''(z) = H(z)\,(z - \beta)^n$$

and $G(z)$, $H(z) \in \mathfrak{R}(z)$. We put

$$A' = A_1'P + A_2'Q , \qquad A'' = A_1''P + A_2''Q$$

and $B_1 = P(U^{-1} - \alpha^{-1}I)^m + Q(U - \beta I)^n$. From the part of the proof already given it follows that $\operatorname{im} A'' \subset \operatorname{im} B_1$ and thus also $\operatorname{im} A' \subset \operatorname{im} B_1$. In particular, for any $\varphi \in \operatorname{im} P$ there exists an element $f \in X$ such that

$$A'\varphi = \sum_{j=0}^{m-1} a_j P(U^{-1} - \alpha^{-1}I)^j \varphi = P(U^{-1} - \alpha^{-1}I)^m f . \tag{4.5}$$

First we assume $a_0 \neq 0$. Then we can easily see that (4.5) implies

$$\operatorname{im} P = \operatorname{im} P(U^{-1} - \alpha^{-1}I)\,P . \tag{4.6}$$

By condition (1) the equality (4.6) shows the invertibility of the operator $P(U^{-1} - \alpha^{-1}I)\,P \mid \operatorname{im} P$. But this contradicts Theorem 1.3, Chapter 2. Hence $a_0 = 0$.

Now let

$$a_j = 0 \quad (j = 0, \dots, l-1), \qquad a_l \neq 0 \quad (l \leq m-1).$$

Then it follows from equation (4.5) that

$$\operatorname{im} P(U^{-1} - \alpha^{-1}I)^l P = \operatorname{im} P(U^{-1} - \alpha^{-1}I)^{l+1} P.$$

Since $\dim \ker P(U^{-1} - \alpha^{-1}I) = 0$, this implies relation (4.6). Hence $a_j = 0$, $j = 0, 1, \dots, m-1$. Analogously we can prove that $b_j = 0$, $j = 0, 1, \dots, n-1$.

By assertion 5° of 2.3.4 the part of the proof already given shows that the functions $A_1(z)$ and $A_2(z)$ have the representations (4.3) with $A_0(z), B_0(z) \in \Re(z)$. Hence relation (4.4) holds. Since $A \in \Phi_{\pm}(X, \overline{X})$ we have $PA_0' + QB_0 \in \Phi_{\pm}(X)$. From this the assertion follows by Theorem 2.7, Chapter 2.

The index formula follows from formula (4.4) and Theorem 4.2, Chapter 2.

Remark. If in the conditions (4.3) all points α_j and β_k are distruct different, then the invertibility of the operator $A \in \mathcal{L}(X, \overline{X})$ corresponds to the number $\sum_1^r m_j - \operatorname{ind}(B_0(z)/A_0(z))$.

By Theorem 4.2, Chapter 2, and the following lemma this assertion follows immediately from the representation $PA_1 + QA_2 = B(PA_0' + QB_0)$.

Lemma 4.4. *Let A, B be arbitrary operators from the algebra $\Re(U)$ whose symbols have no common zeros on the unit circle. Then*

$$\dim \ker (AB + BQ) = \dim \ker (PA + QB). \tag{4.7}$$

Proof. Since $\Re(U)$ is a commutative Banach algebra and the functions $A(z)$, $B(z)$ have no common zeros on the unit circle, there exist two operators $R_1, R_2 \in \Re(U)$ with

$$AR_1 + BR_2 = I. \tag{4.8}$$

Now, let $(AP + BQ)\,x = 0$, i.e.

$$APx = -BQx. \tag{4.9}$$

We put $y = R_2Px - R_1Qx$. Using formulae (4.8) and (4.9) we obtain

$$Ay = -Qx, \qquad By = Px. \tag{4.10}$$

Hence $(PA + QB)\,y = 0$. Moreover, relations (4.10) show that $y = 0$ implies $x = 0$. Thus

$$\dim \ker (AP + BQ) \leq \dim \ker (PA + QB). \tag{4.11}$$

Now we prove the reverse inequality. Let $(PA + QB)\,x = 0$, i.e. $PAx = -QBx$. The relations $PQ = QP = 0$ show that

$$PAx = QBx = 0.$$

Consider the element $y = Bx - Ax$. Obviously, $Py = Bx$ and $Qy = -Ax$. This implies $(AP + BQ)\,y = 0$. If $y = 0$, then $Ax = Bx = 0$, and from (4.8) we conclude that $x = 0$. Thus we obtain the inequality

$$\dim \ker (PA + QB) \leq \dim \ker (AP + BQ). \tag{4.12}$$

Equation (4.7) now follows from (4.11) and (4.12).

4.4.2. As in Section 4.3 we can now consider the operator $D = A_+P + A_-Q$. We suppose that D satisfies condition (2′) of 4.1.1. Thus the space $\widetilde{X} = \widetilde{X}(\varrho_+, \varrho_-)$ is defined.

Lemma 4.5. *If the hypotheses of Lemma* 4.1 *hold, then* $[P, A_0] \in \mathcal{K}(\widetilde{X}, X)$.

Using the relations

$$[P, U^{-1}] = \frac{1}{\alpha} QU^{-1}P(U - \alpha I)\, P - \frac{1}{\alpha} QU^{-l+1}P \qquad (|\alpha| = 1, l = 1, 2, \ldots)$$

we can prove Lemma 4.5 analogously to Lemma 4.1.

It follows from Lemma 4.5 together with the representation

$$A_0D = DA_0 + (A_+ - A_-)\,[A_0, P]$$

that each operator $A_0 \in \mathfrak{R}(U)$ with a symbol $A_0(z) \in \mathfrak{R}(\boldsymbol{\alpha}, \boldsymbol{m}; \boldsymbol{\beta}, \boldsymbol{n})$ admits a continuous extension to the space $\widetilde{X}$. Since

$$PD - DP \quad \text{and} \quad QD = DQ\,,$$

the operators P, Q also possess continuous extensions to $\widetilde{X}$. Furthermore, the Theorem 4.1 remains valid if $\overline{X}$ is replaced by $\widetilde{X}$.

It follows from the proof of Lemma 4.4 that equation (4.7) also holds for the operators $AP + BQ$ and $PA + QB$ considered in the space $\widetilde{X}$ if the symbols $A(z)$, $B(z)$ belong to the algebra $\mathfrak{R}(\boldsymbol{\alpha}, \boldsymbol{m}; \boldsymbol{\beta}, \boldsymbol{n})$ and have no common zeros on the unit circle. By a corresponding modification of the considerations in the proof of the Theorem 4.2 we see that this theorem also remains true if the pair of spaces $X, \overline{X}$ is replaced by $\widetilde{X}, X$. Under the hypothesis (4.3) the index of the operator $A \in \mathfrak{L}(\widetilde{X}, X)$ can be calculated by the formula

$$\operatorname{Ind} A = \varkappa\beta(\hat{U})\,, \qquad \varkappa = \operatorname{ind}\big(B_0(z)/A_0(z)\big) + \sum_{k=1}^{s} n_k\,.$$

If the points α_j and β_k are all different, then

$$\dim\ker A = \max(\varkappa, 0)\,\beta(\hat{U})\,, \qquad \dim\operatorname{coker} A = \max(-\varkappa, 0)\,\beta(\hat{U})\,.$$

In the sequel we maintain the notations (3.7). By combining the results of this and the preceding subsection and taking into consideration the representations for the paired operators used already in Subsection 4.3.5 we obtain the following theorem which, for symbols from the algebra $\mathfrak{R}(\boldsymbol{\alpha}, \boldsymbol{m}; \boldsymbol{\beta}, \boldsymbol{n})$. sharpens Theorems 3.5 and 3.6.

Theorem 4.3. *Let* $A_1(z)$, $A_2(z)$ *be functions from the algebra* $\mathfrak{R}(\boldsymbol{\alpha}, \boldsymbol{m}; \boldsymbol{\beta}, \boldsymbol{n})$ *and* $A = A_1P + A_2Q$ $(A = PA_1 + QA_2)$. *For* A *to be a* Φ_+*-operator or a* Φ_-*-operator which acts from* $\widetilde{X}(\varrho'_+, \varrho'_-)$ *into the space* $\overline{X}(\varrho''_+, \varrho''_-)$ *it is necessary and sufficient that the functions* $A_1(z)$ *and* $A_2(z)$ *possess the representation* (4.3). *where* $A_0(z)$ *and* $B_0(z)$ *are functions from the algebra* $\mathfrak{R}(z)$ *which vanish nowhere on the unit circle.*

If conditions (4.3) *are satisfied, then* A *is a continuous* Φ*-operator from* $\tilde{X}(\varrho'_+, \varrho'_-)$ *into* $\overline{X}(\varrho''_+, \varrho''_-)$ *with the index*

$$\operatorname{Ind} A = \lambda\beta(\hat{U}), \qquad \lambda = \operatorname{ind}\left(B_0(z)/A_0(z)\right) + \sum_{k=1}^{s} n'_k - \sum_{j=1}^{r} m''_j .$$

If in the conditions (4.3) *the points* α_j *and* β_k *are all different, then*

$$\dim\ker A = \max(\lambda, 0)\,\beta(\hat{U}), \qquad \dim\operatorname{coker} A = \max(-\lambda, 0)\,\beta(\hat{U}). \quad (4.13)$$

4.4.3. The essential considerations of Subsections 4.4.1 and 4.4.2 can also be carried over to the case in which the symbols of the operators A_1 and A_2 have finitely many zeros of non-integral order.

Let m'_j, m''_j ($j = 1, 2, \ldots, r$) and n'_k, n''_k ($k = 1, 2, \ldots, s$) be arbitrary non-negative real numbers and

$$m_j = m'_j + m''_j, \qquad n_k = n'_k + n''_k .$$

Now, by $\boldsymbol{m}(\boldsymbol{n})$ we denote the integral multiindex with components $-[-m_j]$, $j = 1, 2, \ldots, r$ ($-[-n_k]$, $k = 1, \ldots, s$), where $[-m_j]$ is the integral part of the number $-m_j$. Moreover. we maintain the notation introduced in 4.3.5.

We suppose that the symbols of the operators A_1 and A_2 possess the representations (3.8), where the functions $A''_0(z)$ and $B''_0(z)$ belong to the algebras $\Re(\boldsymbol{\beta}, \boldsymbol{n})$ and $\Re(\boldsymbol{\alpha}, \boldsymbol{m})$, respectively, and vanish nowhere on the unit circle. Then, using the results of Subsection 2.3.4 and the formulae (2.8), Chapter 2, it follows that the operator $A = A_1P + A_2Q$ ($A = PA_1 + QA_2$) has the representation

$$A = B''(A''_0P + B''_0Q + T)\,D'$$

with $T \in \mathcal{K}(X)$. This implies $A \in \Phi(\tilde{X}(\varrho'_+, \varrho'_-), \overline{X}(\varrho''_+, \varrho''_-))$ and the index formula

$$\operatorname{Ind} A = \lambda\beta(\hat{U}), \qquad \lambda = \operatorname{ind} B''_0(z) - \operatorname{ind} A''_0(z).$$

Furthermore, in the present case Lemma 4.1, Lemma 4.5 and thus also the formulae (4.13) remain true (if the points α_j and β_k are all different).

4.5. The Method of Left Regularization

Let $A \in \mathcal{L}(X, Y)$ be a non-normally solvable operator with the representation (1.9). In Section 4.1 we saw that it is possible to construct a Banach space $\overline{Y}$ such that the operator $A : X \to \overline{Y}$ is a Φ-operator.

In this section we consider another method for constructing a pair of spaces with the mentioned properties. It is based on the knowledge of a closed left regularizer of the operator A. At the end both methods are compared with each other.

In the sequel X and Y are Banach spaces.

4.5.1. Let $A(X \to Y)$ be a closed operator. We assume that a closed left regularizer R of the operator A is at hand. In the domain $D(R)$ we introduce a new norm by

$$|y| = \|y\|_Y + \|Ry\|_X \qquad (y \in D(R)), \tag{5.1}$$

where $\|\cdot\|_X$ and $\|\cdot\|_Y$ denote the norms of the spaces X and Y, respectively. By the closedness of the operator R the domain $D(R)$ is a Banach space in the new norm. We denote this space by Y_R.

Furthermore, by R' we denote the operator which acts from Y_R into X and is defined by $R'y = Ry$. Then from 4.1 we obtain the inequality

$$\|R'y\|_X \leqq |y|,$$

i.e. R' is a bounded operator: $R' \in \mathcal{L}(Y_R, X)$.

By A' we denote the operator which acts from X into Y_R and is defined by the equation

$$A'x = Ax, \qquad D(A') = D(A).$$

Then the closedness of the operator A and the relation $\|y\|_Y = |y|$ $(y \in Y_R)$ immediately show that the operator A' is closed.

By $R'A' = RA = I + T$ the operator R' is a left bounded regularizer of A', and thus, by Theorem 4.1, Chapter 1, A' is a Φ_+-operator.

In the case $A \in \mathcal{L}(X, Y)$ also $A' \in \mathcal{L}(X, Y_R)$ holds, and in this case A' is a Φ-operator if the kernel of R is finite dimensional.

In fact, from (5.1) and $RA = I + T$ we obtain the inequality

$$|A'x| = \|Ax\|_Y + \|RAx\|_X \leqq C\,\|x\|_X,$$

i.e. $A' \in \mathcal{L}(X, Y_R)$. Moreover, if $\ker R = \ker R'$ is finite dimensional, then Theorem 5.5, Chapter 1, shows that A' has a finite index. Thus A' is a Φ-operator.

The following theorem yields a simple criterion for the compactness of an operator $T \in \mathcal{L}(X, Y_R)$.

Theorem 5.1. *Let $T \in \mathcal{K}(X, Y)$ be a compact operator with* $\operatorname{im} T \subset D(R)$. *For the operator $T' \in \mathcal{L}(X, Y_R)$ to be compact it is necessary and sufficient that $T_0 = RT$ be a compact operator on X.*

Proof. The necessity of the condition is obvious since $RT = R'T'$, as a product of a continuous and a compact operator, is compact.

Sufficiency. Let T_0 be compact. We consider an arbitrary bounded set $E \subset X$ and choose an arbitrary sequence $\{x_n\} \subset E$. We have to show the existence of a subsequence $\{x_{n_k}\}$ for which the sequence $\{T'x_{n_k}\}$ is convergent in Y_R.

Since the operator T and T_0 are compact, there exists a subsequence $\{x_{n_k}\}$ such that

$$y_{n_k} = Tx_{n_k} \to y_0, \qquad T_0 x_{n_k} \to x_0,$$

where $x_0 \in X$, $y_0 \in Y$. Hence

$$y_{n_k} \in D(R)\,, \qquad y_{n_k} \to y_0\,, \qquad R y_{n_k} \to x_0\,.$$

From this the relations $y_0 \in D(R)$ and $Ry_0 = x_0$ follow since the operator R is closed.

Thus we obtain $y_0 \in Y_R$ and

$$|T' x_{n_k} - y_0| = ||T x_{n_k} - y_0||_Y + ||T_0 x_{n_k} - x_0||_X \to 0\,.$$

This proves the theorem.

4.5.2. Let R again be a closed left regularizer of the closed operator $A(X \to Y)$.

In concrete problems it is often difficult to describe the domain $D(R)$ of the operator R analytically. But sometimes it is possible to find a space Z which is embedded into $D(R)$ and which has a simpler structure than $D(R)$. In this case A can be reduced to a normally solvable operator from X into Z in the following way.

Let Z be a Banach space continuously embedded into the Banach space Y_R. By

$$\underline{A}x = Ax\,, \qquad D(\underline{A}) = \{x \in D(A) : Ax \in Z\}$$

we define an operator $\underline{A}(X \to Z)$. It is easy to check that $\underline{A}$ is closed.

The operator $\underline{A}$ possesses a bounded left regularizer. Hence it is a Φ_+-operator.

In fact, let R'' be the operator which maps Z into X and which is defined by $R''y = Ry$. For all $y \in Z$ the operator R'' satisfies the inequality

$$||R''y||_X \leqq |y| \leqq C\,||y||_Z\,,$$

i.e. $R'' \in \mathscr{L}(Z, X)$. Moreover, since

$$R''\underline{A}x = x + Tx \qquad (x \in D(\underline{A}))\,,$$

R'' is a left regularizer of A.

In the sequel we suppose that $A \in \mathscr{L}(X, Y)$ and that $\alpha(R) = \dim \ker A$ is finite. Then $A' \in \mathscr{L}(X, Y_R)$ is a Φ-operator. Moreover, since A is a bounded right regularizer of the closed operator R, Theorem 4.2, Chapter 1, shows that also R is a Φ-operator.

Theorem 5.2. *If Z is dense in Y_R, then A is a Φ-operator and the following relations hold:*

$$\ker \underline{A} = \ker A'\,, \qquad (\operatorname{im} \underline{A})^{\perp} = (\operatorname{im} A')^{\perp}\,, \qquad \operatorname{Ind} \underline{A} = \operatorname{Ind} A'\,. \quad (5.2)$$

Proof. Obviously, it is sufficient to prove the second of the relations (5.2). The first follows immediately from the definition of the operator $\underline{A}$.

Since Z is continuously and densely embedded into Y_R we have $Y_R^* \subset Z^*$. From this and from the abvious relation $\operatorname{im} A \subset \operatorname{im} A'$ we obtain the inclusion $(\operatorname{im} A')^{\perp} \subset (\operatorname{im} \underline{A})^{\perp}$.

Thus it still remains to show that the difference $(\operatorname{im} \underline{A})^\perp - (\operatorname{im} A')^\perp$ is the empty set. We assume that this set contains a certain functional f_0. Since the number $m = \beta(A')$ is finite, there exists a basis $f_1, f_2, \dots, f_m$ in the subspace $(\operatorname{im} A')^\perp$. Then the functionals $f_0, f_1, \dots, f_m$ from Z^* are linearly independent. Thus there exists an element $y_0 \in Z$ with $f_0(y_0) = 1$ and

$$f_j(y_0) = 0\,, \qquad j = 1, 2, \dots, m\,. \tag{5.3}$$

Since the operator A' is normally solvable, it follows from (5.3) that there exists an element $x_0 \in X$ such that $A'x_0 = y_0$. But since $y_0 \in Z$ we have $x_0 \in D(\underline{A})$ and $\underline{A}x_0 = A'x_0 = y_0$. Taking into consideration that $f_0 \in (\operatorname{im} \underline{A}^\perp)$ we obtain

$$f_0(y_0) = f_0(\underline{A}x)_0 = 0\,.$$

This contradiction proves the assertion.

Remark. Let the Banach space Z be continuously embedded into Y, and let $Z \subset D(R)$. Then Z is also continuously embedded into the space Y_R if for all $z \in Z$ an inequality of the form

$$||Rz||_X \leqq C\,||z||_Z$$

holds, where C is a constant independent of z.

This follows immediately from the inequalities

$$|z| = ||z||_Y + ||Rz||_X \leqq ||z||_Y + C\,||z||_Z \leqq C'\,||z||_Z\,.$$

Finally, for the case $X = Y$ the following lemma yields a sufficient condition for the denseness of a set $Z \subset D(R)$ in the norm of the space $Y_R = X_R$.

Lemma 5.1. *Let $Z \subset D(R)$ be a linear subset dense in the space X such that $Rx \in Z$ $(x \in X)$ implies $x \in Z$. Then Z is dense in the space X_R.*

Proof. Since R is a Φ-operator in X, the space X possesses the following representations with a certain closed subspace X_1 and a finite dimensional subspace X_2 (comp. Section 1.2):

$$X = X_1 \dotplus \ker R\,, \qquad X = \operatorname{im} R \dotplus X_2\,.$$

From this if follows that $D(R) = D_1 \oplus \ker R$, where $D_1 = D(R) \cap X_1$.

By R_1 we denote the restriction of the operator R to the subspace D_1 (comp. Subsection 1.2.5). Then R_1^{-1} is bounded, i.e. for a certain constant $C > 0$ and arbitrary $x \in D_1$ we have

$$||x|| \leqq C\,||R_1x|| = C\,||Rx||\,. \tag{5.4}$$

Each element $x \in D(R)$ can uniquely be represented in the form $x = x' + x''$ with $x' \in D_1$ and $x'' \in \ker R$. Since, by Lemma 2.2, Chapter 1, the set $(\operatorname{im} R) \cap$ $\cap\, Z$ is dense in $\operatorname{im} R$, there exists a sequence of elements $Rx_n' \in Z$ with $x_n' \in D_1$ and $Rx_n' \to Rx'$. Then the inequality (5.4) shows that $x_n' \to x'$.

Now, if we put $x_n = x'_n + x''$, then $x_n \to x$ and $Rx_n \to Rx$. This implies

$$|x_n - x| = ||x_n - x|| + ||Rx_n - Rx|| \to 0 .$$

Since by hypothesis $Rx_n = Rx'_n \in Z$ implies $x_n \in Z$, the assertion is proved.

4.5.3. Now we shall compare the method of left regularization with the method considered in Section 4.1.

Let the operator $A \in \mathscr{L}(X, Y)$ posses the representation $A = BC$, where $B \in \mathscr{L}(Y)$ satisfies condition (1) of 4.1 and $C \in \mathscr{L}(X, Y)$ is a Φ-operator. Obviously, $\operatorname{im} A \subset \overline{Y} = \operatorname{im} B$.

Furthermore, let R be a certain closed left regularizer of the operator A, i.e. $\operatorname{im} A \subset Y_R = D(R)$ and $RA = I + T$, where T is a compact operator on X.

As in 1.2.5. by C_1 we denote the restriction of the operator C to an arbitrary topological complement of $\ker C$ and by Q a continuous projection of the space Y onto $\operatorname{im} C$. Since obviously $B^{-1}y \in \operatorname{im} C$ for any $y \in \operatorname{im} A$ we obtain

$$Ry = (I + T)\, C_1^{-1}B^{-1}y = (I + T)\, C_1^{-1}QB^{-1}y \,. \tag{5.5}$$

Thus by the right-hand side of the formula (5.5) the operator R can be extended to $\operatorname{im} B$. Hence without loss of generality we can assume that $\operatorname{im} B \subset D(R)$.

Obviously, the operator RB is a left regularizer of C. Since C is a Φ-operator, RB must necessarily be a bounded operator (see Theorem 5.4, Chapter 1). Consequently for any $y \in \operatorname{im} B$ the estimate

$$|y| = ||y||_Y + ||Ry||_X \leqq c\, ||B^{-1}y||_Y = c\, ||y||_{\overline{Y}} \tag{5.6}$$

holds, where $c = \text{const}$. Thus we have the continouus embedding $Y \subset \overline{Y}_R$.

If in particular $\operatorname{im} B = D(R)$, then using Banach's theorem on homomorphisms we conclude from (5.6) that the norms $|\ |$ and $||\ ||_{\overline{Y}}$ defined on $\operatorname{im} B$ are equivalent. Thus in this case the spaces $\overline{Y}$ and Y_R coincide. This holds in particular if the regularizer R has the form $R = MB^{-1}$, where M is a regularizer of the Φ-operator C. Consequently the results of 4.5.2 also remain true if we put $Y_R = \operatorname{im} B$.

4.5.4. As an illustration we consider two simple examples.

Example 1. Let A be the operator defined on the space $C[0, 1]$ by the equation

$$(A\varphi)\,(x) = x\varphi(x) + \int_0^1 K(x, y)\, \varphi(y)\, dy \,. \tag{5.7}$$

Here $K(x, y)$ is a kernel continuous and continuously differentiable by x in the square $[0, 1] \times [0, 1]$.

We can write the operator A in the form $A = B + T$, where A is the operator of the multiplication by the independent variable x and T is the integral operator with kernel $K(x, y)$. The operator T is compact on $C[0, 1]$. Obviously, the operator B satisfies the condition (1) from 4.1, and it is not normally solvable in $C[0, 1]$.

By R we denote the operator defined by

$$(R\varphi)(x) = \frac{\varphi(x) - \varphi(0)}{x}. \tag{5.8}$$

The domain (DR) will be the set of all functions $\varphi \in C[0, 1]$ for which the right-hand side of (5.8) also belongs to $C[0, 1]$. Obviously, $RB = I$, i.e. the operator R is a left inverse and thus also a left regularizer of B. It is easy to see that R is a closed operator.

Introducing the norm

$$|\varphi| = \max_{0 \leq x \leq 1} |\varphi(x)| + \max_{0 \leq x \leq 1} \left| \frac{\varphi(x) - \varphi(0)}{x} \right|$$

on $D(R)$ we obtain the Banach space $C_R[0, 1]$. It is immediately seen that $C_R[0, 1]$ is the extension of the space $\overline{C}[0, 1] = \operatorname{im} B$ for one dimension (all constants).

By 4.5.1 the operator B: $C[0, 1] \to C_R[0, 1]$ is a Φ-operator. The index of this operator is equal to one since a function $\varphi \in C_R[0, 1]$ belongs to im B if and only if $\varphi(0) = 0$. Since the operator

$$(RT\varphi)(x) = \int_0^1 \frac{K(x, y) - K(0, y)}{x} \varphi(y)\, dy$$

is compact on $C[0, 1]$, by Theorem 5.1 the operator T: $C[0, 1] \to C_R[0, 1]$ is also compact. Hence the operator

$$A : C[0, 1] \to C_R[0, 1]$$

is a Φ-operator with index one.

By means of the remarks from 4.5.2 it is easily checked that the Banach space $C^{(1)}[0, 1]$ with the norm

$$||\varphi||_{C^{(1)}} = \max_{0 \leq x \leq 1} |\varphi(x)| + \max_{0 \leq x \leq 1} |\varphi'(x)|$$

is continuously and densely embedded into the space $C_R[0, 1]$.

An analogous result can be obtained for the operator

$$(A\varphi)(x) = a(x)\varphi(x) + \int_0^1 K(x, y)\varphi(y)\, dy,$$

where the function $a(x) \in C[0, 1]$ has finitely many zeros of integral orders in the interval $[0, 1]$.

Example 2. Consider the differential equation

$$(A\varphi)(x) = x\varphi'(x) + c(x)\varphi(x) = f(x) \tag{5.9}$$

on the interval [0, 1]. Here the function $c(x) \in C[0, 1]$ is continuously differentiable in a certain neighborhood (on the right) of the point $x = 0$ and $c(0) > -1$. We maintain the notation introduced at the example 1.

a) First we consider the case $c(x) \equiv \gamma = \text{const}$.

If $\gamma \neq 0$, then it can easily be checked that for any function $f(x) \in C_R[0, 1]$ the equation (5.9) possess the unique solution

$$\varphi(x) = \frac{1}{x^\gamma} \int_0^x t^\gamma g(t)\, dt + \frac{f(0)}{\gamma} \quad (x \neq 0)\,, \quad \varphi(0) = \frac{f(0)}{\gamma} \quad \text{with} \quad g(t) = \frac{f(x) - f(0)}{x}\,.$$

If $= 0$, then the equation (5.9) with $f(x) \in C_R[0, 1]$ has a solution $\varphi \in C^{(1)}[0, 1]$ if and only if $f(0) = 0$. This solution is uniquely determined up to a constant.

From this it now follows that in the case $c(x) \equiv \gamma$ the operator A: $C^{(1)}[0, 1] \to C_R[0, 1]$ defined by (5.9) is a Φ-operator with index zero.

b) In the general case $c(x) \neq \text{const}$ the operator A can be written in the form $A = A_0 + T$, where

$$(A_0\varphi)\,(x) = x\varphi'(x) + c(0)\,\varphi(x)\,, \qquad (T\varphi)\,(x) = [c(x) - c(0)]\,\varphi(x)\,.$$

It is easy to see that the operator

$$(RT\varphi)\,(x) = \frac{c(x) - c(0)}{x}\,\varphi(x)\,,$$

is a compact operator from $C^{(1)}[0, 1]$ into $C[0, 1]$. By Theorem 5.1 this shows that the operator $T : C^{(1)}[0, 1] \to C_R[0, 1]$ is also compact. Hence the operator

$$A : C^{(1)}[0, 1] \to C_R[0, 1]$$

is a Φ-operator with index zero.

4.6. Comments and References

4.1. The method considered here is essentially due to G. N. ČEBOTAREV [2—3], M. I. HAIKIN [1], V. B. DYBIN [3] as well as to V. B. DYBIN and N. K. KARAPETJANC [2], where it appears as the *method of normalization* in the more special form $D = I$ in the investigation of Wiener-Hopf integral equations of non-normal type and of its discrete analogues (for this cf. also the Sections 4.5 and 1.5.3). The author [6] (comp. also [7], [10—11]) used a similar method of factorizing of singular integral operators of non-normal type for inversion and for calculation of the indices of such operators. M. I. HAIKIN [1] gave first an example of a factorization of the form (1.1) with $D \neq I$ in connection with the investigation of Wiener-Hopf integral equations of the first kind whose symbol has a zero of fractional order at the point at infinity.

In the general form represented here this method was described simultaneously by A. A. SEMENCUL [1] and the author [13].

4.2—4.5. Most of the results of these sections are published here for the first time.

Theorem 2.1 and 2.2 generalize some results of G. N. ČEBOTAREV [3] on Wiener-Hopf integral equations in the space of bounded functions.

The results of Subsection 4.3.6 were already ealier stated for singular integral equations by the author [10] and B. SILBERMANN [4].

The Lemma 4.4 was proved by S. PRÖSSDORF and B. SILBERMANN [4]. Further problems on the one-sided invertibility of paired operators are treated in the paper of B. SILBERMANN [7].

CHAPTER 5

WIENER-HOPF INTEGRAL EQUATIONS OF NON-NORMAL TYPE AND THEIR DISCRETE ANALOGUE

In this chapter the results from Chapter 4 are applied to discrete Wiener-Hopf equations, Wiener-Hopf integral equations and the corresponding paired equations whose symbols possess finitely many zeros of integral and fractional orders.

The first four sections have uniform structure. At the beginning of each of these sections the kernel of the corresponding operator B from the representation (1.1), Chapter 4, is determined for the type of equation in question, and an analytic description of the spaces $\overline{X} = \operatorname{im} B$ and $\widetilde{X}$ are given. Then we investigate the properties of the equations inquestion in these spaces first in the case of integral orders and then in the case of arbitrary real orders of the zeros of the symbol. Here we succeed in sharpening the general theorems of Chapter 4. The main results of the present chapter are formulated in Theorems 1.5, 2.6, 3.3 to 3.5, 4.2 and 4.3.

Some of the considerations and results of Sections 5.1 and 5.2 (resp. 5.3 and 5.4) are analogous. Therefore, as a rule, they are represented in the discrete case in more detail and in the continuous in caseless detail. For Wiener-Hopf integral equations the case in which the symbol vanishes at the point at infinity (equation of the first kind) is exceptional. A separate subsection deals with this case.

5.1. Discrete Wiener-Hopf Equations

In this section E (resp. E_+) is any one of the sequence spaces l^p $(1 \leq p < \infty)$, m, c or c_0 introduced in Section 3.1. U and P are the operators defined on E by relations (1.1) and (1.2), respectively, of Chapter 3.

5.1.1. The Sequence Spaces Generated by Zeros of the Symbol

1. Let α be an arbitrary point on the unit circle. The operators B_α and D_α defined by relations (2.7_1), Chapter 4, now have the form

$$B_\alpha\xi = \{\xi_{n+1} - \alpha^{-1}\xi_n\}_{-\infty}^{\infty}, \qquad D_\alpha\xi = \{\xi_{n-1} - \alpha\xi_n\}_{-\infty}^{\infty}, \tag{1.1}$$

where $\xi = \{\xi_n\}_{-\infty}^{\infty} \in E$. Obviously, the operators $\hat{B}_\alpha = PB_\alpha | E_+$ and $\hat{D}_\alpha = D_\alpha P \mid E_+$ considered on E_+ are represented by the following Toeplitz matrices:

$$\hat{B}_\alpha = \begin{pmatrix} -\bar{\alpha} & 1 & 0 & 0 & \dots \\ 0 & -\bar{\alpha} & 1 & 0 & \dots \\ 0 & 0 & -\bar{\alpha} & 1 & \dots \\ \dots & \dots & \dots & \dots & \dots \end{pmatrix}, \quad \hat{D}_\alpha = \begin{pmatrix} -\alpha & 0 & 0 & \dots \\ 1 & -\alpha & 0 & \dots \\ 0 & 1 & -\alpha & \dots \\ \dots & \dots & \dots & \dots \end{pmatrix}.$$

By $\hat{G}_\alpha$ we denote the operator which with any number sequence $f = \{f_k\}_0^\infty$ associates the number sequence

$$\hat{G}_\alpha f = \left\{ -\frac{1}{\alpha} \sum_{k=0}^{n} a^{k-n} f_k \right\}_{n=0}^{\infty}. \tag{1.1'}$$

Hence the operator $\hat{G}_\alpha$ is defined by the triangular matrix

$$\hat{G}_\alpha = -\bar{\alpha} \begin{pmatrix} 1 & 0 & 0 & \dots \\ \bar{\alpha} & 1 & 0 & \dots \\ \bar{\alpha}^2 & \bar{\alpha} & 1 & \dots \\ \dots & \dots & \dots & \dots \end{pmatrix}.$$

It is easy to see that the following two formulae hold[1]):

$$\hat{G}_\alpha \hat{D}_\alpha f = f, \qquad \hat{D}_\alpha \hat{G}_\alpha f = f. \tag{1.2}$$

2. Now let $\alpha_1, \alpha_2, \dots, \alpha_r$ be distinct points on the unit circle and m_j', m_j'' $(j = 1, 2, \dots . r)$ positive integers. We put

$$B = \prod_{j=1}^{r} B_{\alpha_j}^{m_j''}, \qquad D = \prod_{j=1}^{r} D_{\alpha_j}^{m_j'}, \qquad \hat{G} = \prod_{j=1}^{r} \hat{G}_{\alpha_j}^{m_j'} \tag{1.3}$$

and denote the restrictions of the operators PBP and PDP to the space E_+ by $\hat{B}$ and $\hat{D}$, respectively (comp. (2.7_2), Chapter 4).

By (1.2) we have

$$\hat{G}\hat{D}x = x, \qquad \hat{D}\hat{G}x = x \qquad \forall\, x \in E_+.$$

Hence the operator $\hat{D}$ satisfies condition (2′) of Section 4.1 with $\hat{D}^{(-1)} = \hat{G}$. Following the considerations of Section 4.2.3 we now introduce the space $\tilde{E}_+(\varrho_+) = \hat{G}(E_+)$ in which the norm is defined by

$$\|f\|_{\tilde{E}_+} = \|\hat{D}f\|_{E_+} \qquad (f \in \tilde{E}_+(\varrho_+)).$$

Here $\varrho_\pm(z)$ are the functions defined by equations (2.6), Chapter 4:

$$\varrho_+(z) = \prod_{j=1}^{r} (z - \alpha_j)^{m_j'}, \qquad \varrho_-(z) = \prod_{j=1}^{r} (z^{-1} - \alpha_j^{-1})^{m_j''}. \tag{1.4}$$

Obviously, $\varrho_+(z) \in W^+$ $(\varrho_-(z) \in W^-)$ is the symbol of the operator $D(B)$.

[1]) With respect to the second formula in (1.2) we remark that the operator $\hat{D}_\alpha$ is defined in a natural way on *any* sequence of the form $f = \{f_k\}_0^\infty$.

Remark. It is easy to determine the order of increase of the terms of a sequence $\eta \in \widetilde{E}_+(\varrho_+)$. In the case of the space $E_+ = m_+$ we have

$$\eta_n = O(n^{m'}), \quad m' = \max\{m_1', \dots, m_r'\}.$$

This is obtained by using the relation

$$\hat{G}_{\alpha_2}\hat{G}_{\alpha_1} = \frac{1}{\alpha_1 - \alpha_2}(\hat{G}_{\alpha_1} - \hat{G}_{\alpha_2}) \quad (a_1 \neq \alpha_2).$$

3. Now we investigate the null space of the operator $\hat{B}$.

Lemma 1.1. *If E_+ is one of the spaces l_+^p $(1 \leqq p < \infty)$ or c_+^0, then* $\ker \hat{B} = \{0\}$.

Proof. $\hat{B}_\alpha \xi = 0$ $(\xi \in E_+;\ j = 1, 2, \dots, r)$ implies

$$\xi_{n+1} = \frac{1}{\alpha}\xi_n \qquad (n = 0, 1, \dots),$$

i.e. $\xi = C\{\alpha^{-n}\}_{n=0}^{\infty}$ with $C = \xi_0 = \text{const}$. But this sequence belongs to one of the spaces l_+^p or c_+^0 only if $C = 0$. Hence $\dim \ker \hat{B}_\alpha = 0$. Using relation (2.7_2), Chapter 4, we obtain the assertion.

Lemma 1.2. *The kernel of the operator $\hat{B}$ considered in the space m_+ is r-dimensional. The sequences $e_{\alpha j}^- = \{\alpha_j^{-n}\}_{n=0}^{\infty}$ $(j = 1, 2, \dots, r)$ form a basis in it.*

Proof. As in the proof of Lemma 1.1 $\hat{B}_{\alpha_j}\xi = 0$ $(\xi \in m_+)$ implies $\xi = Ce_{\alpha j}^-$, where now $e_{\alpha j}^- \in m_+$. We obtain the assertion by Theorem 2.1, Chapter 4, and the remark following this Theorem.

Remark. The preceding considerations show immediately that in the case of the space c_+ the kernel of $\hat{B}$ depends essentially on the situation of the points α_j. It consists only of the zero element if $\alpha_j \neq 1$ $(j = 1, 2, \dots, r)$ and of all stationary sequences if $\alpha_j = 1$ for some j.

4. We further follow the considerations of Section 4.2.3 and introduce the space $\overline{E}_+(\varrho_-)$ as the image space of the operator $\hat{B} = PB \mid E_+$.

First, let E_+ be one of the spaces l_+^p $(1 \leqq p < \infty)$ or c_+^0. Because of Lemma 1.1 the norm of an element $\eta \in \overline{E}_+(\varrho_-)$ is defined by

$$||\eta||_{\overline{E}_+} = ||\xi||_{E_+} \qquad (\eta = \hat{B}\xi, \quad \xi \in E_+). \tag{1.5}$$

For the analytic description of the space $\overline{E}_+(\varrho_-)$ we introduce the following notation: If for the number sequence $f = \{f_k\}_0^\infty$ the corresponding series $\sum_{k=0}^{\infty} f_k$ is convergent, then by $f^{(-1)}$ we denote the sequence

$$f^{(-1)} = \left\{\sum_{k=n}^{\infty} f_k\right\}_{n=0}^{\infty}.$$

If for this sequence the corresponding series is also convergent, then we put $f^{(-2)} = [f^{(-1)}]^{(-1)}$ etc. By $E_+^{(m)}$ (m a natural number) we denote the collection of all sequences $f \in E_+$ for which the elements $f^{(-k)}$ ($k = 1, 2, \dots, m$) exist and also belong to E_+. We equip the set $E_+^{(m)}$ with the norm

$$||f||_0 = \sum_{k=0}^{m} ||f^{(-k)}||_{E_+} . \tag{1.6}$$

Theorem 1.1. *Let E_+ be one of the spaces l_+^p $(1 \leqq p < \infty)$ or c_+^0. For the sequence $f = \{f_n\}_0^\infty \in E_+$ to belong to the space $\overline{E}_+(\varrho_-)$ it is necessary and sufficient that $f_{\alpha_j} = \{\alpha_j^n f_n\}_{n=0}^\infty \in E_+^{(m_j'')}$ $(j = 1. 2. \dots . r)$. The norm (1.5) is equivalent to the norm $\sum_{j=1}^{r} ||f_{\alpha_j}||_0$.*

Proof. By Theorem 2.2, Chapter 4, it is only necessary to prove the assertion for $r = 1$. Thus we assume $r = 1$ and put $m = m_1''$, $\alpha = \alpha_1$. The assertion is proved by complete induction.

Necessity. First, we consider the case $m = 1$. Thus let $f = \hat{B}_\alpha \xi$ $(\xi \in E_+)$. Then $f_n = \xi_{n+1} - \alpha^{-1}\xi_n$ $(n = 0, 1, \dots)$. Since $\xi \in E_+$ this implies

$$\sum_{k=n}^{\infty} \alpha^k f_k = - \alpha^{n-1} \xi_n .$$

Hence $f_\alpha^{(-1)} = \dfrac{-1}{\alpha} \xi_\alpha \in E_+$.

Now we assume that the necessity of the conditions is proved for $m - 1$. Let $f = \hat{B}_\alpha^m \xi$ $(\xi \in E_+)$. Putting $\eta = \hat{B}_\alpha^{m-1}\xi$ we obtain $f \in \hat{B}_\alpha \eta$ and consequently

$$f_\alpha^{(-1)} = \frac{-1}{\alpha} \eta_\alpha .$$

By the assumption we have

$$f_\alpha^{(-k-1)} = \frac{-1}{\alpha} \eta_\alpha^{(-k)} \in E_+ \qquad (k = 0, 1, \dots, m - 1) ,$$

i.e. $f_\alpha \in E_+^{(m)}$.

Sufficiency. Let $f_\alpha \in E_+^{(m)}$. We show that the sequence $\xi = \{\xi_n\}_0^\infty \in E_+$ with

$$\xi_n = \frac{(-1)^m}{\alpha^{n-m}} [f_\alpha^{(-m)}]_n \qquad (n = 0, 1, \dots) \tag{1.7}$$

satisfies the equation $\hat{B}_\alpha^m \xi = f$.

In fact, in the case $m = 1$ we have

$$\xi_{n+1} - \alpha^{-1}\xi_n = \frac{1}{\alpha^n} \left\{ \sum_{k=n}^{\infty} \alpha^k f_k - \sum_{k=n+1}^{\infty} \alpha^k f_k \right\} = f_n .$$

We assume that the assertion is proved for $m - 1$. Let ξ_n $(n = 0, 1, \ldots)$ be defined by (1.7). We put

$$\eta_n = \frac{-1}{\alpha^{n-1}} [f_\alpha^{(-1)}]_n \qquad (n = 0, 1, \ldots)$$

and $\eta = \{\eta_n\}_0^\infty \in E_+$. Then $\eta_\alpha = -\alpha f_\alpha^{(-1)}$, and using the earlier part of the proof we obtain $\hat{B}_\alpha \eta = f$.

Further, we obviously have

$$\xi_n = \frac{(-1)^{m-1}}{\alpha^{n-m+1}} [\eta_\alpha^{(-m+1)}]_n \qquad (n = 0, 1, \ldots) .$$

Hence it follows from the assumption that $\hat{B}_\alpha^{m-1}\xi = \eta$, and thus $\hat{B}_\alpha^m \xi = \hat{B}_\alpha \eta = f$ holds.

It remains to show the equivalence of the norms. First, by (1.7) for $f = \hat{B}_\alpha^m \xi \in \overline{E}_+(\varrho_-)$ we obtain the estimate

$$||f||_{\overline{E}_+} = ||\xi||_{E_+} = ||f_\alpha^{(-m)}||_{E_+} \leqq ||f_\alpha||_0 . \tag{1.8}$$

On the other hand, the inequalities

$$||f_\alpha^{(-k)}||_{E_+} \leqq C||\xi||_{E_+} \qquad (k = 0, 1, \ldots, m) \tag{1.9}$$

hold for a certain constant C.

In fact, for $m = 1$ this follows immediately from

$$f = \hat{B}_\alpha \xi , \qquad f_\alpha^{(-1)} = \frac{-1}{\alpha} \xi_\alpha .$$

For arbitrary m the inequalities (1.9) are proved by complete induction in the same way as the necessity of the conditions formulated in the theorem. (1.8) and (1.9) immediately show the equivalence of the norms $||f||_{\overline{E}_+}$ and $||f_\alpha||_0$. This proves the theorem.

The preceding proof (see formula (1.7)) immediately yields

Corollary 1.1. *The inverse of the operator $\hat{B}_\alpha$ has the form*

$$\hat{B}_\alpha^{-1} f = \left\{ -\alpha^{-n+1} \sum_{k=n}^\infty \alpha^k f_k \right\}_{n=0}^\infty \qquad (f = \{f_n\}_0^\infty \in \operatorname{im} \hat{B}_\alpha) .$$

5. Now we introduce a sufficiently large class of sequences which is contained in the space $\overline{E}_+(\varrho_-)$ and which is dense in this space.

Let E be an any one of the spaces l^p $(1 \leqq p < \infty)$ or c_0 and m an arbitrary real number. By E^m we denote the space of all number sequences of the form $f = \{(1 + |n|)^{-m} \xi_n\}_{n=-\infty}^\infty$, where $\xi = \{\xi_n\} \in E$. With the norm

$$|f| = ||\xi||_E$$

E^m is a Banach space isometric and isomorphic with E.

In the case of a positive integral m the sequence $f = \{f_n\}_{-\infty}^{\infty}$ belongs to E^m if and only if

$$\eta^{(k)} = \{n^k f_n\}_{n=-\infty}^{\infty} \in E\,, \qquad k = 0, 1, \dots, m\,.$$

In this case $|f|$ is equivalent to the norm

$$||f||_{E^m} = \sum_{k=0}^{m} ||\eta^{(k)}||_E\,.$$

By E_+^m we denote the corresponding spaces of sequences of the form $\{f_n\}_{n=0}^{\infty}$.

Since $(l^p)^* = l^q$ $(q = p/(p-1)$ for $1 < p < \infty$ and $q = \infty$ for $p = 1)$, $l^{q,-m}$ is the conjugate space of $l^{p,m}$ for $1 \leqq p < \infty$ and any real number m. The corresponding statement holds for $l_+^{p,m}$.

Theorem 1.2. *Let m be a positive integer, ε an arbitrary positive number and $1 \leqq p < \infty$. For any sequence $f \in l_+^{p,m}$ $(c_+^{0,m+\varepsilon})$ the sequences $f^{(-k)} \in l_+^p$ (c_+^0), $k = 0, 1, \dots, m$ exist. The space $l_+^{p,m}$ $(c_+^{0,m+\varepsilon})$ is continuously and densely embedded into the space $l_+^{p(m)}$ $(c_+^{0(m)})$.*

Proof. First of all, we consider the case $m = 1$.

If $f = \{f_n\}_{n=0}^{\infty} \in l_+^{1,1}$, then we obtain

$$||f^{(-1)}||_{l_+} \leqq \sum_{n=0}^{\infty} \sum_{k=n}^{\infty} |f_k| = \sum_{k=0}^{\infty} (k+1)\,|f_k| < \infty\,.$$

Hence $f^{(-1)} \in l_+$ and

$$||f||_0 = \sum_{k=0}^{1} ||f^{(-k)}||_{l_+} \leqq 2\,||f||_{l_+^{1,1}}\,.$$

Now let $f \in l_+^{p,1}$, $1 < p < \infty$. We define $f_{-1} = 0$ and

$$\varphi(s, t) = t\,|f_{[s\cdot t]-2}| \qquad (1 \leqq s, t < \infty)\,,$$

where $[s]$ is the integral part of the number s. The integrals of this function with respect to the variables s and t satisfy the estimates

$$\int_1^{\infty} \varphi(s, t)\, ds = \int_t^{\infty} |f_{[z]-2}|\, dz \geqq \int_{[t]+1}^{\infty} |f_{[z]-2}|\, dz = \sum_{n=[t]-1}^{\infty} |f_n| \geqq |f_{[t]-1}^{(-1)}|\,,$$

$$\left(\int_1^{\infty} [\varphi(s, t)]^p\, dt\right)^{1/p} = \frac{1}{s^{(p+1)/p}} \left[\int_s^{\infty} x^p\, |f_{[x]-2}|^p\, dx\right]^{1/p}$$

$$\leqq \frac{1}{s^{(p+1)/p}} \left(\sum_{n=1}^{\infty} (n+1)^p\, |f_{n-2}|^p\right)^{1/p} \leqq \frac{C}{s^{(p+1)/p}}\, ||f||_{l_+^{p,1}}\,,$$

where the constant C depends only on p. By the known inequality[1])

$$\left[\int_1^{\infty} \left\{\int_1^{\infty} \varphi(s, t)\, ds\right\}^p dt\right]^{1/p} \leqq \int_1^{\infty} \left\{\int_1^{\infty} [\varphi(s, t)]^p\, dt\right\}^{1/p} ds$$

[1]) See G. H. Hardy, J. E. Littlewood and G. Polya [1], Theorem 202.

we then obtain the estimate

$$\|f^{(-1)}\|_{l_+^p} \leqq Cp\,\|f\|_{l_+^{p,1}}\,.$$

Hence we have $f^{(-1)} \in l_+^p$ and also the continuous embedding $l_+^{p,1} \subset l_+^{p(1)}$.

Finally, in the case $f \in c_+^{0}$ we can conclude that

$$\|f^{(-1)}\|_{c_+^0} \leqq \sum_{n=0}^{\infty} |f_n| \leqq \|f\|_{c_+^{0,1+\varepsilon}} \sum_{n=0}^{\infty} \frac{1}{(1+n)^{1+\varepsilon}}\,.$$

Thus the assertion is proved for $m = 1$.

For $m \geqq 2$ we obtain the assertion by successive application of the preceding considerations and by using the relation

$$f_k^{(-2)} = (1-k)\,f_k^{(-1)} + \sum_{n=k}^{\infty} n f_n\,, \qquad f^{(-2)} = \{f_k^{(-2)}\}_{k=0}^{\infty}\,,$$

which is easily checked. The assertion of the theorem about the dense embedding can easily be proved if we use the fact that each sequence with only finitely many non-zero terms belongs to $l_+^{p,m}$ $(c_+^{0,m+\varepsilon})$.

Corollary 1.2. *Let* E_+ *be one of the spaces* l_+^p $(1 \leqq p < \infty)$ *or* c_+^0, $\varrho_\pm(z)$ *the functions defined by the equations* (1.4) *and*

$$m' = \max\{m_1', \dots, m_r'\}\,, \qquad m'' = \max\{m_1'', \dots, m_r''\}\,.$$

Then we have

$$E_+^{m''+\varepsilon} \subset \overline{E}_+(\varrho_-) \qquad (\varepsilon > 0 \text{ for } E_+ = c_+^0\,,\ \varepsilon = 0 \text{ for } E_+ = l_+^p)\,,$$

$$\widetilde{E}_+(\varrho_+) \subset E_+^{-m'-\varepsilon} \qquad (\varepsilon > 0 \text{ for } E_+ = l_+\,,\ \varepsilon = 0 \text{ for } E_+ = l_+^p\,,\ 1 < p < \infty\,, \text{ or } E_+ = c_+^0)$$

in the sense of a continuous embedding.

The first of these embeddings follows immediately from Theorem 1.1 and Theorem 1.2. The second can be proved analogously.

Remark 1. In the case of the space c_+^0 the assertion of the Theorem 1.2 (and thus also of the Corollary 1.2) does not hold for $\varepsilon = 0$.

For instance, the sequence $\left\{\dfrac{1}{n \ln n}\right\}_{n=2}^{\infty}$ belongs to the space $c_+^{0,1}$ but the corresponding series is divergent.

Remark 2. It is easily seen that the estimate

$$f_n = O(n^{m'-1})$$

holds for any sequence $\{f_n\}_0^\infty \in \tilde{l}_+^1(\varrho_+)$.

6. Now we come to the analytic description of the space $\overline{E}_+(\varrho)_-$ in the case where E_+ is one of the sequence spaces m_+ or c_+. Taking into consideration Lemma 1.2 and the remark following it we define the norm of an element

$\eta \in \overline{E}_+(\varrho_-)$ in the present case according to Section 4.2.3 by

$$||\eta||_{\overline{E}_+} = \inf_{PB\xi=\eta} ||\xi||_{E_+}. \tag{1.10}$$

For the description of the space $\overline{E}_+(\varrho_-)$ we use the well-known Cesàro summability method for series (see G. M. FICHTENHOLZ [1], Vol. II, p. 420). We recall that a series

$$x_0 + x_1 + \dots + x_n + \dots$$

is said to be *C_1-summable to the C_1-sum S*, in symbols $(C_1) \sum_{k=0}^{\infty} x_k = S$, if the sequence of the arithmetic means of its partial sums $S_n = \sum_{k=0}^{\infty} x_k$, i.e. the sequence

$$\sigma_n = \frac{S_0 + S_1 + \dots + S_n}{n+1},$$

converges to S. If this condition is satisfied, then in the sequel we call the sequence $x = \{x_n\}_0^\infty$ is also to be C_1-summable, and by $x^{(-1)}$ we denote the sequence

$$x^{(-1)} = \{x_n^{(-1)}\}_0^\infty, \qquad x_n^{(-1)} = (C_1) \sum_{k=n}^{\infty} x_k.$$

If the sequence $x^{(-1)}$ is also C_1-summable, then we put $x^{(-2)} = [x^{(-1)}]^{(-1)}$ etc.

In the case of the space $E_+ = m_+$ we introduce $E_+^{(m)}$ (m an arbitrary natural number) as the collection of all sequences $x = \{x_n\}_0^\infty \in E_+$ which have the following properties:

(1) $x^{(-k)}$ is C_1-summable for $k = 0, 1, \dots, m-2$.
(2) $x^{(-k)} \in E_+$ for $k = 0, 1, \dots, m-1$.
(3) $\left\{\sum_{k=0}^{n} x_k^{(-m+1)}\right\}_{n=0}^{\infty} \in E_+$.

We define the norm of an element $x \in E_+^{(m)}$ by the equation

$$||x||_0 = \sum_{l=0}^{m-1} ||x^{(-l)}||_{m_+} + \inf_C \left\| C + \sum_{k=0}^{n} x_k^{(-m+1)} \right\|_{m_+},$$

where the infimum is taken over all constants C.

Theorem 1.3. *For the sequence $f = \{f_n\}_0^\infty \in m_+$ to belong to the space $\overline{m}_+(\varrho_-)$ it is necessary and sufficient that $f_{\alpha_j} = \{\alpha_j^n f_n\}_{n=0}^\infty \in E_+^{(m_j'')}$ ($j = 1, 2, \dots, r$). The norm (1.10) in $\overline{m}_+(\varrho_-)$ is equivalent to the norm $\sum_{j=1}^{r} ||f_{\alpha_j}||_0$.*

A corresponding theorem holds in the case of the space c_+.

Theorem 1.4. *For the sequence $f = \{f_n\}_0^\infty \in c_+$ to belong to the space $\overline{c}_+(\varrho_-)$ it is necessary and sufficient that for every $j = 1, 2, \dots, r$ the following three* *conditions are satisfied:*

a) $f_{\alpha_j}^{(-k)}$ *is* C_1*-summable for* $k = 0, 1, \dots, m_j'' - 2$.

b) $[f_{\alpha_j}^{(-k)}]_{\bar{\alpha}_j} \in c_+$ *for* $k = 0, 1, \dots, m_j'' - 1$.

c) *There exists a constant* $C = C_j$ *such that*[1])

$$\left\{\alpha_j^{-n}\left[\sum_{k=0}^{n} f_{\alpha_j,k}^{(-m_j''+1)} + C\right]\right\}_{n=0}^{\infty} \in c_+ .$$

The norm in $\bar{c}_+(\varrho_-)$ is equivalent to the norm $\sum_{j=1}^{r} \|f_{\alpha_j}\|_0$, where

$$\|f_{\alpha_j}\|_0 = \sum_{l=0}^{m_j''-1} \|f_{\alpha_j}^{(-l)}\|_{c_+} + \left\|C + \sum_{k=0}^{n} f_{\alpha_j,k}^{(-m_j''+1)}\right\|_{c_+} \tag{1.11}$$

if $\alpha_j \neq 1$. If $\alpha_j = 1$, then on the right-hand side of (1.11) the infimum over all constants C is to be taken.

The proofs of Theorems 1.3 and 1.4 are left to the reader. They are completely analogous to the proof of Theorem 1.1.

5.1.2. Symbols with Zeros of Integral Orders

1. We consider the discrete Wiener-Hopf operator $\hat{A}$ defined on the space E_+ by the matrix $\{a_{j-k}\}_{j,k=0}^{\infty}$, where the numbers a_k $(k = 0, \pm 1, \dots)$ satisfy the condition

$$\sum_{k=0}^{\infty} |a_k| < \infty .$$

By Section 3.1 we have $\hat{A} = PAP|E_+$, where A is the operator represented in the space E by the matrix $\{a_{j-k}\}_{j,k=-\infty}^{\infty}$.

We suppose that the symbol

$$a(z) = \sum_{k=-\infty}^{\infty} a_k z^k \qquad (|z| = 1)$$

of the operator $\hat{A}$ admits the representation

$$a(z) = \prod_{j=1}^{r} (z - \alpha_j)^{m_j}\, b(z) . \tag{1.12}$$

Here $\alpha_1, \alpha_2, \dots, \alpha_r$ are distinct points of the unit circle, $m_1, m_2, \dots, m_r$ positive integers and $b(z)$ $(|z| = 1)$ a function which can be expanded into an absolutely convergent Fourier series: $b(z) \in W$.[2])

[1]) Obviously, the constant C in the condition c) is uniquely determined if and only if $\alpha_j \neq 1$.

[2]) Necessary and sufficient conditions for the representability of the function $a(z)$ in the form (1.12) will be given in no. 3 of this subsection.

Following the considerations of Section 4.2.3 we introduce non-negative integers m'_j and m''_j with

$$m'_j + m''_j = m_j \qquad (j = 1, 2, \dots, r),$$

and we represent the symbol $a(z)$ in the form

$$a(z) = \varrho_-(z)\, c(z)\, \varrho_+(z)\,. \tag{1.13}$$

Here $c(z) = \prod_{j=1}^{r} (-\alpha_j z)^{m''_j}\, b(z)$, and $\varrho_\pm(z)$ are the functions introduced in the preceding subsection (see (1.4)). By $\hat{C}$ we denote the discrete Wiener-Hopf operator with the symbol $c(z) \in W$ defined in the space E_+.

Now the following theorem holds.

Theorem 1.5. *For $\hat{A}$ as an operator from $\tilde{E}_+(\varrho_+)$ into $\bar{E}_+(\varrho_-)$ to be a Φ_+-operator or a Φ_--operator it is necessary and sufficient that the function $c(z)$ in the representation* (1.13) (*or, what is the same, the function $b(z)$ in* (1.12)) *vanishes nowhere on the unit circle.*

If this condition is satisfied, then the invertibility of the operator $\hat{A}$ corresponds to the number $\varkappa - \delta$ and

$$\dim\ker \hat{A} = \max(\delta - \varkappa, 0)\,, \qquad \dim \operatorname{coker} \hat{A} = \max(\varkappa - \delta, 0)\,.$$

Here

$$\varkappa = \operatorname{ind} c(z) = \operatorname{ind} b(z) + \sum_{j=1}^{r} m''_j\,,$$

$$\delta = \begin{cases} 0 & \textit{für} \quad E_+ = l^p_+ (1 \leqq p < \infty) \quad \textit{oder} \quad E_+ = c^0_+\,, \\ r'' & \textit{für} \quad E_+ = m_+\,, \end{cases}$$

and r'' is the number of the positive m''_j $(j = 1, 2, \dots, r)$. In the case $E_+ = c_+$ there holds

$$\delta = \begin{cases} 1, & \textit{if} \quad \alpha_{j_0} = 1 \quad \textit{and} \quad m''_{j_0} > 0 \qquad (1 \leqq j_0 \leqq r)\,, \\ 0 & \textit{otherwise}\,. \end{cases}$$

Proof. The first part of the theorem follows immediately from Theorem 2.3, Chapter 4.

Let $b(z) \neq 0$ $(|z| = 1)$. Then Theorem 2.3, Chapter 4, Lemma 1.1 and Lemma 1.2 imply that $\hat{A}$ is a Φ-operator from $\tilde{E}_+(\varrho_+)$ into $\bar{E}_+(\varrho_-)$. For the index of this operator we have

$$\operatorname{Ind} \hat{A} = \delta - \varkappa\,. \tag{1.14}$$

In the case of the spaces l^p_+ or c^0_+ we obtain $\ker B = \{0\}$ by Lemma 1.1. Hence the operator $\hat{A} = \hat{B}\hat{C}\hat{D}$ is invertible if and only if $\hat{C}$ is invertible, i.e. the invertibility of the operator $\hat{A}$ corresponds to the number $\varkappa$.

Now we consider the case of the space m_+. We factorize the function $c(z) \in W$ which is unequal to zero (comp. Theorem 3.1, Chapter 2):

$$c(z) = c_-(z)\, z^\varkappa\, c_+(z) \qquad (|z| = 1)\,.$$

By $b_0, b_1, b_2, \ldots$ we denote the Fourier coefficients of the function $c_-\left(\frac{1}{z}\right) \in W_+$.

In the case $\varkappa \leqq 0$ we have dim coker $\hat{C} = 0$. Let $\varkappa > 0$. By Theorem 1.3, Chapter 3, the transposed homogeneous equation system

$$\sum_{k=0}^{\infty} c_{k-j}\psi_k = 0 \qquad (j = 0, 1, \ldots) \tag{1.15}$$

($c_0, c_{\pm 1}, \ldots$ Fourier coefficients of the function $c(z)$) belonging to the operator $\hat{C}$ has exactly $\varkappa$ linearly independent solutions

$$\xi^{(m)} = \{\underbrace{0, \ldots, 0}_{m-1}, b_0, b_1, b_2, \ldots\} \qquad (m = 1, 2, \ldots, \varkappa)$$

which belong to all spaces E_+. Hence the general solution of the system (1.15) has the form

$$\psi = \sum_{m=1}^{\varkappa} \gamma_m \xi^{(m)} = \{\psi_0, \psi_1, \psi_2, \ldots\}\,, \tag{1.16}$$

where γ_m are arbitrary constants. Lemma 1.2, Chapter 4, Lemma 1.2 and Theorem 1.3, Chapter 3, show that for any sequence $\eta = \{\eta_j\}_0^\infty \in \widetilde{E}_+(\varrho_-)$ the inhomogeneuos equation system

$$\sum_{k=0}^{\infty} a_{j-k}\xi_k = \eta_j \qquad (j = 0, 1, \ldots) \tag{1.17}$$

has a solution in the space $\widetilde{E}_+(\varrho_-)$ if and only if for any sequence $\zeta = \{\zeta_+\}_0^\infty \in m_+$ with $\hat{B}\zeta = \eta$ the condition

$$\sum_{k=0}^{\infty} \zeta_k \psi_k = 0 \tag{1.18}$$

is satisfied for each sequence of the form (1.16) satisfying the conditions

$$\sum_{k=0}^{\infty} \psi_k \alpha_j^{-k} = 0 \qquad (j = 1, \ldots, r\colon m_j'' > 0)\,. \tag{1.19}$$

After some simple calculations and using the relation $c_-(z) \neq 0$ ($|z| = 1$) we see that (1.19) is equivalent to the following system of equations for the determining the coefficients γ_m:

$$\sum_{m=1}^{\varkappa} \gamma_m \alpha_j^{-m+1} = 0 \qquad (j = 1, \ldots, r\colon m_j'' > 0)\,. \tag{1.20}$$

Thus the following statement holds: For the solvability of the system (1.17) it is necessary and sufficient that conditions (1.18) are satisfied for each sequence of the form (1.16), where the coefficients γ_m are to be determined from (1.20). The number of coefficients γ_m which can be chosen arbitrarily in (1.20) and thus also the number of solvability conditions (1.18) is equal to the greater one of the two numbers $\varkappa - r''$ and 0. Hence

$$\dim \operatorname{coker} \hat{A} = \max (\varkappa - r'', 0) . \tag{1.21}$$

The formulae (1.14) and (1.21) yield the assertion for the case of the space m_+.

For the space c_+ the proof is similar. In this case the system (1.20) has the form

$$\sum_{m=1}^{\varkappa} \gamma_m = 0 \tag{1.20'}$$

if for a certain j_0 $(1 \leqq j_0 \leqq r)$ the relations $\alpha_{j_0} = 1$ and $m''_{j_0} > 0$ hold. If, on the contrary, $\alpha_j \neq 1$ $(j = 1, \ldots, r)$, then all coefficients γ_m can be arbitrarily chosen. In this case the solvability conditions (1.18) have the form

$$\sum_{k=m-1}^{\infty} \zeta_k b_{k+1-m} = 0 . \tag{1.18'}$$

The last statement holds also in the case of the spaces l^p_+ and c^0_+.

Thus Theorem 1.5 is completely proved.

We remark that in the preceding proof necessary and sufficient conditions for the solvability of the inhomogeneous system of equations (1.17) have also been stated. These conditions have the form (1.18) and (1.20) (or (1.20′)) resp. (1.18′).

Remark. The hypothesis $b(z) \in W$ assumed above can be weakened. In the case of the space l^2_+ the function $b(z)$ may be an arbitrary continuous function. For the space l^p_+ $(1 \leqq p < \infty)$ and c^0_+ it is sufficient to demand that $b(z) \in \Re(z)$. The assertions of Theorem 1.5 hold also under these more general hypotheses.

2. If E_+ is one of the spaces l^p_+ $(1 \leqq p < \infty)$ or c^0_+, then under the hypotheses of this subsection it is possible to formulate a corresponding analogue of Theorem 1.3 of Chapter 3.

For this we first remark that for the transpose operator $\hat{A}'$ the representation $\hat{A}' = \hat{D}'\hat{C}'\hat{B}'$ holds. Here $\hat{D}'$ and $\hat{B}'$ have symbols $\varrho'_-(z) = \varrho_+\left(\frac{1}{z}\right) \in W^-$ and $\varrho'_+(z) = \varrho_-\left(\frac{1}{z}\right) \in W^+$, respectively. Now let $c(z) \neq 0$ $(|z| = 1)$. Then by Theorem 1.5 $\hat{A}'$ is a Φ-operator from $\widetilde{E}_+(\varrho'_+)$ into $\overline{E}_+(\varrho'_-)$, where

$$\dim \ker \hat{A}' = \max (\varkappa, 0) , \qquad \dim \operatorname{coker} \hat{A}' = \max (-\varkappa, 0) .$$

Every solution ψ of the system (1.15) can be represented in the form $\psi = \hat{B}'\tilde{\psi}$, where $\tilde{\psi}$ belongs to all spaces $\widetilde{E}_+(\varrho'_+)$. Now the solvability conditions (1.18) have the form

$$0 = \sum_{k=0}^{\infty} \zeta_k \psi_k = \sum_{k=0}^{\infty} (\hat{B}^{-1}\eta)_k \, (\hat{B}'\tilde{\psi})_k .$$

By the right-hand side of the last equation the element $\tilde{\psi}$ defines a linear functional on $\overline{E}_+(\varrho_-)$. This is easy to see by means of the Hölder inequality. Therefore we define

$$\langle \eta, \tilde{\psi} \rangle \equiv \sum_{k=0}^{\infty} \eta_k \tilde{\psi}_k = \sum_{k=0}^{\infty} (\hat{B}^{-1}\eta)_k \, (\hat{B}'\tilde{\psi})_k \qquad (\eta \in \overline{E}_+(\varrho_-)) .$$

It is easy to see that in the case $E_+ = l_+^p$ $(1 \leqq p < \infty)$ every linear functional on the space $\overline{l}_+^p(\varrho_-)$ is of this form.

Thus the following theorem holds.

Theorem 1.6. *Let E_+ be one of the spaces l_+^p $(1 \leqq p < \infty)$ or c_+^0, and let the symbol $a(z)$ have the form* (1.13), *where $c(z) \neq 0$ $(|z| = 1)$.*

If $\varkappa = \operatorname{ind} c(z) \leqq 0$, then the inhomogeneous system (1.17) *has a solution $\xi \in \tilde{E}_+(\varrho_+)$ for any sequence $\{\eta_k\}_0^\infty \in \overline{E}_+(\varrho_-)$ and the corresponding homogeneous system has exactly $-\varkappa$ linearly independent solutions.*

In the case $\varkappa > 0$ the homogeneous system

$$\sum_{k=0}^{\infty} a_{k-j} \tilde{\psi}_k = 0 \qquad (j = 0, 1, \ldots)$$

transposed to (1.17) *has $\varkappa$ linearly independent solutions $\tilde{\psi}^{(l)} = \{\tilde{\psi}_k^{(l)}\}_{k=0}^\infty$ $(l = 1, \ldots, \varkappa)$ which belong to all spaces $\tilde{E}_+(\varrho'_+)$. The system* (1.17) *has a solution $\xi \in \tilde{E}_+(\varrho_+)$ if and only if*

$$\sum_{k=0}^{\infty} \eta_k \tilde{\psi}_k^{(l)} = 0 \qquad (l = 1, \ldots, \varkappa) .$$

3. Now we want to clarify under what conditions about the sequence of the Fourier coefficients of the symbol $\{a_k\}_{-\infty}^{\infty}$ the representation (1.12) holds.

Let α be an arbitrary point on the unit circle $(|\alpha| = 1)$. For the sequences $\xi = \{\xi_n\}_{-\infty}^{\infty} \in l^1$ we introduce the operator R_α defined by

$$R_\alpha \xi = \left\{ \begin{array}{l} \left\{ \alpha^{-j-1} \sum\limits_{n=j+1}^{\infty} \alpha^n \xi_n \right\}_{j=0}^{\infty} \\ \left\{ -\alpha^{-j-1} \sum\limits_{n=-\infty}^{j} \alpha^n \xi_n \right\}_{j=-\infty}^{-1} \end{array} \right. . \tag{1.22}$$

By $l_{\alpha,m}^1$ ($m \geqq 0$ and integer) we denote the set of all sequences $\xi \in l^1$ for which

$$R_\alpha^k \xi \in l^1 , \qquad k = 0, 1, \ldots, m .$$

Then the algebra $W(\alpha, m)$ (comp. 2.3.4) can be characterized as follows.

Theorem 1.7. *Let $a(z)$ be a function from the algebra W and a_n $(n = 0, \pm 1, \ldots)$ its Fourier coefficients. Then $a(z) \in W(\alpha, m)$ if and only if $\{a_n\}_{-\infty}^{\infty} \in l_{\alpha,m}^1$.*

Proof. First we consider the case $m = 1$, and we put $b(z) = (z - \alpha)^{-1} \cdot [a(z) - a(\alpha)]$.

Let $a(z) \in W(\alpha, 1)$. Then $b(z) = \sum_{-\infty}^{\infty} b_n z^n \in W$, $a(z) = \sum_{-\infty}^{\infty} a_n z^n \in W$ and consequently

$$a_n = b_{n-1} - \alpha b_n \qquad (n = \pm 1, \pm 2, \dots) .$$

The last equations show at once that $R_\alpha\{a_n\} = \{b_n\} \in l^1$.

Now let conversely $\{a_n\} \in l^1_{\alpha,1}$. We put $\{b_n\} = R_\alpha\{a_n\}$ ($\in l^1$). A simple calculation yields

$$\sum_{-\infty}^{\infty} b_n z^n = \frac{a(z) - a(\alpha)}{z - \alpha} = b(z) ,$$

i.e. $b(z) \in W$ and hence $a(z) \in W(\alpha, 1)$.

Thus the theorem is proved for $m = 1$. For $m > 1$ the assertion is easily obtained by successive application of the preceding considerations.

From Theorems 5° and 7° of Subsection 2.3.4 we obtain

Corollary 1.3. *For the function $a(z) \in W$ to have the representation* (1.12) *with $b(z) \in W$ it is necessary and sufficient that the following two conditions both hold:*

a) $a(z) \in W(\boldsymbol{\alpha}, \boldsymbol{m})$ *with* $\boldsymbol{\alpha} = \{\alpha_1, \alpha_2, \dots, \alpha_r\}$, $\boldsymbol{m} = \{m_1, m_2, \dots, m_r\}$.

b) $a^{\{k\}}(\alpha_j) = 0$, $k = 0, 1, \dots, m_j - 1$, $j = 1, 2, \dots, r$.

In the sequel we denote the algebra of all functions $a(z) = \sum_{-\infty}^{\infty} a_n z^n \in W$ with $\{a_n\}_{-\infty}^{\infty} \in l^{1,m}$ by W^m. From the proof of the Theorem 1.2 it follows easily that $l^{1,m} \subset l^1_{\alpha,m}$ and thus $W^m \subset W(\alpha, m)$ for arbitrary α ($|\alpha| = 1$). As is well-known, every m times differentiable function $a(z)$ whose derivatives belong to W belongs to W^m (cf. G. M. Fichtenholz [1], Vol. III, no. 504).

Corollary 1.4. *If the function $a(z) \in W^m$ has a zero of order $m_j \leqq m$ ($j = 1, 2, \dots, r$) in the point α_j($|\alpha_j| = 1$), then $a(z)$ admits the representation* (1.12) *with $b(z) \in W$.*

5.1.3. Symbols with Zeros of Non-Integral Orders

Now we assume that m, m_j' and m_j'' ($j = 1, 2, \dots, r$) are arbitrary non-negative real numbers. By B and D we denote the operators from the algebra $\mathfrak{R}(U)$ with the symbols $\varrho_-(z)$ and $\varrho_+(z)$, respectively, where $\varrho_-(z)$ and $\varrho_+(z)$ are defined by (1.4). $B_\alpha^m(D_\alpha^m)$ will be the operator from $\mathfrak{R}(U)$ with symbol $(z^{-1} - \alpha^{-1})^m$ (resp. $(z - \alpha)^m$). Then for B and D the representations (1.3) also hold.

From the expansion of the function $(z - \alpha)^m \in W^+$:

$$(z-\alpha)^m = (-\alpha)^m\left[1 - m\bar{\alpha}z - \frac{m(1-m)}{2!}\bar{\alpha}^2z^2 - \frac{m(1-m)(2-m)}{3!}\bar{\alpha}^3z^3 - \dots\right]$$

it follows by the definition of the operators from the algebra $\Re(U)$ (see Section 2.1) that the operators $\hat{B}_\alpha^m = PB_\alpha^m | E_+$ and $\hat{D}_\alpha^m = D_\alpha^m P | E_+$ are represented in the space E_+ by the Toeplitz matrices

$$\hat{B}_\alpha^m = (-\bar{\alpha})^m \begin{pmatrix} 1 & -m\alpha & -\frac{m(1-m)}{2!}\alpha^2 & \dots \\ 0 & 1 & -m\alpha & \dots \\ 0 & 0 & 1 & \dots \\ \dots & \dots & \dots & \dots \end{pmatrix}$$

and

$$\hat{D}_\alpha^m = (-\alpha)^m \begin{pmatrix} 1 & 0 & 0 & \dots \\ -m\bar{\alpha} & 1 & 0 & \dots \\ -\frac{m(1-m)}{2!}\bar{\alpha}^2 & -m\bar{\alpha} & 1 & \dots \\ \dots & \dots & \dots & \dots \end{pmatrix}$$

respectively. For the operators $\hat{B}$ and $\hat{D}$ the relations (2.7_2), Chapter 4 also hold.

Now by $\hat{G}_\alpha^m$ we denote the operator which is defined for any sequence $f = \{f_n\}_0^\infty$ by the triangular matrix

$$\hat{G}_\alpha^m = (-\bar{\alpha})^m \begin{pmatrix} 1 & 0 & 0 & \dots \\ m\bar{\alpha} & 1 & 0 & \dots \\ \frac{m(m+1)}{2!}\bar{\alpha}^2 & m\bar{\alpha} & 1 & \dots \\ \dots & \dots & \dots & \dots \end{pmatrix}.$$

We define the operator $\hat{G}$ again as in (1.3). It is easily checked that

$$\hat{G}_\alpha^m \hat{D}_\alpha^m f = f, \qquad \hat{D}_\alpha^m \hat{G}_\alpha^m f = f.$$

Hence the space $\widetilde{E}_+(\varrho_+) = \hat{G}(E_+)$ can be introduced in the same way as in 5.1.1.

Lemma 1.1 and Lemma 1.2 remain true in the present case. Again we equip the image space $\overline{E}_+(\varrho_-) = \operatorname{im} \hat{B}$ with the norm defined in 4.2.3.

It is immediately clear that the inverse of the operator $\hat{B}_\alpha^m$ is given in the spaces $l_+^p (1 \leq p < \infty)$ and c_+^0 by the Toeplitz matrix

$$(\hat{B}_\alpha^m)^{-1} = (-\alpha)^m \begin{pmatrix} 1 & m\alpha & \frac{m(m+1)}{2!}\alpha^2 & \dots \\ 0 & 1 & m\alpha & \dots \\ 0 & 0 & 1 & \dots \\ \dots & \dots & \dots & \dots \end{pmatrix}.$$

Thus by Theorem 2.2, Chapter 4, in the present case the analytic description of the space $\overline{E}_+(\varrho_-)$ is also possible.

Now we consider the discrete Wiener-Hopf operator $\hat{A}$ whose symbol $a(z)$ ($|z| = 1$) has the form (1.13) in the space E_+. For the operator $\hat{A}$ the Theorem 1.5 also holds. This follows immediately from the proof of this theorem given above.

5.2. Wiener-Hopf Integral Equations

In this section E is any one of the function spaces (2.4) from Section 3.2 on the real line. E_+ is the corresponding space on the positive half-line. P and U are the operators defined on E by relations (2.11) and (2.12), respectively, of Chapter 3.

In Section 3.2 we saw that for Wiener-Hopf integral equations the point at infinity plays a special role. This will be seen still more clearly in this section, where, besides equations of the second kind whose symbol may vanish at finite points, integral equations of the first kind are also considered. These equations have no discrete analogue.

5.2.1. The Function Spaces Generated by Zeros of the Symbol

1. If the unit circle is mapped onto the real line by stereographic projection, then to the symbol $z^{\pm 1} - \tilde{\alpha}^{\pm 1}$ ($|z| = |\tilde{\alpha}| = 1$) there corresponds the symbol $\dfrac{\lambda - \alpha}{\lambda \pm i}$ $(-\infty < \lambda,\ \alpha < \infty)$ if $\tilde{\alpha} \neq 1$ and the symbol $\dfrac{1}{\lambda \pm i}$ if $\tilde{\alpha} = 1$ (comp. 3.2.2).

Thus we now consider the Wiener-Hopf integral operators B_α, D_α and B_∞, D_∞ which are defined on the space E by the following equations (α an arbitrary point on the real line):

$$(B_\alpha\varphi)(t) = \varphi(t) - (1 + i\alpha)\, e^t \int_t^\infty e^{-s}\varphi(s)\, ds\,,$$

$$(D_\alpha\varphi)(t) = \varphi(t) - (1 - i\alpha)\, e^{-t} \int_{-\infty}^t e^{s}\varphi(s)\, ds\,,$$

$$(B_\infty\varphi)(t) = ie^t \int_t^\infty e^{-s}\varphi(s)\, ds\,,$$

$$(D_\infty\varphi)(t) = -\, ie^{-t} \int_{-\infty}^t e^{s}\varphi(s)\, ds\,.$$

The symbols of the operators B_α, B_∞, D_α, D_∞ have the form

$$\frac{\lambda - \alpha}{\lambda - i},\ \frac{1}{\lambda - i},\ \frac{\lambda - \alpha}{\lambda + i},\ \frac{1}{\lambda + i} \qquad (-\infty \leqq \lambda \leqq \infty)\,,$$

respectively.

Obviously, B_α and B_∞ are the operators of the convolution with the functions[1])

$$\delta - (1 + i\alpha)\, e^t(t < 0)\,, \qquad ie^t(t < 0)\,;$$

belonging to $\widetilde{L}^-$ and L^-, respectively. D_α and D_∞ are the operators of convolution with the functions

$$\delta - (1 - i\alpha)\, e^{-t} \qquad (t > 0)\,, \qquad -ie^{-t} \qquad (t > 0)$$

from $\widetilde{L}^+$ and L^+, respectively. If m is a natural number, then it is easily seen, that B_∞^m and D_∞^m are the operators of convolution with the functions

$$-\frac{(-i)^m}{(m-1)!} e^t t^{m-1} \qquad (t < 0)\,, \qquad \frac{(-i)^m}{(m-1)!} e^{-t} t^{m-1} \qquad (t > 0)\,,$$

respectively. The symbols of these operators are $(\lambda - i)^{-m}$ and $(\lambda + i)^{-m}$, respectively.

By $\hat{D}_\alpha$ and $\hat{D}_\infty$ we denote the restriction of the operators $PD_\alpha P$ and $PD_\infty P$, respectively, to the space E_+. $\hat{B}_\alpha$ and $\hat{B}_\infty$ have a corresponding meaning. Furthermore, let $\hat{G}_\alpha$ and $\hat{G}_\infty$ be the operators which with any function f locally integrable on $(0, \infty)$ (i.e. integrable on each finite interval $(0, a)$) associate the functions

$$(\hat{G}_\alpha f)(t) = f(t) + (1 - i\alpha)\, e^{-i\alpha t} \int_0^t e^{i\alpha s} f(s)\, ds \qquad (t > 0)\,,$$

$$(\hat{G}_\infty f)(t) = i[f(t) + f'(t)] \qquad (t > 0)\,,$$

respectively. Here $f'(t)$ is the (generalized) derivative of the function $f(t)$ (continued to the negative half-line by zero) in the sense of distribution theory (comp. I. M. Gelfand and G. E. Šilov [1]). By a simple calculation it is easily seen that the relations (1.2) hold (for arbitrary finite α and also for $\alpha = \infty$).

2. Now let $\alpha_1, \alpha_2, \dots, \alpha_r$ be distinct points on the real line, $m'_\infty \geqq 0$, $m'_1, \dots, m'_r \geqq 0$ and $m''_\infty \geqq 0$, $m''_1, \dots, m''_r > 0$ integers. We put

$$B = B_\infty^{m''_\infty} \prod_{j=1}^r B_{\alpha_j}^{m''_j}\,, \qquad D = D_\infty^{m'_\infty} \prod_{j=1}^r D_{\alpha_j}^{m'_j}\,, \qquad \hat{G} = \hat{G}_\infty^{m'_\infty} \prod_{j=1}^r \hat{G}_{\alpha_j}^{m'_j}\,. \tag{2.1}$$

By $\hat{B}$ and $\hat{D}$ we denote the restrictions of the operators PB and DP, respectively, to the space E_+ (comp. formula (2.7_2), Chapter 4). Furthermore we define

$$\varrho_+(\lambda) = (\lambda + i)^{-m'_\infty} \left(\frac{\lambda - \alpha_1}{\lambda + i}\right)^{m'_1} \cdots \left(\frac{\lambda - \alpha_r}{\lambda + i}\right)^{m'_r}, \qquad (-\infty \leqq \lambda \leqq \infty)$$

$$\varrho_-(\lambda) = (\lambda - i)^{-m''_\infty} \left(\frac{\lambda - \alpha_1}{\lambda - i}\right)^{m''_1} \cdots \left(\frac{\lambda - \alpha_r}{\lambda - i}\right)^{m''_r}.$$

[1]) As in Section 3.2 we will here also identify a function defined on the positive (negative) half-line with the function extended to the whole real line by zero.

Obviously, $\varrho_+(\lambda) \in \mathfrak{L}^+(\varrho_-(\lambda) \in \mathfrak{L}^-)$ is the symbol of the operator $D(B)$. Now we can introduce the space $\widetilde{E}_+(\varrho_+) = \widehat{G}(E_+)$ in a way completely, analogous to 5.1.1, no 2.

We remark that especially in the particular case $m_j' = 0$ $(j = 1, \dots, r)$, i.e.

$$\varrho_+(\lambda) = (\lambda + i)^{-m_\infty'},$$

the space $\widetilde{E}_+(\varrho_+)$ is obtained from E_+ by adding the generalized derivatives up to the m_∞'th order of all functions from E_+. Here each element $\varphi \in \widetilde{E}_+(\varrho_+)$ has the form

$$\varphi(t) = \left(\frac{d}{dt} + 1\right)^{m_\infty'} f(t) \qquad (f \in E_+)\,.$$

3. Next we shall investigate the kernel of the operator $\widehat{B}$. The results are completely analogous to the corresponding results for discrete Wiener-Hopf equations.

Lemma 2.1. *If E_+ is one of the spaces L_+^p $(1 \leqq p < \infty)$ or C_+^0, then* $\ker \widehat{B} = \{0\}$.

Proof. Let

$$\widehat{B}_{\alpha_j}\varphi = \varphi(t) - (1 + i\alpha_j)\, e^t \int_t^\infty e^{-s}\varphi(s)\, ds = 0 \qquad (t > 0)$$

with $\varphi \in E_+$. Then φ is absolutely continuous. By differentiation of the last equation we obtain $\varphi'(t) = -\, i\alpha_j\varphi(t)$, and this implies $\varphi(t) = Ce^{-i\alpha_j t}$ $(C = \text{const})$. Since $\varphi \in E_+$, it follows that $C = 0$, i.e. $\varphi = 0$.

If $\widehat{B}_\infty\varphi = 0$ $(\varphi \in E_+)$, then we multiply this equation by e^{-t}. The following differentiation immediately yields $\varphi = 0$. From this we obtain the assertion by using formula (2.7_2), Chapter 4.

Lemma 2.2. $\ker B$ *is r-dimensional in the space M_+, and the functions*

$$e^{-i\alpha_1 t},\ e^{-i\alpha_2 t},\ \dots,\ e^{-i\alpha_r t} \qquad (t > 0) \tag{2.2}$$

form a basis in it.

Proof. It follows from the proof of Lemma 2.1 that the function $e^{-i\alpha_j t}$ $(t > 0$ $(j = 1, \dots, r)$ is the unique solution of the equation $\widehat{B}_{\alpha_j}\varphi = 0$ in the space M_+. Also in M_+ the equation $\widehat{B}_\infty\varphi = 0$ has only the trivial solution. Thus by Theorem 2.1, Chapter 4, it is sufficient to show that $e^{-i\alpha_j t} \notin \operatorname{im} \widehat{B}_{\alpha_j}$.

We assume the opposite, i.e. for a certain function $\varphi \in M_+$ let

$$\varphi(t) - (1 + i\alpha_j)\, e^t \int_t^\infty e^{-s}\varphi(s)\, ds = e^{-i\alpha_j t} \qquad (t > 0)\,.$$

By differentiation of the last equation we obtain

$$\varphi'(t) + i\alpha_j\varphi(t) = -\,(1 + i\alpha_j)\, e^{-i\alpha_j t}$$

and hence

$$\varphi(t) = -\left((1 + i\alpha_j)\, t + C\right) e^{-i\alpha_j t} \qquad (t > 0)\,. \tag{2.3}$$

But the function (2.3) does not belong to M_+. This contradiction proves the assertion.

Corollary 2.1. *ker $\hat{B}_\infty = \{0\}$ holds in every space E_+.*

We remark that the functions (2.2) also belong to the space M_+^u. But the function $e^{-i\alpha t}$ $(t > 0)$ belongs to C_+ only if $\alpha = 0$. Hence we have:

The kernel of the operator $\hat{B} = PB \mid C_+$ ($\hat{B}$ considered in the space C_+) consists only of the zero element if $\alpha_j \neq 0$ for all $j = 1, 2, \ldots, r$ and of all constants if $\alpha_j = 0$ for some j.

According to Section 4.2.3 we denote the image space of the operator $\hat{B} = PB \mid E_+$ ($\hat{B}$ considered in the space E_+) by $\overline{E}_+(\varrho_-)$. Using Lemma 2.1 and Lemma 2.2 we define the norm of a function $f \in \overline{E}_+(\varrho_-)$ by one of the following relations:

$$\|f\|_{\overline{E}_+} = \|\hat{B}^{-1} f\|_{E_+}$$

if E_+ is one of the spaces L_+^p $(1 \leqq p < \infty)$ or C_+^0, and by

$$\|f\|_{\overline{E}_+} = \inf_{\hat{B}\varphi = f} \|\varphi\|_{E_+}$$

if $E_+ = M_+$. A corresponding definition holds for the space C_+.

Our next task is the analytic description of the space $\overline{E}_+(\varrho_-)$.

4. First we consider the image space

$$E_+^{(\infty)} = \overline{E}_+\{(\lambda - i)^{-m''_\infty}\} = \operatorname{im} \hat{B}_\infty^{m''_\infty}$$

with the norm

$$\|f\|_{E_+^{(\infty)}} = \|\varphi\|_{E_+} \qquad (f = \hat{B}_\infty^{m''_\infty} \varphi)\,.$$

Theorem 2.1. *$f \in E_+^{(\infty)}$ holds if and only if $f^{(k)}(t)$ $(t \geqq 0)$ is absolutely continuous (on each finite interval) for $k = 0, 1, \ldots, m''_\infty - 1$ and $f^{(k)}(t) \in E_+$ for $k = 0, 1, \ldots, m''_\infty$. The norm $\|f\|_{E_+^{(\infty)}}$ is equivalent to the norm $\|f\|_\infty = \sum_{k=0}^{m''_\infty} \|f^{(k)}\|_{E_+}$.*

Proof. First we consider the case $m''_\infty = 1$.

Let $f = \hat{B}_\infty \varphi = i e^t \int_t^\infty e^{-s} \varphi(s)\, ds$ $(t \geqq 0,\ \varphi \in E_+)$. Obviously, $f(t)$ is absolutely continuous, $f \in E_+$ and $f'(t) = -i\hat{B}_0 \varphi \in E_+$. Furthermore, $\|f\|_{E_+} + \|f'\|_{E_+} \leqq c\|\varphi\|_{E_+}$.

Conversely, if $f(t)$ is absolutely continuous and $f, f' \in E_+$, then it is easily checked that the function $\varphi = i(f' - f) \in E_+$ satisfies the equation $f = \hat{B}_\infty \varphi$. Hence $f \in E_+^{(\infty)}$. Moreover, the estimate

$$\|f\|_{E_+^{(\infty)}} = \|\varphi\|_{E_+} \leqq \|f\|_{E_+} + \|f'\|_{E_+}$$

holds. Thus the theorem is proved for $m''_\infty = 1$.

For arbitrary m''_∞ the assertion is easily obtained by complete induction.

Corollary 2.2. *The following formula holds:*

$$(\hat{B}_\infty)^{-1} = i\left(\frac{d}{dt} - 1\right).$$

Remark. $f = \hat{B}^m_\infty \varphi$ ($\varphi \in E_+$, m natural number) implies $f^{(m)} = (-i)^m \hat{B}^m_0 \varphi$.

5. Next we consider the image space

$$E_+^{(j)} = \overline{E}_+\left\{\left(\frac{\lambda - \alpha_j}{\lambda - i}\right)^{m''_j}\right\} = \text{im } \hat{B}^{m''_j}_{\alpha_j} \qquad (j = 1, \dots, r).$$

First let E_+ be one of the spaces L^p_+ $(1 \leqq p < \infty)$ or C^0_+. Then the norm of a function $f \in E_+^{(j)}$ is defined by

$$\|f\|_{E_+^{(j)}} = \|\varphi\|_{E_+}, \qquad (f = \hat{B}^{m''_j}_{\alpha_j}\varphi).$$

Now we introduce the following notation. Let $f(t) \in L(0, \tau)$ for any $\tau > 0$, and let the improper integral

$$\int_0^\infty f(s)\, ds = \lim_{\tau \to \infty} \int_0^\tau f(s)\, ds$$

exist. Then we put

$$f^{(-1)}(t) = \int_t^\infty f(s)\, ds \qquad (t \geqq 0).$$

Obviously, $f^{(-1)}(t)$ is absolutely continuous, and $[f^{(-1)}(t)]' = -f(t)$. Furthermore, we put $f^{(-2)}(t) = [f^{(-1)}(t)]^{(-1)}$ if this function exists etc. Finally, let $f^{(0)}(t) = f(t)$.

Theorem 2.2. *$f \in E_+^{(j)}$ holds if and only if $[e^{-i\alpha_j t} f(t)]^{(-k)} \in E_+$ $(k = 0, 1, \dots, m''_j)$. The norm $\|f\|_{E_+^{(j)}}$ is equivalent to the norm*

$$\|f\|_j = \sum_{k=0}^{m''_j} \|[e^{i\alpha_j t} f(t)]^{(-k)}\|_{E_+}.$$

Proof. First we consider the case $m''_j = 1$.

Let $f \in E_+^{(j)}$, i.e. $f = B_{\alpha_j}\varphi$ with $\varphi \in E_+$. Multiplying both sides of the last equation by $e^{i\alpha_j s}$ and then integrating over the interval (t, τ) $(0 < t < \tau)$ we obtain

$$\int_t^\tau e^{i\alpha_j s} f(s)\, ds = \int_t^\tau e^{i\alpha_j s}\varphi(s)\, ds - (1 + i\alpha_j)\int_t^\tau e^{(1+i\alpha_j)s}\, ds \int_s^\infty e^{-x}\varphi(x)\, dx.$$

Partial integration of the second integral on the right-hand side yields

$$\int_t^\tau e^{i\alpha_j s} f(s)\, ds = e^{(1+i\alpha_j)t}\int_t^\infty e^{-s}\varphi(s)\, ds - e^{(1+i\alpha_j)\tau}\int_\tau^\infty e^{-s}\varphi(s)\, ds. \tag{2.4}$$

By a simple estimate (use of the Hölder inequality in the case $E_+ = L^p_+$) it is easily seen that the last term in (2.4) tends to zero for $\tau \to \infty$. Hence

$$[e^{i\alpha_j t} f(t)]^{(-1)} = e^{(1+i\alpha_j)t} \int_t^\infty e^{-s} \varphi(s)\, ds \in E_+ .$$

The last equation can also be written in the form

$$\hat{B}_\infty \varphi = i e^{-i\alpha_j t} \int_t^\infty e^{i\alpha_j s} f(s)\, ds . \tag{2.5'}$$

Applying the operator $i\left(\frac{d}{dt} - 1\right)$ to both sides of this equation and using Corollary 2.2 we now find

$$\varphi(t) = f(t) + (1 + i\alpha_j)\, e^{-i\alpha_j t} \int_t^\infty e^{i\alpha_j s} f(s)\, ds . \tag{2.5}$$

From the equations $f = \hat{B}_{\alpha_j \varphi}$ and (2.5'), (2.5) it follows immediately that the norms $\|f\|_{E^{(j)}_+}$ and $\|f\|_j$ are equivalent.

Conversely, let now $f(t) \in E_+$ and $[e^{i\alpha_j t} f(t)]^{(-1)} \in E_+$. Then the function $\varphi(t)$ defined by (2.5) also belongs to E_+. It is easy to check that $\hat{B}_{\alpha_j}\varphi = f$ and consequently we have $f \in E^{(j)}_+$.

Thus the theorem is proved for $m''_j = 1$. For arbitrary $m''_j \geqq 1$ the assertion is easily obtained by complete induction.

Along with formula (2.5) we have simultaneously constructed the inverse of the operator $\hat{B}_{\alpha_j} = PB_{\alpha_j} | E_+$.

Corollary 2.3. *The following formula holds:*

$$\hat{B}_0^{-1} f = f(t) + f^{(-1)}(t) \qquad (f \in \operatorname{im} \hat{B}_0) .$$

Corollary 2.4. *For arbitrary* α $(-\infty < \alpha < \infty)$ *the formula*

$$\hat{B}_\alpha^{-1} f = f(t) + (1 + i\alpha)\, e^{-i\alpha t} [e^{i\alpha t} f(t)]^{(-1)}$$

holds.

6. Now we show that the spaces $E^{(j)}_+$ introduced in no. 5 contain sufficiently many functions.

By $L^{p,m}$ ($1 \leqq p \leqq \infty$, m an arbitrary integer) we denote the space of all functions of the form $f(t) = (t + i)^{-m} \varphi(t)$ with $\varphi(t) \in L^p$. With the norm

$$|f|_m = \|\varphi\|_{L^p}$$

$L^{p,m}$ is a Banach space isometrically isomorphic to the space L^p. For $m = 0$ we have $L^{p,0} = L^p$.

Obviously, in the case $m > 0$ a function $f(t)$ belongs to the space $L^{p,m}$ if and only if

$$t^k f(t) \in L^p \qquad (k = 0, 1, \dots, m) ,$$

and the norm $|f|_m$ is equivalent to the norm

$$||f||_{L^{p,m}} = \sum_{k=0}^{m} ||t^k f(t)||_{L^p} .$$

By $L_+^{p,m}$ we again denote the corresponding space of functions on the positive half-line $[0, +\infty)$.

Using the equality $(L^p)^* = L^q$ ($q = \dfrac{p}{p-1}$ for $1 < p < \infty$ and $q = \infty$ for $p = 1$) we can easily see that $L^{q,-m}$ is the conjugate space of $L^{p,m}$ for $1 \leqq p < \infty$ and any integer m. The corresponding statement holds for $L_+^{p,m}$.

Theorem 2.3. *For any $f(t) \in L_+^{p,m}$ $(1 \leqq p < \infty,\ m \geqq 0)$ there exist the functions $f^{(-k)}(t)$, and $f^{(-k)}(t) \in L_+^p$ $(k = 0, 1, \dots, m)$.*

The space $L_+^{p,m}$ is continuously and densely embedded into the space $E_+^{(\alpha)} = \overline{L_+^p}\left\{\left(\dfrac{\lambda-\alpha}{\lambda-i}\right)^m\right\}$.

Proof. We prove the theorem by complete induction. First we assume $m = 1$.

For $p = 1$ partial integration yields

$$\int_0^\infty |f^{(-1)}(t)|\, dt \leqq \int_0^\infty dt \int_t^\infty |f(s)|\, ds = \int_0^\infty t|f(t)|\, dt$$

since by $f \in L_+^{1,1}$ the relation

$$t\int_t^\infty |f(s)|\, ds \leqq \int_t^\infty s\,|f(s)|\, ds \underset{t\to+\infty}{\to} 0$$

holds. Hence we obtain $f^{(-1)}(t) \in L_+^1$, and Theorem 2.2 implies

$$c||f||_{E_+^{(\alpha)}} \leqq \sum_{k=0}^{1} ||f^{(-k)}(t)||_{L_+^1} \leqq ||f||_{L_+^{1,1}} \qquad (c = \text{const} > 0)\,.$$

Now consider the case $p > 1$. First we remark that the function $f(t) \in L_+^{p,1}$ is summable. For, by the Hölder inequality, we have

$$\int_1^\infty |f(t)|\, dt \leqq (q-1)^{-1/q}\left[\int_1^\infty |t\, f(t)|^p\, dt\right]^{1/p} .$$

Consequently the function $f^{(-1)}(t)$ exists. Now we show that $f^{(-1)}(t) \in L_+^p$.

To this end we introduce the function

$$\varphi(s, t) = t\,|f(st)| \qquad (s \geqq 1, t \geqq 1)\,.$$

For this function we calculate the integrals

$$\int_1^\infty \varphi(s, t)\, ds = \int_t^\infty |f(x)|\, dx, \quad \left\{\int_1^\infty [\varphi(s, t)]^p\, dt\right\}^{1/p} = \frac{1}{s^{1+1/p}}\left\{\int_s^\infty |xf(x)|^p\, dx\right\}^{1/p} .$$

By the well-known inequality[1])

$$\left[\int_1^\infty \left\{\int_1^\infty \varphi(s,t)\,ds\right\}^p dt\right]^{1/p} \leqq \int_1^\infty \left\{\int_1^\infty [\varphi(s,t)]^p\,dt\right\}^{1/p} ds$$

we then obtain the estimate

$$\left\{\int_1^\infty |f^{(-1)}(t)|^p\,dt\right\}^{1/p} \leqq p\left\{\int_1^\infty |xf(x)|^p\,dx\right\}^{1/p}.$$

Moreover, since for any $t \geqq 0$ the inequality

$$|f^{(-1)}(t)| \leqq \int_0^\infty |f(t)|\,dt \leqq c_0[\|f\|_{L_+^p} + \|t\,f(t)\|_{L_+^p}]$$

holds, we have

$$\|f^{(-1)}\|_{L_+^p} \leqq c[\|f\|_{L_+^p} + \|t\,f(t)\|_{L_+^p}]\,.$$

This implies the continuity of the embedding $L_+^{p,m} \subset E_+^{(\alpha)}$ in the case $m = 1$.

For $m \geqq 2$ we obtain the last assertion by successive application of the considerations stated above if we use the fact that for any function $f(t) \in L_+^{p,2}$ the relation

$$f^{(-2)}(t) = -\,tf^{(-1)}(t) + \int_t^\infty sf(s)\,ds$$

holds. The validity of this equation can be shown by partial integration.

Now we prove that the set $L_+^{p,m}$ is dense in the space $E_+^{(\alpha)}$. Let $f = \hat{B}_\alpha^m \varphi$ ($\varphi \in L_+^p$) be an arbitrary function from $E_+^{(\alpha)}$ and $\varepsilon > 0$. As is well-known, there exists a continuous function ψ with a compact support such that $\|\varphi - \psi\|_{L_+^p} < \varepsilon$. Then the function $g = \hat{B}_\alpha^m \psi$ also has compact support and thus belongs to $L_+^{p,m}$. By the definition of the norm in $E_+^{(\alpha)}$ we have $\|f-g\|_{E_+^{(\alpha)}} = \|\varphi - \psi\|_{L_+^p} < \varepsilon$.

This proves the theorem.

Remark. By the same means we can prove that the continuous embedding

$$\tilde{L}_+^q \left\{\left(\frac{\lambda - \alpha}{\lambda + i}\right)^m\right\} \subset L_+^{q,-m} \qquad (1 < q \leqq \infty\,, \quad m \geqq 0)$$

holds.

7. Now we consider the space $E_+^{(j)} = \operatorname{im} \hat{B}_{\alpha_j}^{m_j''}$ in the case where E_+ is the space M_+. By Lemma 2.2 in this case the norm of a function $f \in E_+^{(j)}$ is given by the formula

$$\|f\|_{E_+^{(j)}} = \inf_C \|\varphi(t) + Ce^{-i\alpha_j t}\|_{M_+}\,, \tag{2.6}$$

where $f = \hat{B}_{\alpha_j}^{m_j''} \varphi$ ($\varphi \in M_+$).

---

1) See G. H. Hardy, J. E. Littlewood and G. Polya [1], Theorem 202.

For the analytic description of the space $E_+^{(j)} = M_+^{(j)}$ we use the Cesàro summability method for improper integrals (see e.g. A. ZYGMUND [1], Chapter III). Let $f(t)$ be a measurable function on $(0, +\infty)$. The improper integral $\int_0^\infty f(t)\,dt$ is said to be C_1*-summable to the value* S, in symbols $(C_1)\int_0^\infty f(t)\,dt = S$, if

$$t^{-1}\int_0^t dx \int_0^x f(\xi)\,d\xi = t^{-1}\int_0^t (t-x)\,f(x)\,dx \underset{t\to+\infty}{\to} S\,.$$

If this condition is satisfied, then the function $f(t)$ is also said to be C_1-summable, and we put

$$f^{(-1)}(t) = (C_1)\int_t^\infty f(s)\,ds \qquad (t \geqq 0)\,.$$

If the function $f^{(-1)}(t)$ is also C_1-summable, then we put $f^{(-2)}(t) = [f^{(-1)}(t)]^{(-1)}$ etc.

In the sequel let $f(t) \in M_+$. Then $(C_1)\int_0^\infty f(s)\,ds = \int_0^t f(s)\,ds + f^{(-1)}(t)$. Consequently $f^{(-1)}(t)$ is absolutely continuous, and $[f^{(-1)}(t)]' = -f(t)$.

Theorem 2.4. *$f \in E_+^{(j)} = M_+^{(j)}$ holds if and only if the function $f_{\alpha_j}(t) = e^{i\alpha_j t}f(t)$ satisfies the following conditions:*

(1) $f_{\alpha_j}^{(-k)}(t)$ is C_1-summable for $k = 0, 1, \dots, m_j'' - 2$.
(2) $f_{\alpha_j}^{(-k)}(t) \in M_+$ for $k = 0, 1, \dots, m_j'' - 1$.
(3) $g(t) = \int_0^t f_{\alpha_j}^{(-m_j''+1)}(s)\,ds \in M_+$.

The norm (2.6) *is equivalent to the norm*

$$||f_j|| = ||f_{\alpha_j}||_{M_+} + \dots + ||f_{\alpha_j}^{(-m_j''+1)}||_{M_+} + \inf_C ||C + g||_{M_+}\,.$$

The proof of Theorem 2.4 is analogous to the proof of Theorem 2.2. We have only to observe that the subspace of all constants from M_+ is an invariant subspace of the operators PB_∞ and PA_-.

A result analoguos to Theorem 2.4 can be stated also for the space C_+ (comp. also Theorem 1.4). But we shall desist from its formulation.

8. Theorems 2.1 to 2.4 and Theorem 2.2 of Chapter 4 yield the analytic description of the space $\overline{E}_+(\varrho_-)$.

Theorem 2.5. *The formula $\overline{E}_+(\varrho_-) = E_+^{(1)} \cap \dots \cap E_+^{(r)} \cap E_+^{(\infty)}$ holds, and the norm $||f||_{\overline{E}_+}$ is equivalent to the norm $|f| = ||f||_1 + \dots + ||f||_r + ||f||_\infty$.*

Together with Theorems 2.1 to 2.3 this implies:

Corollary 2.5. *Let $E_+ = L_+^p$ $(1 \leqq p < \infty)$ and $m = \max\{m_1'', \dots, m_r''\}$. Then $E_+^{(\infty)} \cap L_+^{p,m} \subset \overline{E}_+(\varrho_-)$ holds in the sense of a continuous embedding.*

Remark. In the case $m'_\infty = 0$ we have the continuous embedding

$$\widetilde{L}^p_+(\varrho_+) \subset L^{p,\,-m'}_+ , \quad 1 < p \leqq \infty , \quad m' = \max\{m'_1, \dots, m'_r\} ;$$

$$\widetilde{L}^1_+(\varrho_+) \subset L^{1,\,-(m'+\varepsilon)}_+ \quad (\varepsilon > 0) .$$

Moreover, it is easily seen that for a function $f(t)$ belonging to all spaces $\widetilde{E}_+(\varrho_+)$ the following estimate holds for $t \to \infty$:

$$f(t) = O(t^{m'-1}) .$$

5.2.2. Symbols with Zeros of Integral Order

In the space E_+ we consider the Wiener-Hopf integral equation

$$(\hat{A}\varphi)(t) = c\varphi(t) - \int_0^\infty k(t-s)\,\varphi(s)\,ds = f(t) \qquad (0 \leqq t < \infty) , \tag{2.7}$$

where c is a constant and $k(t) \in L$. We assume that the symbol

$$\mathcal{A}(\lambda) = c - \int_{-\infty}^{\infty} e^{i\lambda t} k(t)\,dt \qquad (-\infty \leqq \lambda \leqq \infty)$$

has the representation

$$\mathcal{A}(\lambda) = (\lambda + i)^{-m_\infty} \left(\frac{\lambda - \alpha_1}{\lambda + i}\right)^{m_1} \cdots \left(\frac{\lambda - \alpha_r}{\lambda + i}\right)^{m_r} \mathcal{B}(\lambda) . \tag{2.8}$$

Here $\alpha_1, \alpha_2, \dots, \alpha_r$ are distinct points on the real line, $m_\infty \geqq 0$, $m_1, \dots, m_r > 0$ are integers and $\mathcal{B}(\lambda) \in \mathfrak{L}$.[1])

We remark that the case $m_\infty > 0$ corresponds to the equation of the first kind ($c = 0$). In the case of an equation of the second kind ($c \neq 0$) we have $m_\infty = 0$ and $\mathcal{B}(\infty) \neq 0$.

1. Similarily as in 5.1.2. we introduce non-negative integers m'_j and m''_j with

$$m'_j + m''_j = m_j \qquad (j = 1, \dots, r) , \qquad m'_\infty + m''_\infty = m_\infty ,$$

and we represent the symbol $\mathcal{A}(\lambda)$ in the form

$$\mathcal{A}(\lambda) = \varrho_+(\lambda)\,\mathcal{C}(\lambda)\,\varrho_-(\lambda) . \tag{2.9}$$

Here $\varrho_\pm(\lambda)$ are the functions introduced in the preceding subsection, and

$$\mathcal{C}(\lambda) = \left(\frac{\lambda - i}{\lambda + i}\right)^{m''_1 + \dots + m''_r + m''_\infty} \mathcal{B}(\lambda) .$$

Theorem 2.6. *For the operator $\hat{A}$ defined by* (2.7) *to be a Φ_+-operator or a Φ_--operator from $\widetilde{E}_+(\varrho_+)$ into $\overline{E}_+(\varrho_-)$ it is necessary and sufficient that $\mathcal{C}(\lambda) \neq 0$, $-\infty \leqq \lambda \leqq \infty$ (or, what is the same, $\mathcal{B}(\lambda) \neq 0$).*

---

[1]) Necessary and sufficient conditions for the representability of the function $\mathcal{A}(\lambda)$ in the form (2.8) will be given in no. 3 of this subsection.

If this condition is satisfied, then the invertibility of the operator $\hat{A}$ corresponds to the number $\varkappa - \delta$, and the formulae

$$\dim \ker \hat{A} = \max(\delta - \varkappa, 0), \qquad \dim \operatorname{coker} \hat{A} = \max(\varkappa - \delta, 0)$$

hold. Here

$$\varkappa = \operatorname{ind} \mathscr{C}(\lambda) = \operatorname{ind} \mathscr{B}(\lambda) + m''_\infty + \sum_{j=1}^{r} m''_j ,$$

$$\delta = \begin{cases} 0 & \textit{for} \quad E_+ = L_+^p \ (1 \leqq p < \infty) \quad \textit{or} \quad E_+ = C_+^0 , \\ r'' & \textit{for} \quad E_+ = M_+ , \end{cases}$$

and r'' is the number of the positive m''_j ($j = 1, 2, \ldots, r$). In the case $E_+ = C_+$ we have

$$\delta = \begin{cases} 1 , & \textit{if} \quad \alpha_{j_0} = 0 \quad \textit{and} \quad m''_{j_0} > 0 \qquad (1 \leqq j_0 \leqq r) , \\ 0 & \textit{otherwise} . \end{cases}$$

Proof. The proof is analogous to the proof of Theorem 1.5. We obtain the first part of the theorem easily from Theorem 2.3. Chapter 4. The same holds for the second part of the theorem in the case of the spaces L_+^p and C_+^0.

We shall prove the second assertion of the theorem for the space M_+. Thus let $\mathscr{C}(\lambda) \neq 0$ $(-\infty \leqq \lambda \leqq \infty)$. By Theorem 2.3, Chapter 4, and Lemma 2.2 the operator $\hat{A}$ is a Φ-operator from $\tilde{E}_+(\varrho_+)$ into $\overline{E}_+(\varrho_-)$ with the index

$$\operatorname{Ind} \hat{A} = r'' - \varkappa . \tag{2.10}$$

By Theorem 3.3, Chapter 2, the function $\mathscr{C}(\lambda) \in \mathfrak{L}$ can be factorized:

$$\mathscr{C}(\lambda) = \mathscr{C}_-(\lambda) \left(\frac{\lambda - i}{\lambda + i}\right)^{\varkappa} \mathscr{C}_+(\lambda), \tag{2.11}$$

where $\mathscr{C}_\pm(\lambda) \in \mathfrak{L}^\pm$ and $\mathscr{C}_\pm^{-1}(\lambda) \in \mathfrak{L}^\pm$ holds.

By $\hat{C}$ we denote the Wiener-Hopf integral operator defined on the space M_+ with symbol $\mathscr{C}(\lambda)$ and by $\hat{C}'$ the transpose operator to $\hat{C}$. For the symbol $\mathscr{C}(-\lambda)$ of the operator $\hat{C}'$ we then obtain the factorization

$$\mathscr{C}(-\lambda) = \mathscr{C}_+(-\lambda) \left(\frac{\lambda - i}{\lambda + i}\right)^{-\varkappa} \mathscr{C}_-(-\lambda) .$$

Here obviously $\mathscr{C}_\mp(-\lambda) \in \mathfrak{L}^\pm$.

In the case $\varkappa \leqq 0$ we have $\dim \operatorname{coker} \hat{C} = \dim \ker \hat{C}' = 0$.

Now let $\varkappa > 0$. By Theorem 2.3, Chapter 3, the Fourier transform of the general solution of the equation $\hat{C}'\psi = 0$ has the form

$$\Psi(\lambda) = \mathscr{C}_-^{-1}(-\lambda) \frac{P_{\varkappa-1}(\lambda)}{(\lambda + i)^\varkappa} , \tag{2.12}$$

where $P_{\varkappa-1}(\lambda)$ is an arbitrary polynomial in λ of a degree $\leqq \varkappa - 1$. Using Lemma 2.2 and Theorem 2.4. Chapter 3, we conclude from Lemma 1.2, Chapter 4, that for any function $f(t) \in \overline{E}_+(\varrho_-)$ equation (2.7) has a solution $\varphi \in \in \widetilde{E}_+(\varrho_+)$ if and only if the condition

$$\int_0^\infty h(t)\, \psi(t)\, dt = 0 \tag{2.13}$$

is satisfied for each function $\psi(t)$ whose Fourier transform is given by formula (2.12) and vanishes at the point $-\alpha_j$:

$$\Psi(-\alpha_j) = \int_{-\infty}^{\infty} \psi(t)\, e^{-i\alpha_j t}\, dt = 0 \qquad (j = 1, \dots, r : m_j'' > 0)\,. \tag{2.14}$$

Here $h(t) \in M_+$ is an arbitrary function which satisfies the conditionn $\hat{B}h = f$.

Without loss of generality we can assume that $m_j'' > 0$ for $j = 1, \dots, r''$. Then conditions (2.14) are equivalent to the equations $P_{\varkappa-1}(-\alpha_j) = 0$ $(j = 1, \dots, r'')$. Hence we obtain

$$\Psi(\lambda) = \frac{\mathscr{C}_-^{-1}(-\lambda)}{(\lambda + i)^\varkappa}\, (\lambda + \alpha_1) \dots (\lambda + \alpha_{r''})\, P_{\varkappa - r'' - 1}(\lambda)\,, \tag{2.15}$$

where $P_{\varkappa-r''-1}(\lambda)$ is an arbitrary polynomial of degree $\leqq \varkappa - r'' - 1$ $(P_{\varkappa-r''-1}(\lambda) \equiv 0$ for $\varkappa \leqq r'')$. Hence for the solvability of equation (2.7) it is necessary and sufficient that all functions $\psi(t)$ whose Fourier transform is given by the formula (2.15) satisfy condition (2.13). Obviously, the number of obtained conditions is equal to

$$\dim \operatorname{coker} \hat{A} = \max\,(\varkappa - r'', 0). \tag{2.16}$$

Now the assertion follows from formulae (2.10) and (2.16).

In the case of the space C_+ we obtain by the same arguments that conditions (2.13) are the solvability conditions for equation (2.7), where the Fourier transform of the function $\psi(t)$ is given by the formula

$$\Psi(\lambda) = \frac{\mathscr{C}_-^{-1}(-\lambda)}{(\lambda + i)^\varkappa}\, \lambda P_{\varkappa-2}(\lambda)$$

if $\alpha_{j_0} = 0$ and $m_{j_0}'' > 0$ for a certain j_0, $1 \leqq j_0 \leqq r$ and by formula (2.12) in all other cases.

Thus Theorem 2.6 is proved.

2. Now we shall investigate the *solvability conditions* for equation (2.7), The proof of Theorem 2.6 shows that (2.13) gives necessary and sufficient conditions for the solvability of equation (2.7) in $\widetilde{E}_+(\varrho_+)$ with $f(t) \in \overline{E}_+(\varrho_-)$. Here $h(t) \in E_+$ is an arbitrary function which satisfies the equation $\hat{B}h = f$, and the Fourier transform of the function $\psi(t)$ is given by (2.15) for the space M_+ and by (2.12) for the spaces L_+^p $(1 \leqq p < \infty)$ and C_+^0. In the case of the spaces L_+^p or C_+^0 the solvability conditions (2.13) can be given in the more

explicit form

$$\int_0^\infty h(t) \left[e^{-t} t^l + \int_0^t e^{s-t} (t-s)^l \, c_-(-s) \, ds \right] dt = 0 \tag{2.17}$$

$$(l = 0, 1, \ldots, \varkappa - 1) .$$

Here $c_-(t) \in L^-$ is the function whose Fourier transform is equal to $\mathscr{E}_-^{-1}(\lambda) - 1$. $\mathscr{E}_-(\lambda)$ is determined by (2.11) and the condition $\mathscr{E}_-(\infty) = 1$.

Theorem 2.2, Chapter 3, Lemma 2.1 and Lemma 2.2 imply:

Theorem 2.7. *The homogeneous equation $\hat{A}\varphi = 0$ has the same solutions in all spaces $\tilde{E}_+(\varrho_+)$, where E_+ is any one of the spaces L_+^p $(1 \leqq p < \infty)$ or C_+^0. If $m_j'' = 0$ $(j = 1, \ldots, r)$, then this holds also for all the other spaces E_+.*

For the cases mentioned in Theorem 2.7 (i.e. E_+ is one of the spaces L_+^p $(1 \leqq p < \infty)$, C_+^0 or $m_j'' = 0$, $j = 1, \ldots, r$ holds) an analogue of the Theorem 2.4 from Chapter 3 can also be formulated. To this end, in addition the functions $\varrho_\pm(\lambda)$ introduced obove we still consider the functions

$$\varrho'_+(\lambda) = \varrho_-(-\lambda) \in \mathfrak{L}^+ , \qquad \varrho'_-(\lambda) = \varrho_+(-\lambda) \in \mathfrak{L}^- .$$

Obviously, $\varrho'_+(\lambda)$ $(\varrho'_-(\lambda))$ is the symbol of the transposed operator $\hat{B}'$ $(\hat{D}')$. For the symbol $\mathscr{A}(-\lambda)$ of the operator $\hat{A}'$ transpose to $\hat{A}$ we obtain the representation

$$\mathscr{A}(-\lambda) = \varrho'_-(\lambda) \, \mathscr{E}(-\lambda) \, \varrho'_+(\lambda) .$$

If $\mathscr{E}(\lambda) \neq 0$ $(-\infty \leqq \lambda \leqq \infty)$ holds, then by Theorem 2.6 $\hat{A}'$ is a Φ-operator from $\tilde{E}_+(\varrho'_+)$ into $\bar{E}_+(\varrho'_-)$, where

$$\dim \ker \hat{A}' = \max(\varkappa, 0) , \qquad \dim \operatorname{coker} \hat{A}' = \max(-\varkappa, 0) .$$

Each element $\psi \in \tilde{L}_+^q(\varrho'_+)$ $(1 \leqq q \leqq \infty)$ can be regarded as a linear functional on the space $\bar{L}_+^p(\varrho_-)$ $\left(p = \dfrac{q}{q-1}\right)$ if the value of the functional ψ on the element $f \in \bar{L}_+^p(\varrho_-)$ is defined by the formula

$$\langle f, \psi \rangle \equiv \int_0^\infty f(t) \, \psi(t) \, dt = \int_0^\infty \hat{B}^{-1} f \cdot \hat{B}' \psi \, dt . \tag{2.18}$$

The continuity of the functional (2.18) follows from the estimate

$$|\langle f, \psi \rangle| \leqq \|\hat{B}^{-1} f\|_{L_+^p} \, \|\hat{B}' \psi\|_{L_+^q} = \|f\|_{\bar{L}_+^p(\varrho_-)} \, \|\psi\|_{\tilde{L}_+^q(\varrho'_+)} \cdot$$

For $1 \leqq p < \infty$ every linear functional on the space $\bar{L}_+^p(\varrho_-)$ has the form (2.18). This follows easily from $(L_+^p)^* = L_+^q$.

Obviously, by formula (2.18) every element ψ belonging to all spaces $\tilde{E}_+(\varrho'_+)$ can be regarded as a linear functional on the space $\bar{E}_+(\varrho_-)$.

We remark that the integral relation (2.18) holds for any function $\psi(t) \in L_+^q$ (cf. the formula (2.23) of Chapter 3). The validity of the formula (2.18) can also be proved if $m_\infty = 0$ and $f(t) \in L_+^{p,m}$ $(1 \leqq p < \infty)$ with $m = \max\{m_1'', \dots, m_r''\}$.

In fact, under the hypotheses in question we have $f \in \bar{L}_+^p(\varrho_-)$ by Corollary 2.5. We choose a sequence of continuous functions with compact supports $f_n(t)$ converging to the function $f(t)$ with respect to the norm in $L_+^{p,m}$. Then $\hat{B}^{-1}f_n$ converges to $\hat{B}^{-1}f$ with respect to the norm in L^p (cf. Corollary 2.4). Using the remark after Theorem 2.3 and formula (2.23) of Chapter 3 we can easily prove the relation

$$\int_0^\infty f_n(t)\,\psi(t)\,dt = \int_0^\infty \hat{B}^{-1}f_n \cdot \hat{B}'\psi\,dt\,.$$

The formula (2.18) is now obtained by passing to the limit for $n \to \infty$.

Now, Theorem 2.4, Chapter 3, implies

Theorem 2.8. *Let E_+ be one of the spaces L_+^p $(1 \leqq p < \infty)$ or C_+^0, and let the symbol $\mathcal{A}(\lambda)$ possess the form* (2.8), *where $\mathcal{B}(\lambda) \neq 0$ $(-\infty \leqq \lambda \leqq \infty)$.*

In the case $\varkappa = \operatorname{ind} \mathcal{C}(\lambda) \leqq 0$ equation (2.7) *has a solution $\varphi \in \widetilde{E}_+(\varrho_+)$ for any right-hand side $f(t) \in \overline{E}_+(\varrho_-)$, and the corresponding homogeneous equation has exactly $-\varkappa$ linearly independent solutions.*

For $\varkappa > 0$ the homogeneous equation

$$c\psi(t) - \int_0^\infty k(s-t)\,\psi(s)\,ds = 0$$

transpose to (2.7) *has exactly $\varkappa$ linearly independent solutions $\psi_l(t)$ $(l = 0, 1, \dots, \varkappa - 1)$ belonging to all spaces $\widetilde{E}_+(\varrho_+')$. In this case equation* (2.7) *possesses a (necessarily unique) solution $\varphi \in \widetilde{E}_+(\varrho_+)$ if and only if*

$$\int_0^\infty f(t)\,\psi_l(t)\,dt = 0 \qquad (l = 0, 1, \dots, \varkappa - 1)\,.$$

If $m_j'' = 0$ $(j = 1, 2, \dots, r)$ then the theorem holds for any space E_+.

3. Now we want to clarify under what conditions on the kernel $k(t)$ in equation (2.7) the symbol $\mathcal{A}(\lambda)$ has the representation (2.8).

Analogously to 5.1.2, no. 3, we first introduce an operator R_α $(-\infty < \alpha < \infty)$ which for a function $\varphi \in L = L^1(-\infty, \infty)$ is defined by

$$(R_\alpha\varphi)(t) = \begin{cases} ie^{-i\alpha t}\displaystyle\int_t^\infty e^{i\alpha s}\varphi(s)\,ds\,, & t > 0\,, \\ -ie^{-i\alpha t}\displaystyle\int_{-\infty}^t e^{i\alpha s}\varphi(s)\,ds\,, & t < 0\,. \end{cases} \tag{2.19}$$

By $L_{\alpha,m}^1$ ($m \geqq 0$ and integer) we denote the set of all functions $\varphi \in L$ for which

$$R_\alpha^k\varphi \in L\,, \qquad k = 0, 1, \dots, m\,.$$

Using the class $L^1_{\alpha, m}$ we can characterize the algebra $\mathfrak{L}(\alpha, m)$ (cf. 2.3.4) as follows.

Theorem 2.9. *Let*

$$\mathcal{F}(\lambda) = c + \int_{-\infty}^{\infty} e^{i\lambda t} f(t)\, dt \qquad (-\infty < \lambda < \infty)$$

be a function in the algebra $\mathfrak{L}$. *Then* $\mathcal{F}(\lambda) \in \mathfrak{L}(\alpha, m)$ *if and only if* $f(t) \in L^1_{\alpha, m}$.

Proof. First we consider the case $m = 1$. We put $\mathcal{G}(\lambda) = \left(\dfrac{\lambda + i}{\lambda - \alpha}\right)[\mathcal{F}(\lambda) - \mathcal{F}(\alpha)]$.

Let $\mathcal{F}(\lambda) \in \mathfrak{L}(\alpha, 1)$, i.e. $\mathcal{G}(\lambda) \in \mathfrak{L}$. Obviously, $\mathcal{G}(\infty) = c' = c - \mathcal{F}(\alpha)$. By $g(t)$ we denote the inverse Fourier transform of the function $\mathcal{G}(\lambda) - c'$. It is easily seen that

$$f(t) = (D_\alpha g)(t) + c'(-1 + i\alpha)\, e_+(t) \tag{2.20}$$

with $e_+(t) = e^{-t}$ for $t > 0$ and $e_+(t) = 0$ for $t < 0$.

By means of the argument in the proof of Theorem 2.2 we obtain the equation

$$[I + (i + \alpha) R_\alpha] D_\alpha = I. \tag{2.21}$$

(2.20) and (2.21) imply $R_\alpha f \in L$.

Conversely, let $f \in L^1_{\alpha, 1}$. We put $g(t) = f(t) + (i + \alpha)(R_\alpha f)(t)$ and $g_\pm(t) = (i + \alpha)(R_\alpha f)(t)$ $(t \gtrless 0)$. Then $g(t)$ and $g_\pm(t)$ are functions of L. Using the identities

$$\left.\begin{aligned} \int_0^\infty e^{i\lambda t} g_+(t)\, dt &= i\,\frac{g_+(+0)}{\lambda} + \frac{i}{\lambda}\int_0^\infty e^{i\lambda t} g_+'(t)\, dt\,, \\ \int_{-\infty}^0 e^{i\lambda t} g_-(t)\, dt &= -\,i\,\frac{g_-(-0)}{\lambda} + \frac{i}{\lambda}\int_{-\infty}^0 e^{i\lambda t} g_-'(t)\, dt \end{aligned}\right\} \quad (\lambda \neq 0)$$

we obtain

$$\frac{\lambda + i}{\lambda - \alpha}[\mathcal{F}(\lambda) - \mathcal{F}(\alpha)] = (Fg)(\lambda) - (Ff)(\alpha) \in \mathfrak{L}\,,$$

i.e. $\mathcal{F}(\lambda) \in \mathfrak{L}(\alpha, 1)$.

Thus the theorem is proved for $m = 1$. For $m > 1$ the assertion is obtained by successive application of the preceding considerations.

Theorems 5° and 7° of Subsection 2.3.4 imply

Corollary 2.6. *Let* $\mathcal{A}(\lambda) \in \mathfrak{L}$, $\boldsymbol{\alpha} = \{\alpha_1, \alpha_2, \ldots, \alpha_n\}$ *a system of points on the real line and* $\boldsymbol{m} = \{m_1, m_2, \ldots, m_r\}$ *a system of natural numbers. For of*

$$\prod_{j=1}^{r} \left(\frac{\lambda + i}{\lambda - \alpha_j}\right)^{m_j} \mathcal{A}(\lambda) \in \mathfrak{L}\,, \tag{2.22}$$

to hold the two following conditions together are necessary and sufficient:

a) $\mathcal{A}(\lambda) \in \mathfrak{L}(\boldsymbol{\alpha}, \boldsymbol{m})$,

b) $\mathcal{A}^{\{k\}}(\alpha_j) = 0\,, \qquad k = 0, 1, \dots, m_j - 1\,, \qquad j = 1, 2, \dots, r\,.$

In the sequel by $\mathfrak{L}^m$ we denote the algebra of the Fourier transforms of all functions of the form $c\delta + f$ with $f \in L^{1,m}$. In the case $c = 0$ we write $\mathring{\mathfrak{L}}^m$.

From the estimate

$$\int_{-\infty}^{\infty} |(R_\alpha^k \varphi)(t)|\, dt \leqq \frac{1}{k!} \int_{-\infty}^{\infty} |t|^k\, |\varphi(t)|\, dt$$

it follows immediately that $L^{1,m} \subset L^1_{\alpha, m}$ and thus $\mathfrak{L}^m \subset \mathfrak{L}(\alpha, m)$ for any α $(-\infty < \alpha < \infty)$. It is easily seen that the function $\mathcal{F}(\lambda) \in \mathfrak{L}^m$ has continuous derivatives

$$\mathcal{F}^{(k)}(\lambda) = \int_{-\infty}^{\infty} e^{i\lambda t} (it)^k f(t)\, dt\,, \qquad k = 1, 2, \dots, m\,.$$

Corollary 2.7. *If the function $\mathcal{A}(\lambda) \in \mathfrak{L}^m$ possesses a zero of order $m_j \leqq m$ $(j = 1, 2, \dots, r)$ in the point $\alpha_j \in (-\infty, \infty)$, then relation (2.22) holds.*

Theorem 2.10. *Let $\mathcal{F}(\lambda) = Ff \in \mathring{\mathfrak{L}}$ and m a natural number. Then $(\lambda + i)^m \cdot \mathcal{F}(\lambda) \in \mathfrak{L}$ if and only if the function $f(t)$ satisfies the following three conditions:*

(1) *$f(t), f'(t), \dots, f^{(m-2)}(t)$ are absolutely continuous on $(-\infty, \infty)$.*

(2) *$f^{(m-1)}(t)$ is absolutely continuous on $(-\infty, 0]$ and on $[0, \infty)$.*

(3) *$f^{(k)}(t) \in L$, $k = 0, 1, \dots, m$.*

Proof. First we consider the case $m = 1$ and we prove the necessity of the formulated conditions.

Let $\mathcal{G}(\lambda) = (\lambda + i)\, \mathcal{F}(\lambda) \in \mathfrak{L}$, i.e. $\mathcal{G}(\lambda) = c + (Fg)(\lambda)$ $(g \in L)$. Since $(\lambda + i)^{-1}$ is the Fourier transform of the function $\varphi(t) = -ie^{-t}$ $(t > 0)$, $\mathcal{F}(\lambda)$ is the Fourier transform of the convolution

$$f(t) = (c\delta + g) * \varphi = c\varphi(t) - i \int_{-\infty}^{t} e^{s-t} g(s)\, ds\,.$$

The last relation shows that the function $f(t)$ is absolutely continuous on $(-\infty, 0]$ and on $[0, \infty)$ and that $f(t)$ has a jump $-ic$ with $c = \lim_{\lambda\to\infty} \lambda\, \mathcal{F}(\lambda)$ at $t = 0$. Furthermore, (for $t \neq 0$) we have

$$f'(t) = -c\varphi(t) - ig(t) \in L\,.$$

For the proof of sufficiency we assume that $f(t)$ satisfies conditions (2) and (3) for $m = 1$. For $a < 0 < b$ and $-\infty < \lambda < \infty$ we then obtain

$$\int_a^b \frac{d}{dt}[e^{i\lambda t} f(t)]\, dt = f(+0) - f(-0) + e^{i\lambda b} f(b) - e^{i\lambda a} f(a) =$$

$$= i\lambda \int_a^b e^{i\lambda t} f(t)\, dt + \int_a^b e^{i\lambda t} f'(t)\, dt .$$

For $a \to -\infty$ the last two integrals possess a finite limit. Hence the finite limit $\lim_{a\to-\infty} e^{i\lambda a} f(a)$ exists which since $e^{i\lambda t} f(t) \in L$, must be equal to zero. Also we have $\lim_{b\to+\infty} e^{i\lambda b} f(b) = 0$. Passing to the limit in the last equation yields the relation

$$i\lambda \mathcal{F}(\lambda) = c - \int_{-\infty}^{\infty} e^{i\lambda t} f'(t)\, dt \in \mathfrak{L}$$

which we had to prove. Thus the assertion is proved for $m = 1$.

For $m \geqq 2$ we obtain the assertion by successive application of the preceding considerations, where we must observe that e.g. in the case $(\lambda + i)^2 \mathcal{F}(\lambda) \in \in \mathfrak{L}$ the relation $c = \lim_{\lambda\to\infty} \lambda \mathcal{F}(\lambda) = 0$ holds and thus $f(t)$ is absolutely continuous on the whole real line.

Remark. Under the hypotheses of Theorem 2.10 we have

$$\lim_{\lambda\to\infty} \lambda^m \mathcal{F}(\lambda) = i^m [f^{(m-1)}(+0) - f^{(m-1)}(-0)] .$$

Thus the continuity of the $(m - 1)$-th derivative of the function $f(t)$ at the point $t = 0$ is equivalent to the statement that $(\lambda + i)^m \mathcal{F}(\lambda) \in \mathring{\mathfrak{L}}$.

5.2.3. Equations of the First Kind

For $c = 0$ (2.7) yields the Wiener-Hopf integral equation of the first kind

$$\hat{A}\varphi = \int_0^\infty k(t - s)\, \varphi(s)\, ds = f(t) \qquad (0 \leqq t < \infty) . \tag{2.23}$$

As always we assume also here that $k(t) \in L$. Under the hypothesis that the symbol

$$\mathcal{A}(\lambda) = \int_{-\infty}^{\infty} e^{i\lambda t} k(t)\, dt \qquad (-\infty \leqq \lambda \leqq \infty) \tag{2.24}$$

has the representation (2.8) the solvability conditions for the equation (2.23) are described by Theorems 2.6 to 2.8.

In this subsection we discuss those cases in which the function $\mathcal{A}(\lambda)$ has a zero of a not necessarily integral order at the point $\lambda = \infty$.

1. First we consider the following special case of representation (2.8): $m_1 = \dots = m_r = 0$, i.e.

$$\mathcal{A}(\lambda) = (\lambda + i)^{-m_\infty}\, \mathcal{B}(\lambda) \qquad (-\infty \leqq \lambda \leqq \infty)\,. \tag{2.24'}$$

Here m_∞ is a positive integer and $\mathcal{B}(\lambda) \in \mathfrak{L}$ a function which satisfies the condition $\mathcal{B}(\lambda) \neq 0$ for all λ $(-\infty \leqq \lambda \leqq \infty)$. By Theorem 2.10 representation (2.24′) is possible if and only if the function $k(t)$ satisfies conditions (1)—(3) of that theorem with $m = m_\infty$.

Let m_∞' and m_∞'' be two arbitrary non-negative integers whose sum is equal to m_∞. We put

$$\varrho_+(\lambda) = (\lambda + i)^{-m_\infty'}\,, \qquad \varrho_-(\lambda) = (\lambda - i)^{-m_\infty''}\,, \qquad \varrho_+'(\lambda) = \varrho_-(-\lambda)\,.$$

Then $E_+^{(\infty)} = \overline{E}_+(\varrho_-)$ is the space described in Theorem 2.1. The spaces $\widetilde{E}_+(\varrho_+)$ and $\widetilde{E}_+(\varrho_+')$ consist of all generalized functions of the form

$$\left(\frac{d}{dt} + 1\right)^{m_\infty'} f(t) \quad \text{bzw.} \quad \left(\frac{d}{dt} + 1\right)^{m_\infty''} f(t) \qquad (f \in E_+)\,.$$

(The derivative $\frac{df}{dt}$ is to be understand in the sense of distribution theory, cf. 5.2.1, no. 2).

Under the hypotheses assumed above Theorems 2.6 to 2.8 imply the following properties of equation (2.23):

1°. *The operator $\hat{A}$ is a Φ-operator from $\widetilde{E}_+(\varrho_+)$ into $E_+^{(\infty)}$. The invertibility of $\hat{A}$ corresponds to the number*

$$\varkappa = \operatorname{ind} \mathcal{B}(\lambda) + m_\infty''\,,$$

and

$$\dim \ker \hat{A} = \max(-\varkappa, 0)\,, \qquad \dim \operatorname{coker} \hat{A} = \max(\varkappa, 0)\,.$$

2°. *The homogeneous equation*

$$\int_0^\infty k(t - s)\, \varphi(s)\, ds = 0 \qquad (0 \leqq t < \infty) \tag{2.25}$$

has the same solutions in all spaces $\widetilde{E}_+(\varrho_+)$.

3°. *If $\varkappa \leqq 0$, then equation* (2.23) *has a solution $\varphi \in \widetilde{E}_+(\varrho_+)$ for any right-hand side $f(t) \in E_+^{(\infty)}$.*

If $\varkappa > 0$, then the homogeneous equation transpose to (2.23)

$$\int_0^\infty k(s - t)\, \psi(s)\, ds = 0$$

has exactly $\varkappa$ linearly independent solutions $\psi_l(t)$ $(l = 0, 1, \dots, \varkappa - 1)$ which belong to all spaces $\widetilde{E}_+(\varrho_+')$. In this case the equation (2.23) *(with $f(t) \in E_+^{(\infty)}$*

possesses a (necessarily unique) solution $\varphi \in \widetilde{E}_+(\varrho_+)$ *if and only if*

$$\int_0^\infty f(t)\,\psi_l(t)\,dt = 0 \qquad (l = 0, 1, \ldots, \varkappa - 1)\,.$$

2. In the sequel we investigate equation (2.23) in the case where the symbol (2.24) has a zero of non-integral order at the point at finity.

Let μ and ν be positive numbers which are not integers. We introduce the functions

$$b(t) = \frac{e^{i\frac{\pi\mu}{2}}}{\Gamma(\mu)}\,e^t(-t)^{\mu-1} \qquad (t < 0)$$

and

$$d(t) = \frac{e^{-i\frac{\pi\nu}{2}}}{\Gamma(\nu)}\,e^{-t}t^{\nu-1} \qquad (t > 0)\,,$$

where Γ is the Gamma function. The Fourier transforms of the functions $b(t) \in L^-$ and $d(t) \in L^+$ are the functions

$$\varrho_-(\lambda) = \frac{1}{(\lambda - i)^\mu}\left(\varrho_-(0) = e^{i\frac{\pi\mu}{2}}\right), \tag{2.26$_1$}$$

$$\varrho_+(\lambda) = \frac{1}{(\lambda + i)^\nu}\left(\varrho_+(0) = e^{-i\frac{\pi\nu}{2}}\right), \tag{2.26$_2$}$$

respectively.

By $B^\mu_\infty(D^\nu_\infty)$ we denote the operator of convolution with the function $b(t)$ $(d(t))$ in the space E. Then the operators $\hat{B}^\mu_\infty = PB^\mu_\infty$ and $\hat{D}^\nu_\infty = D^\nu_\infty P$ are defined on the space E_+ by

$$(\hat{B}^\mu_\infty\varphi)\,(t) = \frac{e^{i\frac{\pi\mu}{2}}}{\Gamma(\mu)}\int_t^\infty e^{t-s}(s-t)^{\mu-1}\,\varphi(s)\,ds\,,$$

$$(t > 0,\ \varphi \in E)$$

$$(\hat{D}^\nu_\infty\varphi)\,(t) = \frac{e^{i\frac{\pi\nu}{2}}}{\Gamma(\nu)}\int_0^t e^{s-t}(t-s)^{\nu-1}\,\varphi(s)\,ds\,.$$

The kernels of the operators $\hat{B}^\mu_\infty$ and $\hat{D}^\nu_\infty$ consist only of the zero element of the space E_+. This follows immediately from the following general lemma.

Lemma 2.3. *Let the function* $k(t) \in L^-$ *satisfy the following three conditions: (1) There exists a number* $\varepsilon > 0$ *with* $e^{-\varepsilon t}k(t) \in L$. *(2) The Fourier transform* $\mathcal{K}(\lambda)$ *of the function* $k(t)$ *is different from zero for all* λ *with* $\operatorname{Im}\lambda \leqq \varepsilon$. *(3) There exist an integer* $m > 0$ *and a positive number* γ *such that the estimate* $|\lambda^m\mathcal{K}(\lambda)| \geqq \gamma$ *holds for all sufficiently great* $|\lambda|$ *with* $\operatorname{Im}\lambda \leqq \varepsilon$.

Then equation (2.25) possesses only the zero solution in any space E_+.

Proof. We remark that condition (1) guarantees the analyticity of the function $\mathcal{K}(\lambda)$ in the half-plane $\operatorname{Im}\lambda < \varepsilon$ (cf. E. C. TITCHMARSH [1]).

First let $\varphi(t) \in L_+$ be a solution of equation (2.25). Then

$$\int_0^\infty k(t-s)\,\varphi(s)\,ds = f_-(t) \qquad (-\infty < t < \infty)$$

with $f_-(t) \in L^-$. From this it follows by the convolution theorem that

$$\Phi_+(\lambda) = \mathcal{K}^{-1}(\lambda)\,\mathcal{F}_-(\lambda) \tag{2.27}$$

($\Phi_+(\lambda)$ is the Fourier transform of the function $\varphi(t)$ continued to the negative half-line by zero).

The left-hand (right-hand) side of the equation (2.27) is an analytic function in the upper (lower) half-plane and is continuous on the real line. Hence (2.27) defines an entire function which tends to zero for $\lambda \to \infty$, $\operatorname{Im}\lambda \geqq 0$ and does not increase more quickly than $|\lambda|^m$ for $\lambda \to \infty$, $\operatorname{Im}\lambda < 0$. But such a function can only be the zero function, i.e. $\Phi_+(\lambda) = 0$ and hence $\varphi = 0$.

If $\varphi(t)$ is a solution of equation (2.25) which belongs to any other space E_+, then the function $\psi(t) = e^{-\varepsilon_0 t}\,\varphi(t) \in L_+$ $(0 < \varepsilon_0 < \varepsilon)$ is a solution of the equation $\int_t^\infty l(t-s)\,\psi(s)\,ds = 0$ $(t > 0)$, where the kernel $l(t) = e^{-\varepsilon_0 t}\,k(t)$ also satisfies conditions (1)—(3). Then by the part of the proof already given we obtain $\psi = 0$ and thus $\varphi = 0$.

This proves the lemma.

Obviously, the functions $b(t)$ and $d(-t)$ $(t < 0)$ satisfy the conditions of Lemma 2.3.

In the present case the spaces $\overline{E}_+(\varrho_-) = \operatorname{im}\hat{B}^\mu_\infty$ and $\widetilde{E}_+(\varrho_+)$ $[\hat{D}^\nu_\infty(\widetilde{E}_+) = E_+]$ can be described by means of the concept of derivative of fractional order (comp. I. M. GELFAND and G. E. ŠILOV [1], Kap. 1. § 5).

Let $-\infty \leqq a \leqq +\infty$ and

$$v(t) = \frac{1}{\Gamma(\mu)} \int_a^t (t-s)^{\mu-1}\,u(s)\,ds\,,$$

where we put $(t-s)^{\mu-1} = |t-s|^{\mu-1} \cdot e^{i\pi(\mu-1)}$ for $t < s$. As is well-known, $v(t)$ is called the *μ-fold integral* of the function $u(t)$ over the interval (a, t). The function $u(t)$ is called the *derivative of order* μ of the function $v(t)$, in symbols $u(t) = \frac{d^\mu v}{dt^\mu}[a]$. Now we obtain:

The space $\overline{E}_+(\varrho_-)$ consists of all functions $f(t) \in E_+$ with the property that for $t > 0$

$$e^t \frac{d^\mu}{dt^\mu}\left(e^{-t} f(t)\right)[+\infty] \in E_+$$

holds. The space $\widetilde{E}_+(\varrho_+)$ *consists of all generalized functions of the form*

$$\varphi(t) = e^{-t}\frac{d^\nu}{dt^\nu}(e^t\psi(t))\,[0]\,, \qquad \psi(t) \in E_+\,.$$

For $0 < \mu < 1$, $0 < \nu < 1$ the spaces $\overline{E}_+(\varrho_-)$ and $\widetilde{E}_+(\varrho_+)$ can also be characterized directly. Using the known property of the Euler integrals of the first kind (see G. M. Fichtenholz [1], Vol. II, no. 529)

$$\int\limits_0^1 x^{\mu-1}(1-x)^{-\mu}\,dx = \frac{\pi}{\sin\mu\pi} \qquad (0 < \mu < 1)$$

we check easily that the equation $\hat{B}^\mu_\infty\varphi = f(\varphi \in E_+)$ is equivalent to the equation

$$\int\limits_t^\infty e^{t-s}(s-t)^{-\mu}\,f(s)\,ds = \frac{\pi e^{i\frac{\pi\mu}{2}}}{\Gamma(\mu)\sin\mu\pi}\int\limits_t^\infty e^{t-s}\varphi(s)\,ds \qquad (t > 0)\,.$$

Then by Theorem 2.1 we have: *The function* $f(t) \in E_+$ *belongs to* $\overline{E}_+(\varrho_-)$ *if and only if the function*

$$\int\limits_t^\infty e^{t-s}(s-t)^{-\mu}\,f(s)\,ds\ (\in E_+)$$

is absolutely continuous for $t \geqq 0$ *and its derivative belongs to* E_+.

Obviously, the space $\widetilde{E}_+(\varrho_+)$ consists of the generalized solutions of the so-called *Abel integral equation*

$$\int\limits_0^t (t-s)^{\nu-1}\,e^s\varphi(s)\,ds = e^t\,\psi(t) \qquad (t > 0,\ \psi(t) \in E_+)\,.$$

These solutions have the form

$$\varphi(t) = \frac{\sin\nu\pi}{\pi}\,e^{-t}\frac{d}{dt}\int\limits_0^t (t-s)^{-\nu}\,e^s\,\psi(s)\,ds$$

(see F. D. Gahov [1]).

Now we consider the integral equation (2.23) under the hypothesis that the symbol (2.24) permits a representation of the form (2.9) with a function $\mathscr{C}(\lambda) \in \mathfrak{L}$ and with the functions $\varrho_\pm(\lambda)$ defined by (2.26). From the results of Subsection 4.2.4 we obtain the following theorem.

Theorem 2.11. *For the operator* $\hat{A}$ *as an operator from* $\widetilde{E}_+(\varrho_+)$ *into* $\overline{E}_+(\varrho_-)$ *to be a* Φ_+*-operator or a* Φ_-*-operator it is necessary and sufficient that* $\mathscr{C}(\lambda) \neq 0$ $(-\infty \leqq \lambda \leqq \infty)$.

If the last condition is satisfied, then the invertibility of the operator $\hat{A}$ corresponds to the number $\varkappa = \operatorname{ind} \mathcal{C}(\lambda)$, and

$$\dim \ker \hat{A} = \max(-\varkappa, 0), \qquad \dim \operatorname{coker} \hat{A} = \max(\varkappa, 0).$$

3. In an analogous way we can treat equations of the first kind whose symbol has a certain logarithmic behaviour at the point at infinity.

We consider the kernels

$$b(t) = e^{2t} \int_0^\infty \frac{(-t)^{x-1}}{\Gamma(x)}\, dx \qquad (t < 0),$$

$$d(t) = e^{-2t} \int_0^\infty \frac{t^{x-1}}{\Gamma(x)}\, dx \qquad (t > 0),$$

whose Fourier transforms are the functions

$$\varrho_-(\lambda) = \frac{1}{\ln(2 + i\lambda)}, \qquad \varrho_+(\lambda) = \frac{1}{\ln(2 - i\lambda)} \qquad (\varrho_-(0) = \varrho_+(0) > 0), \tag{2.28}$$

respectively. It is easily checked that the functions $b(t)$ and $d(-t)$ $(t < 0)$ satisfy the conditions of Lemma 2.3 for $m = 1$ and any positive number $\varepsilon < 1$.

By B and D we denote the operators of convolution with the functions $b(t)$ and $d(t)$, respectively, and by $\hat{B}$, $\hat{D}$ the restrictions of the operators PB and DP, respectively, to the space E_+. For the analytic description of the image space $\overline{E}_+(\varrho_-) = \operatorname{im} \hat{B}$ we consider the function $k(t) = e^t \ln(-t)$ $(t < 0)$. Its Fourier transform is the function

$$\mathcal{K}(\lambda) = \frac{i}{\lambda - i}\,[-\Gamma'(1) + \ln(1 + i\lambda)].$$

It is easily seen that

$$\mathcal{K}(\lambda)\, \varrho_-(\lambda) = \frac{\mathcal{G}(\lambda)}{\lambda - i}, \tag{2.29}$$

where $\mathcal{G}(\lambda) \in \mathfrak{L}^-$, $\mathcal{G}(\lambda) \neq 0$ $(-\infty \leqq \lambda \leqq \infty)$, $\operatorname{ind} \mathcal{G}(\lambda) = 0$.

In fact, we have

$$\ln(\lambda - i) - \ln(2 + i\lambda) = \ln \frac{\lambda - i}{\lambda - 2i} - \ln i,$$

where $\ln \dfrac{\lambda - i}{\lambda - 2i}$ is the Fourier transform of the function $t^{-1}(e^t - e^{2t})$ $(t < 0)$ bolonging to L^-.

Let K and G be the convolution operators in the space E with the symbols $\mathcal{K}(\lambda)$ and $\mathcal{G}(\lambda)$, respectively. By (2.29) we have $KB = B_\infty G$ and hence

$$PKPB = PB_\infty PG . \qquad (2.30)$$

This implies:

The space $\overline{E}_+(\varrho_-)$ *consists of all functions* $f(t) \in E_+$ *for which the function*

$$\int_t^\infty e^{t-s} \ln (s - t) f(s)\, ds \qquad (\in E_+) \qquad (2.31)$$

is absolutely continuous for $t \geqq 0$ *and possesses a derivative belonging to* E_+.

In fact, by (2.30) $f \in \overline{E}_+(\varrho_-)$ implies $PKf \in \operatorname{im} PB_\infty$. Thus by Theorem 2.1 the function (2.31) has the mentioned properties. Conversely, if the function (2.31) satisfies the mentioned conditions, then $PKf \in \operatorname{im} PB_\infty$. Consequently, since the operator PG is invertible in E_+, there exists a function $\varphi \in E_+$ with $PKf = PB_\infty PG\varphi$. Using (2.30) we obtain $PK(f - PB\varphi) = 0$ and thus $f = PB\varphi \in \overline{E}_+(\varrho_-)$ since by Lemma 2.3 $\ker PK = \{0\}$.

For the description of the space $\widetilde{E}_+(\varrho_+)$ $[D(\widetilde{E}_+) = E_+]$ we introduce the function $n(t) = e^{-t} \ln t$ $(t > 0)$. Then

$$\mathcal{N}(\lambda)\, \varrho_+(\lambda) = \frac{\mathcal{H}(\lambda)}{\lambda + i},$$

where $H(\lambda) \in L^+$, $H(\lambda) \neq 0$ $(-\infty \leqq \lambda \leqq \infty)$, ind $H(\lambda) = 0$. This implies

$$NPDP = D_\infty PHP ,$$

where N and H are the convolution operators in E with the symbols $N(\lambda)$ and $H(\lambda)$, respectively.

By means of the last equation the following assertion can easily be proved.

The space $\widetilde{E}_+(\varrho_+)$ *consists of all functions of the form*

$$\hat{G}_\infty f = i[(f(t) + f'(t)]$$

with $f(t) = \int_0^t e^{-t+s} \ln (t - s)\, h(s)\, ds$ $\big(h(s) \in E_+\big)$.

5.2.4. Symbols with Zeros of Non-Integral Order

In the preceding subsection we have considered symbols possessing only one zero of non-integral order in the point $\lambda = \infty$. In this subsection we shall investigate the more general case that the symbol has an arbitrary finite number of zeros of non-integral order. We restrict ourselves to investigations in the space $E_+ = L^p_+$ $(1 \leqq p < \infty)$.

1. Let t_0 be an arbitrary point on the real line and $\mu \geqq 0$ a real number. Then we have

$$\left(\frac{\lambda - t_0}{\lambda + i}\right)^\mu \in \mathfrak{L}^+ , \qquad \left(\frac{\lambda - t_0}{\lambda - i}\right)^\mu \in \mathfrak{L}^- .$$

In fact, using the stereographic projection $z = \frac{\lambda - i}{\lambda + i}$ we obtain $\frac{\lambda - t_0}{\lambda + i} = \frac{z - z_0}{1 - z_0}$ with $z_0 = \frac{t_0 - i}{t_0 + i}$. Since the function $\left(\frac{z - z_0}{1 - z_0}\right)^\mu$ $(|z| = 1)$ belongs to the subalgebra W^+ (cf. 4.2.4), it can be expanded into a power series in a neighborhood of any point z_1 $(|z_1| = 1)$:

$$\left(\frac{z - z_0}{1 - z_0}\right)^\mu = \sum_{n=0}^{\infty} a_n (z - z_1)^n \qquad (|z| = 1) .$$

Thus we obtain

$$\left(\frac{\lambda - t_0}{\lambda + i}\right)^\mu = \sum_{n=0}^{\infty} a_n [X(\lambda) - X(\lambda_1)]^n$$

for a neighborhood of any point λ_1 $(-\infty \leqq \lambda_1 \leqq \infty)$, where $X(\lambda) = \frac{\lambda - i}{\lambda + i} \in \mathfrak{L}$ holds. Hence the function $\left(\frac{\lambda - t_0}{\lambda + i}\right)^\mu$ satisfies the hypotheses of Theorem 2, § 35 (cf. also § 34) of I. M. Gelfand, D. A. Raikov and G. E. Šilov [1]. Consequently this function belongs to the algebra $\mathfrak{L}^+$. The assertion for $\left(\frac{\lambda - t_0}{\lambda - i}\right)^\mu$ can be proved analogously.

By $B_{t_0}^\mu$ we denote the convolution operator in the space L^p $(1 \leqq p < \infty)$ with the symbol $\left(\frac{\lambda - t_0}{\lambda - i}\right)^\mu$ and by Δ_-^μ the operator of the generalized differentiation defined by[1])

$$\Delta_-^\mu v = e^{-i\left(t_0 t - \frac{\pi\mu}{2}\right)} \frac{d^\mu}{dt^\mu} \left(e^{it_0 t} v(t)\right) [+ \infty] .$$

Then

$$B_{t_0}^\mu = \Delta_-^\mu B_\infty^\mu . \tag{2.32}$$

The last formula is a generalization of the relation $B_{t_0} = i e^{-it_0 t} \frac{d}{dt} (e^{it_0 t} B_\infty)$ which we used already in 5.2.1.

For the proof of formula (2.32) we remark that Δ_-^μ is the operator of convolution with the generalized function $e^{-i\left(t_0 t + \frac{\pi\mu}{2}\right)} \frac{t_-^{-\mu-1}}{\Gamma(-\mu)}$[2]). The Fourier trans-

---

[1]) For notation see 5.2.3, no. 2.
[2]) Cf. I. M. Gelfand and G. E. Šilov [1], Kapitel 1, §§ 3 und 5.

form of this generalized function has the form

$$(\lambda - t_0)^\mu \left((-i)^\mu = e^{-i\frac{\pi\mu}{2}} \right).^{1)}$$

Thus $\Delta_-^\mu B_\infty^\mu$ is the operator of convolution in L^p with the function whose Fourier transform is $\left(\frac{\lambda - t_0}{\lambda - i}\right)^\mu$. This proves (2.32).

(2.32) implies $B_\infty^\mu = \Delta_-^{-\mu} B_{t_0}^\mu$, where according to 5.2.3, no. 2, the operator $\Delta_-^{-\mu}$ is defined by the equation

$$\Delta_-^{-\mu} u = \frac{e^{-i\left(t_0 t - \frac{\pi\mu}{2}\right)}}{\Gamma(\mu)} \int_t^\infty |t - s|^{\mu - 1} e^{it_0 s} u(s)\, ds .$$

Obviously, since $P\Delta_-^{-\mu} = P\Delta_-^{-\mu}P$, we obtain

$$PB_\infty^\mu = P\,\Delta_-^{-\mu} P B_{t_0}^\mu . \tag{2.33}$$

By means of similar considerations as in 5.2.3, no. 3, we conclude from (2.33) that $f \in \operatorname{im} PB_{t_0}^\mu$ *holds if and only if* $P\,\Delta_-^{-\mu} f$ *belongs to* $\operatorname{im} PB_\infty^\mu$. We described the image space $\operatorname{im} PB_\infty^\mu = \operatorname{im} \hat{B}_\infty^\mu$ in 5.2.3, no. 2.

Now let $\nu \geqq 0$ be an arbitrary real number and $D_{t_0}^\nu$ the convolution operator in L^p with symbol $\left(\frac{\lambda - t_0}{\lambda + i}\right)^\nu$. By Δ_+^ν we denote the generalized differentiation operator by the formula

$$\Delta_+^\nu v = e^{-i\left(t_0 t - \frac{\pi\nu}{2}\right)} \frac{d^\nu}{dt^\nu} \left(e^{it_0 t} v(t)\right) [-\infty] .$$

Then

$$D_{t_0}^\nu = \Delta_+^\nu D_\infty^\nu . \tag{2.34}$$

The proof is carried out in the same way as in the case of the formula (2.32). We have only to consider that Δ_+^ν is the operator of the convolution with the generalized function $e^{-i\left(t_0 t - \frac{\pi\nu}{2}\right)} \frac{t_+^{-\nu-1}}{\Gamma(-\nu)}$ whose Fourier transform has the form¹)

$$(\lambda - t_0)^\nu \left(i^\nu = e^{i\frac{\pi\nu}{2}} \right).$$

(2.34) implies

$$D_\infty^\nu P = \Delta_+^{-\nu} P D_{t_0}^\nu P \tag{2.35}$$

with

$$\Delta_+^{-\nu} u = \frac{e^{-i\left(t_0 t + \frac{\pi\nu}{2}\right)}}{\Gamma(\nu)} \int_{-\infty}^t (t - s)^{\nu - 1} e^{it_0 s} u(s)\, ds .$$

---

¹) Cf. I. M. Gelfand and G. E. Šilov [1], Kapitel 2, 2.3, formulae (3) and (9).

By means of (2.35) we easily can prove the following assertion:

The space $\widetilde{L}_+^p[\widehat{D}_\infty^\nu(\widetilde{L}_+^p) = L_+^p]$ *consists of all generalized functions of the form*

$$\varphi(t) = e^{-t} \frac{d^\nu}{dt^\nu} (e^t \psi(t)) \, [0] \tag{2.36}$$

with $\psi(t) = \Delta_+^{-\nu} u$ $(u \in E_+)$.

In the sequel by $\widehat{G}_\infty^\nu$ and $\widehat{G}_{t_0}^\nu$ we denote the mappings $\widehat{G}_\infty^\nu \colon \psi \to \varphi$ and $\widehat{G}_{t_0}^\nu \colon u \to \varphi$, respectively, defined by (2.36).

2. Now let $\widehat{A}$ be the Wiener-Hopf integral operator of the form (2.7) whose symbol $\mathcal{A}(\lambda)$ $(-\infty \leqq \lambda \leqq \infty)$ possesses the representation (2.9). Here $\mathcal{C}(\lambda) \in \mathfrak{L}$, and $\varrho_\pm(\lambda)$ are the functions introduced in 5.2.1. no. 2, with arbitrary non-negative real numbers m_j' and m_j''. The operators $\widehat{B}$ and $\widehat{G}$ have the same meaning as in 5.2.1, no. 2. Analogously to 5.2.1 we introduce the spaces $\overline{L}_+^p = \operatorname{im} \widehat{B}$ and $\widetilde{L}_+^p = \widehat{G}(L_+^p)$. By Theorem 2.2, Chapter 4, we have

$$\overline{L}_+^p = (\operatorname{im} \widehat{B}_\infty^{m''_\infty}) \cap (\operatorname{im} \widehat{B}_{\alpha_1}^{m''_1}) \cap \ldots \cap (\operatorname{im} \widehat{B}_{\alpha_r}^{m''_r}) \,.$$

The results from 4.2.4 show immediately, that for the operator $\widehat{A} \colon \widetilde{L}_+^p \to \overline{L}_+^p$ $(1 \leqq p < \infty)$ the Theorem 2.11 holds.

5.2.5. Examples

Example 1. We consider the Wiener-Hopf integral equation of the second kind

$$\varphi(t) - \int_0^\infty k(t - s) \, \varphi(s) \, ds = f(t) \qquad (0 \leqq t < \infty) \tag{2.37}$$

with $k(t) = e^{-t}(2 - t)$ $(t > 0)$. The symbol is

$$\mathcal{A}(\lambda) = \frac{\lambda^2}{(\lambda + i)^2} \,.$$

Consequently it has the form (2.8), where

$$m_\infty = 0 \,, \qquad r = 1 \,, \qquad \alpha_1 = 0 \,, \qquad m_1 = 2 \,, \qquad \mathcal{B}(\lambda) \equiv 1 \,.$$

We solve the equation (2.37) in the two limiting cases

$$1^\circ.\ m_1' = 2 \,, \qquad m_1'' = 0 \qquad (\varrho_+(\lambda) = \mathcal{A}(\lambda), \varrho_-(\lambda) \equiv 1)$$

and

$$2^\circ.\ m_1' = 0 \,, \qquad m_1'' = 2 \qquad \left(\varrho_+(\lambda) \equiv 1 \,, \quad \varrho_-(\lambda) = \frac{\lambda^2}{(\lambda - i)^2}\right).$$

Case 1°. Let $f(t) \in L_+^p$ $(1 \leqq p < \infty)$.

The equation (2.37) possesses a unique solution $\varphi(t) \in \widetilde{L}^p_+(\varrho_+) = \widehat{G}^2_0(L^p_+)$. This solution has the form

$$\varphi(t) = (\widehat{G^2_0}f)\,(t) = f(t) + (2+t)\int_0^t f(s)\,ds - \int_0^t sf(s)\,ds\,.$$

Case 2°. By (2.9) we now have

$$\mathscr{C}(\lambda) = \left(\frac{\lambda - i}{\lambda + i}\right)^2, \qquad \varkappa = 2\,.$$

Let $f(t) \in \overline{L}^p_+(\varrho_-)$. By Theorem 2.6 equation (2.37) possesses a solution $\varphi(t) \in L^p_+$ if and only if the two conditions

$$\int_0^\infty h(t)\,e^{-t}\,dt = 0\,, \qquad \int_0^\infty h(t)\,e^{-t}t\,dt = 0 \tag{2.38}$$

are satisfied. Here $h \in L^p_+$ is the solution of the equation $\widehat{B}^2_0 h = f$. i.e.

$$h(t) = \widehat{B}_0^{-2}f = f(t) + 2f^{(-1)}(t) + f^{(-2)}(t)\,.$$

If the conditions (2.38) are satisfied, then the (unique) solution $\varphi(t) \in L^p_+$ of equation (2.38) can be represented by the following formula (cf. Section 3.2):

$$\varphi(t) = \widehat{C}^{(-1)}h = h(t) - 4\int_0^\infty e^{t-s}(t - s + 1)\,h(s)\,ds \qquad (t > 0)\,. \tag{2.39}$$

In the sequel we suppose in particular that $f(t) \in L^{p,2}_+$ $(1 \leqq p < \infty)$. Then using relations (1.2) and (2.18) with $\widehat{B} = \widehat{B}^2_0$ and $\widehat{B}' = \widehat{D}^2_0$ we can write the solvability conditions (2.38) in the form

$$\int_0^\infty f(t)\,(\widehat{G^2_0}\chi_l)\,(t)\,dt = 0 \qquad (l = 0, 1)\,,$$

where $\chi_l(t) = e^{-t}t^l$. A simple calculation yields

$$\widehat{G^2_0}\chi_0 = 1 + t\,, \qquad \widehat{G^2_0}\chi_1 = t\,.$$

Hence for a function $f(t) \in L^{p,2}_+$ the solvability conditions (2.38) have the form

$$\int_0^\infty f(t)\,dt = 0\,, \qquad \int_0^\infty tf(t) = 0\,.$$

If the last two conditions are satisfied, then it is easily checked that the solution (2.39) of equation (2.37) can be represented in the form

$$\varphi(t) = f(t) - (2+t)\int_t^\infty f(s)\,ds + \int_t^\infty sf(s)\,ds \qquad (t > 0)\,.$$

Example 2. We consider the equation

$$\int_0^\infty k(t-s)\,\varphi(s)\,ds = f(t) \qquad (0 \leqq t < \infty) \tag{2.40}$$

with the kernel $k(t) = \frac{1}{2} e^{-|t|}$. In this case the symbol of equation (2.40) is the function

$$\mathcal{A}(\lambda) = \frac{1}{1+\lambda^2} = \frac{1}{(\lambda - i)(\lambda + i)} .$$

By 5.2.3, no. 1, we have the following solvability statements for equation (2.40).

(a) Let $f(t) \in E_+$. Then there exist exactly two linearly independent solutions of equation (2.40) in the space of all distributions of the form $\psi(t) + 2\psi'(t) + \psi''(t)$ with $\psi(t) \in E_+$.

(b) If $f(t) \in E_+$ is absolutely continuous and $f'(t) \in E_+$, then (2.40) has a unique solution in the space of all distributions of the form $\psi(t) + \psi'(t)$ with $\psi(t) \in E_+$.

(c) If the function $f(t) \in E_+$ together with its derivative $f'(t)$ is absolutely continuous and if the functions $f'(t)$ and $f''(t)$ also belong to E_+, then (2.40) has a solution $\varphi(t) \in E_+$ if and only if

$$\int_0^\infty [f''(t) - 2f'(t) + f(t)]\, e^{-t}\, dt = 0 .$$

Example 3. We consider the equation (2.40) with kernel $k(t) = \frac{1}{\pi} K_0(|t|)$, where $K_0(z)$ is the modified Hankel function of order zero

$$K_0(z) = \frac{\pi i}{2} H_0^{(1)}(iz) .$$

The Fourier transform of $k(t)$ is the function (see I. S. Gradstein and I. M. Ryshik [1], formula 3.773, 6)

$$\mathcal{A}(\lambda) = \frac{1}{\sqrt{1+\lambda^2}} .$$

We put

$$\varrho_-(\lambda) = \frac{1}{\sqrt{\lambda - i}}, \qquad \varrho_+(\lambda) = \frac{1}{\sqrt{\lambda + i}}, \qquad \mathcal{C}(\lambda) = 1 .$$

By 5.2.3, no. 2, equation (2.40) has a unique solution in the space of all distributions of the form

$$\varphi(t) = e^{-t} \frac{d^{1/2}}{dt^{1/2}} (e^t \psi(t))\,[0] \qquad (\psi(t) \in E_+)$$

if the right-hand side $f(t) \in E_+$ is such that the function

$$g(t) = \int_t^\infty e^{t-s} \frac{f(s)}{\sqrt{s-t}}\, ds$$

is absolutely continuous for $t \geqq 0$ and its derivative $g'(t)$ belongs to E_+.

5.3. Paired Discrete Wiener-Hopf Equations

In this and in the next section we apply the results from 4.3—4.4 and the preceding two sections to paired Wiener-Hopf equations.

5.3.1. Sequence Spaces Generated by the Zeros of the Symbol

1. In the sequel E (resp. E_+) is any one of the sequence spaces introduced in Section 3.1. U, P are the operators defined on E by relations (1.1), (1.2), respectively, of Chapter 3.

Moreover, in the space E we introduce the linear continuous operators P' and P'' by the equations

$$P'\{\xi_n\} = \{\eta_n\}, \qquad \eta_n = \begin{cases} \xi_{-n-1}, & n = 0, 1, \dots \\ 0, & n = -1, -2, \dots \end{cases}$$

and

$$P''\{\xi_n\} = \{\zeta_n\}, \qquad \zeta_n = \begin{cases} 0, & n = 0, 1, \dots \\ \xi_{-n-1}, & n = -1, -2, \dots \end{cases}$$

Obviously,

$$\operatorname{im} P' = \ker P' = \operatorname{im} P, \qquad \operatorname{im} P'' = \ker P'' = \operatorname{im} Q,$$
$$P'P'' = P, \qquad P''P' = Q. \tag{3.1}$$

The operator $P'(P'')$ maps $\operatorname{im} Q$ ($\operatorname{im} P$) isometrically onto $\operatorname{im} P$ ($\operatorname{im} Q$).

Let $\alpha_1, \alpha_2, \dots \alpha_r, \beta_1, \beta_2, \dots, \beta_s$ and $\gamma_1, \gamma_2, \dots \gamma_q$ be certain (not necessarily different) points of the unit circle and $m_1, m_2, \dots, m_r$, $n_1, n_2, \dots, n_s$ and $l_1, l_2, \dots, l_q$ positive integers. By B_α, D_α ($|\alpha| = 1$) we denote the operators defined on E by relations (1.1). It is easily seen that

$$B_\alpha Q = P''D_{\bar\alpha}P', \qquad QD_\alpha = P''B_{\bar\alpha}P'. \tag{3.2}$$

In the sequel we use the notation introduced in 4.3:

$$A_+ = \prod_{j=1}^{r} D_{\alpha_j}^{m_j}, \qquad A_- = \prod_{j=1}^{s} B_{\beta_j}^{n_j}, \qquad F = \prod_{j=1}^{q} D_{\gamma_j}^{l_j}, \tag{3.3}$$
$$B = PA_- + QA_+, \qquad D = A_+P + A_-Q. \tag{3.4}$$

The operators A_+, A_- and F have the symbols

$$\varrho_+(z) = \prod_{j=1}^{r} (z - \alpha_j)^{m_j}, \qquad \varrho_-(z) = \prod_{j=1}^{s} (z^{-1} - \beta_j^{-1})^{n_j},$$
$$\varrho(z) = \prod_{j=1}^{q} (z - \gamma_j)^{l_j}, \tag{3.3'}$$

respectively. By $\hat G_\alpha$ we denote the operator defined by formula (1.1') for any sequence $\{f_n\}_{-\infty}^{\infty}$ with $f_n = 0$ for $n < 0$. Now we put

$$G = \prod_{j=1}^{r} \hat G_{\alpha_j}^{m_j}, \qquad H = \prod_{k=1}^{s} \hat G_{\beta_k}^{n_k}, \qquad D^{(-1)} = PGP + P''HP'. \tag{3.5}$$

Using relations (1.2) and (3.2) we can easily see that the operator D satisfies condition (2′) of 4.11, i.e.

$$D^{(-1)}Dx = x\,, \qquad DD^{(-1)}x = x \quad \forall\, x \in E\,. \tag{3.6}$$

Hence we can introduce the space $\widetilde{E}(\varrho_+, \varrho_-) = D^{(-1)}(E)$ with the norm

$$\|f\|_{\widetilde{E}(\varrho_+, \varrho_-)} = \|Df\|_E$$

(cf. 4.3.4).

2. Using the first relation (3.4), Chapter 4, we obtain easily the following lemma from Lemma 1.1 and Lemma 1.2.

Lemma 3.1. (a) *If E is one of the spaces l^p $(1 \leqq p < \infty)$ or c^0, then* $\ker B = \{0\}$.

(b) *In the space m the kernel of the operator B has dimension $r + s$, and the sequences $\xi^{(j)} = \{\xi_n^{(j)}\}$ $(j = 1, \dots, r)$ and $\eta^{(k)} = \{\eta_n^{(k)}\}$ $(k = 1, \dots, s)$ with*

$$\xi_n^{(j)} = \begin{cases} \beta_j^{-n}, & n \geqq 0 \\ 0, & n < 0, \end{cases} \qquad \eta_n^{(k)} = \begin{cases} 0, & n \geqq 0 \\ \alpha_k^{-n}, & n > 0 \end{cases}$$

form a basis in $\ker B$.

The truth of the following Lemma is easily seen.

Lemma 3.2. (a) *If E is one of the spaces l^p $(1 \leqq p < \infty)$ or c^0, then* $\ker F = \{0\}$.

(b) *In the space m the kernel of the operator F has the dimension q, and the sequences $\{\gamma_j^{-n}\}_{n=-\infty}^{\infty}$ $(j = 1, \dots, q)$ form a basis in* $\ker F$.

The formulations of Lemma 3.1 and Lemma 3.2 for the case of the space c are left to the reader.

Since the kernels of the operators B and F are finite dimensional, we can introduce the image spaces $\overline{E}(\varrho_+, \varrho_-) = \operatorname{im} B$ and $\overline{E}(\varrho) = \operatorname{im} F$ with the norms defined in 4.3.4.

3. The following theorem shows that the analytic description of the space $\overline{E}(\varrho_+, \varrho_-)$ can be reduced to the analytic description of the image space $\operatorname{im} PA_-$. This description is given for the different spaces E by the Theorems 1.1, 1.3, and 1.4. Here A_- is the operator defined by the second relation (3.3). Furthermore, we put

$$A'_- = \prod_{j=1}^{r} B_{\alpha_j^{-1}}^{m_j}\,.$$

Theorem 3.1. *For the element $f \in E$ to belong to the space $\overline{E}(\varrho_+, \varrho_-)$ it is necessary and sufficient that*

$$(1)\ Pf \in \operatorname{im} PA_-\,, \qquad (2)\ P'f \in \operatorname{im} PA'_-\,.$$

Proof. By the second relation (3.4), Chapter 4, $f \in \overline{E}(\varrho_+, \varrho_-)$ holds if and only if the conditions (1) and (2′) $Qf \in \operatorname{im} QA_+$ are satisfied.

By Theorems 2.2 and 3.4, Chapter 4, we have

$$\operatorname{im} QA_+ = \bigcap_{j=1}^{r} \operatorname{im} QD_{\alpha_j}^{m_j}, \qquad \operatorname{im} PA'_- = \bigcap_{j=1}^{r} \operatorname{im} PB_{\alpha_j^{-1}}^{m_j}. \tag{3.7}$$

Furthermore, by (3.1) and (3.2) the equation $Qf = QD_\alpha^m \varphi$ is equivalent to the equation $P'f = PB_{\alpha^{-1}}^m P'\varphi$. This together with (3.7) shows the equivalence of the conditions (2) and (2′). Hence the theorem is proved.

Corollary 3.1. *Let E be one of the spaces l^p $(1 \leq p < \infty)$ or c^0. Then the inverse $B^{-1}\colon \overline{E}(\varrho_+, \varrho_-) \to E$ of the operator B has the form*

$$B^{-1} = P\hat{A}_-^{-1}P + P''(\hat{A}'_-)^{-1} P'. \tag{3.8}$$

Here $\hat{A}_-$ and $\hat{A}'_-$ are the restrictions of the operators PA_- and PA'_-, respectively, to the subspace im P.

We remark that the inverse $\hat{B}_\alpha^{-1}$ was constructed in 5.1.1. If we use the Corollary 1.1, then for the inverses of the operators

$$B_1 = PB_\beta + Q, \qquad B_2 = P + QD_\alpha$$

from (3.8) we obtain

$$B_1^{-1}\{f_n\} = P\left\{-\beta^{-n+1} \sum_{k=n}^{\infty} \beta^k f_k\right\} + Q\{f_n\},$$

$$B_2^{-1}\{f_n\} = Q\left\{-\alpha^{-n-1} \sum_{k=-\infty}^{n} \alpha^k f_k\right\} + P\{f_n\}. \tag{3.8′}$$

Corollary 3.2. *Let E be one of the spaces l^p or c^0 and $m = \max\{m_1, \dots, m_r, n_1, \dots, n_s\}$. Then*

$$E^{m+\varepsilon} \subset \overline{E}(\varrho_+, \varrho_-) \ (\varepsilon > 0 \ \text{ for } \ E = c^0;\ \varepsilon = 0 \ \text{ for } \ E = l^p)$$

$$\widetilde{E}(\varrho_+, \varrho_-) \subset E^{-m-\varepsilon} \ (\varepsilon > 0 \ \text{ for } \ E = l^1;\ \varepsilon = 0 \ \text{ for } \ E = l^p, 1 < p < \infty, \ \text{or} \ E = c^0)$$

holds in the sense of a continuous embedding.

This follows immediately from Theorem 3.1, Corollary 3.1 and Corollary 1.2.

4. From the considerations in points 4—6 of 5.1.1 it is easy to see that the analytic description of the space $\overline{E}(\varrho)$ is completely analogous to that of the space $\overline{E}_+(\varrho_-)$. As an example we formulate the analogue of the Theorem 1.1. Here R_γ is the operator defined by relations (1.22).

Theorem 3.2. *Let E be one of the spaces l^p $(1 \leq p < \infty)$ or c^0. For the sequence $f = \{f_n\}_{-\infty}^{\infty} \in E$ to belong to the space $\overline{E}(\varrho)$ it is necessary and sufficient that*

$$R_{\gamma_j}^k f \in E \qquad (k = 1, \dots, l_j;\ j = 1, \dots, q).$$

The norm in $\overline{E}(\varrho)$ is equivalent to the norm $\sum_{j=1}^{q} \|R_{\gamma_j}^{l_j} f\|_E$.

In the proof of Theorem 3.2 we use Lemma 2.1, Chapter 4, and the relation $D_\gamma^{-1} = R_\gamma$.

5. Using the results of 5.1.3 we can easily see from the preceding considerations that the results of points 1—3 of this subsection are also true in the case of non-integers m_j, n_j and l_j.

5.3.2. Symbols with Zeros of Arbitrary Finite Orders

1. Let $a(z)$ and $b(z)$ be functions on the unit circle wich can be expanded into an absolutely convergent Fourier series: $a(z), b(z) \in W$. We assume that the functions $a(z)$, $b(z)$ have the form[1])

$$a(z) = \varrho_+(z)\,\varrho(z)\,a_0(z)\,, \qquad b(z) = \varrho_-(z)\,\varrho(z)\,b_0(z) \tag{3.9}$$

with $a_0(z), b_0(z) \in W$. The functions $\varrho_\pm(z)$, $\varrho(z)$ are defined by (3.3′), where m_j, n_j, l_j are arbitrary positive real numbers.

By a_n, b_n $(n = 0, \pm 1, \ldots)$ we denote the Fourier coefficient of the functions $a(z)$, $b(z)$ and by $A_1, A_2 \in \Re(U)$ the operators with symbols $a(z)$, $b(z)$, respectively. Then the paired operator $A = A_1 P + A_2 Q$ is defined on the space E by the equations

$$\sum_{k=0}^{\infty} a_{j-k}\xi_k + \sum_{k=-\infty}^{-1} b_{j-k}\xi_k = \eta_j \qquad (j = 0, \pm 1, \ldots) \tag{3.10}$$

with $\{\xi_k\}_{-\infty}^{\infty}, \{\eta_k\}_{-\infty}^{\infty} \in E$. For A by 4.3.2—4.3.3 we obtain the representation $A = FCD$ with $C = A_0 P + B_0 Q$, where $A_0, B_0 \in \Re(U)$ are the operators with the symbols $a_0(z)$, $b_0(z)$. F and D are given by the formula (3.3) to (3.4) (cf. 5.1.3).

Theorem 3.3. *For the operator A as an operator from $\widetilde{E}(\varrho_+, \varrho_-)$ into $\overline{E}(\varrho)$ to be a Φ_+-operator or a Φ_--operator it is necessary and sufficient that the functions $a_0(z)$ and $b_0(z)$ in (3.9) vanish nowhere on the unit circle.*

If this condition is satisfied, then the invertibility of the operator A corresponds to the number $\varkappa - \delta$, and

$$\dim \ker A = \max(\delta - \varkappa, 0)\,, \qquad \dim \operatorname{coker} A = \max(\varkappa - \delta, 0)\,.$$

Here

$$\varkappa = \frac{1}{2\pi}\,[\arg a_0(e^{i\varphi})/b_0(e^{i\varphi})]_{\varphi=0}^{2\pi}\,, \tag{3.11}$$

$$\delta = \begin{cases} 0 & \text{for} \quad E = l^p\ (1 \leq p < \infty) \quad \text{or} \quad E = c^0\,, \\ q & \text{for} \quad E = m\,. \end{cases}$$

[1]) Cf. the remark in 4.3.3.

Proof. The first part of the proof is an immediate consequence of Theorem 3.2, Chapter 4. The proof of the second part is a modification of the proof of Theorem 1.5.

Let $a_0(z) \neq 0$, $b_0(z) \neq 0$ ($|z| = 1$). For the spaces l^p ($1 \leq p < \infty$) and c^0 the assertion is obvious since in this case by Lemma 3.2 we have dim ker $F = 0$.

We consider the case $E = m$. By Theorem 3.1, Chapter 2, the function $c(z) = a_0(z)/b_0(z)$ admits a factorization in the algebra W:

$$c(z) = c_-(z)\, z^{\varkappa} c_+(z) \qquad (|z| = 1)\,.$$

Let $\varkappa > 0$. By Theorem 3.3, Chapter 3, the homogeneous transposed equation

$$(PB_0 + QA_0)\,\xi = 0 \tag{3.12}$$

has exactly $\varkappa$ linearly independent solutions $\xi^{(m)}$ ($m = 1, 2, \ldots, \varkappa$) belonging to all spaces E. From the formula (cf. (4.18_2))

$$PB_0 + QA_0 = (PC_-^{-1} + QC_+)\,(PU^{-\varkappa} + Q)\, C_+^{-1} A_0$$

($C_\pm \in \mathfrak{R}(U)$ are the operators with symbols $c_\pm(z)$) it is easily seen that we can assume $\xi^{(m)} = \{d_{k-m+1}\}_{k=-\infty}^{\infty}$ ($m = 1, 2, \ldots, \varkappa$). Here d_n ($n = 0, \pm 1, \ldots$) are the Fourier coefficients of the function $c_+(z)/a_0(z)$ (cf. also the remark from 2.4.2). Hence the general solution of equation (3.12) has the form

$$\sum_{m=1}^{\varkappa} c_m \xi^{(m)} = \{\psi_k\}_{k=-\infty}^{\infty}\,. \tag{3.13}$$

Repeating the corresponding considerations in the proof of Theorem 1.5 and using Theorem 3.3, Chapter 3, and Lemma 3.2 (b) we obtain:

For any sequence $\eta \in \overline{m}(\varrho)$ the system (3.10) has a solution $\xi \in \widetilde{m}(\varrho_+, \varrho_-)$ if and only if the conditions

$$\sum_{k=-\infty}^{\infty} \zeta_k \psi_k = 0$$

are satisfied for every sequence of the form (3.13). Here $\zeta = \{\zeta_k\}_{-\infty}^{\infty} \in m$ is an arbitrary sequence with $F\zeta = \eta$ and the coefficients c_m are to be determined from the equation system

$$\sum_{m=1}^{\varkappa} c_m \gamma_j^{-m+1} = 0 \qquad (j = 1, 2, \ldots, q)\,.$$

This now implies

$$\dim \operatorname{coker} A = \max(\varkappa - q, 0)\,.$$

On the other hand, since by Theorem 3.2 the index satisfies the equation

$$\operatorname{Ind} A = q - \varkappa\,,$$

the theorem is proved.

Remark. For the space $E = c$ the assertion of the theorem holds with

$$\delta = \begin{cases} 1, & \text{if } \ y_{j_0} = 1 \qquad (1 \leqq j_0 \leqq q) \\ 0 & \text{otherwise} . \end{cases}$$

2. Now we considered the paired operator $A' = PA_2 + QA_1$ transpose to A which is defined by the equation system

$$\sum_{k=-\infty}^{\infty} b_{j-k}\xi_k = \eta_j \qquad (j = 0, 1, \dots),$$

$$\sum_{k=-\infty}^{\infty} a_{j-k}\xi_k = \eta_j \qquad (j = -1, -2, \dots).$$

Without loss of generality we suppose that $\varrho(z) \equiv 1$ in the representations (3.9). By 4.3.2—4.3.3 the Operator A' is representable in the form $A' = BC'$ with $C' = PB_0 + QA_0$.

Theorem 3.4. *For the operator A' as an operator from E into $\overline{E}(\varrho_+, \varrho_-)$ to be a Φ_+-operator a Φ_--operator it is necessary and sufficient that the functions $a_0(z)$ and $b_0(z)$ in* (3.9) *vanish nowhere on the unit circle.*

If this condition is satisfied, then the invertibility of the operator A' corresponds to the number $-(\varkappa + \sigma)$, and

$$\dim \ker A' = \max (\varkappa + \sigma, 0), \qquad \dim \operatorname{coker} A' = \max (-\varkappa - \sigma, 0),$$

with

$$\sigma = \begin{cases} 0 & \text{for } \ E = l^p \ (1 \leqq p < \infty) \ \text{ or } \ E = c^0, \\ r + s & \text{for } \ E = m . \end{cases}$$

The proof is analogous to the proof of the Theorem 3.3.

Remark. Theorem 3.4 is also true in the case of the space $E = c$, where we have to put $\sigma = 0$ if $\alpha_j \neq 1$ $(j = 1, \dots, r)$ and $\beta_k \neq 1$ $(k = 1, \dots, s)$ and $\sigma = 1$ in all other cases.

3. Now we apply the results from Section 4.4 to the paired operators $A_1P + A_2Q$ and $PA_1 + QA_2$.

We suppose that the symbols $a(z)$ and $b(z)$ of the operators A_1 and A_2 belong to the algebra $W(\boldsymbol{\alpha}, \boldsymbol{m}; \boldsymbol{\beta}, \boldsymbol{n})$ (cf. 2.3.4):

$$a(z), b(z) \in \bigcap_{j=1}^{r} W(\alpha_j, m_j) \cap \bigcap_{k=1}^{s} W(\beta_k, n_k).$$

For simplicity we assume that the numbers m_j and n_k are integers.

Now let m_j', m_j'' $(j = 1, \dots, r)$ and n_k', n_k'' $(k = 1, \dots, s)$ be non-negative integers such that

$$m_j' + m_j'' = m_j, \qquad n_k' + n_k'' = n_k$$

for all j and k. By $\varrho_\pm'(z)$ and $\varrho_\pm''(z)$ we denote the functions defined by formulae (3.7), Chapter 4. We introduce the spaces $\widetilde{E}(\varrho_+', \varrho_-')$ and $\overline{E}(\varrho_+'', \varrho_-'')$ according to 5.3.1.

Theorem 4.3. Chapter 4, implies

Theorem 3.5. *For the operator* $A = A_1P + A_2Q$ $(A = PA_1 + QA_2)$ *to be a* Φ_+*-operator or a* Φ_-*-operator acting from* $\widetilde{E}(\varrho'_+, \varrho'_-)$ *into the space* $\overline{E}(\varrho''_+, \varrho''_-)$ *it is necessary and sufficient that the functions*

$$a_0(z) = a(z)/\varrho_+(z) \quad \textit{and} \quad b_0(z) = b(z)/\varrho_-(z)$$

belong to the algebra W and vanish nowhere on the unit circle.

If these conditions are satisfied, then A is a Φ*-operator with the index*

$$\operatorname{Ind} A = -\lambda\,, \qquad \lambda = \varkappa + \sum_{j=1}^{r} m''_j + \sum_{k=1}^{s} n''_k\,,$$

where the number $\varkappa$ *is determined by equation* (3.11).

If the points α_j *and* β_k *are distinct, then the invertibility of the operator A corresponds to the number* λ.

5.4. Paired Wiener-Hopf Integral Equations

5.4.1. The Function Spaces Generated by Zeros of the Symbol

1. In this section E is any one of the function spaces (2.4) of Section 3.2, and P, U are the operators defined on the space E by relations (2.11), (2.12), respectively, of Chapter 3. Analogously to 5.3.1 we define linear continuous operators P' and P'' by the equations

$$(P'\varphi)(t) = \begin{cases} \varphi(-t)\,, & t > 0 \\ 0\,, & t < 0\,, \end{cases} \qquad (P''\varphi)(t) = \begin{cases} 0\,, & t > 0 \\ \varphi(-t)\,, & t < 0\,. \end{cases}$$

Obviously, again relations (3.1) hold.

By B_α, D_α, B_∞, D_∞ $(-\infty < \alpha < \infty)$ we denote the operators introduced in 5.2.1. It is easily seen that

$$\begin{aligned} B_\alpha Q &= P''D_{-\alpha}P'\,, & QD_\alpha &= P''B_{-\alpha}P'\,, \\ B_\infty Q &= -\,P''D_\infty P'\,, & QD_\infty &= -\,P''B_\infty P'\,. \end{aligned} \tag{4.1}$$

Now let $\alpha_1, \alpha_2, \dots, \alpha_r$, $\beta_1, \beta_2, \dots, \beta_s$ and $\gamma_1, \gamma_2, \dots, \gamma_q$ be (not necessarily different) points on the real line and $m_\infty, m_1, \dots, m_r$, $n_\infty, n_1, \dots, n_s$ $l_\infty, l_1, \dots, l_q$ positive integers. For the sequel we introduce the following notation

$$A_+ = D_\infty^{m_\infty} \prod_{j=1}^{r} D_{\alpha j}^{m_j}\,, \qquad A_- = B_\infty^{n_\infty} \prod_{j=1}^{s} B_{\beta_j}^{n_j}\,, \qquad F = D_\infty^{l_\infty} \prod_{j=1}^{q} D_{\gamma_j}^{l_j}\,,$$

$$G = \hat{G}_\infty^{m_\infty} \prod_{j=1}^{r} \hat{G}_{\alpha j}^{m_j}\,, \qquad H = (-\hat{G}_\infty)^{n_\infty} \prod_{j=1}^{s} \hat{G}_{-\beta_j}^{n_j}\,.$$

Here $\hat{G}_\alpha$ and $\hat{G}_\infty$ are the operators defined in 5.2.1 for any locally integrable function $f(t)$ on $(-\infty, \infty)$ with $f(t) = 0$ $(t < 0)$.

The operators A_+, A_- and $F(\in \mathfrak{R}(U))$ possess the symbols

$$\varrho_+(\lambda) = (\lambda + i)^{-m_\infty} \prod_{j=1}^{r} \left(\frac{\lambda - \alpha_j}{\lambda + i}\right)^{m_j},$$

$$\varrho_-(\lambda) = (\lambda - i)^{-n_\infty} \prod_{j=1}^{s} \left(\frac{\lambda - \beta_j}{\lambda - i}\right)^{n_j},$$

$$\varrho(\lambda) = (\lambda + i)^{-l_\infty} \prod_{j=1}^{q} \left(\frac{\lambda - \gamma_j}{\lambda + i}\right)^{l_j},$$

respectively. Furthermore, by B, D and $D^{(-1)}$ we denote the operators defined by (3.4) and (3.5).

In the same way as in 5.3.1 we can show that in the present case the equations (3.6) also hold. As in the preceding section we introduce the spaces $\widetilde{E}(\varrho_+, \varrho_-) = D^{(-1)}(E)$, $\overline{E}(\varrho_+, \varrho_-) = \operatorname{im} B$ and $\overline{E}(\varrho) = \operatorname{im} F$.

2. Analogously to Lemma 3.1 and the Lemma 3.2 we can easily prove the following assertions.

Lemma 4.1. (a) *If E is one of the spaces L^p $(1 \leqq p < \infty)$ or C°, then* $\ker B = \{0\}$.

(b) *In the space M the kernel of the operator B has dimension $r + s$. It is spanned by the functions*

$$e^{-i\beta_j t}\ (t > 0;\ j = 1, \ldots, r)\,, \qquad e^{-i\alpha_k t}\ (t < 0;\ k = 1, \ldots, s)\,.$$

Lemma 4.2. (a) *If E is one of the spaces L^p $(1 \leqq p < \infty)$ or C°, then* $\ker F = \{0\}$.

(b) *In the space M the kernel of the operator F has dimension q. It is spanned by the functions $e^{-i\gamma_j t}$ $(j = 1, 2, \ldots, q)$.*

3. Putting

$$A'_- = (-1)^{m_\infty} B_\infty^{m_\infty} \prod_{j=1}^{r} B_{-\alpha_j}^{m_j}$$

and repeating the proofs from 5.3.1 we now can see that Theorem 3.1 and Corollary 3.1 maintain their validity in the present case. Thus the analytic description of the space $\overline{E}(\varrho_+, \varrho_-)$ is reduced to the description of the image space $\operatorname{im} PA_-$ already given in Theorems 2.1, 2.2, 2.4 and 2.5.

In the sequel we again introduce the operator

$$B^{(\infty)} = PB_\infty^{n_\infty} + QD_\infty^{m_\infty}\,.$$

Obviously, $B^{(\infty)}$ is a special case of the more general operator B considered above. From Theorems 3.1 and 2.3 we obtain:

Corollary 4.1. *Let $m = \max\{m_1, \ldots, m_r, n_1, \ldots, n_s\}$ and $E = L^p$ $(1 \leqq p < \infty)$. Then*

$$(\operatorname{im} B^{(\infty)}) \cap L^{p,m} \subset \overline{E}(\varrho_+, \varrho_-)$$

holds in the sense of a continuous embedding. For $m_\infty = n_\infty = 0$ *the relation*

$$\widetilde{E}(\varrho_+, \varrho_-) \subset L^{p, -m'}$$

holds with $m' = m$ *for* $p > 1$ *and* $m' > m$ *for* $p = 1$.

We remark that by Corollaries 2.2 and 2.4 for the inverses of the operators $B^{(\infty)}$ and

$$B = PB_\beta^n + QD_\alpha^m$$

formula (3.8) implies that

$$[B^{(\infty)}]^{-1} f(t) = \begin{cases} i^{n_\infty}\left(\dfrac{d}{dt} - 1\right)^{n_\infty} f(t)\,, & t > 0 \\ i^{m_\infty}\left(\dfrac{d}{dt} + 1\right)^{m_\infty} f(t)\,, & t < 0 \end{cases}$$

and

$$(B^{-1}f)(t) = \begin{cases} [I + (\beta - i)\, R_\beta]^n f\,, & t > 0 \\ [I + (\alpha + i)\, R_\alpha]^m f\,, & t < 0\,. \end{cases}$$

Here R_α is the operator defined by equations (2.19).

4. The analytic description of the space $\overline{E}(\varrho)$ is again completely analogous to the description of the space $\overline{E}_+(\varrho_-)$ (cf. 5.2.1). As an example we formulate the corresponding analogue to Theorems 2.1 and 2.2.

Theorem 4.1. *Let* E *be one of the spaces* L^p $(1 \leqq p < \infty)$ *or* C°. *For the function* $f(t) \in E$ *to belong to the space* $\overline{E}(\varrho)$ *it is necessary and sufficient that*

(1) $f^{(l)}(t)$ $(l = 0, 1, \ldots, l_\infty - 1)$ *is absolutely continuous and* $f^{(l)}(t) \in E$ $(l = 0, 1, \ldots, l_\infty)$.

(2) $R_{\gamma_j}^k f \in E$ $(k = 1, \ldots, l_j;\ j = 1, \ldots, q)$.

The norm in $\overline{E}(\varrho)$ is equivalent to the norm

$$\sum_{l=0}^{l_\infty} \|f^{(l)}\|_E + \sum_{j=1}^{q} \sum_{k=0}^{l_j} \|R_{\gamma_j}^k f\|_E\,.$$

For the proof of Theorem 4.1 one uses Lemma 2.1, Chapter 4, and the relation (cf. also (2.21))

$$D_\gamma^{-1} = I + (i + \gamma)\, R_\gamma\,.$$

In the case of the spaces M or C the second condition of Theorem 4.1 is to be modified in accordance with Theorem 2.4.

5.4.2. Symbols with a Finite Number of Zeros

1. In the space E we consider the integral equation with two kernels of the form

$$c\varphi(t) - \int_0^\infty k_1(t - s)\, \varphi(s)\, ds - \int_{-\infty}^0 k_2(t - s)\, \varphi(s)\, ds = f(t) \quad (-\infty < t < \infty)\,, \tag{4.2}$$

where $k_j(t) \in L$ $(j = 1, 2)$, $c = c_1$ $(t > 0)$, $c = c_2$ $(t < 0)$. The operator defined by the left-hand side of equation (4.2) has the form $A = A_1P + A_2Q$, where $A_1, A_2 \in \mathfrak{R}(U)$ are the operators with the symbols

$$\mathcal{A}(\lambda) = c_1 - \int_{-\infty}^{\infty} e^{i\lambda t}\, k_1(t)\, dt\,, \qquad \mathcal{B}(\lambda) = c_2 - \int_{-\infty}^{\infty} e^{i\lambda t}\, k_2(t)\, dt\,, \tag{4.3}$$

respectively.

We assume that the functions $\mathcal{A}(\lambda)$, $\mathcal{B}(\lambda)$ admit the representation[1])

$$\mathcal{A}(\lambda) = \varrho_+(\lambda)\, \varrho(\lambda)\, \mathcal{A}_0(\lambda)\,, \qquad \mathcal{B}(\lambda) = \varrho_-(\lambda)\, \varrho(\lambda)\, \mathcal{B}_0(\lambda) \tag{4.4}$$

with the functions $\varrho_\pm(\lambda)$, $\varrho(\lambda)$ defined in 5.4.1 and $\mathcal{A}_0(\lambda), \mathcal{B}_0(\lambda) \in \mathfrak{L}$.

Under the hypotheses assumed above the following theorem holds.

Theorem 4.2. *For A as an operator from $\widetilde{E}(\varrho_+, \varrho_-)$ into $\overline{E}(\varrho)$ to be a Φ_+-operator or a Φ_--operator it is necessary and sufficient that the functions $\mathcal{A}_0(\lambda)$ and $\mathcal{B}_0(\lambda)$ are different from zero for all real values λ $(-\infty \leqq \lambda \leqq \infty)$.*

If this condition is satisfied, then the invertibility of the operator A corresponds to the number $\varkappa - \delta$ and

$$\dim \ker A = \max(\delta - \varkappa, 0)\,, \qquad \dim \operatorname{coker} A = \max(\varkappa - \delta, 0)\,.$$

Here

$$\varkappa = \frac{1}{2\pi}\left[\arg \mathcal{A}_0(\lambda)/\mathcal{B}_0(\lambda)\right]_{\lambda=-\infty}^{+\infty}, \tag{4.5}$$

$$\delta = \begin{cases} 0 & \textit{for} \quad E = L^p \ (1 \leqq p < \infty) \quad \textit{or} \quad E = C^0\,, \\ q & \textit{for} \quad E = M\,. \end{cases}$$

The proof is completely analoguos to the proofs of the Theorems 3.3 and 2.6.

2. We consider the operator $A' = PA_2 + QA_1$ transpose to A which is given by the equations

$$A'\varphi = \begin{cases} c_2\varphi(t) - \displaystyle\int_{-\infty}^{\infty} k_2(t-s)\, \varphi(s)\, ds\,, & 0 < t < \infty\,, \\ c_1\varphi(t) - \displaystyle\int_{-\infty}^{\infty} k_1(t-s)\, \varphi(s)\, ds\,, & -\infty < t < 0\,. \end{cases}$$

By means of the same considerations as in 5.3.2 we then obtain a theorem completely analogous to Theorem 3.4.

Remark. For the space $E = L^p$ $(1 \leqq p < \infty)$ by 4.3.7 the hitherto obtained results of this subsection can be carried over also to the case that the zeros of the functions $\mathcal{A}(\lambda)$ and $\mathcal{B}(\lambda)$ have not necessarily integral orders m_j $(j = 1, \dots, r)$ and n_k $(k = 1, \dots, s)$. Using the results from 5.2.3 and 5.2.4 and the corresponding Theorem 3.1 we can give the analytic description of the space $\widetilde{L^p}(\varrho_+, \varrho_-)$ and $\overline{L^p}(\varrho_+, \varrho_-)$ analogously to the integral case.

---

1) Without loss of generality we can assume $\varrho(\lambda) \equiv 1$ and thus $\overline{E}(\varrho) = E$.

3. Analogously to no. 3 of Subsection 5.3.2 we now apply the results of Sections 4.3.5 resp. 4.4 to the operator A defined by the left-hand side of equation (4.2). We restrict ourselves to the case in which E is one of the spaces L^p $(1 \leqq p < \infty)$ or C^0. Furthermore, we put $\varrho(\lambda) \equiv 1$ in (4.4).

Let m_j', m_j'' $(j = 1, \dots, r, \infty)$ and n_k', n_k'' $(k = 1, \dots, s, \infty)$ be non-negative integers such that

$$m_j' + m_j'' = m_j\,, \qquad n_k' + n_k'' = n_k$$

for all relevant values of j and k. Furthermore, we put

$$\varrho_+'(\lambda) = \frac{\prod_{j=1}^{r}\left(\frac{\lambda - \alpha_j}{\lambda + i}\right)^{m_j'}}{(\lambda + i)^{m_\infty'}}\,, \qquad \varrho_-'(\lambda) = \frac{\prod_{k=1}^{s}\left(\frac{\lambda - \beta_k}{\lambda - i}\right)^{n_k'}}{(\lambda - i)^{n_\infty'}}\,,$$

$$\varrho_+''(\lambda) = \frac{\prod_{k=1}^{s}\left(\frac{\lambda - \beta_k}{\lambda + i}\right)^{n_k''}}{(\lambda + i)^{n_\infty''}}\,, \qquad \varrho_-''(\lambda) = \frac{\prod_{j=1}^{r}\left(\frac{\lambda - \alpha_j}{\lambda - i}\right)^{m_j''}}{(\lambda - i)^{m_\infty''}}\,.$$

Lemma 4.3. *Let the function $k(t)$ $(-\infty < t < \infty)$ have the following properties*[1]):

a) $k(t) \in \bigcap_{j=1}^{r} L^1_{\alpha_j, m_j''} \cap \bigcap_{k=1}^{s} L^1_{\beta_k, n_k''}$.

b) *$k(t)$ satisfies conditions* (1)—(3) *of Theorem* 2.10 *for* $m = \max\,\{m_\infty'', n_\infty''\}$.

Let K be the operator of convolution with the function $k(t)$. Then the commutator $T = [P, K]$ is a operator from E into the space $\overline{E}(\varrho_+'', \varrho_-'')$.

Proof. Lemma 4.1, Chapter 4, and Theorem 2.9 imply that under the condition a) T is a compact operator from E into the space im B_2 with

$$B_2 = PB_{\alpha_j}^{m_j''} + QD_{\beta_k}^{n_k''} \qquad (j = 1, \dots, r;\; k = 1, \dots, s)\,.$$

Now we show that T is also compact from E into im $B_2^{(\infty)}$, where $B_2^{(\infty)} = PB_\infty^{m_\infty''} + QD_\infty^{n_\infty''}$. By Theorem 3.4, Chapter 4, this yields the assertion.

By the hypothesis b) and Theorem 2.10 the Fourier transform $\mathcal{K}(\lambda)$ of the function $k(t)$ can be represented as follows:

$$\mathcal{K}(\lambda) = \frac{\mathcal{K}_1(\lambda)}{(\lambda - i)^{m_\infty''}} = \frac{\mathcal{K}_2(\lambda)}{(\lambda + i)^{n_\infty''}} \tag{4.6}$$

with $\mathcal{K}_l(\lambda) \in \mathfrak{L}$ $(l = 1, 2)$. Let K_l be the convolution operator in E with the symbol $\mathcal{K}_l(\lambda)$. Then to the representations (4.6) there correspond the following representations of the operator K:

$$K = B_\infty^{m_\infty''} K_1 \quad \text{und} \quad K = D_\infty^{n_\infty''} K_2\,.$$

[1]) For notation see 5.2.2, no. 3.

Thus we have

$$T = PKQ - QKP = PB_{\infty}^{m''_{\infty}}PK_1Q - QD_{\infty}^{n''_{\infty}}QK_2P =$$
$$= B_2^{(\infty)}(PK_1Q - OQK_2P)\,.$$

The operator enclosed in parantheses is compact on the space E by Lemma 2.1, Chapter 2. Thus T is a compact operator from E into im $B_2^{(\infty)}$.

This proves the lemma.

In the sequel we suppose that the kernels $k_{10}(t)$ and $k_{20}(t)$ which correspond to the functions $\mathcal{A}_0(\lambda)$, $\mathcal{B}_0(\lambda)$ from (4.4) by relations (4.3) satisfy condition b) of Lemma 4.3. Furthermore, let

$$k_{10}(t) \in \bigcap_{k=1}^{s} L^1_{\beta_k, n''_k}\,, \qquad k_{20}(t) \in \bigcap_{j=1}^{r} L^1_{\alpha_j, m''_j}\,.$$

Using Lemma 4.3 and Theorem 2.9 we obtain the following result from Theorem 3.5, Chapter 4.

Theorem 4.3. *For the operator A defined by the left-hand side of* (4.2) *to be a Φ_+-operator or a Φ_--operator from $\widetilde{E}(\varrho'_+, \varrho'_-)$ into $\overline{E}(\varrho''_+, \varrho''_-)$ it is necessary and sufficient that the functions $\mathcal{A}_0(\lambda)$ and $\mathcal{B}_0(\lambda)$ are different from zero for all real values λ $(-\infty \leqq \lambda \leqq \infty)$.*

If the last condition is satisfied, then A is a Φ-operator, and for its index the formula

$$-\operatorname{Ind} A = \varkappa + m''_{\infty} + n''_{\infty} + \sum_{j=1}^{r} m''_j + \sum_{k=1}^{s} n''_k$$

holds, where the number $\varkappa$ is determined by (4.5).

Remark 1. Theorem 4.3 also holds for the operator $A = PA_1 + QA_2$. Moreover, an analogue to Theorem 3.5 can be formulated.

Remark 2. Putting $\varrho(\lambda) \equiv 1$ in (4.4) and proceeding similarly as in 5.2.2 no. 2, we can obtain a corresponding analogon to the Theorem 2.8.

Remark 3. The results from 4.3.6 can be applied to the operator A.

5.5. Non-Bounded Regularization of Paired Wiener-Hopf Equations

We consider the integral equation (4.2) in the space L^p $(1 \leqq p < \infty)$, where $f(t) \in L^p$ is a given function and $\varphi(t) \in L^p$ is the function sought. With respect to the kernels of these equations we state the following hypotheses:

a) The functions (4.3) have the form

$$\mathcal{A}(\lambda) = \varrho_-(\lambda)\, \mathcal{A}_0(\lambda)\,, \qquad \mathcal{B}(\lambda) = \varrho_+(\lambda)\, \mathcal{B}_0(\lambda) \tag{5.1}$$

with the functions $\varrho_\pm(\lambda)$ introduced in 5.4.1, $\mathcal{A}_0(\lambda)$, $\mathcal{B}_0(\lambda) \in \mathfrak{L}$ and $\mathcal{A}_0(\lambda) \neq 0$, $\mathcal{B}_0(\lambda) \neq 0$ $(-\infty \leqq \lambda \leqq \infty)$.

b) The kernels $k_1(t)$ and $k_2(t)$ satisfy conditions (1)—(3) of Theorem 2.10 for $m = \max\{m_\infty, n_\infty\}$.

c) $k_j(t) \in \bigcap_{j=1}^{r} L^1_{\alpha j, mj} \cap \bigcap_{k=1}^{s} L^1_{\beta k, nk} \quad (j = 1, 2)$.

In the sequel by A we denote the operator defined on the space L^p by the left-hand side of the equation (4.2) and by A_0, B_0 the convolution operators in L^p with the symbols $\mathcal{A}_0(\lambda)$, $\mathcal{B}_0(\lambda)$, respectively. Furthermore, we use the notations introduced in 5.4.

By (5.1) we have

$$A_1 = A_- A_0 , \qquad A_2 = A B_0 .$$

This implies

$$A = B(PA_0 + QB_0) + T_0$$

with

$$B = PA_- + QA_+ , \qquad T_0 = [A_1, P] + [A_2, Q]$$

By hypotheses b), c) and Lemma 4.3 T_0 is a compact operator from L^p into the image space $\overline{L^p}\,(\varrho_+, \varrho_-) = \operatorname{im} B$. The inverse B^{-1} of the operator B is given by formula (3.8).

We remark that by Theorem 3.1 and Corollary 2.5 e.g. every function $\varphi(t)$ satisfying the following conditions belongs to the domain $D(B^{-1}) = \operatorname{im} B$:

(1) $\varphi^{(k)}(t)$ is absolutely continuous on the half-lines $[0, +\infty)$ and $(-\infty, 0]$ for $k = 0, 1, \ldots, m - 1$, and the derivatives $\varphi^{(k)}(t)$, $k = 0, 1, \ldots, m$ belong to $L^p(0, \infty)$ resp. $L^p(-\infty, 0)$. Here $m = \max\{m_\infty, n_\infty\}$.

(2) $\varphi(t) \in L^{p,l}$ for $l = \max\{m_1, \ldots, m_r, n_1, \ldots, n_s\}$.

Since the set of these functions is dense in the space L^p, the operator B^{-1} is defined in L^p on a dense set.

For the solvability of equation (4.2) it is necessary that $f(t) \in D(B^{-1})$. We assume that this condition is satisfied.

From the preceding considerations we can now immediately conclude that the integral equation of non-normal type (4.2) is equivalent to the integral equation of normal type

$$(PA_0 + QB_0)\,\varphi + T_1\varphi = B^{-1}f .$$

Here $T_1 = B^{-1}T_0$ is a compact operator in L^p.

If T is an arbitrary compact operator from L^p into the space $\overline{L^p}\,(\varrho_+, \varrho_-)$, then we obtain analogously that the equation

$$A\varphi + T\varphi = f \tag{5.2}$$

is equivalent to the equation of normal type

$$(PA_0 + QB_0)\,\varphi + T_2\varphi = B^{-1}f$$

with the operator $T_2 = T_1 + B^{-1}T$ compact in L^p.

Now by R we denote the bounded regularizer of the Φ-operator $PA_0 + QB_0$ constructed in 3.3.2. Then the operator RB^{-1} defined on $D(B^{-1}) = \operatorname{im} B$

is a non-bounded left regularizer of the operator $A + T$, and each solution of equation (5.2) is also a solution of the Fredholm integral equation

$$R(PA_0 + QB_0)\,\varphi + RT_2\varphi = RB^{-1}f\,. \tag{5.3}$$

But the equivalence of equations (5.2) and (5.3) (for any $f(t)$) holds in general only if for the number $\varkappa$ defined by (4.5) we have $\varkappa \leqq 0$.

The above results together with the results of Section 1.5 immediately show:

Corollary 5.1. *Under the above assumptions the operator A (resp. $A + T$) considered in the space L^p has a finite index and is not normally solvable*[1]).

We remark that the condition of the compactness of T: $\overline{L^p} \to L^p(\varrho_+, \varrho_-)$ is satisfied in the case $p > 1$ e.g. for all integral operators of the form

$$(T\varphi)\,(t) = \int_{-\infty}^{\infty} T(t, s)\,\varphi(s)\,ds\,, \tag{5.4}$$

where the kernel $T(t, s)$ satisfies the following conditions:

α) $T(t, s)$ is m-times differentiable by t for arbitrary s.

β) $(T\varphi)^{(k)}\,(t) = \int_{-\infty}^{\infty} T_t^{(k)}(t, s)\,\varphi(s)\,ds \qquad (\forall\, \varphi \in L^p) \qquad (k = 0, 1, \dots, m)$;

γ) $\displaystyle\int_{-\infty}^{\infty}\int_{-\infty}^{\infty} |(t + i)^l\, T(t, s)|^{p'}\,dt\,ds < \infty \quad \left(\frac{1}{p} + \frac{1}{p'} = 1\right)$.

Here m and l have the same meaning as in the above-mentioned conditions (1) and (2).

We can easily prove this statement using the fact that an integral operator of the form (5.4) is compact on L^p if the condition γ) is satisfied for $l = 0$.

Finally we remark that the arguments of this section can also be applied to paired discrete Wiener-Hopf equations.

5.6. Comments and References

A Wiener-Hopf integral equation of non-normal type was probably considered for the first time by V. A. Fok [1]. He stated sufficient solvability conditions for the right-hand side of such an equation in the case of a symmetric kernel $k(t) = 0(e^{c|t|})$ $(c > 0)$. The solution of paired integral equations of non-normal type (first and second kind) in certain classes of integrable functions was first studied by F. D. Gahov and V. I. Smagina [1] using the Riemann boundary value problem for the real line.

5.1. The most important results of this section are due to G. N. Čebotarev. He formulated Theorems 1.1 (for $p = 1$), 1.3 and 1.5 (without proof) in [3], [4]. Parts of Theorem 1.5 were simultaneously and independently proved by V. B.

---

[1]) Cf. the remark to Theorem 2.1, Chapter 3.

Dybin and N. K. Karapetjanc [2]. The proof of Theorem 1.2 given here was communicated to the author by A. Pomp. It is the discrete analogue of the proof of the Theorem 2.3 (cf. Section 5.2). The embedding $l_+^{p,m+\varepsilon} \subset l_+^{p(m)}$ can be proved for $\varepsilon > 0$, $1 < p < \infty$ simply by using the Hölder inequality (cf. the German edition of this book).

The results of 5.1.2, no. 3 were proved by Š. I. Galiev [3].

The results of Subsection 5.1.3 are partly due to N. K. Karapetjanc [2].

5.2. The results of Subsections 5.2.1 and 5.2.2 with the exception of Theorems 2.3, 2.8 and 2.9 and without consideration of the spaces $\widetilde{E}_+$ are also due to G. N. Čebotarev [1—4]. For equations of the second kind and the spaces L_+^p $(1 \leqq p < \infty)$, C_+^0 Theorem 2.6 was simultaneously stated by V. B. Dybin [3], [4] (cf. also [5]). Theorems 2.3 and 2.9 were actually proved by the author in the paper [11] (with the methods used here). Similar considerations as in no. 2 of 5.2.2 can be found by M. I. Haikin [1] and V. B. Dybin [4].

In Section 5.2.3 we essentially follow the paper of M. I. Haikin [1]. The example 3 from 5.2.5 is also taken from the same paper. The results of Subsection 5.2.4 are new.

Here we refer also to the paper of A. Voigtländer [1], where by the methods of Section 6.5 and by using the Fourier transform some general cases of degenerations of Wiener-Hopf integral equations of the second kind are investigated in the space L^p.

Wiener-Hopf integral equations of the first kind with certain symbols possessing a zero of infinite order at the point $\lambda = \infty$ were recently considered by Š. I. Galiev [1], [3] (cf. also Chapter 10).

Using probability theory the homogeneous Wiener-Hopf integral equation of the second kind was investigated by F. Spitzer [1] under the hypothesis that the kernel of the equation is a probability density, i.e. $k(t) \geqq 0$ and $\int_{-\infty}^{\infty} k(t)\, dt = 1$ (the symbol of such an equation vanishes at the point $\lambda = 0$). The problem of finding the asymptotes of the solution of such an equation depending on the real zeros of the symbol was investigated by H. Widom [1]. Here we could not treat these and similar questions since they are beyond the scope of this book.

5.3. Theorem 3.4 was stated by N. K. Karapetjanc [2] (in the case of zeros of arbitrary finite order).

5.4. The continuous analogue of Theorem 3.4 (see 5.4.2, no. 2) was found for the spaces L^p $(1 \leqq p < \infty)$ and C_0 by V. B. Dybin [3]. This Present author also stated the formula for the inverse B^{-1} (cf. 5.4.1, no. 3).

5.5 Here some earlier results of the author (Thesis, Leningrad 1966) are generalized.

CHAPTER 6

SINGULAR INTEGRAL EQUATIONS OF NON-NORMAL TYPE

Using the methods considered in Chapter 4 and Section 1.5 we shall now investigate singular integral equations with continuous coefficients on finite closed curve systems. The corresponding symbol shall have finitely many zeros of integral or fractional orders. The structure of this chapter is similar to the preceding chapter. First we state some lemmas. Then we describe the spaces in which the singular operators are considered. The most important part of this chapter is formed by Theorems 3.1 (3.2), 3.4 and 3.5, in which we formulate the properties of the singular integral equations in several cases of degeneration of the symbol. At the end the question of the existence and construction of a regularizer and equivalent regularizer of the singular integral operator of non-normal type in the space L^p is considered.

6.1. Lemmas

6.1.1. Notation

1. Let Γ be a smooth oriented plane Jordan curve and s the arc length measured from a fixed point on Γ. The equation of Γ has the form $t(s) = x(s) + iy(s)$, $0 \leqq s \leqq \gamma$. Here the real valued functions $x(s)$, $y(s)$ are continuously differentiable in the interval in question, and their derivatives $x'(s)$, $y'(s)$ cannot vanish simultaneously.

We say that Γ belongs to the class C^m (m a natural number) if the function $t(s)$ has continuous derivatives up to the order m. If in addition the derivative $t^{(m)}(s)$ satisfies a Hölder condition with the exponent $\lambda (0 < \lambda \leqq 1)$, then we write $\Gamma \in C^{m,\lambda}$. Curves of the class $C^{1,\lambda}$ are called *Ljapunov curves*. If $\Gamma \in C^m$ for any m, then Γ is called a curve of class C^∞.

A curve system Γ consisting of finitely many closed or open smooth curves without common points is simply called a *smooth curve system*[1]. We say that it belongs to the class C^m (resp. $C^{m,\lambda}$) if all curves of Γ belong to this class.

2. Let Γ be a smooth curve system. By $C^m(\Gamma)$ we denote the collection of all (complex valued) functions m-times continuously differentiable on Γ

[1]) In the sequel we shall consider only such curve systems.

and by $C^{m,\lambda}(\Gamma)$ $(0 < \lambda \leqq 1)$ the set of the functions $f(t) \in C^m(\Gamma)$ whose m-th derivative $f^{(m)}(t)$ belongs to $H^\lambda(\Gamma)$. The notation $C^{m,\lambda}(\Gamma_1 \times \Gamma_2)$ for functions of two variables defined on $\Gamma_1 \times \Gamma_2$ has the corresponding meaning. $C^m(\Gamma)$ and $C^{m,\lambda}(\Gamma)$ are Banach spaces with the norms

$$||f||_{C^m(\Gamma)} = \sum_{k=0}^{m} \max_{t\in\Gamma} |f^{(k)}(t)|$$

and

$$||f||_{C^{m,\lambda}(\Gamma)} = ||f||_{C^m(\Gamma)} + \sup_{t,\tau\in\Gamma} \frac{|f^{(m)}(t) - f^{(m)}(\tau)|}{|t-\tau|^\lambda}, \tag{1.1}$$

respectively. For $\lambda = 0$ we put $D^{m,0}(\Gamma) = C^m(\Gamma)$ and $H^0(\Gamma) = C(\Gamma)$.

If $\Gamma \in C^{m,\lambda}$, then $C^{m,\lambda}(\Gamma)$ is equal to the set of all functions $f(t)$ defined on Γ for which the function $f(s) = f(t(s))$ of the real variable s has continuous derivatives up to order m in the interval $[0, \gamma]$ and $f^{(m)}(s)$ satisfies a Hölder condition with the exponent λ. Moreover, the norm (1.1) is equivalent to the norm

$$||f||_{m,\lambda} = \sum_{k=0}^{m} \max_{0\leqq s\leqq\gamma} |f^{(k)}(s)| + \sup_{0\leqq s,\,\sigma\leqq\gamma} \frac{|f^{(m)}(s) - f^{(m)}(\sigma)|}{|s-\sigma|^\lambda}.$$

3. As usual, by $W_p^{(m)}(\Gamma)$ $(1 < p < \infty)$ we denote the Sobolev space of all functions $f(t) \in L^p(\Gamma)$ having generalized derivatives $f^{(k)}(t) \in L^p(\Gamma)$, $k = 1, 2, \ldots, m$ (see S. L. Sobolev [1] or V. I. Smirnov [1]). The norm of a function $f \in W_p^{(m)}(\Gamma)$ is defined by the formula

$$||f||_{W_p^{(m)}(\Gamma)} = \sum_{k=0}^{m} ||f^{(k)}||_{L^p(\Gamma)}. \tag{1.2}$$

$W_p^{(m)}(\Gamma)$ is the completion in the norm (1.2) of the set $C^\infty(\Gamma)$ of all functions infinitely differentiable on Γ.

4. Let Γ be a smooth curve system. $t_1, \ldots, t_r$ distinct points on Γ and $\Gamma_{t_j} \subset \Gamma$ $(j = 1, 2, \ldots, r)$ an arc containing t_j as an interior point.

By $C^\lambda(\Gamma; (\Gamma_{t_j}, m_j)_1^r)$ $(0 \leqq \lambda \leqq 1)$ we denote the collection of all functions $f(t) \in H^\lambda(\Gamma)$ whose restrictions to Γ_{t_j} belong to the space $C^{m_j,\lambda}(\Gamma_{t_j})$. Correspondingly, $W_p(\Gamma; (\Gamma_{t_j}, m_j)_1^r)$ denotes the collection of all functions $f(t) \in L^p(\Gamma)$ whose restrictions to Γ_{t_j} belong to $W_p^{(m_j)}(\Gamma_{t_j})$. In the norms

$$||f||_{C^\lambda} = ||f||_{H^\lambda(\Gamma)} + \sum_{j=1}^{\lambda} ||f||_{C^{m_j,\lambda}(\Gamma_{t_j})}$$

and

$$||f||_{W_p} = ||f||_{L^p(\Gamma)} + \sum_{j=1}^{r} ||f||_{W_p^{(m_j)}(\Gamma_{t_j})}$$

$C^\lambda(\Gamma; (\Gamma_{t_j}, m_j)_1^r)$ and $W_p(\Gamma; (\Gamma_{t_j}, m_j)_1^r)$, respectively, are Banach spaces. When the concrete form of the arcs Γ_{t_j} is immaterial we shall replace Γ_{t_j} simply by t_j in the notation of these spaces.

5. If for a function $f(t) \in C(\Gamma)$ and certain arcs Γ_{t_j} $(j = 1, 2, \dots, r)$ the restrictions of $f(t)$ to Γ_{t_j} belong to $C^{m_j,\lambda}(\Gamma_{t_j})$, then we will write $f(t) \in C(\Gamma; (t_j, m_j + \lambda)_1^r)$.

6.1.2. Some Inequalities

Let $f(t) \in C^0(\Gamma; (\Gamma_{t_j}, m_j - 1)_1^r)$. By Hf we denote the Hermite interpolation polynomial of the function $f(t)$ with the interpolation knots t_j satisfying the conditions

$$(Hf)^{(k)}(t_j) = f^{(k)}(t_j) \qquad (k = 0, \dots, m_j - 1;\ j = 1, \dots, r)\,. \tag{1.3}$$

Hf is the uniquely determined polynomial in t of degree $\sum_{j=1}^{r} m_j - 1$ satisfying conditions (1.3). The Hermite interpolation polynomial has the form (see I. S. Beresin and N. P. Shidkov [1])

$$(Hf\,(t) = \sum_{j=1}^{r} \sum_{k=0}^{m_j-1} \frac{1}{k!} f^{(k)}(t_j)\,(t - t_j)^k\, g_{jk}(t)$$

with

$$g_{jk}(t) = h_j(t) \sum_{l=0}^{m_j-k-1} \frac{1}{l!} \left[\frac{1}{h_j(t)}\right]^{(l)}_{t=\alpha_j} (t - t_j)^l\,,$$

$$h_j(t) = (t - t_j)^{-m_j} \prod_{i=1}^{r} (t - t_i)^{m_i}\,.$$

Obviously, for $r = 1$ we have $g_{jk}(t) \equiv 1$.

Now we put

$$F(t) = \frac{f(t) - (Hf)\,(t)}{\prod_{j=1}^{r} (t - t_j)^{m_j}} \qquad (t \in \Gamma)$$

with $F(t_j) = \lim_{t \to t_j} F(t)$, and we prove some properties of the function $F(t)$ which will be important in the sequel.

Lemma 1.1. *Let $f(t) \in C^\lambda(\Gamma; (\Gamma_{t_j}, m_j)_1^r)$ $(0 \leq \lambda \leq 1)$ and $\Gamma \in C^{m,\lambda}$ with $m = \max_j m_j$. Then $F(t) \in H^\lambda(\Gamma)$ and the mapping $f \to F$ of $C(\Gamma; (\Gamma_{t_j}, m_j)_1^r)$ into $H^\lambda(\Gamma)$ is continuous.*

Proof. By successive partial integration it is easily seen that any function $g(x) \in C^m[0, 1]$ satisfies the relation

$$g(1) - \sum_{k=0}^{m-1} \frac{1}{k!} g^{(k)}(0) = \frac{1}{(m-1)!} \int_0^1 (1 - x)^{m-1} g^{(m)}(x)\, dx\,. \tag{1.4}$$

We choose an arbitrary but fixed index j, $1 \leqq j \leqq r$. Let $t = t(s)$ $(0 \leqq s \leqq \leqq \gamma_j)$ be the equation of the subcurve of the system Γ which contains the point t_j. Furthermore, let $t_j = t(s_j)$ and $f(s) = f(t(s))$. Putting $m = m_j$ and

$$g(x) = f[s_j + x(s - s_j)]$$

in the formula (1.4) we obtain

$$\frac{f(s) - \sum_{k=0}^{m_j-1} \frac{1}{k!} f^{(k)}(s_j)\,(s - s_j)^k}{(s - s_j)^{m_j}} =$$

$$= \frac{1}{(m_j - 1)!} \int_0^1 (1 - x)^{m_j - 1} f^{(m_j)}[s_j + x(s - s_j)]\, dx\,.$$

By the remarks made in 6.1.1, no. 2, this shows immediately that the function

$$F_j(t) = \frac{f(t) - \sum_{k=0}^{m_j-1} \frac{1}{k!} f^{(k)}(t_j)\,(t - t_j)^k}{(t - t_j)^{m_j}}$$

satisfies the relations $F_j(t) \in H^\lambda(\Gamma)$ and

$$\|F_j\|_{H^\lambda(\Gamma)} \leqq c\, \|f\|_{C^\lambda} \qquad (c = \text{const})\,. \tag{1.5}$$

On the other hand, for any j the representations

$$F(t) = \frac{1}{h_j(t)}\,[F_j(t) - H_1(t) - H_2(t)] \tag{1.6}$$

hold with

$$H_1(t) = \sum_{l \neq j} \sum_{k=0}^{m_l-1} \frac{1}{k!} f^{(k)}(t_l)\,(t - t_l)^k\, \frac{g_{lk}(t)}{(t - t_j)^{m_j}}\,,$$

$$H_2(t) = \sum_{k=0}^{m_j-1} \frac{1}{k!} f^{(k)}(t_j)\,(t - t_j)^k\, \frac{g_{jk}(t) - 1}{(t - t_j)^{m_j}}\,.$$

It is easily seen that the functions $(t - t_j)^{-m_j}\, g_{lk}(t)$ $(l \neq j)$ and $(t - t_j)^{-m} \times \times [g_{jk}(t) - 1]$ are infinitely differentiable. Hence the assertion of the lemma now follows from (1.5) and (1.6).

Lemma 1.2. *Let $f(t) \in W_p(\Gamma; (\Gamma_{t_j}, m_j)_1^r)$ and $\Gamma \in C^m$ with $m = \max m_j$. Then $F(t) \in L^p(\Gamma)$, and the mapping $f \to F$ of $W_p(\Gamma; (\Gamma_{t_j}, m_j)_1^r)$ into $L^p(\Gamma)$ is continuous.*

First we remark that by the Sobolev embedding theorems (comp. S. L. Sobolev [1]) we have

$$W_p(\Gamma; (\Gamma_{t_j}, m_j)_1^r) \subset C^0(\Gamma; (\Gamma_{t_j}, m_j - 1)_1^r) \tag{1.7}$$

in the sense of a continuous embedding. Thus under the hypotheses of the Lemma 1.2 the existence of the polynomial Hf is secured.

For the proof of the Lemma 1.2 we need some inequalities which we shall state in the following two lemmas.

Lemma 1.3. *Let (a, b) be a finite interval, $f(x) \in L^p(a, b)$ $(1 < p < \infty)$ and*

$$F(x) = \int_a^x |f(y)|\, dy\,, \qquad G(x) = \int_x^b |f(y)|\, dy \qquad (a < x < b)\,.$$

Then

$$\left\|\frac{F(x)}{x-a}\right\|_{L^p(a,b)} \leqq \frac{p}{p-1} \|f\|_{L^p(a,b)}\,, \tag{1.8}$$

$$\left\|\frac{G(x)}{b-x}\right\|_{L^p(a,b)} \leqq \frac{p}{p-1} \|f\|_{L^p(a,b)}\,. \tag{1.9}$$

Proof. First we prove the inequality (1.8) for $a = 0, b = 1$. To this end we introduce the auxiliary function

$$\varphi(x, y) = |f(xy)| \qquad (0 \leqq x, y \leqq 1)\,,$$

and we calculate the following integrals:

$$\int_0^1 \varphi(x, y)\, dx = \frac{F(y)}{y}\,, \qquad \left\{\int_0^1 [\varphi(x, y)]^p\, dy\right\}^{1/p} \leqq \left\{\frac{1}{x}\int_0^1 |f(t)|^p\, dt\right\}^{1/p}.$$

By the well-known inequality[1])

$$\left[\int_0^1 \left\{\int_0^1 \varphi(x, y)\, dx\right\}^p dy\right]^{1/p} \leqq \int_0^1 \left\{\int_0^1 [\varphi(x, y)]^p\, dy\right\}^{1/p} dx$$

we then obtain the estimate

$$\left\|\frac{F(y)}{y}\right\|_{L^p(0,1)} \leqq \frac{p}{p-1} \|f\|_{L^p(0,1)}\,.$$

By the substitution $x = a + (b - a)\, y$ this easily yields the inequality (1.8).

The inequality (1.9) can be proved analogously by using the function $\varphi(x, y) = |f[1 + (1 - y)\,(x - 1)]|$.

Lemma 1.4. *Let $f(x) \in W_p^{(n)}(a, b)$, $a \leqq x_0 \leqq b$ and $f^{(k)}(x_0) = 0$ for $k = 0, 1, \ldots, m - 1$ $(m \leqq n)$.*

Then $g(x) = (x - x_0)^{-m} f(x)$ is a function from $W_p^{(n-m)}(a, b)$ and

$$\|g\|_{W_p^{(n-m)}(a,b)} \leqq \left(\frac{p}{p-1}\right)^m \|f\|_{W_p^{(n)}(a,b)}\,.$$

[1]) See G. H. Hardy, J. E. Littlewood and G. Polya [1], Theorem 202.

Proof. By partial integration it is easily seen that for any $k \leqq n - 1$ the relation

$$g^{(k)}(x) = \frac{1}{x - x_0} \int_{x_0}^{x} \left(\frac{y - x_0}{x - x_0}\right)^k f^{(k+1)}(y)\, dy$$

holds. Thus we obtain by inequalities (1.8) and (1.9) that

$$\|g^{(k)}\|_{L^p(a,b)} \leqq \left\{\int_a^b \left|\frac{1}{x - x_0} \int_{x_0}^{x} |f^{(k+1)}(y)|\, dy\right|^p dx\right\}^{1/p} \leqq \frac{p}{p-1} \|f^{(k+1)}\|_{L^p(a,b)} .$$

From this the assertion follows in the case $m = 1$.

For $m > 1$ we obtain the assertion of the lemma easily by induction. Here we have to use the fact that the function $h(x) = (x - x_0)^{1-m} f(x)$ satisfies the relations $g(x) = (x - x_0)^{-1} h(x)$ and

$$h(x_0) = \lim_{x \to x_0} (x - x_0)^{1-m} f(x) = \frac{f^{(m-1)}(x_0)}{(m-1)!} = 0 .$$

Now we come to the proof of the Lemma 1.2.

For the function $F_j(t)$ introduced at the proof of Lemma 1.1 by Lemma 1.4 we have $F_j(t) \in L^p(\Gamma)$ and

$$\|F_j\|_{L^p(\Gamma)} \leqq c \|f\|_{W_p} .$$

Using the representation (1.6) and the estimates

$$|f^{(k)}(t_j)| \leqq c \|f\|_{W_p} \qquad (k = 0, 1, \dots, m_j - 1;\ j = 1, \dots, r)$$

which hold by (1.7) from this we obtain the assertion of Lemma 1.2.

Remark. The statements of Lemmas 1.1 and 1.2 can be generalized as follows: Let $f(t) \in C^\lambda(\Gamma; (\Gamma_{t_j}, n_j)_1^r)$ $(0 \leqq \lambda \leqq 1)$, $\Gamma \in C^{n,\lambda}$ with $n = \max_j n_j$ and $n_j \geqq m_j$ $(j = 1, 2, \dots, r)$. Then $F(t) \in C^\lambda(\Gamma; (\Gamma_{t_j}, n_j - m_j)_1^r)$ and the mapping $f \to F$ of $C^\lambda(\Gamma; (\Gamma_{t_j}, n_j)_1^r)$ into $C^\lambda(\Gamma_{t_j}, n_j - m_j)_1^r)$ is continuous. Analogously the corresponding generalization of the Lemma 1.2 can be formulated.

This follows easily from the preceding proofs.

6.1.3. Further Properties of the Singular Integral

For the sequel we need some properties of the singular integral operator S, introduced in Section 3.4, on the spaces $W_p(\Gamma; (\Gamma_{t_j}, m_j)_1^r)$ and $C^\lambda(\Gamma; (\Gamma_{t_j}, m_j)_1^r)$.

Lemma 1.5. *Let Γ be an open Ljapunov curve with the initial point a and end point b. If $\varphi(t) \in W_p^{(1)}(\Gamma)$ $(1 < p < \infty)$, then*

$$\frac{d}{dt} \int_\Gamma \frac{\varphi(\tau)}{\tau - t}\, d\tau = \frac{\varphi(a)}{a - t} - \frac{\varphi(b)}{b - t} + \int_\Gamma \frac{\varphi'(\tau)}{\tau - t}\, d\tau \qquad (1.10)$$

for all interior points t of the curve Γ.

Proof. First we suppose that $\varphi(t) \in C^{1,\lambda}(\Gamma)$ $(0 < \lambda < 1)$. We make a cut in the complex plane which joins the point $t \in \Gamma$ $(t \neq a, t \neq b)$ with the point at infinity and which lies on the right-hand side of Γ. In the complex plane cut up in this way $\ln(z - t)$ is a unique function of z.

Let $\varepsilon > 0$ be an arbitrary sufficiently small number. Now on Γ we consider the two arcs t_1t and tt_2 lying on different sides of t and each of them having the arc length ε. By Γ_ε we denote both arcs taken together. Furthermore, let $\psi_\varepsilon(t)$ be the function

$$\psi_\varepsilon(t) = \int_{\Gamma - \Gamma_\varepsilon} \varphi'(\tau) \ln(\tau - t)\, d\tau .$$

By partial integration we obtain

$$\psi_\varepsilon(t) = -\int_{\Gamma-\Gamma_\varepsilon} \frac{\varphi(\tau)}{\tau - t}\, d\tau + \varphi(b)\ln(b - t) - \varphi(a)\ln(a - t) +$$
$$+ \varphi(t_1)\ln(t_1 - t) - \varphi(t_2)\ln(t_2 - t) .$$

For $\varepsilon \to 0$ the integral on the right-hand side tends to the singular integral

$$-\int_\Gamma \frac{\varphi(\tau)}{\tau - t}\, d\tau .$$

The two following terms do not depend on ε. The last two summands can be transformed as follows:

$$\varphi(t_1)\ln(t_1 - t) - \varphi(t_2)\ln(t_2 - t) = \varphi(t)\,[\ln(t_1 - t) - \ln(t_2 - t)] +$$
$$+ [\varphi(t_1) - \varphi(t)]\ln(t_1 - t) - [\varphi(t_2) - \varphi(t)]\ln(t_2 - t) .$$

The first term of the right-hand side of the last equation tends to $\pi i\varphi(t)$ for $\varepsilon \to 0$. The last two terms converge to zero since $\lim_{x\to 0} x^\lambda \ln x = 0$. Thus we obtain

$$\lim_{\varepsilon\to 0} \psi_\varepsilon(t) = \pi i\varphi(t) + \varphi(b)\ln(b - t) - \varphi(a)\ln(a - t) - \int_\Gamma \frac{\varphi(\tau)}{\tau - t}\, d\tau . \tag{1.11}$$

Now we differentiate the function $\psi_\varepsilon(t)$. Using the differentiation rule for parametric integrals with variable limits of integration we find

$$\psi_\varepsilon'(t) = -\int_{\Gamma-\Gamma_\varepsilon} \frac{\varphi'(\tau)}{\tau - t}\, d\tau + [\varphi'(t_1)\ln(t_1 - t)t'(s - \varepsilon) -$$
$$- \varphi'(t_2)\ln(t_2 - t)\, t'(s + \varepsilon)]\, \overline{t'(s)} .$$

By the already given part of the proof the right-hand side converges to the function

$$-\int_\Gamma \frac{\varphi'(\tau)}{\tau - t}\, d\tau + \pi i\varphi'(t) ,$$

for $\varepsilon \to 0$. This convergence is uniform with respect to $t \in \Gamma_0$, where $\Gamma_0 \subset \Gamma$ is an arbitrary closed arc which does not contain the points a and b (cf. N. I. Mushelišvili [1], p. 40). By a well-known theorem from analysis we then obtain

$$\left[\lim_{\varepsilon\to 0} \psi_\varepsilon(t)\right]' = \lim_{\varepsilon\to 0} \psi_\varepsilon'(t) = \pi i \varphi'(t) - \int_\Gamma \frac{\varphi'(\tau)}{\tau - t} d\tau .$$

From this and (1.11) it follows that formula (1.10) holds for the case that $\varphi(t) \in C^{1,\lambda}(\Gamma)$.

Now let $\varphi(t) \in W_p^{(1)}(\Gamma)$. Then there exists a sequence of functions $\varphi_n(t) \in C^{1,\lambda}(\Gamma)$ with

$$\|\varphi - \varphi_n\|_{W_p^{(1)}(\Gamma)} \to 0 .$$

We put

$$\psi(t) = \int_\Gamma \frac{\varphi(\tau)}{\tau - t} d\tau , \qquad \psi_n(t) = \int_\Gamma \frac{\varphi_n(\tau)}{\tau - t} d\tau .$$

By $\chi(t)$ we denote the right-hand side of (1.10).

Since for the functions $\varphi_n(t)$ the formula (1.10) is already proved and since S is a continuous operator on the space $L^p(\Gamma)$ (cf. 3.4.4), we obtain

$$\|\psi - \psi_n\|_{L^p(\Gamma)} \to 0 , \qquad \|\chi - \psi_n'\|_{L^p(\Gamma_0)} \to 0 , \qquad n \to \infty .$$

Since the generalized differentiation operator is closed (see V. I. Smirnov [1]), this implies $\psi(t) \in W_p^{(1)}(\Gamma_0)$ and $\psi'(t) = \chi(t)$, q.e.d.

The following corollary follows easily from the preceding proof.

Corollary 1.1. *Let $\varphi(t) \in W_p^{(1)}(\Gamma)$ and t an interior point of the curve Γ. Then the following rule of partial integration holds:*

$$\int_\Gamma \varphi'(\tau) \ln (\tau - t)\, d\tau = \pi i \varphi(t) + \varphi(b) \ln (b - t) -$$

$$- \varphi(a) \ln (a - t) - \int_\Gamma \frac{\varphi(\tau)}{\tau - t} d\tau .$$

In the sequel by Γ we denote a curve system consisting of finitely many closed Ljapunov curves without common points. Furthermore, let $t_1, t_2, \dots, t_r$ be pairwise different points on Γ and Γ_{t_j}, Γ_{t_j}' $(j = 1, 2, \dots, r)$ arcs on Γ which contain t_j as an interior point and which are such that the closed curve $\overline{\Gamma_{t_j}'}$ (i.e. the curve Γ_{t_j}' including the end points) consists only of interior points of Γ_{t_j}.

From Lemma 1.5 we immediately obtain

Corollary 1.2. *The singular operator S given by*

$$(S\varphi)(t) = \frac{1}{\pi i} \int_\Gamma \frac{\varphi(\tau)}{\tau - t} d\tau \qquad (t \in \Gamma)$$

maps the space $W_p(\Gamma; (\Gamma_{t_j}, m_j)_1^r)$ *continuously into the space* $W_p(\Gamma; (\Gamma'_{t_j}, m_j)_1^r)$ $(1 < p < \infty)$.

The corresponding statement is valid for $C^\lambda(\Gamma; (\Gamma_{t_j}, m_j)_1^r)$ $(0 < \lambda < 1)$.

Corollary 1.3. *The singular operator* S *is a homeomorphism of the space* $W_p^{(m)}(\Gamma)$ $(1 < p < \infty)$ *onto itself. For any* $\varphi \in W_p^{(m)}(\Gamma)$ *we have*

$$\frac{d^k}{dt^k} \int_\Gamma \frac{\varphi(\tau)}{\tau - t}\, d\tau = \int_\Gamma \frac{\varphi^{(k)}(\tau)}{\tau - t}\, d\tau \qquad (k = 1, 2, \dots, m)\,.$$

The corresponding statement is valid for the space $C^{m,\lambda}(\Gamma)$ $(0 < \lambda < 1)$.

For the proof of Corollary 1.3 it is sufficient to remark that in the case of a closed curve $\Gamma (a = b)$ the first two terms of the right-hand side of (1.10) vanish and moreover $S^2 = I$ holds.

Lemma 1.6. *Let* $a(t) \in C(\Gamma; (t_j, m_j + \lambda)_1^r)$ $(0 < \lambda \leqq 1)$ *and* $\Gamma \in C^{m,\lambda}$ *with* $m = \max_j m_j$. *Then the commutator* $T_a = Sa - aS$ *maps the space* $L^p(\Gamma)$ $(1 < p < \infty)$ *compactly into the space* $W_p(\Gamma; (\Gamma_{t_j}, m_j)_1^r)$.

Proof. Let $\varphi \in L^p(\Gamma)$. First we show that the function $\psi = T_a\varphi$ belongs to the space $W_p(\Gamma; (\Gamma_{t_j}, m_j)_1^r)$. Since $\psi \in L^p(\Gamma)$, it is sufficient to consider the restriction of the function ψ to Γ_{t_j}.

We put

$$\psi(t) = \int_\Gamma T(t, \tau)\, \varphi(\tau)\, d\tau\,, \qquad T(t, \tau) = \frac{1}{\pi i} \frac{a(\tau) - a(t)}{\tau - t}$$

and consider only points $t \in \Gamma_{t_j}$ for fixed j, $1 \leqq j \leqq r$. Then the partial derivatives of the function $T(t, \tau)$ with respect to t (resp. with respect to $\tau \in \Gamma_{t_j}$) exist up to the $(m_j - 1)$-th order. These derivatives belong to the class $H^\lambda(\Gamma_{t_j} \times \Gamma)$ (comp. Lemma 1.1). The m_j-th derivative is representable in the form

$$T_t^{(m_j)}(t, \tau) = \frac{\partial^{m_j} T(t, \tau)}{\partial t^{m_j}} = \frac{T_0(t, \tau)}{|t - \tau|^\nu}\,, \tag{1.12}$$

where $1 - \lambda < \nu < 1$ and $T_0(t, \tau) \in H^{\lambda+\nu-1}(\Gamma_{t_j} \times \Gamma)$. The number ν can be chosen arbitrarily in the relevant interval (see N. I. Mushelišvili [1], § 7). Furthermore, we have

$$\psi^{(k)}(t) = \int_\Gamma T_t^{(k)}(t, \tau)\, \varphi(\tau)\, d\tau \qquad (t \in \Gamma_{t_j};\ k = 1, \dots, m_j)\,. \tag{1.13}$$

In fact, let $k = 1, \dots, m_j - 1$. Since the derivatives $T_t^{(k)}(t, \tau)$ are continuous, the formula (1.13) is a well-known property of parametric integrals (see e.g. G. M. Fichtenholz [1], Vol. II). We now prove the formula (1.13) for $k = m_j$. If $\varphi(t) \in C(\Gamma)$, then we obtain the assertion easily by differentiating the integral

$$\int_{\Gamma - \Gamma_\varepsilon} T_t^{(m_j-1)}(t, \tau)\, \varphi(\tau)\, d\tau$$

analogously to the proof of Lemma 1.5 and then passing to the limit for $\varepsilon \to 0$. Now let $\varphi(t) \in L^p(\Gamma)$. We choose a sequence of continuous functions $\varphi_n(t)$ with $||\varphi - \varphi_n||_{L^p(\Gamma)} \to 0$. For $t \in \Gamma_{t_j}$ we put

$$\Phi_n(t) = \int_\Gamma T_t^{(m_j-1)}(t, \tau)\, \varphi_n(\tau)\, d\tau\,,$$

$$\chi(t) = \int_\Gamma T_t^{(m_j)}(t, \tau)\, \varphi(\tau)\, d\tau\,.$$

From the part of the proof given just before we obtain

$$\Phi_n'(t) = \int_\Gamma T_t^{(m_j)}(t, \tau)\, \varphi_n(\tau)\, d\tau\,.$$

Let ν be a fixed number arbitrarily chosen in the interval $1 - \lambda < \nu < 1$. Then with $q = p > 1$, $\sigma = 1$ and $1 < r < \dfrac{1}{\nu}$ by (1.12) all conditions of Theorem 3 (2.X), from the book L. V. Kantorovič and G. P. Akilov [1] are satisfied. Hence the integral operators defined by the right-hand sides of the equations (1.13) are compact from $L^p(\Gamma)$ into $L^p(\Gamma_{t_j})$. Thus

$$||\psi^{(m_j-1)} - \Phi_n||_{L^p(\Gamma_{t_j})} \leqq C_1\, ||\varphi - \varphi_n||_{L^p(\Gamma)} \to 0\,,$$

$$||\chi - \Phi_n'||_{L^p(\Gamma_{t_j})} \leqq C_2\, ||\varphi - \varphi_n||_{L^p(\Gamma)} \to 0\,.$$

This implies $\psi^{(m_j-1)}(t) \in W_p^{(1)}(\Gamma_{t_j})$ and $\psi^{(m_j)}(t) = \chi(t)$ since the generalized differentiation operator is closed. Hence formula (1.13) is also true for $k = m_j$.

Using the compactness of the rigth-hand sides of (1.13) we obtain the assertion of the lemma immediately from these formulae.

Remark. Lemma 1.6 can also be proved for the case $\lambda = 0$ (comp. A. P. Calderon [1]).

Lemma 1.7. *Let* $a(t) \in C^\lambda(\Gamma; (\Gamma_{t_j}, m_j)_1^r)$ $(0 < \lambda \leqq 1)$, $\Gamma \in C^{m,\lambda}$ *with* $m = \max m_j$ *and* $0 < \mu < \lambda$. *Then the commutator* T_a *maps the space* $H^\nu(\Gamma)$ $(0 < \nu \leqq 1)$ *compactly into the space* $C^\mu(\Gamma; (\Gamma_{t_j}, m_j)_1^r)$.

This assertion follows also from formulae (1.13). We have only to observe that by Corollary 4.3, Chapter 3, the right-hand sides of (1.13) define compact operators from $H^\nu(\Gamma)$ into $H^\mu(\Gamma_{t_j})$.

6.1.4. Mean Derivatives

1. Let $[a, b]$ be a finite interval and $f(x)$ a function measurable and Lebesgue integrable on this interval: $f(x) \in L(a, b)$. We put $f(x) = 0$ for all $x \notin [a, b]$.

For any $h > 0$ by $f_h(x)$ we denote the *Steklov mean function*

$$f_h(x) = \frac{1}{2h} \int_{x-h}^{x+h} f(t)\,dt \quad \text{for} \quad a < x < b\,,$$

$$f_h(a) = \frac{1}{h} \int_{a}^{a+h} f(t)\,dt\,, \qquad f_h(b) = \frac{1}{h} \int_{b-h}^{a} f(t)\,dt\,.$$

Let $x_0 \in [a, b]$ be an arbitrary point. If the limit $\alpha = \lim_{h\to 0} f_h(x_0)$ exists, then α is called the *mean limit* of the function f in the point x_0. We write

$$\alpha = \widetilde{\lim_{x\to x_0}} f(x)\,.$$

Lemma 1.8. *If $f(x)$ possesses an (ordinary) limit at the point x_0, then there the mean limit in this point also exist, and both coincide.*

Proof. First, let $\alpha = \lim_{x\to x_0} f(x)$ be finite and $a < x_0 < b$. Then for any $\varepsilon > 0$ there exists a number $\delta = \delta(\varepsilon) > 0$ such that

$$|t - x_0| < \delta\,, \qquad a \leqq t \leqq b\,, \qquad t \neq x_0$$

implies $|f(t) - \alpha| < \varepsilon$. Hence for $h < \delta$ we obtain the inequality

$$|f_h(x_0) - \alpha| \leqq \frac{1}{2h} \int_{x_0-h}^{x_0+h} |f(t) - \alpha|\,dt < \varepsilon\,,$$

i.e. $\widetilde{\lim_{x\to x_0}} f(x) = \alpha$.

For $x_0 = a$, $x_0 = b$ or $\alpha = \infty$ the proof is analogous.

As is well-known, the function $f(x) = \sin\frac{1}{x}$ $(-1 \leqq x \leqq 1)$ has no limit at the point $x = 0$. But $\widetilde{\lim_{x\to 0}} f(x) = 0$ holds.

We now introduce the following notations:

If $f^{\{0\}}(x_0) = \widetilde{\lim_{x\to x_0}} f(x)$ is finite, then we put

$$f^{\{1\}}(x_0) = \widetilde{\lim_{x\to x_0}} \frac{f(x) - f^{\{0\}}(x_0)}{x - x_0}\,.$$

If $f^{\{n-1\}}(x_0)$ is already defined and finite, then we put

$$f^{\{n\}}(x_0) = n!\,\widetilde{\lim_{x\to x_0}} \frac{f(x) - \sum_{k=0}^{n-1} \frac{1}{k!} f^{\{k\}}(x_0)\,(x - x_0)^k}{(x - x_0)^n}\,.$$

If the limit $f^{\{n\}}(x_0)$ exists, then $f^{\{n\}}(x_0)$ is called the *mean derivative of the order* n of the function f in the point x_0.

By Lemma 1.8 we have: If the n-th Taylor derivative (comp. 2.3.4) of the function f exists in the point x_0, then the mean derivative $f^{\{n\}}(x_0)$ also exists, and both derivatives coincide. In particular $f^{\{n\}}(x_0) = f^{(n)}(x_0)$ holds if the ordinary derivative $f^{(n)}(x_0)$ exists.

The mean derivative can also exist if the ordinary (or the Taylor) derivative does not exist. E.g. the functions

$$f(x) = x \sin \frac{1}{x} \, (f(0) = 0) \, ; \qquad f(x) = |x| \, , \qquad -1 \leqq x \leqq 1$$

have the mean derivative $f^{\{1\}}(0) = 0$ in the point $x_0 = 0$.

Lemma 1.9. *Let* $g(x) \in L^p(a, b)$ $(1 \leqq p < \infty)$, *m a non-negative integer and* $x_0 \in [a, b]$. *Then for* $\mu \geqq \frac{1}{p}$ *the function* $f(x) = (x - x_0)^{m+\mu} \cdot g(x)$ *has the mean derivatives* $f^{\{k\}}(x_0) = 0$ $(k = 0, 1, \dots, m)$ *in the point* x_0.

Proof. Using the Hölder inequality we easily obtain the estimate

$$|f_h(x_0)| \leqq \frac{1}{\sqrt[p]{2}} h^{m+\mu-\frac{1}{p}} \left[\int_{x_0-h}^{x_0+h} |g(t)|^p \, dt \right]^{1/p} .$$

Since the integral is absolutely continuous, the assertion follows from this estimate for $k = 0$. For $k = 1, 2, \dots, m$ the assertion is proved by induction.

Remark. The definition of the mean derivative can easily be carried over to functions of several variables. There we have to replace the Steklov mean functions by Sobolev mean functions.

2. A further property of the mean derivatives consists in the following. We consider the operator A defined by the equation

$$(A\varphi)(x) = \frac{\varphi(x) - \sum_{j=0}^{k-1} \frac{1}{j!} \varphi^{\{j\}}(x_0) (x - x_0)^j}{(x - x_0)^k}$$

in the space $L^p(a, b)$. Here k is a fixed natural number and $x_0 \in [a, b]$ a fixed point. The domain $D(A)$ consists of all functions $\varphi \in L^p(a, b)$ for which $(A\varphi)(x) \in L^p(a, b)$. $D(A)$ is dense in $L^p(a, b)$ since by Lemma 1.1 $C^\infty[a, b] \subset \subset D(A)$.

Lemma 1.10. *The operator A is closed.*

Proof. Let $\varphi_n \in D(A)$, $\varphi_n \to \varphi$ and $A\varphi_n \to \psi$ (in the sense of the norm of $L^p(a, b)$). We show that $\varphi \in D(A)$ and $A\varphi = \psi$.

Since the function system $(x - x_0)^j$ $(j = 0, 1, \dots, k - 1)$ is linearly independent, from the stated hypotheses it is easily seen that the finite limits

$$\lim_{n\to\infty} \varphi_n^{\{j\}}(x_0) = c_j \qquad (j = 0, 1, \dots, k - 1)$$

exist. We put

$$\chi(x) = \frac{\varphi(x) - \sum_{j=0}^{k-1} \frac{1}{j!} c_j (x - x_0)^j}{(x - x_0)^k} \qquad (x \neq x_0)$$

and show that $\chi(x) = \psi(x)$ holds almost everywhere in (a, b). By Lemma 1.9 this implies $\varphi^{\{j\}}(x_0) = c_j$ $(j = 0, 1, \dots, k - 1)$. This proves the assertion.

For arbitrary $m = 1, 2, \dots$ we put $I_m = [a, b] - \left(x_0 - \frac{1}{m}, x_0 + \frac{1}{m}\right)$ and $I = [a, b] - \{x_0\}$. Since $I = \bigcup_{m=1}^{\infty} I_m$, it is sufficient to show that $\chi(x) = \psi(x)$ holds almost everywhere on I_m $(m = 1, 2, \dots)$. But this follows immediately from the extimates

$$\int_{I_m} |\psi(x) - \chi(x)|\, dx \leqq \int_{I_m} |\psi(x) - (A\varphi_n)\ (x)|\, dx + \int_{I_m} |(A\varphi_n)\ (x) - \chi(x)|\, dx$$

and

$$\int_{I_m} |\psi(x) - (A\varphi_n)\ (x)|\, dx \leqq C_m' \, ||\psi - A\varphi_n||_{L^p_{(a,\, b)}} ,$$

$$\int_{I_m} |\chi(x) - (A\varphi_n)\ (x)|\, dx \leqq C_m'' \, ||\varphi - \varphi_n||_{L^p_{(a,\, b)}} + \sum_{j=0}^{k-1} |c_j - \varphi_n^{\{j\}}(x_0)|$$

since the right-hand sides of the last two inequalities tend to zero for $n \to \infty$. This proves the lemma.

3. Now let Γ be a sufficiently smooth Jordan curve and $t = t(s)$ $(0 \leqq s \leqq \gamma)$ the equation of Γ, where s is the arc length on Γ. We say that the function $f(t) \in L(\Gamma)$ possesses the mean derivative $f^{\{1\}}(t_0)$ in the point $t_0 = t(s_0) \in \Gamma$ if the function $f(s) = f(t(s))$ of the real variable s has the mean derivative $f^{\{1\}}(s_0)$ at the point $s_0 \in [0, \gamma]$, and we put

$$f^{\{1\}}(t_0) = \overline{t'(s_0)}\, f^{\{1\}}(s_0) \qquad \left(t'(s) = \frac{dt(s)}{ds}\right).$$

Finally by induction we define

$$f^{\{k+1\}}(t) = \overline{t'(s)} \frac{d}{ds} f^{\{k\}}(t) ,$$

where on the right-hand side $\frac{d}{ds} f^{\{k\}}(t)$ is to be calculated formally by the product rule and in the expression obtained we put $\frac{d}{ds} f^{\{l\}}(s) = f^{\{l+1\}}(s)$.

Obviously, all properties of the mean derivative $f^{\{k\}}(t)$ mentioned in no. 1 and no. 2 remain true.

6.2. The Spaces Generated by Zeros of the Symbol

6.2.1. Let Γ be a curve system in the complex plane which consists of finitely many closed Ljapunov curves without common points and which divides the plane into two regions: a connected finite region D_+ containing the origin of the coordinate system and a region D_- containing the point at infinity. The positive orientation on Γ is chosen as usual (cf. 3.4.1). As in Section 3.4 by P and Q we denote the projections

$$P = \tfrac{1}{2}(I + S), \qquad Q = \tfrac{1}{2}(I - S)$$

defined on the space $L^p(\Gamma)$ $(1 < p < \infty)$.

Furthermore, let $\alpha_1, \alpha_2, \dots, \alpha_r, \beta_1, \beta_2, \dots, \beta_s$ and $\gamma_1, \gamma_2, \dots, \gamma_q$ be points on Γ and $m_1, m_2, \dots, m_r, n_1, n_2, \dots, n_s$ and $l_1, l_2, \dots, l_q$ non-negative integers. We put

$$\varrho_+(t) = \prod_{j=1}^{r} (t - \alpha_j)^{m_j}, \qquad \varrho_-(t) = \prod_{j=1}^{s} (t^{-1} - \beta_j^{-1})^{n_j}, \tag{2.1}$$

$$\varrho(t) = \prod_{j=1}^{q} (t - \gamma_j)^{l_j} \qquad (t \in \Gamma).$$

Since obviously $\varrho_+(t) \in C^+(\Gamma)$, $\varrho_-(t) \in C^-(\Gamma)$, by Theorem 5° of 3.4.4 we obtain

$$\varrho_+ P = P\varrho_+ P, \qquad P\varrho_- = P\varrho_- P. \tag{2.2}$$

We now introduce the following operators

$$B = P\varrho_- + Q\varrho_+, \qquad D = \varrho_+ P + \varrho_- Q, \qquad F = \varrho I. \tag{2.3}$$

6.2.2. We first assume that the points α_j and β_k $(j = 1, \dots, r;\ k = 1, \dots, s)$ are pairwise different. Under these hypotheses we prove the following lemma.

Lemma 2.1. *Let $\varrho_+(t)$, $\varrho_-(t)$ be the functions defined by (2.1), $f_+(t) \in L^p_+(\Gamma)$ $(= \operatorname{im} P)$ and $f_-(t) \in L^p_-(\Gamma)$ $(= \operatorname{im} Q)$. Then*

$$\varrho_+(t)\, f_-(t) = \varrho_-(t)\, f_+(t) \qquad (t \in \Gamma) \tag{2.4}$$

implies

$$f_+(t) = f_-(t) = 0.$$

Proof. We put

$$m = \sum_{j=1}^{r} m_j, \qquad n = \sum_{j=1}^{s} n_j. \tag{2.5}$$

For $m = n = 0$ the assertion is obvious since $L^p_+(\Gamma) \cap L^p_-(\Gamma) = \{0\}$. Let $m + n > 0$. First we show that $f_+(t)$ and $f_-(t)$ are polynomials. To this end we introduce the polynomials $g_-(t) = t^{-m}\varrho_+(t)$ and $g_+(t) = t^n\varrho_-(t)$, and we write the equation (2.4) in the form

$$t^{m+n} g_-(t)\, f_-(t) = g_+(t)\, f_+(t). \tag{2.6}$$

Applying the projection P to both sides of the last equation we find that $h_+(t) = g_+(t)\, f_+(t)$ is a polynomial of degree $\leqq m + n - 1$. This polynomial is divisible (in the ring of polynomials) by the polynomial $g_+(t)$ since otherwise, in view of $\beta_k \in \Gamma$ the function $h_+(t)/g_+(t) = f_+(t)$ would not belong to $L_+^p(\Gamma)$. Hence $f_+(t)$ is a polynomial. Similar arguments show that the polynomial $h_+(t)$ is also divisible by the polynomial $t^{m+n}g_-(t) = t^n \varrho_+(t)$. Thus by (2.6) $f_-(t) = h_+(t)/t^{m+n}g_-(t)$ is also a polynomial.

Since by hypothesis the zeros of the polynomials $\varrho_+(t)$ and $\varrho_-(t)$ are different, by (2.4) $f_+(t)/\varrho_+(t) = f_-(t)/\varrho_-(t)$ is a polynomial and thus we have $f_+(t) = f_-(t) = 0$.

Corollary 2.1. *Let $f_+(t) \in L_+^p(\Gamma)$ $(f_-(t) \in L_-^p(\Gamma))$. If the function*

$$\varphi_1(t) = f_+(t)/\varrho_+(t) \qquad (\varphi_2(t) = f_-(t)/\varrho_-(t))$$

belongs to $L^p(\Gamma)$, then necessarily $\varphi_1(t) \in L_+^p(\Gamma)$ $(\varphi_2(t) \in L_-^p(\Gamma))$ holds.

Proof. Let e.g. $\varphi_1(t) \in L^p(\Gamma)$. From the equality $P\varphi_1 = \varphi_1 - Q\varphi_1$ we obtain

$$\varrho_+(t)\, P\varphi_1 - f_+(t) = - \varrho_+(t)\, Q\varphi_1\,,$$

i.e. an equation of the form

$$\varphi_+(t) = \varrho_+(t)\, \varphi_-(t) \qquad (t \in \Gamma)$$

with $\varphi_\pm(t) \in L_\pm^p(\Gamma)$. From this it follows by Lemma 2.1 that $\varphi_+(t) = \varphi_-(t) = 0$. i.e. $Q\varphi_1 = 0$ and thus $\varphi_1 = P\varphi_1 \in L_+^p(\Gamma)$.

The assertion for the function $\varphi_2(t)$ can be proved analogously.

In the sequel by $D^{(-1)}$ we denote the operator defined for any $f \in L^p(\Gamma)$ by the formula

$$D^{(-1)}f = \varrho_+^{-1}Pf + \varrho_-^{-1}Qf\,. \tag{2.7}$$

The collection of all functions of the form (2.7) with $f \in L^p(\Gamma)$ is denoted by $\widetilde{L}^p(\varrho_+, \varrho_-)$.

Obviously, by (2.2) for any $g \in L^p(\Gamma)$ we have

$$D^{(-1)}Dg = g\,. \tag{2.8}$$

Hence $L^p(\Gamma) \subset \widetilde{L}^p(\varrho_+, \varrho_-)$. We now extend D to an operator $\widetilde{D}$ defined on $\widetilde{L}^p(\varrho_+, \varrho_-)$ by putting

$$\widetilde{D}\varphi = \varrho_+\widetilde{P}\varphi + \varrho_-\widetilde{Q}\varphi \qquad (\varphi \in \widetilde{L}^p(\varrho_+, \varrho_-))\,.$$

Here $\widetilde{P}$ and $\widetilde{Q}$ are operators defined for the function $\varphi = \varrho_+^{-1}Pf + \varrho_-^{-1}Qf \in \widetilde{L}^p(\varrho_+, \varrho_-)$ by the equations

$$\widetilde{P}\varphi = \varrho_+^{-1}Pf\,, \qquad \widetilde{Q}\varphi = \varrho_-^{-1}Qf\,.$$

We remark that the operators $\widetilde{P}$ and $\widetilde{Q}$ are uniquely defined since by Lemma 2.1 the function $f \in L^p(\Gamma)$ is uniquely determined by $\varphi \in \widetilde{L}^p(\varrho_+, \varrho_-)$.

Furthermore, by (2.8) for any $g \in L^p(\Gamma)$ the equations $\widetilde{P}g = Pg$, $\widetilde{Q}g = Qg$ and consequently $\widetilde{D}g = Dg$ hold. Obviously,

$$\widetilde{D}D^{(-1)}f = f \qquad (f \in L^p(\Gamma)) .$$

Thus the operator D satisfies condition (2′) of 4.1.1. Hence in the norm

$$||\varphi||_{\widetilde{L}^p(\varrho_+, \varrho_-)} = ||\widetilde{D}\varphi||_{L^p(\Gamma)} = ||f||_{L^p(\Gamma)}$$

$\widetilde{L}^p(\varrho_+, \varrho_-)$ becomes a Banach space which the operator $\widetilde{D}$ maps isometrically onto $L^p(\Gamma)$.

6.2.3. Obviously, dim ker $F = 0$. Following Subsection 4.1.1 by $\overline{L}^p(\varrho) =$ $=$ im F we denote the Banach space consisting of all functions of the form

$$\varphi(t) = \varrho(t)\, f(t) \qquad (f \in L^p(\Gamma)) ,$$

where the norm is defined by the formula

$$||\varphi||_{\overline{L}^p(\varrho)} = ||f||_{L^p(\Gamma)} .$$

Correspondingly let $\widetilde{L}^p(\Gamma)$ be the Banach space of all functions of the form

$$\psi(t) = \varrho^{-1}(t)\, f(t) \qquad (f \in L^p(\Gamma))$$

with the norm

$$||\psi||_{\widetilde{L}^p(\varrho)} = ||f||_{L^p(\Gamma)} .$$

The operator F maps the space $L^p(\Gamma)$ (resp. $\widetilde{L}^p(\Gamma)$) isometrically onto $\overline{L}^p(\varrho)$ (resp. $L^p(\Gamma)$).

Obviously, the considerations of this subsection can also be carried over to the more general case where $\varrho(t)$ is an arbitrary measurable bounded function vanishing on Γ on at most a set of Lebesgue measure zero.

6.2.4. Now let $\alpha_1, \alpha_2, \dots, \alpha_r$ and $\beta_1, \beta_2, \dots, \beta_s$ be arbitrary (not necessarily different) points on Γ.

Lemma. 2.2. *We have* ker B $= \{0\}$.

Proof. Let $B\varphi = 0$ $(\varphi \in L^p(\Gamma))$. Applying the projection P to both sides of this equation, from (2.2) we obtain $P\varrho_- P\varphi = 0$ or

$$P\varphi = \varrho_-^{-1} Q \varrho_- P\varphi .$$

By Lemma 2.1 this implies $P\varphi = 0$. Analogously we obtain $Q\varrho_+ Q\varphi = 0$ or

$$Q\varphi = \varrho_+^{-1} P \varrho_+ Q\varphi .$$

Thus by Lemma 2.1 we obtain also $Q\varphi = 0$ and hence $\varphi = 0$.

In accordance with 4.3.4 by $\overline{L}^p(\varrho_+, \varrho_-)$ we denote the image space im B with the norm

$$||f||_{\overline{L}^p(\varrho_+, \varrho_-)} = ||B^{-1}f||_{L^p(\Gamma)} .$$

Then $\overline{L^p}(\varrho_+, \varrho_-) \subset L^p(\Gamma)$ holds in the sense of a continuous embedding, and the operator B maps $L^p(\Gamma)$ isometrically onto the Banach space $\overline{L^p}(\varrho_+, \varrho_-)$.

We now give an analytic description of the space $\overline{L^p}(\varrho_+, \varrho_-)$.

Theorem 2.1. *Let m and n be the integers defined by* (2.5). *For the function $f(t) \in L^p(\Gamma)$ to belong to the space $\overline{L^p}(\varrho_+, \varrho_-)$ the following conditions are jointly necessary and sufficient:*

(1) *There exists a polynomial* $q_-(t) = \dfrac{a_1}{t} + \dots + \dfrac{a_n}{t_n}$ *such that* $\varphi_1(t) \equiv (Pf - q_-(t))/\varrho_-(t) \in L^p(\Gamma)$.

(2) *There exists a polynomial* $q_+(t) = b_1 + \dots + b_m t^{m-1}$ *such that* $\varphi_2(t) \equiv (Qf - q_+(t))/\varrho_+(t) \in L^p(\Gamma)$.

If the conditions (1) *and* (2) *are satisfied, then* $B^{-1}f = \varphi_1 + \varphi_2$ *and the norm in the space* $\overline{L^p}(\varrho_+, \varrho_-)$ *is equivalent to the norm*

$$||f||' = ||\varphi_1||_{L^p(\Gamma)} + ||\varphi_2||_{L^p(\Gamma)} .$$

Proof. Necessity. Let $f(t) \in \overline{L^p}(\varrho_+, \varrho_-)$, i.e. $f = B\varphi$ with $\varphi(t) \in L^p(\Gamma)$. Applying the operators P and Q to both sides of the last equation and using (2.2) we find

$$P\varrho_- P\varphi = Pf , \qquad Q\varrho_+ Q\varphi = Qf .$$

From this, after some elementary transformations, we obtain

$$t^n f_-(t) = t^n Pf - t^n \varrho_-(t)\, P\varphi \tag{2.9}$$

and

$$Qf - \varrho_+(t)\, Q\varphi = f_+(t) \tag{2.10}$$

with $f_\pm(t) \in L^p_\pm(\Gamma)$. The right-hand sides of the last equations belong to the subspace $L^p_+(\Gamma)$. After subtraction of certain polynomials of the form $q(t) = a_n + a_{n-1}t + \dots + a_1 t^{n-1}$ for (2.9) and of the form $q_+(t) = b_1 + b_2 t + \dots + b_m t^{m-1}$ for (2.10) the left-hand sides belong to the subspace $L^p_-(\Gamma)$. Applying the operator Q to both sides of (2.9) and the operator P to both sides of (2.10) we thus obtain

$$t^n f_-(t) = q(t) , \qquad f_+(t) = q_+(t) . \tag{2.11}$$

It follows immediately from (2.9)—(2.11) that the conditions (1) and (2) are satisfied.

Sufficiency. Conversely, let conditions (1) and (2) be fulfilled. Then by Corollary 2.1 we have $\varphi_1(t) \in L^p_+(\Gamma)$ and $\varphi_2(t) \in L^p_-(\Gamma)$. Hence for the function $\varphi = \varphi_1 + \varphi_2$ there holds $P\varphi = \varphi_1$, $Q\varphi = \varphi_2$. Thus using the relations (2.2) we obtain

$$B\varphi = P\varrho_-\varphi_1 + Q\varrho_+\varphi_2 = P(Pf - q_-) + Q(Qf - q_+) = Pf + Qf = f .$$

The assertion about the equivalence of the norms follows immediately from the estimates

$$||\varphi_1||_{L^p(\Gamma)} \leqq ||P|| \, ||\varphi||_{L^p(\Gamma)}, \qquad ||\varphi_2||_{L^p(\Gamma)} \leqq ||Q|| \, ||\varphi||_{L^p(\Gamma)}$$

and the obvious inequality $||\varphi||_{L^p(\Gamma)} \leqq ||f||'$. This completes the proof of the theorem.

Corollary 2.2. *Let $\Gamma \in C^{\overline{m}}$ with $\overline{m} = \max\{m_1, \dots, m_r, n_1, \dots, n_s\}$. Then*

$$W_p(\Gamma; (\Gamma_{\alpha_j}, m_j)_1^r, (\Gamma_{\beta_k}, n_k)_1^s) \subset \overline{L}^p(\varrho_+, \varrho_-), \tag{2.12}$$

holds in the sense of a continuous embedding.

Proof. Let $f(t)$ be a function which belongs to the space on the left in (2.12). By $q(t)$ we denote the Hermite interpolation polynomial (in t) of the function $t^n Pf$ with knots β_k $(k = 1, \dots, s)$ and by $q_+(t)$ the corresponding interpolation polynomial of Qf with the knots α_j $(j = 1, \dots, r)$. Furthermore, we put $q_-(t) = t^{-n} q(t)$. By Lemma 1.2 and Corollary 1.2 for the functions φ_1 and φ_2 mentioned in Theorem 2.1 we then obtain the estimates

$$||\varphi_1||_{L^p(\Gamma)} \leqq C||f||_{W_p}, \qquad ||\varphi_2||_{L^p(\Gamma)} \leqq C||f||_{W_p}$$

with $C = \text{const}$. From this the assertion follows by Theorem 2.1.

Corollary 2.3. *Let $a(t) \in C(\alpha_j, m_j) \cap C(\beta_k, n_k)$ $(j = 1, \dots, r;\ k = 1, \dots, s)$. The function $a(t)$ is a multiplier in the space $\overline{L}^p(\varrho_+, \varrho_-)$, i.e. for any $f(t) \in \overline{L}^p(\varrho_+, \varrho_-)$ there holds*

$$a(t)\, f(t) \in \overline{L}^p(\varrho_+, \varrho_-), \qquad ||af||_{\overline{L}^p(\varrho_+, \varrho_-)} \leqq c||f||_{\overline{L}^p(\varrho_+, \varrho_-)}.$$

Proof. Let $f = B\varphi$ $(\varphi \in L^p(\Gamma))$. Then

$$a(t)\, f(t) = Ba\varphi + (aB - Ba)\, \varphi,$$

and the assertion follows from Lemma 4.1, Chapter 4.

Taking into consideration Theorem 3.4, Chapter 4, and Lemma 1.9 it is easily seen that Theorem 2.1 can also be formulated as follows.

Theorem 2.2. *For the function $f(t) \in L^p(\Gamma)$ to belong to the space $\overline{L}^p(\varrho_+, \varrho_-)$ it is necessary and sufficient that for all $j = 1, 2, \dots, s$ and $k = 1, 2, \dots, r$ the following conditions are satisfied:*

(1) *There exist mean derivatives $b_{jl} = (Pf)^{\{l\}}(\beta_j)$, $l = 0, 1, \dots, n_j - 1$ such that*

$$\frac{Pf - \sum_{l=0}^{n_j-1} \frac{1}{l!} b_{jl}(t - \beta_j)^l}{(t - \beta_j)^{n_j}} \in L^p(\Gamma).$$

(2) *There exist mean derivatives* $a_{kl} = (Qf)^{\{l\}}(\alpha_k)$, $l = 0, 1, \ldots, m_k - 1$ *such that*

$$\frac{Qf - \sum_{l=0}^{m_k-1} \frac{1}{l!} a_{kl}(t - \alpha_k)^l}{(t - \alpha_k)^{m_k}} \in L^p(\Gamma) .$$

Corollary 2.4. *The inverse of the operator*

$$B = P(t^{-1} - \alpha^{-1})^m + Q(t - \beta)^n$$

is defined by the formula

$$(B^{-1}f)(t) = (t^{-1} - \alpha^{-1})^{-m} \left[Pf - \sum_{k=0}^{m-1} (t^m Pf)^{\{k\}} (\alpha) \frac{(t - \alpha)^k}{k!\, t^m} \right] +$$

$$+ (t - \beta)^{-n} \left[Qf - \sum_{k=0}^{n-1} (Qf)^{\{k\}} (\beta) \frac{(t - \beta)^k}{k!} \right] . \tag{2.13}$$

6.3. Symbols with Zeros of Integral Orders

6.3.1. Let Γ be a curve system of the form described in 6.2.1. In the space $L^p(\Gamma)$ $(1 < p < \infty)$ we consider the singular integral equation

$$A\varphi = a(t)\, \varphi(t) + \frac{b(t)}{\pi i} \int_\Gamma \frac{\varphi(\tau)}{\tau - t}\, d\tau = f(t) \qquad (t \in \Gamma) , \tag{3.1}$$

where $a(t)$ and $b(t)$ are continuous functions on Γ. As in 3.4 we introduce the functions

$$c(t) = a(t) + b(t) , \qquad d(t) = a(t) - b(t) .$$

Using these functions we can represent the operator A in the form

$$A = c(t)\, P + (dt)\, Q .$$

We suppose that the functions $c(t)$ and $d(t)$ admit the representation

$$\left. \begin{aligned} c(t) &= \prod_{j=1}^{r} (t - \alpha_j)^{m_j} \prod_{i=1}^{q} (t - \gamma_i)^{l_i}\, c_0(t) , \\ d(t) &= \prod_{k=1}^{s} (t - \beta_k)^{n_k} \prod_{i=1}^{q} (t - \gamma_i)^{l_i}\, d_1(t) . \end{aligned} \right\} \tag{3.2}$$

Here $c_0(t)$ and $d_1(t)$ are functions from $C(\Gamma)$, α_j, β_k, γ_i points on Γ and m_j, n_k, l_i non-negative integers. By means of the functions $\varrho_\pm(t)$, $\varrho(t)$ defined by (2.1) we can then represent the functions $c(t)$, $d(t)$ in the form

$$c(t) = \varrho_+(t)\, \varrho(t)\, c_0(t) , \qquad d(t) = \varrho_-(t)\, \varrho(t)\, d_0(t) \, . \tag{3.3}$$

where $d_0(t) = t^n d_1(t) \prod_{k=1}^{s} (-\beta_k)^{n_k}$.

We remark that by Lemma 1.1 the function $c(t)$ is representable in the form (3.2) if it belongs to the class $C(\Gamma; (\alpha_j, m_j)_1^r, (\gamma_i, l_i)_1^q)$ and has zeros of the corresponding order at the points α_j and γ_i.

Unless the contrary is stated, without loss of generality we shall assume that the points α_j and β_k ($j = 1, \dots, r$; $k = 1, \dots, s$) in (3.2) are distinct. Then the points γ_i are just the common zeros of the functions $c(t)$ and $d(t)$.

By (2.2) and (3.3) for the operator A we obtain the representation $A = FCD$ with $C = c_0P + d_0Q$. The operators D and F are given by formulae (2.3). Hence using Theorem 3.2, Chapter 4, and the remark following this theorem we immediately obtain the following theorem.

Theorem 3.1. *For the operator A as an operator from $\widetilde{L}^p(\varrho_+, \varrho_-)$ into $\overline{L}^p(\varrho)$ to be a Φ_+-operator or a Φ_--operator it is necessary and sufficient that the functions $c_0(t)$ and $d_0(t)$ from (3.3) do not vanish on Γ.*

If this is the case, then the invertibility of the operator A corresponds to the number

$$\varkappa = \frac{1}{2\pi}\left[\arg\frac{c_0(t)}{d_0(t)}\right]_\Gamma, \tag{3.4}$$

and

$$\dim\ker A = \max(-\varkappa, 0), \qquad \dim\operatorname{coker} A = \max(\varkappa, 0).$$

6.3.2. Now we consider the singular operator $A' = Pd + Qc$ transpose to A. For this operator we analogously obtain the representation $A' = BC'F$, where $C' = Pd_0 + Qc_0$ and B is defined by (2.3).

The following theorem is a realization of Theorem 3.3, Chapter 4.

Theorem 3.2. *For the operator A' as an operator from $\widetilde{L}^p(\varrho)$ into $\overline{L}^p(\varrho_+, \varrho_-)$ to be a Φ_+-operator or a Φ_--operator it is necessary and sufficient that the functions $c_0(t)$ and $d_0(t)$ from (3.3) do not vanish on Γ.*

If this is the case, then the invertibility of the operator A' corresponds to the number $-\varkappa$ and

$$\dim\ker A' = \max(\varkappa, 0), \qquad \dim\operatorname{coker} A' = \max(-\varkappa, 0).$$

Remark. Theorem 3.2 is also still valid if some of the points a_j and β_k coincide for the functions $\varrho_+(t)$, $\varrho_-(t)$. Thus without loss of generality in Theorem 3.2 we can put $\varrho(t) \equiv 1$, i.e. $\widetilde{L}^p(\varrho) = L^p(\Gamma)$.

This follows from the fact that the hypothesis $\alpha_j \neq \beta_k$ was used only for the construction of the space $\widetilde{L}^p(\varrho_+, \varrho_-)$ but not for the construction of the space $\overline{L}^p(\varrho_+, \varrho_-)$ (cf. Subsections 6.2.2 and 6.2.4).

6.3.3. By Theorems 3.1 and 3.2 the singular integral operators A and A' are Noether operators in the pairs of spaces $\widetilde{L}^p(\varrho_+, \varrho_-)$, $\overline{L}^p(\varrho)$ and $\widetilde{L}^p(\varrho)$, $\overline{L}^p(\varrho_+, \varrho_-)$, respectively, if the coefficients $c(t)$ and $d(t)$ possess the representations (3.3), where $c_0(t) \neq 0$, $d_0(t) \neq 0$ ($t \in \Gamma$). Now we prove the converse assertion: If A (resp. A') is a Noether operator on the pair of spaces in question, then the functions $c(t)$ and $d(t)$ necessarily have the form (3.3). In this sense the above constructed spaces exactly correspond to the zeros of the symbol of the operators A and A' considered in this subsection.

Theorem 3.3 *Let the functions $c(t)$ and $d(t)$ be continuous on Γ, and suppose the Taylor derivatives*[1] $c^{\{m_j\}}(\alpha_j)$, $c^{\{l_i\}}(\gamma_i)$, $d^{\{n_k\}}(\beta_k)$, $d^{\{l_i\}}(\gamma_i)$ *exist for all* $j = 1, \dots, r$, $k = 1, \dots, s$ *and* $i = 1, \dots, q$.

For the operator $A = c(t)\,P + d(t)\,Q$ *to be a Φ_+-operator or a Φ_--operator from* $\widetilde{L}^p(\varrho_+, \varrho_-)$ *into* $\overline{L}^p(\varrho)$ *it is necsseary and sufficient that for the functions $c(t)$ and $d(t)$ the representations* (3.3) *hold with certain functions* $c_0(t)$, $d_0(t) \in C(\Gamma)$ *vanishing nowhere on* Γ.

An analogous theorem holds for the transpose operator $A' = Pd + Qc$ and the pair of spaces $\widetilde{L}^p(\varrho)$, $\overline{L}^p(\varrho_+, \varrho_-)$.

Before proving the Theorem 3.3 we state the following lemma.

Lemma 3.1. *Let the functions $c(t)$ and $d(t)$ satisfy the conditions of Theorems* 3.3. *If for any function* $\varphi(t) \in L^p(\Gamma)$ *the function*

$$\frac{c(t)\,P\varphi}{\varrho_+(t)\,\varrho(t)}\left(\frac{d(t)\,Q\varphi}{\varrho_-(t)\,\varrho(t)}\right) \tag{3.5}$$

also belongs to the space $L^p(\Gamma)$, *then* $c(t)$ $(d(t))$ *is representable in the form* (3.3) *with* $c_0(t) \in C(\Gamma)$ $(d_0(t) \in C(\Gamma))$.

Proof. We shall prove the assertion for the function $c(t)$. For $d(t)$ the proof is analogous.

Putting $\varphi \equiv 1$ in (3.5) we obtain $\varrho_+^{-1}\varrho^{-1}c \in L^p(\Gamma)$. Thus in particular the function $c(t)\,(t - \alpha_1)^{-m_1}$ is integrable on Γ. This implies

$$c(t) = (t - \alpha_1)^{m_1}\,\tilde{c}(t) \tag{3.6}$$

with $\tilde{c}(t) \in C(\Gamma)$.

In fact, we can assume that $m_1 \geqq 1$. Thus, since the function $c(t)$ is continuous, necessarily we have $c(\alpha_1) = 0$. Hence

$$\lim_{t\to\alpha_1} \frac{c(t)}{t - \alpha_1} = c^{\{1\}}(\alpha_1)\,.$$

For $t \in \Gamma$ we put

$$c_1(t) = \begin{cases} \dfrac{c(t)}{t - \alpha_1} & \text{wise} \quad t \neq \alpha_1 \\ c^{\{1\}}(\alpha_1) & \text{wise} \quad t = \alpha_1\,. \end{cases}$$

Then obviously $c_1(t)$ is continuous on Γ and $c(t) = (t - \alpha_1)\,c_1(t)$. By induction we easily obtain the representation (3.6).

Analogously we treat the points $\alpha_2, \dots, \alpha_r, \gamma_1, \dots, \gamma_q$. This proves the lemma.

Now we prove Theorem 3.3. In view of Theorem 3.1 we have only to show the necessity of the representations (3.3).

Let the operator A map the space $\widetilde{L}^p(\varrho_+, \varrho_-)$ into the space $\overline{L}^p(\varrho)$ and let A be a Φ_+-operator or a Φ_--operator. Then

$$C\varphi \equiv \varrho^{-1}[\varrho_+^{-1}cP\varphi + \varrho_-^{-1}dQ\varphi] \in L^p(\Gamma)$$

holds for any function $\varphi \in L^p(\Gamma)$. Consequently the hypotheses of the Lemma 3.1 are satisfied, and thus the functions $c(t)$ and $d(t)$ have the form (3.3). Hence

$$C\varphi = c_0P\varphi + d_0Q\varphi\,.$$

---

[1]) For notation see 2.3.4.

Finally by of 4.1.1 the operator C is a Φ_+-operator resp. a Φ_--operator in the space $L^p(\Gamma)$. By Theorem 4.5, Chapter 3, it follows from this that the functions $c_0(t)$ and $d_0(t)$ vanish nowhere on Γ.

This completes the proof of Theorem 3.3.

Remark. Still more than is asserted in Theorem 3.3 holds: If the singular operator A from $\widetilde{L}^p(\varrho_+, \varrho_-)$ into $\overline{L}^p(\varrho)$ is normally solvable, then the representations (3.3) hold, where each of the continuous functions $c_0(t)$ and $d_0(t)$ are either identically equal to zero or nowhere equal to zero on each closed subcurve of Γ.

This follows immediately from the preceding proof and the remark to Theorem 4.5, Chapter 3.

6.3.4. In the sequel we apply the results from Section 4.4 to the singular integral operators (3.1) resp. $Pc + Qd$.

Let α_j and β_k be (not necessarily distinct) points on Γ and m_j, n_k ($j = 1, \dots, r$; $k = 1, \dots, s$) non-negative integers. We assume that the coefficients $a(t)$ and $b(t)$ belong to the subalgebra $C(\boldsymbol{\alpha}, \boldsymbol{m}; \boldsymbol{\beta}, \boldsymbol{n})$ of the algebra $C = C(\Gamma)$ (cf. 2.3.4):

$$a(t), b(t) \in \bigcap_{j=1}^{r} C(\alpha_j, m_j) \cap \bigcap_{k=1}^{s} C(\beta_k, n_k) .$$

Let m_j', m_j'' ($j = 1, \dots, r$) and n_k', n_k'' ($k = 1, \dots, s$) be arbitrary non-negative integers such that

$$m_j' + m_j'' = m_j , \qquad n_k' + n_k'' = n_k$$

for all j and k. We put

$$\begin{aligned} \varrho_+'(t) &= \prod_{j=1}^{r} (t - \alpha_j)^{m_k'} , & \varrho_-'(t) &= \prod_{k=1}^{s} (t^{-1} - \beta_k^{-1})^{n_k'} , \\ \varrho_+''(t) &= \prod_{k=1}^{s} (t - \beta_k)^{n_k''} , & \varrho_-''(t) &= \prod_{j=1}^{r} (t^{-1} - \alpha_j^{-1})^{m_j''} . \end{aligned} \tag{3.7}$$

By $\varrho_\pm(t)$ we again denote the functions defined by (2.1). Theorem 4.3, Chapter 4, implies

Theorem 3.4. *Let the functions $\varrho_+'(t)$ and $\varrho_-'(t)$ have no common zeros. For the singular integral operator $A = c(t)\, P + d(t)\, Q$ (resp. $A = Pc + Qd$) to be a Φ_+-operator or a Φ_--operator acting from $\widetilde{L}^p(\varrho_+', \varrho_-')$ into the space $\overline{L}^p(\varrho_+'', \varrho_-'')$ it is necessary and sufficient that the functions*

$$c_0(t) = c(t)/\varrho_+(t) , \qquad d_0(t) = d(t)/\varrho_-(t) \tag{3.8}$$

are continuous on Γ and vanish nowhere on Γ.

If these conditions are satisfied, then A is a Φ-operator from $\widetilde{L}^p(\varrho_+', \varrho_-')$ into $\overline{L}^p(\varrho_+'', \varrho_-'')$, and for its index the formula

$$\operatorname{Ind} A = -\lambda , \qquad \lambda = \varkappa + \sum_{j=1}^{r} m_j'' + \sum_{k=1}^{s} n_k''$$

holds, where the number $\varkappa$ is determined by (3.4).

If the points α_j and β_k are all distinct, then

$$\dim \ker A = \max(-\lambda, 0), \qquad \dim \operatorname{coker} A = \max(\lambda, 0).$$

6.3.5. Under the hypotheses of Subsection 6.3.4 we now put in particular $m''_j = m_j$ ($j = 1, \dots, r$) and $n''_k = n_k$ ($k = 1, \dots, s$). We assume that the continuous functions (3.8) vanish nowhere on Γ. Then $\widetilde{L}^p(\varrho'_+, \varrho'_-) = L^p(\Gamma)$ and the singular integral operator A is a continuous Φ-operator from $L^p(\Gamma)$ into the space $\overline{L}^p(\varrho''_+, \varrho''_-)$.

Using the notations of subsection 4.5.2 we put

$$X = Y = L^p(\Gamma), \qquad Z = W_p(\Gamma; (\alpha_j, m_j)_1^r, (\beta_k, n_k)_1^s)$$

and $R = (P\varrho''_- + Q\varrho''_-)^{-1}$. In addition to the continuous operator $A: L^p(\Gamma) \to \overline{L}^p(\varrho''_+, \varrho''_-)$ we still consider the closed operator $\underline{A}: L^p(\Gamma) \to Z$ defined by the equations

$$\underline{A}\varphi = A\varphi, \qquad D(\underline{A}) = \{\varphi \in L^p(\Gamma): A\varphi \in Z\}.$$

Obviously, the space $\overline{L}^p(\varrho''_+, \varrho''_-)$ is topologically equivalent to the space X_R introduced in 4.5.2. By Theorem 5.2 and Lemma 5.1 of Chapter 4 $\underline{A}$ is a Φ-operator and

$$\ker \underline{A} = \ker A, \qquad (\operatorname{im} \underline{A})^\perp = (\operatorname{im} A)^\perp, \qquad \operatorname{Ind} \underline{A} = \operatorname{Ind} A.$$

6.3.6. It is easily seen from the results of the Section 6.1 and the considerations of 6.2 and 6.3 that all results of the last two sections can be carried over to the space $H^\lambda(\Gamma)$ ($0 < \lambda < 1$). Their formulation is obtained by replacing $L^p(\Gamma)$ by $H^\lambda(\Gamma)$ and $C(\Gamma)$ by $H^\lambda(\Gamma)$ in the above-mentioned results.

6.4. Symbols with Zeros of Non-Integral Orders

The results of the preceding two sections can be generalized to the case where the symbol has finitely many zeros of not necessarily integral orders. For simplicity we assume the Γ consists of one simple closed Ljapunov curve enclosing the origin of the coordinate system.

6.4.1. Let $\alpha_1, \alpha_2, \dots, \alpha_r, \beta_1, \beta_2, \dots, \beta_s$ and $\gamma_1, \gamma_2, \dots, \gamma_q$ be points on Γ and $m_1, m_2, \dots, m_r$, $n_1, n_2, \dots, n_s$ and $l_1, l_2, \dots, l_q$ arbitrary non-negative real numbers. We assume that the points α_j and β_k ($j = 1, \dots, r$; $k = 1, \dots, s$) are distinct. In the sequel we maintain the notation introduced by (2.1) and (2.3).

Obviously, in the present case we also have $\varrho_\pm(t) \in C^\pm(\Gamma)$. Consequently the relations (2.2) remain true. The same statement holds for Lemma 2.1 and thus also for Corollary 2.1.

In fact, from a relation of the form

$$(t - \alpha_1)^{m_1} f_-(t) = (t^{-1} - \beta_1^{-1})^{n_1} f_+(t)$$

with $f_\pm(t) \in L^p_\pm(\Gamma)$ it follows that

$$(t - \alpha_1)^{[m_1]+1} \varphi_-(t) = (t^{-1} - \beta_1^{-1})^{[n_1]+1} \varphi_+(t) ,$$

where

$$\varphi_+(t) = f_+(t)\,(t - \alpha_1)^{[m_1]+1-m_1} \in L^p_+(\Gamma) ,$$
$$\varphi_-(t) = f_-(t)\,(t^{-1} - \beta_1^{-1})^{[n_1]+1-n_1} \in L^p_-(\Gamma)$$

and $[m]$ is the integral part of the number m. Thus by Lemma 2.1 proved in 6.2.2 we have $\varphi_\pm(t) = 0$ and consequently $f_\pm(t) = 0$. Analogously we can treat the case of finitely many points α_j and β_k.

From the considerations in Section 6.2 it is now easily seen that the spaces $\widetilde{L}^p(\varrho_+, \varrho_-)$, $\widetilde{L}^p(\varrho)$ and $\overline{L}^p(\varrho)$ can be constructed in the same way as in the case of integers m_j, n_k and l_i.

6.4.2. It is easy to see that in the present case the Lemma 2.2 also holds. We equip the image space $\overline{L}^p(\varrho_+, \varrho_-) = \operatorname{im} B$ with the norm introduced in 6.2.4.

In order to describe the space $\overline{L}^p(\varrho_+, \varrho_-)$ analytically we first consider the simplest case $r = s = 1$:

$$\varrho_+(t) = (t - \alpha)^m , \qquad \varrho_-(t) = (t^{-1} - \beta^{-1})^n .$$

We introduce the numbers

$$\mu = \left[m + \frac{1}{p}\right], \qquad \nu = \left[n + \frac{1}{p}\right].$$

Theorem 4.1. *For the function $f(t) \in L^p(\Gamma)$ to belong to the space $\overline{L}^p(\varrho_+, \varrho_-)$ it is necessary and sufficient that the following conditions both hold:*[1])

(1) *There exists a polynomial $q_-(t) = \frac{a_1}{t} + \dots + \frac{a_\nu}{t^\nu}$ such that*

$$\varphi_1(t) \equiv \frac{P[(t^{-1} - \beta^{-1})^{\nu-n} f(t)] - q_-(t)}{(t^{-1} - \beta^{-1})^\nu} \in L^p(\Gamma) .$$

(2) *There exists a polynomial $q_+(t) = b_1 + \dots + b_\mu t^{\mu-1}$ such that*

$$\varphi_2(t) = \frac{Q[(t - \alpha)^{\mu-m} f(t)] - q_+(t)}{(t - \alpha)^\mu} \in L^p(\Gamma) .$$

If the conditions (1) *and* (2) *are satisfied, then $B^{-1}f = \varphi_1 + \varphi_2$.*

Proof. Necessity. Let $f = B\varphi \in \overline{L}^p(\varrho_+, \varrho_-)$. Repeating the considerations from the corresponding part of the proof of Theorem 2.1 we obtain

---

[1]) In the case $\nu = 0$ we have to put $q_-(t) \equiv 0$ in the condition (1). The corresponding statement holds for $\mu = 0$. Another description of the space $\overline{L}^p(\varrho_+, \varrho_-)$ is given in Section 6.5.

relations of the form

$$(t^{-1} - \beta^{-1})^{\nu} P\varphi = (t^{-1} - \beta^{-1})^{\nu - n} [f_-(t) + Pf], \tag{4.1}$$

$$(t - \alpha)^{\mu} Q\varphi = (t - \alpha)^{\mu - m} [f_+(t) + Qf]. \tag{4.2}$$

The left-hand side of (4.1) becomes a function from $L^p_+(\Gamma)$ by addition of a suitable polynomial $q_-(t) = \frac{a_1}{t} + \dots + \frac{a_\nu}{t^\nu}$. It is easily seen that the right-hand side of equation (4.1) belongs to $L^q(\Gamma)$ for a certain $q > 1$. Applying the operator P to both sides of (4.1) and using Corollary 2.1 we obtain

$$(t^{-1} - \beta^{-1})^{\nu} P\varphi + q_-(t) = P(t^{-1} - \beta^{-1})^{\nu - n} Pf = P(t^{-1} - \beta^{-1})^{\nu - n} f.$$

From this condition (1) follows immediately. Analoguosly from (4.2) we obtain condition (2).

Using relations (2.2) and Corollary 2.1 we can give the proof of the sufficiency of the conditions (1) and (2) similarily as in the case of the Theorem 2.1. Here we have to apply the relation $(t - \alpha)^{\mu - m} f(t) \in L^q(\Gamma)$ $(q > 1)$ already used in the proof of necessity.

This completes the proof of Theorem 4.1.

In the case $r \geqq 1$, $s \geqq 1$ by Theorem 3.4, Chapter 4, we have

$$\overline{L}^p(\varrho_+, \varrho_-) = \left[\bigcap_{j=1}^{r} \overline{L}^p(\varrho_+^{(j)}, 1)\right] \cap \left[\bigcap_{k=1}^{s} \overline{L}^p(1, \varrho_-^{(k)})\right],$$

where

$$\varrho_+^{(j)}(t) = (t - \alpha_j)^{m_j}, \qquad \varrho_-^{(k)}(t) = (t^{-1} - \beta_k^{-1})^{n_k}.$$

6.4.3. Now we assume that the continuous functions $c(t)$ and $d(t)$ on Γ are representable in the form (3.3), where $c_0(t)$ and $d_0(t)$ are also functions from $C(\Gamma)$. Then for the operators

$$A = c(t)\, P + d(t)\, Q, \qquad A' = Pd + Qc$$

the Theorems 3.1 and 3.2 hold. If representations of the form

$$c(t) = \varrho'_+(t)\, \varrho''_-(t)\, c_0(t), \qquad d(t) = \varrho'_-(t)\, \varrho''_+(t)\, d_0(t)$$

hold, where the functions $\varrho'_\pm(t)$, $\varrho''_\pm(t)$ are given by equations (3.7) with certain non-negative real (in general not integral) numbers m'_j, m''_j, n'_k, n''_k, then the results from Subsection 4.4.3 can be applied.

6.5. A More General Case of Degeneration

Let Γ be a simple closed and sufficiently smooth Ljapunov curve enclosing the origin of the coordinate system. In this section we consider a class of singular integral operators of non-normal type of the form

$$A = a(t)\, P + b(t)\, Q$$

in the space $L^p(\Gamma)$ $(1 < p < \infty)$. In the coefficients $a(t)$ and $b(t)$ we allow the following degenerations:

$$a(t) = R_1(t)\, c(t)\,, \qquad b(t) = R_2(t)\, d(t) \tag{5.1}$$

with

$$\begin{aligned} R_1(t) &= t^{-\eta} \prod_{i=1}^{q} \left(\frac{1}{\alpha_i} - \frac{1}{t}\right)^{n_{1i}+r_{1i}} (t-\alpha_i)^{n_{3i}+r_{3i}}\,, \\ R_2(t) &= t^{\xi} \prod_{i=1}^{q} (t-\alpha_i)^{n_{2i}+r_{2i}} \left(\frac{1}{\alpha_i} - \frac{1}{t}\right)^{n_{4i}+r_{4i}}\,. \end{aligned} \tag{5.2}$$

Here α_i $(i = 1, \dots, q)$ are points on Γ, n_{ji} $(j = 1, \dots, 4)$ non-negative integers, $0 \leqq r_{ji} < 1$ and

$$\eta = \sum_{i=1}^{q} (n_{3i} - [-r_{3i}])\,, \qquad \xi = \sum_{i=1}^{q} (n_{4i} - [-r_{4i}])\,.$$

The functions $c(t)$ and $d(t)$ are continuous on Γ.

We remark that e.g. coefficients with zeros of the form $|t-\alpha|^{\mu}$ ($\mu > 0$ arbitrary real number) can be represented in the form (5.1). Namely, we have

$$|t-\alpha|^{2\mu} = (t-\alpha)^{\mu} \left(\frac{1}{t} - \frac{1}{\alpha}\right)^{\mu} c(t)\,,$$

where the function $c(t) \in C^{\infty}(\Gamma)$ vanishes nowhere on Γ and moreover has index zero (cf. N. I. Mushelišvili [1]. § 6). Obviously, in the case of the unit circle we have $c(t) \equiv 1$.

For further considerations sometimes we still need the hypothesis

$$\min\{n_{1i} + n_{3i} + r_{1i} + r_{3i}\,, \quad n_{2i} + n_{4i} + r_{2i} + r_{4i}\} = 0 \qquad (i = 1, \dots, q)\,. \tag{5.3}$$

Obviously, the condition (5.3) means that the functions $R_1(t)$ and $R_2(t)$ have no common zeros on Γ.

6.5.1. We begin with the study of operators of the form $R_1P + R_2Q$.

Theorem 5.1. *Under the condition* (5.3) *the formula*

$$\dim \ker (R_1P + R_2Q) = -\sum_{i=1}^{q} \left([-r_{3i}] + [-r_{4i}] + \left[r_{3i} + \frac{1}{p'}\right] + \left[r_{4i} + \frac{1}{p'}\right]\right)$$

holds with $\dfrac{1}{p'} = 1 - \dfrac{1}{p}$.

Proof. We put

$$N_1 = \prod_{i=1}^{q} \left(\frac{1}{\alpha_i} - \frac{1}{t}\right)^{n_{1i}+r_{1i}}\,, \qquad N_2 = \prod_{i=1}^{q} (t-\alpha_i)^{n_{2i}+r_{2i}}\,,$$

$$M_1 = \prod_{i=1}^{q} (t-\alpha_i)^{n_{3i}+r_{3i}}\,, \qquad M_2 = \prod_{i=1}^{q} \left(\frac{1}{\alpha_i} - \frac{1}{t}\right)^{n_{4i}+r_{4i}}\,.$$

Since obviously $M_1 \in C^+(\Gamma)$ and $M_2 \in C^-(\Gamma)$, using Theorem 5° of 3.4.4 we obtain the representation

$$R_1P + R_2Q = (t^{-\eta}N_1P + t^{\xi}N_2Q)(M_1P + M_2Q).$$

By the condition (5.3) the functions N_1 and N_2 have no common zeros on Γ, and Lemma 4.4, Chapter 4, yields

$$\dim\ker(t^{-\eta}N_1P + t^{\xi}N_2Q) = \dim\ker(Pt^{-\eta}N_1 + Qt^{\xi}N_2) =$$
$$= \dim\ker(Pt^{-\eta}N_1P \mid \operatorname{im} P) + \dim\ker(Qt^{\xi}N_2Q \mid \operatorname{im} Q).$$

As is well-known, the following relations hold:

$$Pt^{-\eta}N_1P = PN_1PPt^{-\eta}P, \qquad Qt^{\xi}N_2Q = QN_2QQt^{\xi}Q,$$
$$\dim\ker(Pt^{-\eta}P \mid \operatorname{im} P) = \eta, \qquad \dim\ker(Qt^{\xi}Q \mid \operatorname{im} Q) = \xi.$$

Further, Lemma 2.1 (see also 6.4.1) implies

$$\dim\ker(PN_1P \mid \operatorname{im} P) = \dim\ker(QN_2Q \mid \operatorname{im} Q) = 0.$$

Thus we have $\dim\ker(t^{-\eta}N_1P + t^{\xi}N_2Q) = \eta + \xi$.

Let P_σ be an arbitrary polynomial in t of a degree $\sigma \leqq \eta + \xi - 1$. Then obviously each function ψ of the form

$$\psi = N_2P_\sigma - t^{-\eta-\xi}N_1P_\sigma, \tag{5.4}$$

is an element of $\ker(t^{-\eta}N_1P + t^{\xi}N_2Q)$. Since there exist exactly $\eta + \xi$ linearly independent functions of the form (5.4), the kernel of the operator $t^{-\eta}N_1P + t^{\xi}N_2Q$ consists only of functions of the form (5.4).

Now let $\varphi \in \ker(R_1P + R_2Q)$. Then $(M_1P + M_2Q)\varphi \in \ker(t^{-\eta}N_1P + t^{\xi}N_2Q)$, where $(M_1P + M_2Q)\varphi \neq 0$ for $\varphi \neq 0$. Thus $(M_1P + M_2Q)\varphi$ is a function of the form (5.4). This implies

$$M_1P\varphi = N_2P_\sigma, \qquad M_2Q\varphi = -t^{-\eta-\xi}N_1P_\sigma.$$

Since by condition (5.3) the functions M_1 and N_2 (resp. M_2 and N_1) have no common zeros on Γ and the function $\left(\frac{1}{\alpha} - \frac{1}{t}\right)^{-r}$ (resp. $(t-\alpha)^{-r}$) with $\alpha \in \Gamma$, $r \geqq 0$ belongs to $L^p(\Gamma)$ if and only if $\left[r + \frac{1}{p'}\right] = 0$, we immediately obtain the assertion of the theorem.

Corollary 5.1. *If the condition* (5.3) *holds, then*

$$\dim\ker(R_1P + R_2Q) = \dim\ker(PR_1P + QR_2Q). \tag{5.5}$$

Proof. Obviously, we have

$$\dim\ker(PR_1P + QR_2Q) = \dim\ker(PR_1P + Q) + \dim\ker(P + QR_2Q),$$
$$R_1P + Q = (PR_1P + Q)(I + QR_1P),$$
$$P + R_2Q = (P + QR_2Q)(I + PR_2Q).$$

Since the operators $I + QR_1P$ and $I + PR_2Q$ are invertible (see Chapter 2), the assertion follows from Theorem 5.1.

For what follows we introduce the image spaces $\hat{L}^p(R_1, R_2) = \operatorname{im}(R_1P + R_2Q)$ and $\overline{L}^p(R_1, R_2) = \operatorname{im}(PR_1P + QR_2Q)$. We equip both spaces with the corresponding factor norms (cf. Section 4.1.3). Then $\hat{L}^p(R_1, R_2)$ and $\overline{L}^p(R_1, R_2)$ are Banach spaces.

Remark. If

$$n_{3i} + r_{3i} = n_{4i} + r_{4i} = 0 \qquad (i = 1, \dots, q),$$

then $\overline{L}^p(R_1, R_2)$ coincides with the space introduced in Section 6.2.

Our next task is the analytic description of the spaces $\hat{L}^p(R_1, R_2)$ and $\overline{L}^p(R_1, R_2)$.

Theorem 5.2. *If condition* (5.3) *holds, then* $\hat{L}^p(R_1, R_2) = \overline{L}^p(R_1, R_2)$ (*i.e. both spaces are topologicaly equivalent*).

Proof. We put

$$\begin{aligned} R_3 &= \prod_{i=1}^{q} (t - \alpha_i)^{1 - r_{3i} - [1 - r_{3i}]}, \\ R_4 &= \prod_{i=1}^{q} \left(\frac{1}{\alpha_i} - \frac{1}{t}\right)^{1 - r_{4i} - [1 - r_{4i}]}. \end{aligned} \tag{5.6}$$

By condition (5.3) the function $R_1(R_2)$ is differentiable sufficiently often in a certain neighborhood of the zeros of the function $R_2R_4(R_1R_3)$. From the results of Section 6.1.2 we obtain the following representation possibilities for the functions R_1 and R_2:

$$R_1 = R_2R_4g_1 + h_+, \qquad R_2 = R_1R_3g_2 + h_-$$

with $g_1, g_2 \in C(\Gamma)$ and $h_\pm \in C^\pm(\Gamma)$. From this it follows together with

$$R_3, R_2R_4 \in C^+(\Gamma); \qquad R_4, R_1R_3 \in C^-(\Gamma)$$

that

$$R_1P + R_2Q = (PR_1P + QR_2Q)(I + T) \tag{5.7}$$

with $T = QR_4Qg_1P + PR_3Pg_2Q$. By Lemma 4.3, Chapter 3, T is a compact operator on $L^p(\Gamma)$. By k we denote the number (5.5). By Theorem 3.6, Chapter 1, it follows from (5.7) that $R_1P + R_2Q \in \Phi(L^p(\Gamma), \overline{L}^p(R_1, R_2))$ and $\operatorname{Ind}(R_1P + R_2Q) = k$. Since moreover $\dim\ker(R_1P + R_2Q) = k$ holds, $R_1P + R_2Q$ yields a continuous mapping of $L^p(\Gamma)$ onto $\overline{L}^p(R_1, R_2)$.

The mapping $R_1P + R_2Q \in \mathcal{L}(L^p(\Gamma), \hat{L}^p(R_1, R_2))$ has the same properties. From this it follows at once that there exists a continuous linear one-to-one mapping of $\overline{L}^p(R_1, R_2)$ onto $\hat{L}^p(R_1, R_2)$. Hence these spaces are topologicaly equivalent. This completes the proof of the Theorem 5.2.

Remark. If the condition (5.3) is not satisfied, then $\hat{L}^p(R_1, R_2) \neq \bar{L}^p(R_1, R_2)$. Namely, in this case we obviously have $1 \notin \hat{L}^p(R_1, R_2)$, but $1 \in \bar{L}^p(R_1, R_2)$ (cf. Theorem 5.4).

In the case $r_{1i} = r_{2i} = r_{3i} = r_{4i} = 0$ $(i = 1, \dots, q)$ we have $\hat{L}^p(R_1, R_2) \subset \subset \bar{L}^p(R_1, R_2)$. This follows easily from the proof of Theorem 5.2.

Now we put

$$G(t) = t^{\eta+\xi_1} \prod_{i=1}^{q} \left(\frac{1}{\alpha_i} - \frac{1}{t}\right)^{-[-r_{1i}]-r_{1i}} (t - \alpha_i)^{-[-r_{2i}]-r_{2i}}$$

with $\xi_1 = \sum_{i=1}^{q} (n_{1i} - [-r_{1i}])$, $\eta_1 = \sum_{i=1}^{q} (n_{2i} - [-r_{2i}])$ and $\sigma = \eta + \eta_1 + \xi + \xi_1$.

Theorem 5.3. *Let the condition* (5.3) *be satisfied. A function* $f \in L^p(\Gamma)$ *belongs to* $\bar{L}^p(R_1, R_2)$ *if and only if there exists a ploynomial* $P_{\sigma-1}$ (*in* t) *of degree* $\leq \sigma - 1$ *such that*

$$\varphi_1 \equiv \frac{PGf - P_{\sigma-1}}{GR_1} \in L^p(\Gamma)\,, \quad \varphi_2 = \frac{QGf + P_{\sigma-1}}{GR_2} \in L^p(\Gamma)\,. \tag{5.8}$$

Proof. By Theorem 5.2 we have $\bar{L}^p(R_1, R_2) = \hat{L}^p(R_1, R_2)$. Let f be an element of this space, i.e. for a certain $\varphi \in L^p(\Gamma)$ the formula

$$R_1 P\varphi + R_2 Q\varphi = f$$

holds. By multipling this equation by G we obtain

$$GR_1 P\varphi + GR_2 Q\varphi = Gf\,.$$

The function $GR_1P\varphi$ belongs to the subspace $L^p_+(\Gamma) = \operatorname{im} P$. After subtracting a polynomial $P_{\sigma-1}$ the function $GR_2Q\varphi$ belongs to the subspace $L^p_-(\Gamma) = \operatorname{im} Q$. Hence by application of the operator P resp. Q to both sides of the last equation we obtain

$$GR_1 P\varphi + P_{\sigma-1} = PGf\,, \qquad GR_2 Q\varphi - P_{\sigma-1} = QGf\,.$$

This implies relations (5.8) with $\varphi_1 = P\varphi$, $\varphi_2 = Q\varphi$.

Conversely, let now the conditions (5.8) be satisfied. Then by Corollary 2.1 we have $\varphi_1 \in L^p_+(\Gamma)$ and $\varphi_2 \in L^p_-(\Gamma)$. This implies

$$Gf = GR_1 P\varphi_1 + GR_2 Q\varphi_2\,.$$

Since G has only a finite number of zeros, we obtain $f = R_1 P\varphi + R_2 Q\varphi$ with $\varphi = \varphi_1 + \varphi_2$. This proves Theorem 5.3.

Remark 1. If the condition (5.3) is not satisfied, then the formulae (5.8) yield necessary and sufficient conditions for the function $f \in L^p(\Gamma)$ to belong to the space $\hat{L}^p(R_1, R_2)$.

Remark 2. The space $\bar{L}^p(R_1, R_2)$ can also be described analytically without the hypothesis (5.3). Namely, $f \in \bar{L}^p(R_1, R_2)$ obviously holds if and only if simultaneously $Pf \in \bar{L}^p(R_1, 1)$ and $Qf \in \bar{L}^p(1, R_2)$. But the description of the space $\bar{L}^p(R_1, 1)$ and $\bar{L}^p(1, R_2)$ is given by Theorem 5.3.

In order to generalize Corollary 2.2 we introduce the following notation:

$$l_{1i} = n_{1i} + n_{3i} - [-r_{1i}] - [-r_{3i}]\,, \qquad l_{2i} = n_{2i} + n_{4i} - [-r_{2i}] - [-r_{4i}]$$

and

$$l_i = \max\{l_{1i}, l_{2i}\} \qquad (i = 1, \dots, q)\,.$$

Theorem 5.4. *The following formula holds (in the sense of a continuous embedding):*

$$W_p(\Gamma; (\alpha_i, l_i)_{i=1}^q) \subset \overline{L}^p(R_1, R_2)\,.$$

Proof. Let $f \in W_p(\Gamma; (\alpha_i, l_i)_{i=1}^q)$. By Corollary 2.2 we have $f \in \operatorname{im}(PR_1'P + QR_2'Q)$ with

$$R_1' = \prod_{i=1}^q \left(\frac{1}{\alpha_i} - \frac{1}{t}\right)^{l_{1i}}, \qquad R_2' = \prod_{i=1}^q (t - \alpha_i)^{l_{2i}}\,.$$

It is immediately seen that

$$PR_1'P + QR_2'Q = (PR_1P + QR_2Q)\,(PR_3P + QR_4Q)\,(PN_3P + QN_4Q)\,. \tag{5.9}$$

Here

$$N_3 = \prod_{i=1}^q \left(\frac{1}{\alpha_i} - \frac{1}{t}\right)^{-[-r_{1i}]-r_{1i}},$$

$$N_4 = \prod_{i=1}^q (t - \alpha_i)^{-[-r_{2i}]-r_{2i}}\,.$$

The functions R_3, R_4 are defined by equations (5.6). Now (5.9) implies that $f \in \overline{L}^p(R_1, R_2)$. The continuous embedding follows easily from Corollary 2.2.

Remark. We have not used the hypothesis (5.3) at the proof of Theorem 5.4.

6.5.2. We shall now study the properties of the singular integral operator $aP + bQ$ on the pair of spaces $L^p(\Gamma)$, $\overline{L}^p(R_1, R_2)$ $(1 < p < \infty)$. We assume that the coefficients $a(t)$ and $b(t)$ have the form (5.1), where the functions $R_1(t)$ and $R_2(t)$ are given by the formulae (5.2) and the functions $c(t)$, $d(t)$ belong to the following classes:

$$\begin{aligned} c(t) &\in \bigcap_{i=1}^q C(\alpha_i, l_{2i} - [-r_{3i}]) \\ d(t) &\in \bigcap_{i=1}^q C(\alpha_i, l_{1i} - [-r_{4i}])\,. \end{aligned} \tag{5.10}$$

Under these hypotheses we have:

Theorem 5.5. *Let condition* (5.3) *be satisfied. Then* $A = aP + bQ \in \mathcal{L}(L^p(\Gamma), \overline{L}^p(R_1, R_2))$. *This operator is a* Φ_+*-operator or a* Φ_-*-operator if and only if*

$$c(t) \neq 0\,, \qquad d(t) \neq 0 \qquad (t \in \Gamma)\,. \tag{5.11}$$

If conditions (5.11) *are satisfied, then A is a Φ-operator from $L^p(\Gamma)$ into the space $\overline{L}^p(R_1, R_2)$ with the index*

$$\operatorname{Ind} A = -\varkappa - \sum_{i=1}^{q} \left([-r_{3i}] + [-r_{4i}] + \left[r_{3i} + \frac{1}{p'}\right] + \left[r_{4i} + \frac{1}{p'}\right] \right), \tag{5.12}$$

where $\frac{1}{p'} = 1 - \frac{1}{p}$ *and* $\varkappa = \frac{1}{2\pi}\left[\arg \frac{c(t)}{d(t)}\right]_\Gamma$. *The operator $A \in \mathscr{L}(L^p(\Gamma), \overline{L}^p(R_1, R_2))$ is at least one-sided invertible.*

Proof. We put $R_5 = R_2R_3R_4$, $R_6 = R_1R_3R_4$, where the functions R_3, R_4 are defined by (5.6). Obviously. we have $R_5 \in C^+(\Gamma)$. $R_6 \in C^-(\Gamma)$ and $R_1R_5 = R_2R_6$. Using the conditions (5.10) we easily obtain from the results of Section 2.3.4 that the representations

$$c = R_5c_1 + h_+ , \qquad d = R_6d_1 + h_- \tag{5.13}$$

hold with certain functions c_1, $d_1 \in C(\Gamma)$ and $h_\pm \in C^\pm(\Gamma)$. Then a simple calculation yields

$$A = (R_1P + R_2Q)\,[cP + dQ + (R_5 - R_6)\,(Pd_1Q - Qc_1P)]\,. \tag{5.14}$$

As we already remarked in the proof of Theorem 5.2, it follows from (5.7) that $R_1P + R_2Q \in \Phi(L^p(\Gamma), \overline{L}^p(R_1, R_2))$ with $\operatorname{Ind}(R_1P + R_2Q) = k$, where

$$k = k(R_1, R_2) = -\sum_{i=1}^{q} \left([-r_{3i}] + [-r_{4i}] + \left[r_{3i} + \frac{1}{p'}\right] + \left[r_{4i} + \frac{1}{p'}\right] \right).$$

Taking into consideration the results of Section 1.3 we conclude from the equation (5.14) that $A \in \mathscr{L}(L^p(\Gamma), \overline{L}^p(R_1, R_2))$ and that this operator is a Φ_+-operator or a Φ_--operator if and only if $A_1 = cP + dQ + (R_5 - R_6)$ $(Pd_1Q - Qc_1d) \in \Phi_\pm(L^p(\Gamma))$. But this is equivalent to $A_0 = cP + dQ \in \Phi_\pm(L^p(\Gamma))$, since the operator $A_1 - A_0$ is compact on $L^p(\Gamma)$. Theorem 4.5, Chapter 3, shows that $A_0 \in \Phi_\pm(L^p(\Gamma))$ holds if and only if conditions (5.11) are satisfied. If this is the case, then $A_1 \in \Phi(L^p(\Gamma))$ with $\operatorname{Ind} A_1 = -\varkappa$. Now by Theorem 3.6, Chapter 1, the index formula (5.12) follows from equation (5.14).

Now we prove the last assertion of the theorem. We show that the operators $A \in \mathscr{L}(L^p(\Gamma), \overline{L}^p(R_1, R_2))$ and $cP + dQ \in \mathscr{L}(L^p(\Gamma))$ are simultaneously invertible or not invertible. First we consider the case $k(R_1, R_2) = 0$. Then it is immediately seen that

$$(R_5P + R_6Q)\,A_0 = A_1(R_5P + R_6Q)\,. \tag{5.15}$$

From this it follows that the operator A_1 maps the image space $\operatorname{im}(R_5P + R_6Q)$ into itself. Since obviously $\dim\ker(R_5P + R_6Q) = 0$, in the Banach space $\boldsymbol{L} = \operatorname{im}(R_5P + R_6Q)$ (with the norm $\|f\|_{\boldsymbol{L}} = \|(R_5P + R_6Q)^{-1} f\|_{L^p(\Gamma)}$) the operator A_1 possesses the same properties as the operator $A_0 = cP + dQ$

in the space $L^p(\Gamma)$. By $\boldsymbol{A}_1$ we denote the operator A_1 considered in the space $\boldsymbol{L}$.

Now let A_0 be one-sided invertible in $L^p(\Gamma)$. Then by Theorems 4.4 and 4.5, Chapter 3, we have $A_0 \in \Phi(L^p(\Gamma))$. Hence A_1 (in $L^p(\Gamma)$) and $\boldsymbol{A}_1$ are Φ-operators, where moreover

$$\operatorname{Ind} A_0 = \operatorname{Ind} A_1 = \operatorname{Ind} \boldsymbol{A}_1 . \tag{5.16}$$

The set im $(R_5P + R_6Q)$ is dense in $L^p(\Gamma)$. In fact, the equation $PR_6 + QR_5 = PR_6P + QR_5Q$ and Theorem 5.1 show that dim ker $(PR_6 + QR_5) = 0$. This together with equation (4.56), Chapter 3, yields the assertion. From the density of $\boldsymbol{L}$ in $L^p(\Gamma)$ and from the equation (5.16) we obtain easily that dim ker A_1 = dim ker $\boldsymbol{A}_1$. Furthermore, from dim ker $\boldsymbol{A}_1$ = dim ker A_0 (cf. (5.15)) it follows at once that A_1 is invertible in $L^p(\Gamma)$ on the same side as A_0. Hence also $A = (R_1P + R_2Q) A_1 \in \mathscr{L}(L^p(\Gamma), \overline{L}^p(R_1, R_2))$ is invertible on the same side as A_0.

Now we consider the case $k(R_1, R_2) > 0$. Let $A_0 = cP + dQ$ be one-sided invertible, i.e. $A_0 \in \Phi(L^p(\Gamma))$ and moreover the conditions (5.11) are satisfied. Since by (5.3) R_1 and R_2 have no common zeros on Γ, we obtain (cf. the proof of Lemma 4.4, Chapter 4)

$$\dim \ker (R_1cP + R_2dQ) = \dim \ker (PR_1c + QR_2d) . \tag{5.17}$$

The functions R_1 and R_2 can be represented in the form

$$R_1 = t^{-\theta_1} G_1 G_2 , \qquad R_2 = t^{\theta_2} H_1 H_2$$

with

$$\begin{aligned} G_2 &= \prod_{i=1}^{q} (t - \alpha_i)^{-\left([-r_{3i}] + \left[r_{3i} + \frac{1}{p'}\right]\right) r_{3i}} , \\ \theta_1 &= - \sum_{i=1}^{q} \left([-r_{3i}] + \left[r_{3i} + \frac{1}{p'}\right]\right), \\ H_2 &= \prod_{i=1}^{q} \left(\frac{1}{\alpha_i} - \frac{1}{t}\right)^{-\left([-r_{4i}] + \left[r_{4i} + \frac{1}{p'}\right]\right) r_{4i}} , \\ \theta_2 &= - \sum_{i=1}^{q} \left([-r_{4i}] + \left[r_{4i} + \frac{1}{p'}\right]\right). \end{aligned} \tag{5.18}$$

By these statements the functions G_1 and H_1 are uniquely determined. It is easy to see that G_1, H_1 are functions of the form (5.2) for which $k(G_1, H_1) = 0$. From Theorem 5.1. it follows easily that dim ker $(PH_2 + QG_2) = 0$. Now this together with the results obtained before yields

$$\begin{aligned} &\dim \ker (PR_1c + QR_2d) = \dim \ker (PH_2 + QG_2)(PR_1c + QR_2d) = \\ &= \dim \ker [(Pt^{-\theta_1}G_1c + Qt^{\theta_2}H_1d) G_2H_2] \leq \\ &\leq \dim \ker (Pt^{-\theta_1}G_1c + Qt^{\theta_2}H_1d) = \dim \ker (G_1ct^{-\theta_1}P + H_1dt^{\theta_2}Q) = \\ &= \dim \ker (ct^{-\theta_1}P + dt^{\theta_2}Q) . \end{aligned}$$

By Theorem 4.4, Chapter 3, for the operator $A_2 = ct^{-\theta_1}P + dt^{\theta_2}Q$ we have

$$\dim \ker A_2 = \max \{0, \operatorname{Ind} A_2\} .$$

Thus from (5.17) we obtain

$$\dim \ker A \leqq \max \{0, \operatorname{Ind} A_2\} .$$

Using Theorem 5.1 and equation (5.14) we can easily check that

$$\operatorname{Ind} A_2 = \operatorname{Ind} A_0 + \theta_1 + \theta_2 = \operatorname{Ind} A .$$

From this it follows now that

$$\dim \ker A \leqq \max \{0, \operatorname{Ind} A\} ,$$

and together with the obvious relation $\dim \ker A \geqq \operatorname{Ind} A$ we obtain

$$\dim \ker A = \max \{0, \operatorname{Ind} A\} .$$

With this it has been shown that the operator $A \in \mathcal{L}(L^p(\Gamma), \overline{L}^p(R_1, R_2))$ is at least one-sided invertible.

This completes the proof of Theorem 5.5.

Remark 1. With the exception of the last assertion Theorem 5.5 holds also if instead of (5.3) the following condition is satisfied for the functions (5.2):

$$n_{1i} + r_{1i} = n_{2i} + r_{2i} = 0 \qquad (i = 1, \dots, q) . \tag{5.19}$$

This follows from the preceding considerations since under the hypotheses (5.19) the representation (5.7) also holds.

In fact, in this case namely we have

$$R_1P + R_2Q = PR_1P + QR_2Q + (Qt^{-\eta}P + Pt^{\xi}Q)(Pt^{\eta}R_1P + Qt^{-\xi}R_2Q) .$$

Since $t^{-\eta}$ and t^{ξ} are infinitely differentiable functions, they admit the representations

$$t^{-\eta} = R_2R_4g_1 + q_+ , \qquad t^{\xi} = R_1R_3g_2 + q_-$$

with $g_1, g_2 \in C(\Gamma)$ and $q_\pm \in C^\pm(\Gamma)$. This at once implies

$$(Qt^{-\eta}P + Pt^{\xi}Q) = (PR_1P + QR_2Q)(QR_3Qg_1P + PR_4Pg_2Q)$$

and thus also (5.7).

Remark 2. In Chapter 10 we shall also study operators of the form $Pa + Qb$ under the hypotheses of this section (even under more general hypothese).

6.6. The Solution of a Singular Integral Equation of Non-Normal Type by Means of Factorization

Under the hypotheses of Section 6.5.2 (including the condition (5.3)) we now shall deal with the effective solution of the integral equation

$$A\varphi = c(t)\,P\varphi + d(t)\,Q\varphi = f(t) \qquad (t \in \Gamma) . \tag{6.1}$$

The right-hand side $f(t)$ shall belong to the space $\overline{L}^p(R_1, R_2)$ introduced in 6.5. We search for a solution $\varphi(t) \in L^p(\Gamma)$. In the sequel we shall use some considerations from Section 4.3.6.

For the functions $c(t)$ and $d(t)$ in (5.1) we now assume in addition that they belong to the class $H^\lambda(\Gamma)$ $(0 < \lambda < 1)$ and moreover vanish nowhere on Γ. Then by 3.4.1 the function $d(t)/c(t)$ can be factorized:

$$\frac{d(t)}{c(t)} = a_-(t)\, t^{-\varkappa}\, a_+(t) \qquad (t \in \Gamma),$$

where $a_\pm(t) \in H^\lambda(\Gamma)$. For the sequel we still introduce the following notations:

$$M_0(t) = t^{\theta_1} R_1(t), \qquad N_0(t) = t^{-\theta_2} R_2(t),$$

$$c_0(t) = t^{-\theta_1} c(t), \qquad d_0(t) = t^{\theta_2}\, d(t), \qquad \varrho = \theta_1 + \theta_2 - \varkappa,$$

$$V = M_0 P + N_0 Q, \qquad A_0 = c_0 P + d_0 Q.$$

The numbers θ_1, θ_2 are defined by the relations (5.18). Using the Theorem 5° of 3.4.4 we then obtain the following representation of the operator A:

$$A = c_0 a_+ (M_0 P + t^\varrho N_0 Q)\,(a_+^{-1} P + a_- Q). \tag{6.2}$$

The operator $a_+^{-1} P + a_- Q$ has the inverse $a_+ P + a_-^{-1} Q$ in $L^p(\Gamma)$. From Theorem 5.1 it follows easily that $\dim \ker V = 0$.

By L_0 we denote the subspace of the dimension $\max(\varrho, 0)$ formed of all functions of the form

$$\psi(t) = (N_0 - t^{-\varrho} M_0)\, R_3 R_4 P_{\varrho-1}(t).$$

Here $P_{\varrho-1}(t)$ is an arbitrary polynomial of degree $\leqq \varrho - 1$ ($P_{\varrho-1}(t) \equiv 0$ for $\varrho \leqq 0$). The functions $R_3(t)$ and $R_4(t)$ are determined by (5.6). It is easily seen that the space $L^p(\Gamma)$ can be represented as the direct sum

$$\operatorname{im}(P + t^{-\varrho} Q) \dotplus L_0 = L^p(\Gamma).$$

Repeating the considerations from 4.3.6 we now obtain the following two theorems (cf. also Theorem 5.5).

Theorem 6.1. *The kernel of the operator A consists of all functions $(a_+ P + a_-^{-1} Q)\,\psi$ with $\psi \in L_0$, i.e. of all functions of the form*

$$\varphi_0(t) = (a_+ N_0 - t^{-\varrho} a_-^{-1} M_0)\, R_3 R_4 P_{\varrho-1}(t). \tag{6.3}$$

Theorem 6.2. *The invertibility of the operator $A \in \mathscr{L}(L^p(\Gamma), \overline{L}^p(R_1, R_2))$ corresponds to the number $-\varrho = \varkappa - \theta_1 - \theta_2$ and*

$$\dim \ker A = \max(\varrho, 0), \qquad \dim \operatorname{coker} A = \max(-\varrho, 0).$$

The corresponding one-sided inverse has the form

$$A^{(-1)} = (a_+ P + a_-^{-1} Q)\,(P + t^{-\varrho} Q)\, V^{-1} (c_0 a_+)^{-1} I.$$

Now we return to equation (6.1). The following statement follows immediately from Theorem 6.2:

For $\varrho \geqq 0$

$$\varrho_1 = A^{(-1)}f \tag{6.4}$$

is always a solution of equation (6.1). The general solution has the form $\varphi = \varphi_1 + \varphi_0$, where the functions φ_0 are given by the formula (6.3).

In the case $\varrho < 0$ the function (6.4) is the only possible solution of equation (6.1). This function is a solution of (6.1) if and only if the condition

$$Q(t^{-\varrho}QV^{-1}g) = t^{-\varrho}QV^{-1}g\,, \qquad g = f/c_0a_+$$

is satisfied or, what is the same, the conditions

$$\int_\Gamma t^k(V^{-1}g)\,(t)\,dt = 0 \qquad (k = 0, 1, \dots, -\varrho - 1)$$

are satisfied. The last assertion follows from the fact that in the case $\varrho < 0$ a function $f \in L^p(\Gamma)$ satisfies the equation $A_0A_0^{(-1)}f = f$ if and only if $Qt^{-\varrho}Qf = t^{-\varrho}Qf$ (cf. formula (3.15), Chapter 4).

6.7. Non-Bounded Regularization of Singular Integral Equations of Non-Normal Type

6.7.1. Non-Bounded Regularization

We consider the singular integral equation

$$A\varphi = a(t)\,\varphi(t) + \frac{b(t)}{\pi i}\int_\Gamma \frac{\varphi(\tau)}{\tau - t}\,d\tau + \int_\Gamma T(t,\tau)\,\varphi(\tau)\,d\tau = f(t)\,, \tag{7.1}$$

in the space $L^p(\Gamma)$ $(1 < p < \infty)$. Here Γ is a curve system of the form described in 6.2.1. For the coefficients of the equation (7.1) we state the following hypotheses:

(1) $a(t)$, $b(t)$ and $T(t, \tau)$ satisfy a Hölder condition on Γ and $\Gamma \times \Gamma$, respectively.

(2) The functions

$$c(t) = a(t) + b(t)\,, \qquad d(t) = a(t) - b(t)$$

vanish on Γ at the points $\alpha_1, \alpha_2, \dots, \alpha_r$ and $\beta_1, \beta_2, \dots, \beta_s$, respectively. The positive integers m_j resp. n_k $(j = 1, 2, \dots, r;\ k = 1, \dots, s)$ are the orders of these zeros.

(3) In a neighborhood (on Γ) of the points α_j and β_k $(j = 1, \dots, r;\ k = 1, \dots, s)$ the functions $a(t)$, $b(t)$ and $T(t, \tau)$ possess derivatives with respect to t up to the order m_j resp. n_k, and these derivatives satisfy a Hölder condition with exponent $\lambda > \dfrac{p-1}{p}$.

At last let $\Gamma \in C^{\widetilde{m}, \mu}$ with $\widetilde{m} = \max\{m_1, \ldots, m_r, n_1, \ldots, n_s\}$ and $0 < \mu \leq 1$.

We show that under these hypotheses a non-bounded regularizer of the operator A can explicitly be constructed by means of the coefficients $c(t)$ and $d(t)$. (Here, similarily as at the regularization of the equation (7.1) in the normal case, we make no use of the factorization of the functions in the algebra $H^\lambda(\Gamma)$.)

We first prove the following lemma.

Lemma 7.1. *Let Γ_0 be an arc on Γ, $f(t, \tau) \in H^\lambda(\Gamma_0 \times \Gamma)$ $(0 < \lambda \leq 1)$ and*

$$F(t, \tau) = \int_\Gamma \frac{f(t, t_1)}{t_1 - \tau} dt_1 \qquad (t \in \Gamma_0, \tau \in \Gamma).$$

If the partial derivatives $f_t^{(k)}(t, \tau)$ of the function $f(t, \tau)$ with respect to the variable t exist for $k = 1, \ldots, m$ and belong to $H^\lambda(\Gamma_0 \times \Gamma)$, then

$$F_t^{(k)}(t, \tau) = \int_\Gamma \frac{f_t^{(k)}(t, t_1)}{t_1 - \tau} dt_1 \qquad (k = 1, \ldots, m)$$

and $F_t^{(k)}(t, \tau) \in H^{\lambda'}(\Gamma_0 \times \Gamma)$ for any $\lambda' < \lambda$.

Proof. For sufficiently small $\mu > 0$ and fixed $\tau \in \Gamma$ we put

$$g(t, t_1) = \frac{f(t, t_1) - f(t, \tau)}{|t_1 - \tau|^\mu}, \qquad \varphi(t_1) = \frac{|t_1 - \tau|^\mu}{t_1 - \tau}.$$

Then $g(t, t_1)$ satisfies a Hölder condition on $\Gamma_0 \times \Gamma$ (cf. N. I. MUSHELIŠVILI [1], §§ 5–6), and by a well-known theorem from analysis (see G. M. FICHTENHOLZ [1], vol. II) we now have

$$F_t^{(k)}(t, \tau) = \int_\Gamma g_t^{(k)}(t, t_1)\, \varphi(t_1)\, dt_1 + \pi i f_t^{(k)}(t, \tau) = \int_\Gamma \frac{f_t^{(k)}(t, t_1)}{t_1 - \tau} dt_1$$

$$(k = 0, 1, \ldots, m).$$

From this it follows that $F_t^{(k)}(t, \tau) \in H^{\lambda'}(\Gamma_0 \times \Gamma)$ (comp. N. I. MUSHELIŠVILI [1], § 18).

This proves the lemma.

We now come to the construction of the regularizer. If for the function $\varphi(t) \in L^p(\Gamma)$ the mean derivatives $\varphi^{\{k\}}(\alpha_j)$ $(k = 0, 1, \ldots, m_j - 1; j = 1, \ldots, r)$ exist, then by $H_1\varphi$ we denote the Hermite interpolation polynomial of the function φ with the interpolation knots α_j satisfying the conditions

$$(H_1\varphi)^{(k)}(\alpha_j) = \varphi^{\{k\}}(\alpha_j) \qquad (k = 0, 1, \ldots, m_j - 1; j = 1, \ldots, r).$$

Correspondingly $H_2\varphi$ is defined with β_l $(l = 1, \ldots, s)$ instead of α_j.

We consider the operator B given in the space $L^p(\Gamma)$ by the equation

$$B\varphi = \frac{P\varphi - H_1 P\varphi}{c(t)} + \frac{Q\varphi - H_2 Q\varphi}{d(t)}. \tag{7.2}$$

The domain $D(B)$ consists of all functions $\varphi \in L^p(\Gamma)$ for which the right-hand side of (7.2) is defined and is an element of the space $L^p(\Gamma)$. The set $D(B)$ is dense in $L^p(\Gamma)$ since it contains all infinitely differentiable functions on Γ (cf. 6.1). It is easily seen that B is not a bounded operator.

We show that B is a regularizer of the operator A (cf. the definition from 1.4).

Using the commutation formulae (4.37) and (4.38) from Chapter 3 and the equation

$$\int_\Gamma \frac{d\tau}{\tau - t} = \pi i \qquad (t \in \Gamma)$$

for any $\varphi \in L^p(\Gamma)$ we obtain

$$PA\varphi = c(t)\, P\varphi + \int_\Gamma N(t, \tau)\, \varphi(\tau)\, d\tau\,, \tag{7.3}$$

$$QA\varphi = d(t)\, Q\varphi + \int_\Gamma M(t, \tau)\, \varphi(\tau)\, d\tau\,,$$

where

$$2N(t, \tau) = R_a(t, \tau) + G(t, \tau) + L(t, \tau) + 2T(t, \tau)\,,$$
$$2M(t, \tau) = -\, R_a(t, \tau) - G(t, \tau) - L(t, \tau)$$

and

$$R_a(t, \tau) = \frac{1}{\pi i} \frac{a(t) - a(\tau)}{t - \tau}\,, \qquad G(t, \tau) = \frac{1}{\pi i} \int_\Gamma R_b(t, t_1) \frac{dt_1}{\tau - t_1}\,,$$

$$L(t, \tau) = \frac{1}{\pi i} \int_\Gamma [T(t_1, \tau) - T(t, \tau)] \frac{dt_1}{t_1 - t}\,.$$

By Lemma 1.1 and Lemma 7.1 the functions $N(t, \tau)$ and $M(t, \tau)$ satisfy the above-mentioned condition (3) with $m_j - 1$ instead of m_j and $n_k - 1$ instead of n_k. By Lemma 1.9 we obtain

$$H_1[cP\varphi] = 0\,, \qquad H_2[dQ\varphi] = 0\,,$$

and the equations (7.3) imply that $A\varphi \in D(B)$ and

$$BA\varphi = \varphi(t) + \int_\Gamma K(t, \tau)\, \varphi(\tau)\, d\tau \tag{7.4}$$

with

$$K(t, \tau) = \frac{1}{c(t)} [N(t, \tau) - H_1 N] + \frac{1}{d(t)} [M(t, \tau) - H_2 M]\,,$$

where in $H_1 N$ (resp. in $H_2 M$) we have to apply the operation H_1 (resp. H_2) to the argument t.

For the proof of the assertion that B is a regularizer of the operator A it remains to prove that the integral operator with the kernel $K(t, \tau)$ is compact in the space $L^p(\Gamma)$. We show this for the operator with the kernel

$$K_1(t, \tau) = \frac{1}{c(t)} [N(t, \tau) - H_1 N] .$$

For $K_2(t, \tau) = K(t, \tau) - K_1(t, \tau)$ the proof is analogous.

Let U_j be a sufficiently small neighborhood on Γ of the point α_j $(j = 1, \dots, r)$ and

$$\Gamma' = \bigcup_{j=1}^{r} U_j , \qquad \Gamma'' = \Gamma - \Gamma' .$$

We put

$$V(t,\tau) = \begin{cases} 0 , & t \in \Gamma' \\ K_1(t,\tau) , & t \in \Gamma'' \end{cases}; \qquad W(t,\tau) = K_1(t,\tau) - V(t,\tau) .$$

First we consider the operator V with the kernel $V(t, \tau)$:

$$V\varphi = h(t) \Big[\int_\Gamma N(t,\tau)\, \varphi(\tau)\, d\tau + \int_\Gamma V_0(t,\tau)\, \varphi(\tau)\, d\tau \Big]$$

with

$$V_0(t,\tau) = - H_1 N , \qquad h(t) = \begin{cases} 0 , & t \in \Gamma' \\ \dfrac{1}{c(t)} , & t \in \Gamma'' . \end{cases}$$

The operator with the kernel $V_0(t, \tau)$ is a continuous operator of finite rank and thus compact. Since the integral operators with kernels of potential type are compact (comp. L. V. Kantorovič and G. P. Akilov [1], p. 296), $\int_\Gamma N(t, \tau)\, \varphi(\tau)\, d\tau$ and thus V is a compact operator.

It remains still to show that the operator

$$W\varphi = \int_\Gamma W(t,\tau)\, \varphi(\tau)\, d\tau$$

is compact. Since the derivatives $N_t^{(m_j-1)}(t, \tau)$ satisfy a Hölder condition with the exponent $\nu > p^{-1}(p - 1)$ for $t \in U_j, \tau \in \Gamma$ and $j = 1, \dots, r$, we have

$$|K_1(t,\tau)| \leqq c_0 \prod_{j=1}^{r} |t - \alpha_j|^{\nu - 1} \qquad (t \in \Gamma', \tau \in \Gamma) .$$

Hence, because of $(1 - \nu)\, p < 1$, for $p' = p/(p - 1)$ we obtain that the integral

$$\int_\Gamma \Big[\int_\Gamma |W(t,\tau)|^{p'} |d\tau| \Big]^{p/p'} |dt| = \int_{\Gamma'} \Big[\int_\Gamma |K_1(t,\tau)|^{p'} |d\tau| \Big]^{p/p'} |dt|$$

is finite. Thus a well-known sufficient condition for the compactness of the integral operator with the kernel $W(t, \tau)$ in the space $L^p(\Gamma)$ is satisfied (see L. V. Kantorovič and G. P. Akilov [1], p. 292). Consequently the compactness of the integral operator in (7.4) is proved.

The regularizer B possesses a finite dimensional kernel.

In fact, multiplying both sides of the equation $B\varphi = 0$ by $c(t)\, d(t)$ we obtain an equation of the form

$$A_1\varphi \equiv d(t)\, P\varphi + c(t)\, Q\varphi = d(t)\, \psi_1(t) + c(t)\, \psi_2(t)\,,$$

where $\psi_1(t)$ and $\psi_2(t)$ are certain polynomials whose degrees are not greater then $m - 1$ and $n - 1$, respectively, with

$$m = \sum_{j=1}^{r} m_j\,, \qquad n = \sum_{k=1}^{s} n_k\,. \tag{7.5}$$

By the already given part of the proof the operator A_1 has a non-bounded regularizer and consequently $\dim \ker A_1 < \infty$. Hence

$$\dim \ker B \leqq \dim \ker A_1 + m + n < \infty\,.$$

We formulate the results of this subsection in the following theorem.

Theorem 7.1. *Under the hypotheses* (1) *to* (3) *the operator B defined by* (7.2) *is a non-bounded regularizer of the singular integral operator* (7.1). *The kernel of this regularizer is finite dimensional.*

Corollary 7.1. *The operator A considered in the space $L^p(\Gamma)$ possesses a finite index and is not normally solvable.*

This follows immediately from Theorems 5.4 and 5.5 of Chapter 1.

Remark 1. The hypotheses with respect to the kernel $T(t, \tau)$ in (7.1) can be weakened. It is sufficient to demand the representation

$$T(t, \tau) = \frac{k(t, \tau) - k(t, t)}{\tau - t}$$

instead of condition (1), where $k(t, \tau)$ satisfies the condition (1).

This follows from the fact, that in this case also the commutation formulae (4.37), Chapter 3, hold.

Remark 2. Lemma 1.10 shows that the regularizer B is closed.

Furthermore, by Theorem 2.1 we obtain easily that a function φ belongs to the domain $D(B)$ if and only if $\varphi \in \bar{L}^p(\varrho''_+, \varrho''_-)$, where the functions $\varrho''_\pm(t)$ are defined in the same way as in 6.3.5. Hence the space $\boldsymbol{L} = \bar{L}^p(\varrho''_+, \varrho''_-)$ coincides with the space $\boldsymbol{D}$ obtained by introduction of the norm

$$|\varphi| = ||\varphi||_{L^p} + ||B\varphi||_{L^p} \qquad (\varphi \in D(B))$$

on $D(B)$ (cf. 4.5).

Remark 3.[1] The results of this subsection can be carried over to the case of zeros of non-integral order. In this case the regularizer B also has the form (7.2) but the interpolation polynomial $H_1\varphi$ is now determined by the conditions

$$(H_1\varphi)^{(k)}\,(\alpha_j) = \varphi^{\{k\}}(\alpha_j) \qquad \left(k = 0, 1, \dots, \left[m_j - \frac{1}{p}\right];\; j = 1, \dots, r\right).$$

The corresponding statement holds for $H_2\varphi$.

[1]) Cf. S. Prössdorf and B. Silbermann [1].

6.7.2. Equivalent Non-Bounded Regularization

In general, the regularizer B constructed in the preceding subsection is not an equivalent regularizer of the operator A. But B can easily be changed into such an operator which for Ind $A \geqq 0$ is an equivalent left regularizer of A (for this cf. Theorem 5.6, Chapter 1).

Here we maintain the hypotheses about the coefficients of the equation (7.1) assumed in 6.7.1. But now we demand in addition that $\alpha_j \neq \beta_k$ for all $j = 1, \dots, r$ and $k = 1, \dots, s$.

By Lemma 1.1 we can represent the functions $c(t)$ and $d(t)$ in the form

$$c(t) = R_0(t)\, c_0(t)\,, \qquad d(t) = R_1(t)\, d_1(t) \tag{7.6}$$

with

$$R_0(t) = \prod_{j=1}^{r} (t - \alpha_j)^{m_j}\,, \qquad R_1(t) = \prod_{k=1}^{s} (t - \beta_k)^{n_k}\,,$$

where the functions $c_0(t)$ and $d_1(t)$ satisfy a Hölder condition on Γ and vanish nowhere on Γ. In the sequel we shall assume that the functions $c_0(t)$ and $d_1(t)$ in (7.6) also satisfy condition (3) of 6.7.1. (By the remark made at the end of 6.1.2 for this it is sufficient that the functions $c(t)$ and $d(t)$ belong to the class $C^\lambda(\Gamma; (\alpha_j, 2m_j)_1^r, (\beta_k, 2n_k)_1^s)$.

We now consider the operator $\widetilde{B} = B_1 + B_2$, where

$$B_1 = d_1^{-1} R_0^{-1} (I - \widetilde{H}_1) \frac{d_1}{c_0} P\,, \qquad B_2 = c_0^{-1} R_1^{-1} (I - H_2) \frac{c_0}{d_1} Q\,.$$

Here H_2 has the same meaning as in 6.7.1. But

$$\widetilde{H}_1 \varphi = \sum_{k=1}^{m} \frac{a_k}{t^{n+k}}$$

is the Hermite interpolation polynomial with respect to the basis functions t^{-n-k} $(k = 1, \dots, m)$ determined by the conditions

$$(\widetilde{H}_1 \varphi)^{(l)}(\alpha_j) = \varphi^{\{l\}}(\alpha_j) \qquad (l = 0, 1, \dots, m_j - 1;\ j = 1, \dots, r)\,,$$

m and n are defined by (7.5).

Repeating the arguments in 6.7.1 we obtain easily that $\widetilde{B}$ is also a left regularizer of the operator A. By 1.5.2 $\widetilde{B}$ is an equivalent regularizer if and only if dim ker $\widetilde{B} = 0$. Now we shall investigate when this is the case.

Let $\varphi \in L^p(\Gamma)$ be a solution of the equation $B\varphi = 0$. We factorize the function d_1/c_0:

$$\frac{d_1(t)}{c_0(t)} = a_-(t)\, t^{\varkappa_0} a_+(t)\,; \qquad \varkappa_0 = \frac{1}{2\pi} \left[\arg \frac{d_1(t)}{c_0(t)}\right]_\Gamma .$$

Putting this into the equation $\widetilde{B}\varphi = 0$ after some elementary calculations we obtain a relation of the form

$$R_1 a_+ t^{\varkappa_0} P\varphi - R_0 a_+ t^{\varkappa_0} H_2 \varphi_2 = R_1 a_-^{-1} \widetilde{H}_1 \varphi_1 - R_0 a_-^{-1} Q\varphi$$

with φ_1 and φ_2 from $L^p(\Gamma)$.

Now we assume $\varkappa_0 \geqq m$. Multiplying both sides of the last equation by t^{-m} we obtain that the left-hand side belongs to $L_+^p(\Gamma)$ and the right-hand side to $L_-^p(\Gamma)$. Hence

$$P\varphi = \frac{R_0(t)}{R_1(t)} \sum_{l=0}^{n-1} b_l t^l, \qquad Q\varphi = \frac{R_1(t)}{R_0(t)} \sum_{k=1}^{m} \frac{a_k}{t^{n+k}}.$$

Since $\alpha_j \neq \beta_k$ and $P\varphi, Q\varphi \in L^p(\Gamma)$, from this it follows necessarily that $b_l = 0$ $(l = 0, \dots, n-1)$ and $a_k = 0$ $(k = 1, \dots, m)$. Hence $P\varphi = Q\varphi = 0$, and consequently $\varphi = 0$.

Thus we have proved the following theorem.

Theorem 7.2. *The operator $\widetilde{B}$ is an equivalent left regularizer of the operator A if $\varkappa_0 \geqq m$.*

Remark 1. The result formulated in Theorem 7.2 cannot be sharpened. Namely, it is possible to show that there exists a compact (and even finite dimensional) integral operator T with the properties demanded in 6.7.1 such that

$$\operatorname{Ind} A = \varkappa_0 - m \qquad (\operatorname{Ind} A = \varkappa_0 + n)$$

holds (see S. Prössdorf [3]). Then consequently $\operatorname{Ind} A < 0$ for $\varkappa_0 < m$, and in this case by Theorem 5.6, Chapter 1, the operator A cannot have an equivalent left regularizer.

Remark 2. In the case $\varkappa_0 \leqq m$ the operator $\widetilde{B}$ constructed above is an equivalent right regularizer of the operator $A\colon L^p(\Gamma) \to \bar{L}^p(\varrho_+'', \varrho_-'')$ (cf. the remark at the end of 6.7.1).

This assertion can be proved by the methods of a paper by S. Prössdorf and E. Teichmann [1].

6.8. Comments and References

6.1. In the Subsections 6.1.2 and 6.1.3 for the most part we follow the authors paper [10] Corollary 1.1 is due to S. G. Mihlin [1]. The proof of Lemma 1.5 given here is a slight modification of the proof of Corollary 1.1 given by S. G. Mihlin [1]. The formula for differentiation of the singular integral mentioned in Corollary 1.3 can be found in the book of F. D. Gahov ([1], p. 49).

The results in 6.1.4 were obtained by the author in his thesis (Leningrad 1966) (cf. also [8]).

6.2. The results of Subsections 6.2.2 and 6.2.3 were simultaneously found by the author [13] and A. A. Semencul [1]. The proof of Lemma 2.1 is taken from A. A. Semencul [1]. The results of Subsection 6.2.4 were proved by the author [13—14].

6.3—6.4. Here essentially we present the results of the author [10], [13—14].

In this connection we also refer to the papers of B. Silbermann [1—2], where similar results as in 6.3.5 were stated for the spaces $H^\lambda(\Gamma)$ in the case of zeros of not necessarily integral orders. In the papers of S. MEYER [1—2] some results from Section 6.3 were carried over to the case of zeros of logarithmic type and to some cases where the functions $a \pm b$ are identically equal to zero on whole arcs. Also F. D. GAHOV [3] considered the last-mentioned case in another connection.

In this connection some results of N. E. TOVMASJAN [1] are remarkable. This author considered the equation (7.1) under the following hypotheses about the coefficients:

a) Each of the functions $a(t) + b(t)$ and $a(t) - b(t)$ is either identically equal to zero on Γ or unequal to zero everywhere on Γ.

b) The kernel $T(t, \tau)$ has the form

$$T(t, \tau) = K_1(t, \tau) \ln \left(1 - \frac{\tau}{t}\right) + K_2(t, \tau) \ln \left(1 - \frac{t}{\tau}\right) + K_3(t, \tau) ,$$

where $K_j(t, \tau)$ $(j = 1, 2, 3)$ are infinitely differentiable functions.

Necessary and sufficient conditions for the solvability of the equation (7.1) are formulated and the index of the equation is calculated (this index essentially depends on the functions $K_1(t, t)$ and $K_2(t, t)$). The operator A proves to be Φ-operator on the space $C^\infty(\Gamma)$ (for this comp. Chapter 9).

6.5. With the exception of the last assertion of Theorem 5.5 the results of this section were obtained by S. PRÖSSDORF and B. SILBERMANN [1] together. The last assertion of Theorem 5.5 (about one-sided invertibility) was proved by B. SILBERMANN [8] in the more general case of piecewise continuous coefficients (cf. Chapter 10). Here we essentially follow the paper of B. SILBERMANN [8].

6.6. This section contains generalizations of some results of the author [10] and B. SILBERMANN [4], [2]. Similar results were proved also for the space $H^\lambda(\Gamma)$ by B. SILBERMANN [2].

For the case that the symbol has zeros of integral orders the equation (6.1) was first solved by F. D. GAHOV in his 1941 thesis under suitable smoothness hypotheses on the coefficients and the right-hand side by reduction to a Riemann-Hilbert boundary value problem (cf. F. D. GAHOV [1]). N. P. VEKUA [1] and B. V. HVEDELIDZE [1] studies the solution of this equation in the case where the functions $c(t)$ and $d(t)$ have zeros of an order less than one.

6.7. The results of this section are due to the author [2], [8]. We remark here that in the paper [2] the question of the existence of an equivalent regularizer is also answered positively for a given right-hand side of the equation (7.1): If under the hypotheses of Subsection 6.7.2 the equation (7.1) is solvable for fixed f, then it is possible to construct a left regularizer B such that equation (7.1) is equivalent to the Fredholm integral equation $BA\varphi = Bf$.

Other methods for the reduction of a singular integral equation of non-normal type to a Fredholm integral equation of the second kind were earlier developed by F. D. GAHOV [1] and D. I. ŠERMAN [1]. GAHOV solved this problem using the solution of the equation (6.1) obtained by himself (thus in particular by using the factorization of functions). ŠERMAN [1] mainly uses means of the theory of complex functions (without factorization). But he has to demand for the coefficients of the equation (7.1) (besides some other conditions) that one of the functions $a + b$ or $a - b$ be unequal to zero every-where on the integration curve. The other function can vanish at single points. The results of D. I. ŠERMAN were carried over to the case of zeros of fractional orders by A. E. KOSULIN [1]. The method given in Section 6.7 can in a way be considered as a generalization of the method of ŠERMAN (cf. also the author's paper [1]).

CHAPTER 7

SYSTEMS OF SINGULAR EQUATIONS OF NORMAL TYPE

Here the main results of the Chapters 2 and 3 are generalized to the case of systems of singular equations.

It is relatively easy to obtain necessary and sufficient conditions for the validity of the Noether properties of such a system. We reduce the present case to the scalar case by means of a general theorem on operator matrices which is given in the first section. Rather great difficulties appear in the transfer of quantitative statements from the scalar case to the case of systems. For the solution of these problems we need the factorization of matrix functions in several algebras of continuous functions. Therefore these questions have a central place in this chapter.

7.1. A Theorem on Operator Matrices

7.1.1. In the sequel we shall use the following notation:

If X is an arbitrary vector space, then by X_n (n a natural number) we denote the set of all n-dimensional vectors with the components from X and by $X_{n\times n}$ the set of all quadratic matrices of the order n with elements from X.

If X is an F-space (a normed or unitary space), then X_n can be equipped with a countable system of semi-norms (a norm or a scalar product) in the following way. As the p-th semi-norm (norm or scalar product) of a vector $x = (x_1, \dots, x_n) \in X_n$ we take the sum of the p-th semi-norms (norms or scalar products) of the single components x_j:

$$\|x\|_p = \sum_{j=1}^{n} \|x_j\|_p \quad (p = 1, 2, \dots) .$$

If X is a normed space, then the norm of a matrix $A = \{a_{jk}\}_1^n \in X_{n\times n}$ can be defined by

$$\|A\| = n \max_{j, k} \|a_{jk}\| .$$

If X is a Banach algebra, then with this norm $X_{n\times n}$ is also a Banach algebra.

If X is an arbitrary F-space and $\mathscr{L}(X)$ the set of all linear continuous operators on X, then $\mathscr{L}_{n\times n}(X)$ can be identified with $\mathscr{L}(X_n)$. With other words, each operator $A \in \mathcal{L}(X_n)$ can be written as a matrix $A = \{A_{jk}\}_1^n$, where $A_{jk} \in \mathcal{L}(X)$. The operator A is compact if and only if all the operators A_{jk} are compact.

To the adjoint operator A^* there corresponds the adjoint matrix $A^* = \{A_{kj}^*\}_1^n$.

7.1.2. Theorem 1.1. *Let X be an F-space and $A = \{A_{kj}\}_1^n \in \mathcal{L}(X_n)$, where the operators A_{jk} commute pairwise up to a compact operator. Then A is a Φ_+-operator if and only if* $\det A (\in \mathcal{L}(X))$ *is a Φ_+-operator*[1]).

Proof. Let $B = \{B_{kj}\}$, where B_{jk} is the cofactor of the element A_{jk} in the determinant $\det A$. Then

$$BA = \{\det A \delta_{jk}\}_1^n = AB .$$

If now $\det A$ is a Φ_+-operator in X, then obviously $\{\det A \delta_{jk}\}_1^n$ is a Φ_+-operator in X_n. Hence by Theorem 3.1, Chapter 1, A is a Φ_+-operator.

Conversely, let A be a Φ_+-operator. For any combination $i_1, \dots, i_m$ and $j_1, \dots, j_m$ of m numbers $(1 \leqq m \leqq n)$ from the set $\{1, \dots, n\}$ we put

$$B_{i_1 \dots i_m}^{j_1 \dots j_m} = \det \begin{pmatrix} A_{i_1 j_1} \dots A_{i_1 j_m} \\ \dots\dots\dots \\ A_{i_m j_1} \dots A_{i_m j_m} \end{pmatrix} .$$

For $m = 0$ we define $B_{i_1 \dots i_m}^{j_1 \dots j_m} = I$ (I the identity operator in the space X).

By $\{\|x\|_p\}_{p=1}^\infty$ we denote a generating system of semi-norms in X. We show that for all $m = 0, \dots, n-1$ and $p = 1, 2, \dots$ an estimate of the form

$$C \sum \|B_{i_1 \dots i_m}^{j_1 \dots j_m} x\|_p \leqq \sum \|B_{i_1 \dots i_{m+1}}^{j_1 \dots j_{m+1}} x\|_q + \sum_l \|T_l x\|_q \tag{1.1}$$

holds simultaneously for all $x \in X$. Here T_l are certain (finitely many) compact operators. $C > 0$ is a constant (depending only on m and p) and q an index depending on p. In the first sum of (1.1) it is summed over all combinations $i_1, \dots, i_m$ and $j_1, \dots, j_m$. The corresponding statement holds for the second sum. For $m = 0$ the inequality (1.1) yields

$$C\|x\|_p \leqq \sum_{i,k} \|A_{ik} x\|_q + \sum_l \|T_l x\|_q . \tag{1.2}$$

The assertion of the theorem follows easily from the inequality (1.1). In fact, applying (1.1) successively from $m = 0$ to $m = n - 1$ and using the relation

$$\pm B_{i_1 \dots i_n}^{j_1 \dots j_n} = B_{12 \dots n}^{12 \dots n} = \det A$$

---

[1]) The determinant $\det A$ is defined in the same way as in the case of a number matrix. There the ordering of the factors is unessential since all possible results can only differ by a compact part (cf. Theorem 3.7*, Chapter 1). Thus the operator equations appearing at the proof of the Theorems 1.1 and 1.2 are to be understood as congruences modulo $\mathcal{K}(X)$ (resp. modulo $\mathcal{K}(X_n)$).

we obtain an estimate of the form

$$C\|x\|_p \leqq \|\det Ax\|_q + \sum_l \|T_l x\|_q .$$

By Lemma 3.1, Chapter 1, this implies that det A is a Φ_+-operator.

It remains only to prove the inequality (1.1). We first show that the relation (1.2) holds. By Lemma 3.1, Chapter 1, there exists a compact operator $T = \{T_{jk}\}_1^n \in \mathcal{K}(X_n)$ such that the estimate

$$C\|z\|_p \leqq \|Az\|_q + \|Tz\|_q \qquad (C = C_p = \text{const})$$

is valid for all $z = (z_1, \dots, z_n) \in X_n$ and all $p = 1, 2, \dots$. But this is equivalent to

$$C\left(\sum_{j=1}^n \|z_j\|_p\right) \leqq \sum_{i=1}^n \left\|\sum_{j=1}^n A_{ij} z_k\right\|_q + \sum_{i=1}^n \left\|\sum_{j=1}^n T_{ij} x\right\|_q . \tag{1.3}$$

Putting $z_k = x$ and $z_j = 0$ for $j \neq k$ in (1.3) we obtain

$$C\|x\|_p \leqq \sum_{i=1}^n \|A_{ik} x\|_q + \sum_{i=1}^n \|T_{ik} x\|_q .$$

Summing the last inequality over $k = 1, \dots, n$ we see that (1.2) holds.

Now let m with $1 \leqq m \leqq n - 1$ be a fixed index. Further let the combinations $i_1, \dots, i_m$ and $j_1, \dots, j_{m+1}$ in each case consist of distinct numbers in the set $\{1, \dots, n\}$. In (1.3) we put

$$z_k = \begin{cases} 0 & \text{for} \quad k \notin \{j_1, \dots, j_{m+1}\} \\ (-1)^{r+1} B_{i_1 \dots i_m}^{j_1 \dots j_{r-1} j_{r+1} \dots j_{m+1}} x & \text{for} \quad k = j_r . \end{cases}$$

Then

$$\sum_{j=1}^n A_{ij} z_j = \det \begin{pmatrix} A_{ij_1} & A_{ij_2} & \dots & A_{ij_{m+1}} \\ A_{i_1 j_1} & A_{i_1 j_2} & \dots & A_{i_1 j_{m+1}} \\ A_{i_m j_1} & A_{i_m j_2} & \dots & A_{i_m j_{m+1}} \end{pmatrix} x ,$$

which is easily checked by expansion by the first row. This implies

$$\left\|\sum_{j=1}^n A_{ij} z_j\right\|_q = \begin{cases} \|B_{i_1 \dots i_m i}^{j_1 \dots j_{m+1}} x + T_0 x\|_q & \text{for} \quad i \notin \{i_1, \dots, i_m\} \\ T_0 \|x\|_q & \text{for} \quad i \in \{i_1, \dots, i_m\} . \end{cases}$$

Thus from (1.3) there follows an inequality of the form

$$C \sum_{r=1}^{m+1} \|B_{i_1 \dots i_m}^{j_1 \dots j_{r-1} j_{r+1} \dots j_{m+1}} x\|_p \leqq \sum_{i \notin \{i_1, \dots, i_m\}} \|B_{i_1 \dots i_m i}^{j_1 \dots j_{m+1}} x\|_q + \sum_k \|T_k x\|_q .$$

Summing up in the last inequality over all combinations $i_1, \dots, i_m$ and $j_1, \dots, j_{m+1}$ we obtain the estimate (1.1).

This completes the proof of the theorem.

Remark. If, under the hypotheses of Theorem 1.1, det A is a Φ_--operator, then A is a Φ_--operator.

This follows immediately from the first part of the proof by Theorem 3.3, Chapter 1.

Theorem 1.2. *Let X be a Banach space. Under the hypotheses of the Theorem 1.1 $A \in \mathcal{L}(X_n)$ is a $\Phi_+(\Phi_-)$-operator if and only if* det A *is a $\Phi_+(\Phi_-)$-operator.*

Proof. By Theorem 1.1 and the above stated remark it remains only to show that $A \in \Phi_-(X_n)$ implies det $A \in \Phi_-(X)$.

If $A \in \Phi_-(X_n)$, then $A^* \in \Phi_+(X_+^n)$, and by Theorem 1.1 we have

$$\det A^* = (\det A)^* \in \Phi_+(X^*) .$$

But from this it follows that det $A \in \Phi_-(X)$.

7.2. Factorization of Matrix Functions

In the sequel by Γ we denote a closed smooth (in general multiply connected) oriented curve which divides the complex plane into two regions: a connected bounded region $D_+(\ni 0)$ and an unbounded region $D_-(\ni \infty)$. By $G_\pm = D_\pm \cup \cup \Gamma$ we denote the closure of these regions. Furthermore, in this section we maintain all notations introduced in 2.3.

7.2.1. According to the notation introduced in 7.1.1 $C_{n\times n}(\Gamma)$ is the Banach algebra of all continuous matrix functions on Γ of order n.

Let $A(z) \in C_{n\times n}(\Gamma)$ be a matrix function which is non-singular on the whole curve Γ. A representation of the form

$$A(z) = A_-(z)\, D(z)\, A_+(z) \qquad (z \in \Gamma) \tag{2.1}$$

is called a *right factorization* of $A(z)$ if the matrix functions $D(z)$, $A_+(z)$ and $A_-(z)$ have the following properties: $D(z)$ is a diagonal matrix function of the form

$$D(z) = \{z^{\varkappa_j}\delta_{jk}\}_1^n \qquad (z \in \Gamma) ,$$

where $\varkappa_1 \geqq \varkappa_2 \geqq \ldots \geqq \varkappa_n$ are certain integers. $A_\pm(z)$ is a quadratic $n \times n$ matrix function possessing a continuation which is analytic in the region $D_\pm$ and continuous in $G_\pm$. Furthermore,

$$\det A_+(z) \neq 0 \qquad (z \in G_+) , \qquad \det A_-(z) \neq 0 \qquad (z \in G_-) ,$$

holds.

The factorization of the matrix function $A(z)$ arising from (2.1) by commutation of the factors $A_+(z)$ and $A_-(z)$ is called a *left factorization*. Obviously, each right (left) factorization of the matrix function $A(z)$ generates a left (right) factorization of the transposed matrix function $A'(z)$ and of the inverse matrix function $A^{-1}(z)$.

Theorem 2.1. *If the matrix function $A(z) \in C_{n\times n}(\Gamma)$ admits a right (left) factorization, then the numbers $\varkappa_j = \varkappa_j(A)$ $(j = 1, 2, \ldots, n)$ are uniquely determined by the matrix function $A(z)$.*

Proof. We assume that as well as the factorization (2.1) $A(z)$ possesses a second right factorization of the form

$$A(z) = \widetilde{A}_-(z)\,\widetilde{D}(z)\,\widetilde{A}_+(z)\,, \tag{2.2}$$

where

$$\widetilde{D}(z) = \{z^{\widetilde{\varkappa}_j}\delta_{jk}\}_1^n\,.$$

From (2.1) and (2.2) we obtain the equation

$$B_-(z)\,D(z) = \widetilde{D}(z)\,B_+(z) \tag{2.3}$$

with

$$B_-(z) = \widetilde{A}_-^{-1}(z)\,A_-(z)\,, \qquad B_+(z) = \widetilde{A}_+(z)\,A_+^{-1}(z)\,. \tag{2.4}$$

If we denote the elements of the matrix function $B_\pm(z)$ by $b_{jk}^\pm$, then (2.3) can be written in the form

$$b_{jk}^-(z)\,z^{\varkappa_k} = z^{\widetilde{\varkappa}_j}b_{jk}^+(z) \qquad (j, k = 1, 2, \ldots, n)\,.$$

From this we obtain

$$b_{jk}^-(z) = b_{jk}^+(z) = 0 \qquad (z \in \Gamma) \tag{2.5}$$

for all j, k with $\varkappa_k < \widetilde{\varkappa}_j$. Namely, by the well-known Liouville theorem it follows from the equation

$$b_{jk}^-(z) = z^{\widetilde{\varkappa}_j - \varkappa_k}b_{jk}^+(z)$$

that both sides are constant. But the right-hand side vanishes at the point $z = 0$. Hence we have (2.5).

Contrary to the assertion of the theorem we assume that $\varkappa_r \neq \widetilde{\varkappa}_r$ for a certain $r(1 \leqq r \leqq n)$. Without loss of generality we can assume that $\varkappa_r < \widetilde{\varkappa}_r$. Obviously, we then have

$$\varkappa_k < \widetilde{\varkappa}_j \qquad (j = 1, \ldots, r;\ k = r, \ldots, n)$$

and consequently

$$b_{jk}^+(z) = 0 \qquad (j = 1, \ldots, r;\ k = r, \ldots, n)\,.$$

It follows from the last equations that each minor of the order r formed from the first r rows of the matrix $B_+(z)$ is identically equal to zero. Then Laplace's expansion theorem shows that $\det B_+(z) \equiv 0$. With this contradiction we have proved the theorem.

According to the factorization type the numbers $\varkappa_i (i = 1, 2, \ldots, n)$ are called *right (left) indices* or also *partial indices* of the matrix function $A(z)$. The sum

$$\varkappa = \sum_{i=1}^{n} \varkappa_i$$

is called the *total index (sum index)* of the matrix function $A(z)$. The equation (2.1) shows immediately that

$$\varkappa = \sum_{i=1}^{n} \varkappa_i = \operatorname{ind} \det A(z) . \tag{2.6}$$

In general, the right and left indices of a matrix function do not coincide. But it follows from (2.6) that the total index is independent of the factorization type.

Theorem 2.1 implies: If the matrix function $A(z)$ admits a factorization (2.1) and if $\alpha_\pm$ is an arbitrary (finite) point in $D_\pm$, then $A(z)$ possesses the factorization

$$A(z) = \widetilde{A}_-(z) \left\{ \left(\frac{z - \alpha_+}{z - \alpha_-} \right)^{\varkappa_j} \delta_{jk} \right\}_1^n \widetilde{A}_+(z) . \tag{2.7}$$

Conversely, it is possible to obtain a factorization of the form (2.1) from the factorization (2.7) without changing the indices $\varkappa_j$.

In the sequel $\mathfrak{A}(\Gamma)$ is a Banach algebra of continuous functions on the curve Γ satisfying conditions a) and b) of Section 2.3. Lemma 3.1, Chapter 2, shows immediately: If $A(z) \in \mathfrak{A}_{n\times n}(\Gamma)$ $(\mathfrak{A}^\pm_{n\times n}(\Gamma))$ and $\det A(z) \neq 0$ for $z \in \Gamma(G_\pm)$, then $A^{-1}(z) \in \mathfrak{A}_{n\times n}(\Gamma)$ $(\mathfrak{A}^\pm_{n\times n}(\Gamma))$.

A right factorization of the form (2.1) of a non-singular matrix function $A(z) \in \mathfrak{A}_{n\times n}(\Gamma)$ is said to be *canonical*, if $A_+^{\pm 1}(z) \in \mathfrak{A}^+_{n\times n}(\Gamma)$ and $A_-^\pm \in \mathfrak{A}^-_{n\times n}(\Gamma)$. Analogously the left canonical factorization is defined.

If $\mathfrak{A}(\Gamma)$ is a decomposing algebra (i.e. $\mathfrak{A}(\Gamma) = \mathfrak{A}^+(\Gamma) + \mathring{\mathfrak{A}}^-(\Gamma)$), then we obviously have $\mathfrak{A}_{n\times n}(\Gamma) = \mathfrak{A}^+_{n\times n}(\Gamma) + \mathring{\mathfrak{A}}^-_{n\times n}(\Gamma)$.

Next we shall prove the existence of the canonical factorization for non-singular matrix functions from Banach algebras $\mathfrak{A}(\Gamma)$ with two norms satisfying conditions α)—δ) of Section 2.3.

Theorem 2.2. *Let $\mathfrak{A}(\Gamma)$ be a Banach algebra of continuous functions on Γ satisfying conditions α) to δ) from Section* 2.3.

Then each non-signular matrix function $A(z) \in \mathfrak{A}_{n\times n}(\Gamma)$ admits a right (left) canonical factorization.

For the proof of this theorem we need the following lemma.

Lemma 2.1. *Let $\mathfrak{A}(\Gamma)$ be an arbitrary algebra of continuous functions satisfying conditions* a) *and* b) *of* 2.3.

If the function $a(z) \in \mathfrak{A}^+(\Gamma)$ $(\mathfrak{A}^-(\Gamma))$ assumes the value zero at a point $z_0 \in D_+(D_-)$, then $(z - z_0)^{-1} a(z) \in \mathfrak{A}^+(\Gamma)$ $(\mathfrak{A}^-(\Gamma))$.

Proof. Let e.g. $a(z) \in \mathfrak{A}^+(\Gamma)$ and $a(z_0) = 0$ $(z_0 \in D_+)$. In view of conditions a) and b) we have $(z - z_0)^{-1} a(z) \in \mathfrak{A}(\Gamma)$. By $p_n(z)$ $(n = 1, 2, ...)$ we denote a sequence of functions from $R^+(\Gamma)$ converging uniformly to $a(z)$ on Γ. Obviously, we have

$$r_n(z) = \frac{p_n(z) - p_n(z_0)}{z - z_0} \in R^+(\Gamma) \qquad (n = 1, 2, ...) .$$

Since $p_n(z_0) \to 0$ the sequence $r_n(z)$ converges uniformly on Γ to $(z - z_0)^{-1}a(z)$ and so the function $(z - z_0)^{-1} a(z)$ belongs to $C^+(\Gamma)$ and thus to $\mathfrak{A}^+(\Gamma)$. This proves the lemma.

Furthermore, we remark that the Lemma 3.3 from Chapter 2 also holds for matrix functions if in its formulation $\mathfrak{A}(\Gamma)$ and $\widetilde{\mathfrak{A}}(\Gamma)$ are replaced by $A_{n\times n}(\Gamma)$ and $\widetilde{\mathfrak{A}}_{n\times n}(\Gamma)$, respectively, and $\widetilde{P}$ is the projection which projects $\widetilde{\mathfrak{A}}_{n\times n}(\Gamma)$ onto $\widetilde{\mathfrak{A}}^+_{n\times n}(\Gamma)$ parallel to $\overset{\circ}{\widetilde{\mathfrak{A}}}{}^-_{\times n}(\Gamma)$.

The proof is completely analogous to the case $n = 1$.

Proof of the Theorem 2.2. We carry out the proof for the right canonical factorization. The existence of the left canonical factorization can be proved analogously.

First we consider the case $A(z) \in \mathfrak{A}^+_{n\times n}(\Gamma)$. By $z_1, z_2, \ldots, z_q$ we denote the zeros of the function $\det A(z)$ in the region D_+ and by $m_1, m_2, \ldots, m_q$ their multiplicities. For the sequel it is convenient to put $z_q = 0$. If this point is not a zero of the function $\det A(z)$, then we define $m_q = 0$.

Now let $f_j(z)$ be the j-th row of the matrix function $A(z) = \{a_{jk}(z)\}_1^n$:

$$f_j(z) = \{a_{j1}(z), a_{j2}(z), \ldots, a_{jn}(z)\} \qquad (j = 1, 2, \ldots, n)\,.$$

By p_j $(j = 1, 2, \ldots, n)$ we denote the order of the zero $z = z_1$ of the vector function $f_j(z)$. Obviously $\sum_{j=1}^{n} p_j \leq m_1$.

Withous loss of generality we may assume that $p_1 \geq p_2 \geq \ldots \geq p_n$. Let $\sum_{j=1}^{n} p_j < m_1$. Then there exist complex numbers $c_1, c_2, \ldots, c_l$ $(l \leq n,\ c_l = 1)$ such that the vector function

$$f(z) = \sum_{j=1}^{l} \frac{c_j f_j(z)}{(z - z_1)^{p_j}} \qquad (z \in \Gamma)$$

vanishes at the point $z = z_1$. By Lemma 2.1 the components of this vector function belong to the algebra $\mathfrak{A}^+(\Gamma)$. Obviously, the vector function

$$\hat{f}(z) = f(z)\,(z - z_1)^{p_l} = \sum_{j=1}^{l} \frac{c_j f_j(z)}{(z - z_1)^{p_j - p_l}}$$

then possesses a zero of the order $\hat{p}_l > p_l$ in the point $z = z_1$.

Now we form the matrix function

$$\hat{B}_1(z) = \begin{Vmatrix} 1 & 0 & \ldots\ 0 \ldots\ 0 \\ 0 & 1 & \ldots\ 0 \ldots\ 0 \\ \cdots & \cdots & \cdots \\ \dfrac{c_1}{(z - z_1)^{p_1 - p_l}} & \dfrac{c_2}{(z - z_1)^{p_2 - p_l}} & \ldots\ 1 \ldots\ 0 \\ \cdots & \cdots & \cdots \\ 0 & 0 & \ldots\ 0 \ldots\ 1 \end{Vmatrix} \quad l\text{-th row}\,.$$

Then $\det \hat{B}_1(z) = 1$, $\hat{B}_1^{\pm 1}(z) \in \mathfrak{A}^-_{n\times n}(\Gamma)$.

Since we can obviously obtain the matrix function $\hat{B}_1(z)\,A(z)$ from the matrix function $A(z)$ by replacing the l-th row vector of $A(z)$ by the vector $\hat{f}(z)$, we have $\hat{B}_1(z)\,A(z) \in \mathfrak{A}^+_{n\times n}(\Gamma)$. If we obtain

$$\sum_{j\neq l} p_j + \hat{p}_l < m_1 ,$$

then we can repeat the operation carried out just now with the matrix function $\hat{B}_1(z)\,A(z)$ instead of $A(z)$. After a finite number of steps we then obtain a matrix function $B_1(z)\,A(z)$ whose row vectors possess zeros of the order

$$\hat{p}_1 \geqq \hat{p}_2 \geqq \dots \geqq \hat{p}_n \quad \text{with} \quad \sum_{j=1}^{n} \hat{p}_j = m_1 ,$$

at the point $z = z_1$. Here $\det B_1(z) = 1$, $B_1^{\pm 1} \in \mathfrak{A}^-_{n\times n}(\Gamma)$.

We form the following diagonal matrix function $D_1(z) \in \mathfrak{A}^-_{n\times n}(\Gamma)$:

$$D_1(z) = \left\{\left(\frac{z}{z - z_1}\right)^{\hat{p}_j} \delta_{jk}\right\}_1^n \qquad (z \in \Gamma) .$$

It is easy to see that the determinant of the matrix function $A_1(z) = D_1(z)\,B_1(z)\,A(z)$ does not vanish at the point $z = z_1$. Hence in the region D_+ the function $\det A_1(z)$ has only the zeros $z_2, z_3, \dots, z_q$ whose orders are $m_2, m_3, \dots, m_q + m_1$, respectively.

Similar operations can now be carried out for the matrix function $A_1(z)$ and the zero $z = z_2$. The result is a matrix function $A_2(z) = D_2(z)\,B_2(z)\,A_1(z)$ whose determinant has only the zeros $z_3, z_4, \dots, z_q$ with the multiplicities $m_3, m_4, \dots, m_q + m_1 + m_2$, respectively, in the region D_+. Repeating this process we then come to a matrix function

$$A_{q-1}(z) = D_{q-1}(z)\,B_{q-1}A_{q-2}(z) \in \mathfrak{A}^+_{n\times n}(\Gamma)$$

with $B^{\pm 1}_{q-1}(z)$ and $D^{\pm 1}_{q-1}(z)$ from $\mathfrak{A}^-_{n\times n}(\Gamma)$, where in the region D_+ the function $\det A_{q-1}(z)$ has an only zero at the point $z = 0$ of the order $\sum m_j$.

Finally by means of the above described operation we can construct a matrix function $B(z)$ with

$$B(z)\,A_{q-1}(z) \in \mathfrak{A}^+_{n\times n}(\Gamma) , \qquad B^{\pm 1}(z) \in \mathfrak{A}^-_{n\times n}(\Gamma)$$

such that the function $\det\,[B(z)\,A_{q-1}(z)]$ only has a zero at the point $z = 0$ of the multiplicity $\sum m_j$ and the sum of the orders $\varkappa_1 \geqq \varkappa_2 \geqq \dots \geqq \varkappa_n$ of the zero $z = 0$ of the corresponding row vectors of the matrix function $B(z)\,A_{q-1}(z)$ is equal to $\sum m_j$.

Now we form the diagonal matrix function

$$D(z) = \{z^{\varkappa_j}\delta_{jk}\}_1^n .$$

By Lemma 2.1 we have

$$D^{-1}(z)\,B(z)\,A_{q-1}(z) = A_+(z) \in \mathfrak{A}^+_{n\times n}(\Gamma) , \tag{2.8}$$

where det $A_+(z) \neq 0$ $(z \in G_+)$. From (2.8) we obtain

$$B(z)\,D_{q-1}(z)\,B_{q-1}(z) \dots D_1(z)\,B_1(z)\,A(z) = D(z)\,A_+(z)$$

and thus

$$A(z) = A_-(z)\,D(z)\,A_+(z) \qquad (z \in \Gamma)$$

with

$$A_-(z) = [B(z)\,D_{q-1}(z)\,B_{q-1}(z) \dots D_1(z)\,B_1(z)]^{-1}\,.$$

Here $A_-^{\pm 1}(z) \in \mathfrak{A}^-_{n\times n}(\Gamma)$, $A_+^{\pm 1}(z) \in \mathfrak{A}^+_{n\times n}(\Gamma)$. Hence we have proved the theorem for the case $A(z) \in \mathfrak{A}^+_{n\times n}(\Gamma)$.

Now let $A(z) \in \mathfrak{A}_{n\times n}(\Gamma)$ be an arbitrary non-singular matrix function. By condition β) of Section 2.3 there exists a matrix function $R(z) = \{r_{jk}(z)\}_1^n \in R_{n\times n}(\Gamma)$ with $R^{-1}(z) \in R_{n\times n}(\Gamma)$ such that

$$\|I - A(z)\,R^{-1}(z)\|_{\mathfrak{A}_{n\times n}} < \min\,(\|\widetilde{P}\|^{-1}, \|\widetilde{Q}\|^{-1}, \varrho)\,.$$

By the above stated remark to Lemma 3.3 of Chapter 2 the matrix function $A(z)\,R^{-1}(z)$ admits a factorization of the form

$$A(z)\,R^{-1}(z) = B_-(z)\,B_+(z)\,,$$

where $B_-^{\pm 1} \in \mathfrak{A}^-_{n\times n}(\Gamma)$ and $B_+^{\pm 1}(z) \in \mathfrak{A}^+_{n\times n}(\Gamma)$. Hence

$$B_+(z)\,R(z) = B_-^{-1}(z)\,A(z)\,. \tag{2.9}$$

Let $\lambda_1^+, \lambda_2^+, \dots, \lambda_s^+$ be all poles of the elements $r_{jk}(z)$ in the region D_+ (each pole counted according to its order). We put

$$r(z) = \prod_{i=1}^{s} (z - \lambda_i^+)\,.$$

Then we obviously have

$$\widetilde{A}(z) = r(z)\,B_+(z)\,R(z) \in \mathfrak{A}^+_{n\times n}(\Gamma)$$

and det $\widetilde{A}(z) \neq 0$ $(z \in \Gamma)$. By the already given part of the proof the matrix function $\widetilde{A}(z)$ admits a factorization

$$\widetilde{A}(z) = \widetilde{A}_-(z)\,\widetilde{D}(z)\,\widetilde{A}_+(z) \qquad (z \in \Gamma)\,.$$

Using (2.9) we obtain from this that

$$A(z) = B_-(z)\,\widetilde{A}_-(z)\,\hat{D}(z)\,\widetilde{A}_+(z) \qquad (z \in \Gamma) \tag{2.10}$$

with $\hat{D}(z) = r^{-1}(z)\,\widetilde{D}(z)$.

The matrix $\widetilde{D}(z)$ is of diagonal form. The elements of its principal diagonal $d_j(z)$ $(j = 1, 2, \dots, n)$ belong to $R(\Gamma)$. Hence the functions $d_j(z)$ admit a factorization (see equation (1.12), Chapter 2):

$$d_j(z) = d_j^-(z)\,z^{\varkappa_j} d_j^+(z) \qquad (z \in \Gamma)\,,$$

where $d_j^{\pm}(z) \in R^{\pm}(\Gamma)$, $(j = 1, 2, \ldots, n)$. Without loss of generality we can assume that $\varkappa_1 \geqq \varkappa_2 \geqq \ldots \geqq \varkappa_n$. Thus for $\hat{D}(z)$ we obtain the factorization

$$\hat{D}(z) = D_-(z)\, D(z)\, D_+(z) \qquad (z \in \Gamma) \tag{2.11}$$

with $D(z) = \{z^{\varkappa_j}\delta_{jk}\}_1^n$, $D_+^{\pm 1}(z) \in R_{n\times n}^+(\Gamma)$ and $D_-^{\pm 1}(z) \in R_{n\times n}^-(\Gamma)$.

From (2.10) and (2.11) we obtain the desired factorization for the matrix function $A(z)$:

$$A(z) = A_-(z)\, D(z) A_+(z) \qquad (z \in \Gamma)\,,$$

where

$$A_-(z) = B_-(z)\, \tilde{A}_-(z)\, D_-(z)\,, \qquad A_+(z) = D_+(z)\, \tilde{A}_+(z)\,.$$

This proves the theorem.

Remark. If the algebra $\mathfrak{A}(\Gamma)$ satisfies the conditions α), β) and γ) of Section 2.3, then the condition δ) is necessary for each non-singular matrix function $A(z) \in \mathfrak{A}_{n\times n}(\Gamma)$ to admit a right (left) canonical factorization (cf. S. Prössdorf and G. Unger [1]).

Since every decomposing R-algebra satisfies the conditions of Theorem 2.2, we have the following corollary.

Corollary 2.1. *Let $\mathfrak{A}(\Gamma)$ be a decomposing R-algebra. Then each non-singular matrix function $A(z) \in \mathfrak{A}_{n\times n}(\Gamma)$ admits a right (left) canonical factorization.*

This holds in particular for the Wiener algebra W of all functions on the unit circle which can be expanded into an absolutely convergent Fourier series, i.e. we have:

Theorem 2.3. *Each non-singular matrix function $A(z) \in W_{n\times n}$ admits a right (left) canonical factorization.*

Remark. Theorem 2.3 remains valid if in its formulation the algebra W is replaced by W^m $(1 \leqq m \leqq \infty)$ or $W(\boldsymbol{\alpha}, \boldsymbol{m})$ (see 5.1.2, no. 3).

Furthermore, as an immediate consequence of the Theorem 2.2 and Corollary 4.6, Chapter 3, we obtain:

Theorem 2.4. *Each non-singular matrix function $A(z) \in H_{n\times n}^{\lambda}(\Gamma)$ $(0 < \lambda < 1)$ admits a right (left) canonical factorization.*

The conditions of Theorem 2.2 can be essentially weakened. In particular the following theorem holds.

Theorem 2.2*. *Let $\tilde{\mathfrak{A}}(\Gamma)$, $\mathfrak{A}(\Gamma)$ and $\mathfrak{A}^\circ(\Gamma)$ be (in general non-normed) algebras of continuous functions on the curve Γ satisfying the conditions* a) *and* b) *of Section 2.3 and the following conditions:*

1) $\mathfrak{A}(\Gamma) \subset \mathfrak{A}^\circ(\Gamma) \subset \tilde{\mathfrak{A}}(\Gamma)$.
2) *$\tilde{\mathfrak{A}}(\Gamma)$ is a decomposing Banach algebra.*
3) *$R(\Gamma)$ is dense in $\mathfrak{A}(\Gamma)$ with respect to the norm of $\tilde{\mathfrak{A}}(\Gamma)$.*
4) *For any function $A(z) \in \mathfrak{A}(\Gamma)$ we have:*

$$T_a\colon \tilde{\mathfrak{A}}(\Gamma) \to \mathfrak{A}^\circ(\Gamma) \text{ and } T_a\colon \mathfrak{A}^\circ(\Gamma) \to \mathfrak{A}(\Gamma).$$

where the mapping T_a is defined by the equation

$$T_a\varphi = \widetilde{P}a\varphi - a\widetilde{P}\varphi \qquad (\varphi \in \widetilde{\mathfrak{A}}(\Gamma))$$

and $\widetilde{P}$ is the projection of $\widetilde{\mathfrak{A}}(\Gamma)$ onto $\widetilde{\mathfrak{A}}^+(\Gamma)$.

Then each non-singular matrix function $A(z) \in \mathfrak{A}_{n\times n}(\Gamma)$ admits a right (left) canonical factorization.

Proof. By $\widetilde{\mathscr{P}}$ we denote the projection of $\widetilde{\mathfrak{A}}_{n\times n}(\Gamma)$ onto $\widetilde{\mathfrak{A}}^+_{n\times n}(\Gamma)$ and by $\widetilde{Q}$ the complementary projection of $\widetilde{\mathscr{P}}$. Let $B(z)$ be an arbitrary matrix function from $\mathfrak{A}_{n\times n}(\Gamma)$ with

$$||I - B(z)||_{\widetilde{\mathfrak{A}}_{n\times n}} < \min\,(||\widetilde{\mathscr{P}}||^{-1}, ||\widetilde{Q}||^{-1})\,.$$

By Lemma 3.2, Chapter 2, $B(z)$ admits the factorization

$$B(z) = B_-(z)\, B_+(z)$$

with $B_-^{\pm 1}(z) \in \widetilde{\mathfrak{A}}^-_{n\times n}(\Gamma)$ and $B_+^{\pm 1}(z) \in \widetilde{\mathfrak{A}}^+_{n\times n}(\Gamma)$. Using the condition 4) we conclude by means of the arguments in the proof of Lemma 3.3, Chapter 2, that $B_-^{\pm 1}(z) \in \mathfrak{A}^{0-}_{n\times n}(\Gamma)$ and $B_+^{\pm 1}(z) \in \mathfrak{A}^{0+}_{n\times n}(\Gamma)$. Repeating the same arguments we finally obtain $B_-^{\pm 1}(z) \in \mathfrak{A}^-_{n\times n}(\Gamma)$ and $B_+^{\pm 1}(z) \in \mathfrak{A}^+_{n\times n}(\Gamma)$. The rest of the proof is carried out in the same way as the corresponding part of the proof of Theorem 2.2.

As a consequence of the Theorem 2.2* for instance we obtain the following theorem.

Theorem 2.5.[1]) *Let $t_1, t_2, \ldots, t_r$ be distinct points on Γ, $m_1, m_2, \ldots, m_r$ positive integers and $\mathfrak{A}(\Gamma) = C^\lambda(\Gamma; (t_j, m_j)_1^r)$ $(0 < \lambda < 1)$.*

Each non-singular matrix function $A(z) \in \mathfrak{A}_{n\times n}(\Gamma)$ admits a right (left) canonical factorization.

For the proof of the Theorem 2.5 we remark that the algebras $\mathfrak{A}(\Gamma) = C^\lambda(\Gamma; (t_j, m_j)_1^r)$, $\mathfrak{A}^\circ(\Gamma) = C^\mu(\Gamma; (t_j, m_j)_1^r)$ and $\widetilde{A}(\Gamma) = H^\mu(\Gamma)$ with $0 < \mu < \lambda$ satisfy the conditions of Theorem 2.2*. Namely, by Lemma 1.7, Chapter 6, we have $T_a\colon \widetilde{\mathfrak{A}}(\Gamma) \to \mathfrak{A}^\circ(\Gamma)$. In the following way we can see that $T_a\varphi \in \mathfrak{A}(\Gamma)$ for any $a \in \mathfrak{A}(\Gamma)$ and $\varphi \in \mathfrak{A}^\circ(\Gamma)$. First, for any neighborhood $\Gamma_0 \in \Gamma$ of the point t_j with end points α, γ and

$$(S_0\varphi)\,(t) = \int\limits_{\Gamma_0} \frac{\varphi(\tau)}{\tau - t}\, d\tau \qquad (t \in \Gamma_0)\,, \qquad \psi_0 = (S_0 a - a S_0)\,\varphi$$

---

[1]) For notation see 6.1.1. Theorem 2.5 asserts: If in a neighborhood of the points t_j $(j = 1, \ldots, r)$ the matrix function $A(z) \in H^\lambda(\Gamma)$ has derivatives up to the order m_j satisfying a Hölder condition with the exponent λ, then the factors $A_\pm(z)$ in (2.1) possess the same properties (possibly for a smaller neighborhood of the points t_j).

we have (cf. Lemma 1.5, Chapter 6)

$$\psi_0'(t) = (S_0 a' - a' S_0)\,\varphi + (S_0 a - a S_0)\,\varphi' +$$

$$+ [a(\alpha) - a(t)]\,\frac{\varphi(\alpha)}{\alpha - t} + [a(t) - a(\beta)]\,\frac{\varphi(\beta)}{\beta - t}\,.$$

Using Corollary 4.2, Chapter 3, we obtain easily from the last formula that the assertion holds.

The following theorem is a special case of Theorem 2.5.

Theorem 2.6.[1] *Each non-singular matrix function* $A(z) \in C^{m,\lambda}_{n\times n}(\Gamma)$ *admits a right (left) canonical factorization.*

Remark. Theorem 2.6 remains valid if in its formulation the algebra $C^{m,\lambda}(\Gamma)$ is replaced by $C^\infty(\Gamma)$.

7.2.2. The concept of the *canonical factorization on the real line* is introduced analogously to 2.3.3. The right canonical factorization of a matrix function $A(\lambda) \in \mathfrak{C}_{n\times n}$ ($\mathfrak{C}$ an R-algebra of continuous functions on the real line) is a representation of the form

$$A(\lambda) = A_-(\lambda)\,D(\lambda)\,A_+(\lambda) \tag{2.12}$$

with a diagonal matrix function $D(\lambda)$ of the form

$$D(\lambda) = \left\{\left(\frac{\lambda - i}{\lambda + i}\right)^{\varkappa_j} \delta_{jk}\right\}_1^n,$$

where $\varkappa_1 \geqq \varkappa_2 \geqq \dots \geqq \varkappa_n$ are integers and $A_+^{\pm 1}(\lambda) \in \mathfrak{C}^+_{n\times n}$, $A_-^{\pm 1} \in \mathfrak{C}^-_{n\times n}$.

By stereographic projection of the unit circle onto the closed real line, Corollary 2.1 immediately yields the following theorem.

Theorem 2.7. *Let* $\mathfrak{C}$ *be a decomposing R-algebra of functions on the real line. Then each matrix function* $A(\lambda) \in \mathfrak{C}_{n\times n}$ *with*

$$\det A(\lambda) \neq 0 \qquad (-\infty \leqq \lambda \leqq \infty) \tag{2.13}$$

admits a right (left) canonical factorization.

Finally from Theorem 2.7 we obtain:

Theorem 2.8. *If the matrix function* $A(\lambda) \in \mathfrak{L}_{n\times n}$[2] *satisfies the condition* (2.13), *then* $A(\lambda)$ *admits a right (left) canonical factorization.*

Remark. Theorem 2.8 remains valid if in its formulation $\mathfrak{L}$ is replaced by $\mathfrak{L}^m (1 \leqq m \leqq \infty)$ or $\mathfrak{L}(\boldsymbol{\alpha}, \boldsymbol{m})$ (see 5.2.2, no. 3).

---

[1]) Theorem 2.6 can also be obtained immediately from Theorem 2.2. By Corollary 1.3, Chapter 6, and Corollary 4.2, Chapter 3, we can easily check that the algebra $C^{m,\lambda}(\Gamma)$ satisfies the conditions of the Theorem 2.2.

[2]) For the definition of the algebra $\mathfrak{L}$ see 3.2.1.

7.3. Systems of Abstract Singular Equations

7.3.1. Let $U \in \mathfrak{L}(X)$ be an invertible operator on the Banach space X and $P_1 \in \mathfrak{L}(X)$ a projection. We assume that U and P_1 satisfy the conditions (I), (II) of 2.1.2 and the condition (III) of 2.2.3:

(I) *The spectral radii of U and U^{-1} are equal to one.*

(II) $UP_1 = P_1UP_1$, $\quad UP_1 \neq P_1U$, $\quad P_1U^{-1} = P_1U^{-1}P_1$.

(III) $\beta(\hat{U}) = \dim \operatorname{coker} (U \mid \operatorname{im} P_1) < \infty$.

According to the notations introduced in 7.1.1 let $\mathfrak{R}_{n\times n}(U)$ be the algebra of all quadratic matrices of the order n with elements from $\mathfrak{R}(U)$. With each operator (in X_n) $A = \{A_{jk}\}_1^n \in \mathfrak{R}_{n\times n}(U)$ we associate the continuous matrix function on the unit circle $A(z) = \{A_{jk}(z)\}_1^n$ ($|z| = 1$). Here $A_{jk}(z)$ is the symbol of the operator $A_{jk} \in R(U)$. The matrix function $A(z)$ is called the *symbol* of A.

Furthermore, by P and Q we denote the projections defined in the space X_n by

$$P = \{P_1\delta_{jk}\}_1^n, \qquad Q = I - P.$$

Analogously to the Chapter 2 by $\hat{\mathfrak{R}}_{n\times n}(U)$ we denote the set of all operators of the form $\hat{A} = PAP \mid \operatorname{im} P$ (restriction of the operator A to the subspace $\operatorname{im} P$) with $A \in \mathfrak{R}_{n\times n}(U)$. The symbol $A(z)$ of the operator A is also called the symbol of $\hat{A}$.

Correspondingly, by $\tilde{\mathfrak{R}}_{n\times n}(U)$ we denote the algebra of all operators of the form

$$C = AP + BQ + T \tag{3.1}$$

with $A, B \in \mathfrak{R}_{n\times n}(U)$ and $T \in \mathcal{K}(X_n)$. The matrix function

$$C(z, \theta) = A(z)\frac{1+\theta}{2} + B(z)\frac{1-\theta}{2} \quad (|z| = 1,\ \theta = \pm 1)$$

is called the *symbol* of the operator C. Here $A(z)$ and $B(z)$ are the symbols of the operators A and B, respectively.

Remark. Contrary to the case $n = 1$, $\mathfrak{R}_{n\times n}(U)$ (and thus also $\tilde{\mathfrak{R}}_{n\times n}(U)$) is not a commutative algebra. But the symbol defined here has all the properties $\sigma.1$ to $\sigma.4$ of 2.1.1 (in $\sigma.2$ attention must be paid to the ordering of the factors). This follows easily from the validity of these properties in the case $n = 1$ and from Lemma 2.1, Chapter 2.

Theorem 3.1. *The operator $\hat{A} \in \hat{\mathfrak{R}}_{n\times n}(U)$ is a $\Phi_+(\Phi_-)$-operator if and only if its symbol satisfies the condition*

$$\det A(z) \neq 0 \qquad (|z| = 1). \tag{3.2}$$

Proof. By Lemma 2.1, Chapter 2, all commutators $P_1 A_{jk} - A_{jk} P_1$ $(j, k = 1, \ldots, n)$ are compact. From this it follows easily that

$$\det \hat{A} = P_1 A_1 P_1 + T_1$$

with $T_1 \in \mathcal{K}(X)$, $A_2 \in \mathfrak{R}(U)$ and $A_1(z) = \det A(z)$. To finish the proof it is sufficient to apply Theorems 1.2 (from 7.1), 1.3 (Chapter 2) and 3.7* (Chapter 1).

Theorem 3.2. *The operator* (3.1) *from* $\widetilde{\mathfrak{R}}_{n \times n}(U)$ *is a* $\Phi_+(\Phi_-)$*-operator if and only if*

$$\det A(z) \neq 0\,, \qquad \det B(z) \neq 0 \qquad (|z| = 1)\,. \tag{3.3}$$

If these conditions are satisfied, then $R = A^{-1}P + B^{-1}Q$ *is a two-sided regularizer of the operator* C.

The proof of the first assertion is carried out completely analogously to the proof of the preceding theorem. We only have to take into consideration that

$$\det C = A_1 P_1 + B_1 Q_1 + T_1$$

with $B_1 \in \mathfrak{R}(U)$, $B_1(z) = \det B(z)$ and $Q_1 = I_1 - P_1$ (I_1 the identity operator in X). The second assertion of the theorem is a direct consequence of the Lemma 2.1, Chapter 2.

If condition (3.2) (resp. (3.3)) is satisfied, then we say that the symbol does not *degenerate* and $\hat{A}$ (resp. C) is called an *operator of normal type*. Obviously, the conditions (3.3) are equivalent to

$$\det C(z, \theta) \neq 0 \qquad (|z| = 1, \theta = \pm 1)\,.$$

7.3.2. Now we assume that the operator U satisfies the conditions (I)–(III) and in addition the following condition:

(IV) $\mathfrak{R}(U)$ is an algebra without radical.

Then to each function $A_1(z) \in \mathfrak{R}(z)$ there corresponds exactly one operator $A_1 \in \mathfrak{R}(U)$ whose symbol is equal to $A_1(z)$.

In the sequel by $\mathfrak{R}_{n \times n}(z)$ we denote the algebra of all matrix symbols $A(z)$ with $A \in \mathfrak{R}_{n \times n}(U)$.

Theorem 3.3. *Let the conditions* (3.3) *be satisfied and let the matrix function* $C(z) = B^{-1}(z)\, A(z) \in \mathfrak{R}_{n \times n}(z)$ *admits a right canonical factorization*

$$C(z) = C_-(z)\, D(z)\, C_+(z)\,, \qquad D(z) = \{z^{\varkappa_j} \delta_{jk}\}_1^n\,. \tag{3.4}$$

Then the operator $AP + BQ$ *is a* Φ*-operator and*

$$\dim \ker (AP + BQ) = -\,\beta(\hat{U}) \sum_{\varkappa_j < 0} \varkappa_j\,, \tag{3.5$_1$}$$

$$\dim \operatorname{coker} (AP + BQ) = \beta(\hat{U}) \sum_{\varkappa_j > 0} \varkappa_j\,, \tag{3.5$_2$}$$

$$\operatorname{Ind} (AP + BQ) = -\,\beta(\hat{U}) \operatorname{ind} \det C(z)\,. \tag{3.5$_3$}$$

If the equation

$$(AP + BQ)\,\varphi = f \qquad (f \in X_n) \tag{3.6}$$

is solvable, then one of its solutions is given by the formula

$$\varphi = (AP + BQ)^{[-1]} f$$

with

$$(AP + BQ)^{[-1]} = (C_+^{-1}P + C_-Q)\,(D^{[-1]}P + Q)\,C_-^{-1}B^{-1} \tag{3.7}$$

and

$$D^{[-1]} = \{U^{-\varkappa_j}\delta_{jk}\}_1^n .$$

Here $C_\pm \in \Re_{n\times n}(U)$ are the operators with the symbols $C_\pm(z)$.

The corresponding statement holds for the operator $PA + QB$, where the matrix function $A(z)\,B^{-1}(z)$ plays the role of the matrix function $B^{-1}(z)\,A(z)$.

Proof. By (3.4) the operator $AP + BQ$ can be represented in the following form (cf. also (4.18_1) from Chapter 2):

$$AP + BQ = BC_-(DP + Q)\,(C_+P + C_-^{-1}Q)\,,$$

where $D = \{U^{\varkappa_j}\delta_{jk}\}_1^n$. The two outer operators on the right-hand side of the last equation are continuously invertible. Thus using Theorem 4.2, Chapter 2, we easily obtain the formulae (3.5).

Furthermore, we can easily see that

$$(AP + BQ)\,(AP + BQ)^{[-1]}\,(AP + BQ) = AP + BQ\,.$$

This immediately implies the last assertion of the theorem.

Corollary 3.1. *Under the conditions of the Theorem 3.3 the operator $AP + BQ$ is invertible (left invertible, right invertible) if and only if all right indices of the matrix function $C(z)$ are equal to zero (non-negative, non-positive). If this is the case, then the inverse (the corresponding one-sided inverse) operator is represented by the formula (3.7).*

Corollary 3.2. *The following formula holds:*

$$\ker\,[AP + BQ] = [C_+^{-1}P + C_-Q]\,(\ker\,[DP + Q])\,.$$

Corollary 3.3. *The equation (3.6) is solvable if and only if*

$$QD^{[-1]}PC_-^{-1}B^{-1}f = 0\,. \tag{3.8}$$

Proof. Obviously, the equation (3.6) is solvable if and only if the equation

$$(DP + Q)\,\psi = C_-^{-1}B^{-1}f$$

is solvable. From the last assertion of Theorem 3.3 we easily conclude that this is the case if and only if the equation (3.8) holds.

Putting $B = I$ in Theorem 3.3 we obtain the corresponding statement for the operators $\hat{A} = PAP \mid \operatorname{im} P \in \hat{\Re}_{n\times n}(U)$.

7.4. Systems of Wiener-Hopf Equations

7.4.1. Systems of Discrete Wiener-Hopf Equations

Let E_+ be an arbitrary one of the sequence spaces l_+^p $(1 \leqq p < \infty)$, c_+^0, c_+ or m_+ introduced in Section 3.1 and U, P_1 the operators defined by the equations (1.1) resp. (1.2) from Chapter 3 in the corresponding space E. If the matrix function $A(z)$ belongs to the algebra $W_{n\times n}$ and thus to $\mathfrak{R}_{n\times n}(z)$, then the corresponding operator $\hat{A} \in \hat{\mathfrak{R}}_{n\times n}(U)$ with the symbol $A(z)$ is represented in $(E_+)_n$ by the Toeplitz matrix

$$\hat{A} = \{a_{j-k}\}_{j,k=0}^{\infty} . \tag{4.1}$$

Here (4.1) now is a block matrix: a_j $(j = 0, \pm 1, \ldots)$ is a $n \times n$ matrix whose elements are the j-th Fourier coefficients of the corresponding elements of the matrix function $A(z)$:

$$A(z) = \sum_{j=-\infty}^{\infty} a_j z^j \qquad (|z| = 1) .$$

In this symbols the conditional equation for the operator $\hat{A}$ is given by

$$\hat{A}\xi = \left\{ \sum_{k=0}^{\infty} a_{j-k}\xi_k \right\}_{j=0}^{\infty} .$$

Here $\xi = \{\xi_n\}_0^\infty$ is a sequence of n-dimensional vectors whose components form a number sequence from E_+ for a fixed index i $(i = 1, \ldots, n)$.

By Theorem 2.3 the matrix function $A(z)$ admits a right (left) canonical factorization if $\det A(z) \neq 0$ $(|z| = 1)$.

The next theorem follows immediately from Theorems 3.1 and 3.3 and the equation (1.4), Chapter 2.

Theorem 4.1. *Let $A(z) \in W_{n\times n}$. The operator $\hat{A}$ defined by* (4.1) *on the space $(E_+)_n$ is a $\Phi_+(\Phi_-)$-operator if and only if the condition* (3.2) *is satisfied. If this is the case and*

$$A(z) = A_-(z)\, D(z)\, A_+(z) , \qquad D(z) = \{z^{\varkappa_j}\delta_{jk}\}_1^n$$

yields a right canonical factorization of the matrix function $A(z)$, then

$$\dim \ker \hat{A} = -\sum_{\varkappa_j<0} \varkappa_j , \qquad \dim \operatorname{coker} \hat{A} = \sum_{\varkappa_j>0} \varkappa_j . \tag{4.2$_1$}$$

If the equation $\hat{A}\varphi = f (f \in (E_+)_n)$ is solvable, then one of its solutions is given by the formula

$$\varphi = \hat{A}_+^{-1}\hat{D}^{[-1]}\hat{A}_-^{-1}f . \tag{4.2$_2$}$$

Here $\hat{A}_+^{-1}$, $\hat{D}^{[-1]}$ and $\hat{A}_-^{-1}$ are operators in $(E_+)_n$ of the form (4.1) with the symbols $A_+^{-1}(z)$, $D^{-1}(z)$ and $A_-^{-1}(z)$, respectively.

Remark. Obviously, the first part of both theorems holds for any matrix function $A(z) \in \mathfrak{R}_{n\times n}(z)$.

7.4.2. Systems of Wiener-Hopf Integral Equations

Now let E be an arbitrary one of the function spaces (2.4) from Section 3.2 on the real line, E_+ the corresponding space on the positive half-line and P_1, U the operators defined on E by the equations (2.11) and (2.12), respectively. In this case, all operators defined on the space $(E_+)_n$ by an equation of the form

$$\hat{A}\varphi = c\varphi(t) - \int_0^\infty k(t-s)\,\varphi(s)\,ds = f(t) \qquad (0 \leqq t < \infty) \tag{4.3}$$

belong to $\hat{\mathfrak{R}}_{n\times n}(U)$. Here $k(t)$ is a matrix function of the order n with elements from $L = L^1(-\infty, \infty)$, c a number matrix of the order n and $\varphi(t)$, $f(t)$ are vectors from $(E_+)_n$.

Analogously to Section 3.2 the matrix function

$$\mathcal{A}(\lambda) = c - \int_{-\infty}^{\infty} e^{i\lambda t} k(t)\,dt \qquad (-\infty \leqq \lambda \leqq \infty) \tag{4.4}$$

is also called the *symbol* of the operator A. If the condition

$$\det\left\{c - \int_{-\infty}^{\infty} e^{i\lambda t} k(t)\,dt\right\} \neq 0 \qquad (-\infty \leqq \lambda \leqq \infty) \tag{4.5}$$

is satisfied, then by Theorem 2.8 $\mathcal{A}(\lambda)$ admits a right (left) canonical factorization of the form (2.12).

From the Theorems 3.1 and 3.3 we now obtain:

Theorem 4.2. *Let $k(t) \in \mathfrak{L}_{n\times n}$. The operator $\hat{A}$ defined on the space $(E_+)_n$ by the equation* (4.3) *is a $\Phi_+(\Phi_-)$-operator if and only if the condition* (4.5) *is satisfied.*

If this is the case, then the formulae (4.2_1) *hold, where $\varkappa_j$ $(j = 1, \ldots, n)$ are the right indices of the matrix function* (4.4).

In the case of the solvability of the equation (4.3) *one of its solutions is given by the formula* (4.2_2). *Here $\hat{A}_\pm^{-1}$ and $\hat{D}^{[-1]}$ are integral operators in $(E_+)_n$ of the form* (4.3) *with the symbols $A_\pm^{-1}(\lambda)$ and $\left\{\left(\frac{\lambda - i}{\lambda + i}\right)^{-\varkappa_j} \delta_{jk}\right\}_1^n$, respectively.*

Remark. Theorem 4.2 shows that the operator (4.3) is not a Φ_+-operator and not a Φ_--operator in $(E_+)_n$ if especially the number matrix $c = \mathcal{A}(\infty)$ is singular ($\det c = 0$). In this case (4.3) is called a *system of the first kind.* If $\det c \neq 0$, then the system is called a *system of the second kind.* In this case without loss of generality we can put $c = I$ (I the unit matrix) in (4.3).

7.4.3. Systems of Paired Wiener-Hopf Equations

Completely analogously to the preceding two subsections we can apply Theorem 3.3 to paired system of the form

$$\sum_{k=0}^{\infty} a_{j-k}\xi_k + \sum_{k=-\infty}^{-1} b_{j-k}\xi_k = \eta_j \qquad (j = 0, \pm 1, \ldots) \tag{4.6}$$

or

$$\left.\begin{aligned}\sum_{k=-\infty}^{\infty} a_{j-k}\xi_k &= \eta_j \qquad (j = 0, 1, \ldots)\,, \\ \sum_{k=-\infty}^{\infty} b_{j-k}\xi_k &= \eta_j \qquad (j = -1, -2, \ldots)\,.\end{aligned}\right\} \tag{4.6'}$$

Here a_j and b_j ($j = 0, \pm 1, \ldots$) are matrices of order n formed from the Fourier coefficients of the matrix functions $A(z)$, $B(z) \in W_{n \times n}$. ξ_j and η_j ($j = 0, \pm 1, \ldots$) are sequences of n-dimensional vectors whose components form a sequence from the space E (see 7.4.1) for a fixed index i ($i = 1, 2, \ldots, n$).

The same statement holds for systems of integral equations of the form

$$c\varphi(t) - \int_0^\infty k_1(t-s)\,\varphi(s)\,ds - \int_{-\infty}^0 k_2(t-s)\,\varphi(s)\,ds = f(t) \qquad (-\infty < t < \infty) \tag{4.7}$$

resp.

$$\left.\begin{aligned}c_1\varphi(t) - \int_{-\infty}^{\infty} k_1(t-s)\,\varphi(s)\,ds &= f(t) \qquad (0 < t < \infty) \\ c_2\varphi(t) - \int_{-\infty}^{\infty} k_2(t-s)\,\varphi(s)\,ds &= f(t) \qquad (-\infty < t < 0)\,,\end{aligned}\right\} \tag{4.7'}$$

where $k_j(t)$ ($j = 1, 2$) are matrix functions with elements from L, $\varphi(t)$, $f(t)$ vectors from E_n and c, c_1, c_2 number matrices with

$$c = \begin{cases} c_1\,, & t > 0 \\ c_2\,, & t < 0\,. \end{cases}$$

The Theorems 2.3 and 2.8 show immediately that Theorem 3.3 is applicable to systems of the form (4.6)—(4.7). Under the conditions of this Subsection the formulae (3.5) also hold with $\beta(\hat{U}) = 1$.

7.5. Systems of Singular Integral Equations

We consider a system of singular integral equations of the form

$$\mathfrak{A}\varphi = A(t)\,\varphi(t) + \frac{B(t)}{\pi i}\int_\Gamma \frac{\varphi(\tau)}{\tau - t}\,d\tau + T\varphi = f(t) \qquad (t \in \Gamma) \tag{5.1}$$

in the space $L_n^p(\Gamma)$ ($1 < p < \infty$). Here Γ is a Ljapunov curve system of the form described in 6.2.1, $A(t)$ and $B(t)$ are arbitrary continuous matrix functions on Γ of the order n, T is an arbitrary compact operator in the space $L_n^p(\Gamma)$ and $\varphi(t)$, $f(t)$ are vectors from $L_n^p(\Gamma)$.

According to the definition given in 7.3.1 the matrix function

$$A(t, \theta) = A(t) + \theta B(t) \qquad (t \in \Gamma;\ \theta = \pm 1)$$

is called the *symbol* of the operator A defined on the space $L_n^p(\Gamma)$ by the equation (5.1). We put

$$C(t) = A(t) + B(t)\,, \qquad D(t) = A(t) - B(t) \tag{5.2}$$

and

$$S = \{S_1\delta_{jk}\}_1^n\,, \qquad P = \tfrac{1}{2}(I + S)\,, \qquad Q = \tfrac{1}{2}(I - S)\,.$$

Here S_1 is the scalar singular operator defined in the space $L^p(\Gamma)$ by the equation

$$(S_1 g)(t) = \frac{1}{\pi i}\int_\Gamma \frac{g(\tau)}{\tau - t}\,d\tau\,.$$

Using the above introduced notations we can write the operator (5.1) in the form

$$\mathfrak{A} = CP + DQ + T\,,$$

where C and D are the operators of the multiplication by the matrix functions $C(t)$ and $D(t)$, respectively, on the space $L_n^p(\Gamma)$.

Theorem 5.1. *For the operator $\mathfrak{A}$ defined on the space $L_n^p(\Gamma)$ to be a $\Phi_+(\Phi_-)$-operator it is necessary and sufficient that the conditions*

$$\det C(t) \neq 0\,, \qquad \det D(t) \neq 0 \qquad (t \in \Gamma) \tag{5.3}$$

hold.

If conditions (5.3) *are satisfied, then*

$$\operatorname{Ind} \mathfrak{A} = \frac{1}{2\pi}\left[\arg \frac{\det D(t)}{\det C(t)}\right]_\Gamma \tag{5.4}$$

and the operator $R = C^{-1}P + D^{-1}Q$ is a two-sided regularizer of the operator $\mathfrak{A}$.

Proof. The first part of the theorem is proved in the same way as Theorem 3.2 by using Theorem 1.2, Theorem 4.5, Chapter 3, and Theorem 3.7*, Chapter 1.

By Lemma 4.3, Chapter 3, it follows easily from the formula (2.7), Chapter 2, that R is a regularizer.

Now we prove the formula (5.4). If $C(t)$ and $D(t)$ are matrix functions from $H_{n\times n}^\lambda \Gamma)$ $(0 < \lambda < 1)$, then (5.4) follows from the subsequent Theorem 5.2. Now let $C(t)$, $D(t)$ be arbitrary matrix functions satisfying the conditions (5.3). For an arbitrary sufficiently small number $\varepsilon > 0$ we can find matrix functions $\widetilde{C}(t)$ and $\widetilde{D}(t)$ from $H_{n\times n}^\lambda(\Gamma)$ satisfying the conditions (5.3) and moreover the inequalities

$$\|C(t) - \widetilde{C}(t)\|_{C_{n\times n}(\Gamma)} < \varepsilon\,, \qquad \|D(t) - \widetilde{D}(t)\|_{C_{n\times n}(\Gamma)} < \varepsilon\,.$$

Since the index of a function is an integer, for sufficiently small ε we can obtain that

$$\operatorname{ind} \det C(t) = \operatorname{ind} \det \widetilde{C}(t)\,, \qquad \operatorname{ind} \det D(t) = \operatorname{ind} \det \widetilde{D}(t)\,.$$

Now we introduce the operator

$$\tilde{\mathfrak{A}} = \tilde{C}P + \tilde{D}Q + T\,.$$

From the part of the proof already given it follows that

$$\operatorname{Ind} \tilde{\mathfrak{A}} = \frac{1}{2\pi}\left[\arg \frac{\det D(t)}{\det C(t)}\right]_\Gamma . \tag{5.5}$$

On the other hand, for a suitable choise of the number ε the norm of the operator $\mathfrak{A} - \tilde{\mathfrak{A}}$ is sufficiently small. Hence by Theorem 3.8, Chapter 1, we have $\operatorname{Ind} \mathfrak{A} = \operatorname{Ind} \tilde{\mathfrak{A}}$. This together with (5.5) implies the formula (5.4).

By Theorem 2.4 the next theorem follows immediately from the proof of Theorem 3.3.

Theorem 5.2. *Let $C(t)$ and $D(t)$ be matrix functions from $H^\lambda_{n\times n}(\Gamma)$ satisfying conditions* (5.3). *Furthermore let*

$$D^{-1}(t)\, C(t) = C_-(t)\, D_0(t)\, C_+(t)\,, \qquad D_0(t) = \{t^{\varkappa_j}\delta_{jk}\}_1^n$$

be a right canonical factorization of the matrix function $D^{-1}(t)\, C(t)$.

Then for the operator $\mathfrak{A}_0 = CP + DQ$ we have

$$\dim \ker \mathfrak{A}_0 = -\sum_{\varkappa_j<0} \varkappa_j\,, \qquad \dim \operatorname{coker} \mathfrak{A}_0 = \sum_{\varkappa_j>0} \varkappa_j\,.$$

The equation $\mathfrak{A}_0\varphi = f(f \in L^p_n(\Gamma))$ is solvable if and only if the vector

$$g(t) = C_-^{-1}(t)\, D^{-1}(t)\, f(t) = (g_1(t), \dots, g_n(t))$$

satisfies the conditions

$$\int_\Gamma t^k g_j(t)\, dt = 0 \qquad (k = 0, 1, \dots, \varkappa_j - 1) \tag{5.6}$$

for all $j = 1, 2, \dots, n$ with $\varkappa_j > 0$. If this is the case, then a solution of this equation is given by the formula $\varphi = \mathfrak{A}_0^{[-1]}f$, where

$$\mathfrak{A}_0^{[-1]} = (C_+^{-1}P + C_-Q)\,(D_0^{-1}P + Q)\, C_-^{-1}D^{-1}\,.$$

We remark that the conditions (5.6) are just the solvability conditions (3.8).

From Corollary 3.2 we immediately obtain:

Corollary 5.1. *The general solution of the homogeneous equation $\mathfrak{A}_0\varphi = 0$ is given by the formula*

$$\varphi(t) = [C_+^{-1}(t) - C_-(t)\, D_0(t)]\, P(t)\,, \qquad P(t) = (P_{-\varkappa_1-1}, \dots, P_{-\varkappa_n-1})\,,$$

where $P_{-\varkappa_j-1}(t)$ is an arbitrary polynomial of a degree $\leqq -\varkappa_j - 1$ ($P_{-\varkappa_j-1}(t) \equiv \equiv 0$ for $\varkappa_j \geqq 0$).

Remark. If the matrix functions $C(t)$, $D(t)$ satisfy the conditions of Theorem 5.2 and T is an integral operator whose kernel has the form

$$T(t, \tau) = \frac{K(t, \tau) - K(t, t)}{\tau - t}$$

with $K(t, \tau) \in H^{\mu}_{n \times n}(\Gamma \times \Gamma)$ $(\lambda < \mu \leqq 1)$, then Theorems 5.1 and 5.2 remain valid if in their formulation $L^p_n(\Gamma)$ is replaced by $H^\lambda_n(\Gamma)$.

For the Theorem 5.2 this is obvious. For the Theorem 5.1 we obtain this assertion by means of the same considerations as in 3.4.9.

7.6. Comments and References

7.1. The Theorems 1.1 and 1.2 were proved for the case of a Banach space by U. Köhler in his diploma paper 1971. Here in principle we give the proof of U. Köhler. For Φ-operators the Theorem 1.2 was first found by N. J. Krupnik [1].

Theorem 1.1 (and a corresponding theorem for Φ_--operators) was recently generalized by U. Köhler and B. Silbermann [1—2] to the case of general locally convex topological vector spaces.

7.2. Theorem 2.1 is due to I. C. Gohberg and M. G. Krein [3]. Theorem 2.2 is a certain modification of a Theorem of M. S. Budjanu and I. C. Gohberg ([1—2], Theorem 8.1). The proof of Theorem 2.2 given here is in principle taken from the mentioned papers of these authors (cf. also I. C. Gohberg and I. A. Feldman [1], Kapitel 8). There also Theorem 2.6 was proved for the first time.

Corollary 2.1 and Theorem 2.7 were first proved by I. C. Gohberg [2]. Theorems 2.3 and 2.8 are due to I. C. Gohberg and M. G. Krein [3]. Theorem 2.4 was found by J. Plemelj, N. I. Mushelišvili and N. P. Vekua (for this comp. N. I. Mushelišvili [1]). Theorems 2.2* and 2.5 are probably new.

7.3—7.4. Theorems 3.1 to 3.3 are due to I. C. Gohberg [2]. They are generalizations of the Theorems 4.1 and 4.2 and of the corresponding results for systems of paired Wiener-Hopf equations (cf. 7.4.3) stated ealier by I. C. Gohberg and M. G. Krein [3—4]. In the Sections 7.3 and 7.4 to a great extend we follow I. C. Gohberg and I. A. Feldman ([1], Kapitel 8).

7.5. Theorem 5.2 and the second part of Theorem 5.1 were stated for the space $H^\lambda_n(\Gamma)$ in 1943 by N. I. Mushelišvili [1] and N. P. Vekua [2]. The regularizer of the operator $\mathfrak{A}$ and the solvability conditions (5.6) were already ealier given by G. Giraud (1939). S. G. Mihlin [1] carried over the results of Giraud to the case of the space $L^2_n(\Gamma)$. For the spaces $L^p_n(\Gamma)$ Theorems 5.1 and 5.2 were proved by I. C. Gohberg [1—2], B. V. Hvedelidze [1], G. F. Mandžavidze and B. V. Hvedelidze [1], I. B. Simonenko [1].

CHAPTER 8

SYSTEMS OF SINGULAR EQUATIONS OF NON-NORMAL TYPE

In this chapter the results of Chapters 5 and 6 are generalized to systems of discrete Wiener-Hopf equations, of Wiener-Hopf integral equations, of the corresponding paired equations and of singular integral equations.

It is possible to apply the methods of Chapter 4 since singular matrix symbols whose determinant vanishes only in a finite number of points admit special factorizations (see Section 8.1). Because of this fact we succeed in representing the operator A of the corresponding equation system in the form

$$A = VBCDW .$$

Here C is a Φ-operator, V and W are invertible operators on the space $X = E_n$ and B, D are diagonal operators of the form

$$B = \{B_j \delta_{jk}\}_1^n , \qquad D = \{D_j \delta_{jk}\}_1^n ,$$

where B_j and D_j ($j = 1, \dots , n$) are scalar operators considered in Chapters 5 and 6 and satisfying conditions (1) and (2′) of 4.1.1. Putting

$$\overline{X} = \operatorname{im}(VB) , \qquad \widetilde{X} = G(X) , \qquad G = W^{-1}\{D_j^{(-1)} \delta_{jk}\}_1^n$$

we prove that the operator A is a Φ-operator from $\widetilde{X}$ into the space $\overline{X}$ for which the results of the Section 4.1 remain valid.

The analytic description of the space $\overline{X}$ can easily be reduced to the case $n = 1$ considered in the earlier chapters: Let $f \in E_n$ and $V^{-1}f = (g_1, g_2, \dots , g_n)$. Then $f \in \overline{X}$ holds if and only if

$$g_j \in \operatorname{im} B_j = B_j(E) \qquad (j = 1, \dots , n)$$

is valid.

8.1. Representations of Singular Matrix Functions

8.1.1. Let Γ be a smooth plane (in general multiply connected) Jordan curve and $A(t)$ a matrix function defined on Γ which is singular at the point $\alpha \in \Gamma$, i.e. $\det A(\alpha) = 0$. Let the integer $m > 0$ be the order of the zero α of the determinant $\det A(t)$.

We assume that $A(t) \in \mathfrak{A}_{n\times n}(\Gamma)$, where $\mathfrak{A}(\Gamma)$ is a certain algebra of continuous functions on Γ containing the set $R(\Gamma)$ of all rational functions without poles on Γ. Under weak additional hypotheses about the elements of the matrix function $A(t)$ which in the sequel will be sated explicitly the function $A(t)$ can be represented as the product

$$A(t) = R(t)\, D(t)\, A_0(t)\,. \tag{1.1}$$

Here $D(t)$ is a diagonal matrix function of the form

$$D(t) = \{(t - \alpha)^{\mu_j}\, \delta_{jk}\}_1^n$$

with non-negative integers $\mu_1 \geqq \mu_2 \geqq \dots \geqq \mu_n \geqq 0$. $R(t)$ is a polynomial matrix in t with a constant determinant different from zero. $A_0(t)$ is a matrix function with $A_0(t) \in \mathfrak{A}_{n\times n}(\Gamma)$ and $\det A_0(\alpha) \neq 0$ ($\det A_0(t) \neq 0$ for all $t \in \Gamma$ if $\alpha \in \Gamma$ is the only zero of $\det A(t)$). Then obviously we have

$$m = \sum_{j=1}^{n} \mu_j\,.$$

In order to construct the representation (1.1) we proceed as follows. By $a_j(t)$ ($j = 1, \dots, n$) we denote the j-th row vector of the matrix function $A(t)$. Then we can find non-negative integers $\nu_1 \geqq \nu_2 \geqq \dots \geqq \nu_n$ such that

$$a_j(t) = (t - \alpha)^{\nu_j}\, a_j^0(t)\,, \qquad a_j^0(t) \in \mathfrak{A}_n(\Gamma) \qquad (j = 1, \dots, n)\,.$$

(E.g. the numbers $\nu_j = 0$ satisfy these conditions).

If $\sum_{j=1}^{n} \nu_j = m$, then at once we obtain the representation (1.1) with $R(t) = I$ and $\mu_j = \nu_j$ choosing the matrix formed by the row vectors $a_j^0(t)$ as the matrix $A_0(t)$.

Now let $\sum_1^n \nu_j < m$. Then the vectors $a_j^0(\alpha)$ ($j = 1, \dots, n$) are not linearly independent, i.e.

$$\sum_1^n c_j a_j^0(\alpha) = 0\,, \tag{1.2}$$

where the numbers c_j are not all equal to zero. By ν_l we denote the greatest of all numbers ν_j ($j = 1, \dots, n$) for which $c_j \neq 0$. Now we form the matrix function $T^{(1)}(t)$ whose elements $T_{jk}(t)$ ($j, k = 1, \dots, n$) are defined as follows:

$$T_{jk}(t) = \delta_{jk}\ (j \neq l)\,, \qquad T_{lk}(t) = c_k(t - \alpha)^{\nu_l - \nu_k}\,.$$

If we now denote the j-th row vector of the matrix function $B(t) = T^{(1)}(t) \times \times A(t)$ by $b_j(t)$, then

$$b_j(t) = a_j(t)\ (j \neq l)\,, \qquad b_l(t) = (t - \alpha)^{\nu_l} \sum_1^n c_k a_k^0(t)\,.$$

In the sequel we assume that the components of the vector $\sum c_k a_k^0(t)$ satisfy the following condition:

(A) If $a(\alpha) = 0$, then $(t - \alpha)^{-1} a(t) \in \mathfrak{A}(\Gamma)$.

Then by (1.2) the vector $b_l(t)$ can be represented in the form

$$b_l(t) = (t - \alpha)^{\nu_l + 1} b_l^0(t) , \qquad b_l^0(t) \in \mathfrak{A}_n(\Gamma) .$$

Hence after the multiplication of the matrix $A(t)$ by $T^{(1)}(t)$ the number ν_l is enlarged to $\nu_l + 1$. Here $T^{(1)}(t)$ is a polynomial matrix with det $T^{(1)}(t) \equiv c_l \neq 0$. Multiplying the matrix function $T^{(1)}(t)$ on the left by a constant elementary matrix (whose determinant is different from zero), if necessary, we can fix it so that in the matrix function $B(t)$ the numbers

$$\nu_j^{(1)} = \begin{cases} \nu_j , & j \neq l \\ \nu_l + 1 , & j = l \end{cases}$$

are arranged in a monotonically decreasing sequence.

Repeating the preceding considerations for the matrix $B(t)$ instead of $A(t)$ and continuing this process we finally obtain at last after finitely many steps that

$$T(t)\, A(t) = D(t)\, A_0(t)$$

with $A_0(t) \in \mathfrak{A}_{n \times n}(\Gamma)$ and det $A_0(\alpha) \neq 0$. This yields the representation (1.1) for $R(t) = T^{-1}(t)$.

It follows from the preceding constructions that the representation (1.1) of the matrix function $A(t)$ can effectively be realized if at every step the elements of the new matrix function satisfy condition (A). By 2.3.4 this is the case if the elements of $A(t)$ belong to the algebra $\mathfrak{A}(\alpha, m)$.

Furthermore, we remark: If in the preceding constructions $t - \alpha$ is everywhere replaced by $t^{-1} - \alpha^{-1}$ (here we assume $0 \notin \Gamma$), then in a completely analogous way we obtain the representation

$$A(t) = \overline{R}(t)\, \overline{D}(t)\, \overline{A}_0(t) , \tag{1.3}$$

where $\overline{D}(t)$ is a diagonal matrix function of the form

$$\overline{D}(t) = \{(t^{-1} - \alpha^{-1})^{\mu_j} \delta_{jk}\}_1^n$$

and $\overline{R}(t)$ is a polynomial matrix in t^{-1} with a constant determinant different from zero. $\overline{A}_0(t)$ has the same meaning as $A_0(t)$ in (1.1).

8.1.2. If we finally replace the roles of the rows and columns in the matrix functions at the constructions of the Subsection 8.1.1, then we obtain the representations

$$A(t) = A_1(t)\, D(t)\, R_1(t) \tag{1.4}$$

and

$$A(t) = \overline{A}_1(t)\, \overline{D}(t)\, \overline{R}_1(t) . \tag{1.5}$$

The factors in the representations (1.1) and (1.4) (resp. (1.3) and (1.5)) have the same structure.

Theorem 1.1. *If the matrix function $A(t)$ permits a representation* (1.1), *then the numbers μ_j ($j = 1, \dots, n$) are uniquely determined.*

Proof. In addition to (1.1) let $A(t)$ possess a second representation of the form

$$A(t) = \tilde{R}(t)\,\tilde{D}(t)\,\tilde{A}_0(t)\,, \qquad \tilde{D}(t) = \{(t-\alpha)^{\tilde{\mu}_j}\,\delta_{jk}\}_1^n\,.$$

Then

$$B_1(t)\,D(t) = \tilde{D}(t)\,B_2(t) \tag{1.6}$$

with

$$B_1(t) = \tilde{R}^{-1}(t)\,R(t)\,, \qquad B_2(t) = \tilde{A}_0(t)\,A_0^{-1}(t)\,.$$

Hence $\det B_1(t) \equiv c_1 \neq 0$, $\det B_2(\alpha) \neq 0$.

Assuming that $\mu_r < \tilde{\mu}_r$ for a certain r ($1 \leqq r \leqq n$) by means of similar consideration as at the proof of the Theorem 2.1, Chapter 7, we easily come to the conclusion that (1.6) implies $\det B_1(\alpha) = 0$. This contradiction proves the assertion.

Remark. The proof of Theorem 1.1 shows that the numbers μ_j ($j = 1, \dots, n$) coincide for all representations (1.1) and (1.3)—(1.5). In the sequel these numbers are called the *partial orders of the zero* α of the function $\det A(t)$.

8.1.3. Now we assume that the determinant of the matrix function $A(t) \in \mathfrak{A}_{n\times n}(\Gamma)$ possesses finitely many zeros of the integral orders m_j at the points $\alpha_j \in \Gamma$ ($j = 1, \dots, r$).

We apply the constructions from 8.1.1 to the matrix function $A(t)$ successively for the points $\alpha_1, \alpha_2, \dots, \alpha_r$. We assume that after each step condition (A) of 8.1.1 is satisfied. First, by (1.1) we can write $A(t)$ in the form

$$A(t) = R_1(t)\,D_1(t)\,A_1(t)\,,$$

where $R_1(t)$ is a polynomial matrix in t with $\det R_1(t) \equiv \text{const} \neq 0$ and $D_1(t) = \{(t-\alpha_1)^{\mu_j^{(1)}}\,\delta_{jk}\}_1^n$ is a diagonal matrix. The determinant of the matrix function $A_1(t) \in \mathfrak{A}_{n\times n}(\Gamma)$ possesses only $r-1$ zeros on Γ in the points $\alpha_2, \dots, \alpha_r$.

If we now carry out the constructions from 8.1.1 for $A_1(t)$ at the point α_2 etc., then for the matrix function $A(t)$ at last we obtain the representation

$$A(t) = R_1(t)\,D_1(t) \dots R_r(t)\,D_r(t)\,A_0'(t)\,. \tag{1.7}$$

Here $R_l(t)$ ($l = 1, \dots, r$) are polynomial matrices in t with a constant determinant different from zero. Furthermore, $D_l(t) = \{(t-\alpha_l)^{\mu_j^{(l)}}\delta_{jk}\}_1^n$, the determinant of the matrix function $A_0'(t) \in \mathfrak{A}_{n\times n}(\Gamma)$ is unequal to zero at all points $t \in \Gamma$ and $\sum_{j=1}^{n} \mu_l^{(l)} = m_l$ ($l = 1, \dots, r$). By a well-known theorem on polynomial matrices (see F. R. Gantmaher [1], Satz 3, Kapitel VI) now we have

$$R_1(t)\,D_1(t) \dots R_r(t)\,D_r(t) = R(t)\,D_1(t) \dots D_r(t)\,R'(t)\,, \tag{1.8}$$

where $R(t)$ and $R'(t)$ also are polynomial matrices in t with a constant determinant different from zero. Thus from (1.7) and (1.8) there follows again the representation (1.1) for the matrix function $A(t)$, where now

$$D(t) = \{(t - \alpha_1)^{\mu_j^{(1)}} \dots (t - \alpha_r)^{\mu_j^{(r)}} \delta_{jk}\}_1^n \tag{1.9}$$

and $A_0(t) = R'(t) A_0'(t)$.

Repeating the preceding constructions with $t^{-1} - \alpha_l^{-1}$ instead of $t - \alpha_l$ $(l = 1, \dots, r)$ we obtain the representation (1.3) for the matrix function $A(t)$, where

$$\overline{D}(t) = \{(t^{-1} - \alpha_1^{-1})^{\mu_j^{(1)}} \dots (t^{-1} - \alpha_r^{-1})^{\mu_j^{(r)}} \delta_{jk}\}_1^n . \tag{1.10}$$

In an analogous way the representations (1.4) and (1.5) can be generalized to the case of finitely many zeros of the determinant $\det A(t)$.

Thus taking into consideration the results of 2.3.4 we obtain the following theorem.

Theorem 1.2. *Let* $A(t) \in \mathfrak{A}_{n\times n}(\Gamma)$ *and* $\boldsymbol{\alpha} = \{\alpha_1, \alpha_2, \dots, \alpha_r\}$ *the system of all zeros of the determinant* $\det A(t)$ *of the integral multiplicity* $\boldsymbol{m} = \{m_1, m_2, \dots, m_r\}$.

If $A(t) \in \mathfrak{A}_{n\times n}(\boldsymbol{\alpha}, \boldsymbol{m})$ *then* $A(t)$ *admits the representations* (1.1) *and* (1.3)—(1.5). *Here* $A_k(t)$, $\overline{A}_k(t)$ $(k = 0, 1)$ *are regular matrix functions from* $\mathfrak{A}_{n\times n}(\Gamma)$. $R(t)$ *and* $R_1(t)$ $(\overline{R}(t)$ *and* $\overline{R}_1(t))$ *are polynomial matrices in* t *(in* t^{-1}*) with a constant determinant different from zero. The diagonal matrices* $D(t)$, $\overline{D}(t)$ *have the form* (1.9), (1.10), *respectively, where the integers* $\mu_1^{(l)} \geqq \mu_2^{(l)} \geqq \dots \geqq \mu_n^{(l)} \geqq 0$ $(l = 1, \dots, r)$ *are uniquely determined.*

8.1.4. Finally we still consider the case of the closed real line $\Gamma = \{-\infty, \infty\}$ and $\mathfrak{A}(\Gamma) = \mathfrak{L}$. For this case we obtain analogous results by stereographic projection of the unit circle.

Thus let now $A(\lambda) \in \mathfrak{L}_{n\times n}$ and let the determinant $\det A(\lambda)$ possess zeros of the integral orders m_j at the finite points $\alpha_j \in (-\infty, \infty)$ $(j = 1, \dots, r)$.

If we now apply the constructions from 8.1.1—8.1.3 to the matrix function $A(\lambda)$, where we have to replace $z^{-1} - \alpha^{-1}$ by $\dfrac{\lambda - \alpha}{\lambda - i}$ everywhere in the preceding considerations, then we obtain the representation

$$A(\lambda) = \overline{R}(\lambda)\, \overline{D}(\lambda)\, \overline{A}_0(\lambda) . \tag{1.11}$$

Here $\overline{D}(\lambda)$ is a diagonal matrix of the form

$$\overline{D}(\lambda) = \left\{\left(\frac{\lambda - \alpha_1}{\lambda - i}\right)^{\mu_j^{(1)}} \dots \left(\frac{\lambda - \alpha_r}{\lambda - i}\right)^{\mu_j^{(r)}} \delta_{jk}\right\}_1^n , \tag{1.11'}$$

where $\mu_1^{(l)} \geqq \mu_2^{(l)} \geqq \dots \geqq \mu_n^{(l)}$ $(l = 1, \dots, r)$ are non-negative integers with $\sum_{j=1}^{n} \mu_j^{(l)} = m_l$. $\overline{R}(\lambda)$ $(-\infty \leqq \lambda \leqq \infty)$ is a matrix function with a constant determinant different from zero and with elements which are polynomials with respect to $\frac{1}{\lambda - i}$, and $\overline{A}_0(\lambda) \in \mathfrak{L}_{n\times n}$, $\det \overline{A}_0(\lambda) \neq 0$ $(-\infty \leqq \lambda < \infty)$.

By Subsection 2.3.4 the above-mentioned constructions are always possible if $A(\lambda) \in \mathfrak{L}(\boldsymbol{\alpha}, \boldsymbol{m})$.

Interchanging the roles of the rows and columns in an analogous way we obtain a representation of the form

$$A(\lambda) = A_0(\lambda)\, D(\lambda)\, R(\lambda)\,. \tag{1.12}$$

Here $R(\lambda)$ $(D(\lambda))$ has the same form as the matrix function $\overline{D}(\lambda)$ $(\overline{R}(\lambda))$ with $\frac{\lambda - \alpha_j}{\lambda + i}\left(\frac{1}{\lambda + i}\right)$ instead of $\frac{\lambda - \alpha_j}{\lambda - i}\left(\frac{1}{\lambda - i}\right)$, and $A_0(\lambda) \in \mathfrak{L}_{n\times n}$, $\det A_0(\lambda) \neq 0$ $(-\infty \leqq \lambda \leqq \infty)$. Furthermore. such representations are possible which are obtained from (1.11) resp. (1.12) by inverting the ordering of all factors.

Finally we consider the case where the determinant of the matrix function $A(\lambda) \in \mathfrak{L}_{n\times n}$ has a zero of the integral order m in the point $\lambda = \infty$. Here we still assume that the elements of $A(\lambda)$ satisfy the conditions (1)—(3) of Theorem 2.10, Chapter 5. Using this theorem we then obtain the representation (1.11) by application of the constructions described in 8.1.1 to the matrix function $A(\lambda)$ (with $\frac{1}{\lambda - i}$ instead of $z^{-1} - \alpha^{-1}$), where now $\overline{D}(\lambda)$ is a diagonal matrix of the form

$$\overline{D}(\lambda) = \left\{\left(\frac{1}{\lambda - i}\right)^{\mu_j} \delta_{jk}\right\}_1^n. \tag{1.13}$$

Correspondingly we can obtain to the analogue of the representations (1.12), (1.1) and (1.5).

8.2. Systems of Discrete Wiener-Hopf Equations

8.2.1. The Spaces Generated by Zeros of the Symbol

1. In the sequel by E_+ we denote one of the sequence spaces l_+^p $(1 \leqq p < \infty)$ or c_+^0.

Let α_l $(l = 1, \dots, r)$ be points of the unit circle $(|\alpha_l| = 1)$ and $\mu_j^{(l)}$ $(j = 1, \dots, n)$ arbitrary non-negative integers. By $\overline{D}(t)$ we denote the diagonal matrix (1.10) and by $\overline{R}(t)$ an arbitrary polynomial matrix in t^{-1} with a constant determinant different from zero. In the space $(E_+)_n$ we consider the discrete Wiener-Hopf operator $\hat{B}$ with the symbol (cf. 7.4.1)

$$B_1(t) = \overline{R}(t)\, \overline{D}(t) \qquad (|t| = 1)\,. \tag{2.1}$$

By means of the results of the Section 5.1 the image space $\operatorname{im} \hat{B}$ of this operator can be described as follows.

First, by Theorem 4.1, Chapter 7, the operator $\hat{R}_-$ considered in $(E_+)_n$ with the symbol $\overline{R}(t)$ is invertible, where moreover $\hat{R}_-^{-1} \in \hat{\mathfrak{E}}_{n \times n}(U)$ holds.[1]) From this it follows by Lemma 1.1, Chapter 5, that ker $\hat{B} = \{0\}$. Now let $f \in (E_+)_n$ and

$$g = \hat{R}_-^{-1}f = (g_1, g_2, \ldots, g_n) .$$

Then we have $f \in \text{im } \hat{B}$ if and only if

$$g_j \in \text{im } \prod_{l=1}^r \hat{B}_{\alpha l}^{\mu_j^{(l)}} \qquad (j = 1, \ldots, n) . \tag{2.2}$$

If conditions (2.2) are satisfied, then

$$\hat{B}^{-1}f = (h_1, h_2, \ldots, h_n) . \qquad h_j = \prod_{l=1}^r \hat{B}_{\alpha l}^{-\mu_j^{(l)}} g_j ,$$

where $\hat{B}_\alpha^{-1}$ is the operator defined by Corollary 1.1, Chapter 5, and $\hat{B}_\alpha^{-\mu}$ its μ-th power. With the norm $||f|| = ||\hat{B}^{-1}f||_{(E_+)_n}$ the space im $\hat{B}$ becomes a Banach space which in the sequel we shall denote by $(\overline{E}_+)_n$.

Remark. We put $\mu = \max\limits_{j,l} \mu_j^{(l)}$. Then we have

$$(E_+^{\mu+\varepsilon})_n \subset (\overline{E}_+)_n \qquad (\varepsilon > 0 \quad \text{for} \quad E_+ = c_+^0 , \qquad \varepsilon = 0 \quad \text{for} \quad E_+ = l_+^p)$$

in the sense of a continuous embedding. This follows immediately from Corollary 1.2, Chapter 5.

2. Let also $\tilde{\mu}_j^{(l)}$ $(l = 1, 2, \ldots, r; j = 1, \ldots, n)$ be non-negative integers. Let $D(t)$ be the diagonal matrix

$$D(t) = \{(t - \alpha_1)^{\tilde{\mu}_j^{(1)}} \ldots (t - \alpha_r)^{\tilde{\mu}_j^{(r)}} \delta_{jk}\}_1^n \tag{2.3}$$

and $R(t)$ a polynomial matrix in t with a constant determinant different from zero.

In the space $(E_+)_n$ we consider the discrete Wiener-Hopf operator $\hat{D}$ with the symbol matrix $D_1(t) = D(t) R(t)$. By $\hat{R}_+$ we denote the corresponding operator with the symbol $R(t)$ and by $\hat{G}_l$ $(l = 1, \ldots, r)$ the operator mapping a vector of the form $\varphi = (\varphi_1, \varphi_2, \ldots, \varphi_n)$, where φ_j $(j = 1, \ldots, n)$ are arbitrary number sequences, onto the vector

$$\hat{G}_l\varphi = (\hat{G}_{\alpha l}^{\tilde{\mu}_1^{(l)}}\varphi_1, \hat{G}_{\alpha l}^{\tilde{\mu}_2^{(l)}}\varphi_2, \ldots, \hat{G}_{\alpha l}^{\tilde{\mu}_n^{(l)}}\varphi_n) . \tag{2.4_1}$$

Here $\hat{G}_\alpha$ is the operator introduced in 5.1.1. Now we consider the operator

$$\hat{G} = \hat{R}_+^{-1}\hat{G}_1 \ldots \hat{G}_r \tag{2.4_2}$$

on $(E_+)_n$.

---

[1]) For the notation $\mathfrak{E}(U)$ see Section 2.1.

Then by equations (1.2), Chapter 5, we have

$$\hat{G}\hat{D}\varphi = \varphi . \qquad \hat{D}\hat{G}\varphi = \varphi \qquad \forall\, \varphi \in (E_+)_n . \tag{2.5}$$

Following the considerations of Section 4.2.3 we now introduce the space $(\widetilde{E}_+)_n = \hat{G}(E_+)_n$, and we define the norm on this space by

$$\|f\| = \|\hat{D}f\|_{(E_+)_n} \qquad (f \in (\widetilde{E}_+)_n) .$$

The space $(E_+)_n$ is continuously embedded into the Banach space $(\widetilde{E}_+)_n$. $\hat{D}$ effects an isometric isomorphism of $(\widetilde{E}_+)_n$ onto $(E_+)_n$.

Remark. We put $\tilde{\mu} = \max\limits_{j,l} \tilde{\mu}_j^{(l)}$. By Corollary 1.2, Chapter 5, we have

$$(\widetilde{E}_+)_n \subset (E_+^{-(\tilde{\mu}+\varepsilon)})_n \qquad (\varepsilon > 0 \quad \text{for} \quad E_+ = l_+ ; \qquad \varepsilon = 0 \quad \text{for} \quad E_+ = l_+^p, \; 1 < p < \infty , \quad \text{or} \quad E_+ = c_+^0) .$$

8.2.2. Symbols with Zeros of Integral Orders

In the space $(E_+)_n$ (E_+ one of the sequence spaces l_+^p, $1 \leqq p < \infty$, or c_+^0) we consider the system of discrete Wiener-Hopf equations of the form

$$\hat{A}\xi = \sum_{k=0}^{\infty} a_{j-k}\xi_k = \eta_j \qquad (j = 0, 1, 2, \ldots) , \tag{2.6}$$

where

$$A(t) = \sum_{j=-\infty}^{\infty} a_j t^j \qquad (|t| = 1)$$

is a matrix function from $W_{n\times n}$.

We assume that the determinant $\det A(t)$ has a finite number of zeros of the integral orders m_l on the unit circle at the points α_l ($l = 1, \ldots, r$). Furthermore, let $A(t) \in \bigcap\limits_{l=1}^{r} W(\alpha_l, m_l)$.[1]) By $m_j^{(l)}$ ($j = 1, \ldots, n$) we denote the partial orders of the zero α_l.

Now we choose arbitrary non-negative integers $\mu_j^{(l)}$ and $\tilde{\mu}_j^{(l)}$ such that

$$\mu_j^{(l)} + \tilde{\mu}_j^{(l)} = m_j^{(l)} \qquad (j = 1, \ldots, n; \quad l = 1, \ldots, r) . \tag{2.7}$$

By Theorem 1.2 the matrix function $A(t)$ then admits the representation

$$A(t) = B_1(t)\, A_0(t)\, D_1(t) . \tag{2.8}$$

Here $B_1(t)$ and $D_1(t)$ are the matrix functions defined in 8.2.1, and for $A_0(t)$ we have $A_0(t) \in W_{n\times n}$, $\det A_0(t) \neq 0$ ($|t| = 1$). By $\hat{A}_0$ we denote the Wiener-Hopf operator with the symbol $A_0(t)$ considered in the space $(E_+)_n$. Then to the relation (2.8) there corresponds the following representation of the operator $\hat{A}$:

$$\hat{A} = \hat{B}\hat{A}_0\hat{D} . \tag{2.9}$$

[1]) For notation see 5.1.2, no. 3.

By Theorem 2.3, Chapter 7, the matrix function $A_0(t)$ has a right canonical factorization. Let $\varkappa_j$ $(j = 1, \ldots, n)$ be the right partial indices of $A_0(t)$.

By Theorem 4.1, Chapter 7, and the results of Section 4.1 from (2.9) we obtain the following theorem.

Theorem 2.1. *The operator $\hat{A}$ is a Φ-operator from $(\widetilde{E}_+)_n$ into the space $(\overline{E}_+)_n$, and the following formulae hold::*

$$\dim \ker \hat{A} = - \sum_{\varkappa_j<0} \varkappa_j , \qquad \dim \operatorname{coker} A = \sum_{\varkappa_j>0} \varkappa_j . \tag{2.10}$$

Remark 1. It is easily seen that (2.8) and (2.10) yield the following formula for the index of the operator $\hat{A}$:

$$- \operatorname{Ind} \hat{A} = \operatorname{ind} \det A_0(t) = \mu + \operatorname{ind} \left[\frac{\det A(t)}{\prod_{l=1}^{r} (t - \alpha_l)^{m_l}} \right],$$

where

$$\mu = \sum_{l,j} \mu_j^{(l)} .$$

Remark 2. The equation

$$\hat{A}\varphi = f \qquad (\varphi \in (\widetilde{E}_+)_n , \quad f \in (\overline{E}_+)_n)$$

is equivalent to the equation

$$\hat{A}_0\psi = f_1 \qquad (\psi \in (E_+)_n)$$

with $f_1 = \hat{B}^{-1}f \in (E_+)_n$.

Remark 3. The solvability conditions for the system (2.6) can be stated similarily as in Theorem 1.6, Chapter 5.

8.3. Systems of Wiener-Hopf Integral Equations

8.3.1. The Spaces Generated by Zeros of the Symbol

1. Now let $\alpha_l \in (-\infty, \infty)$ $(l = 1, \ldots, r)$ be finite points on the real line and $\mu_j^{(l)}$ $(j = 1, \ldots, n)$ arbitrary non-negative integers. By $\overline{R}(\lambda)$ and $\overline{D}(\lambda)$ $(-\infty \leqq \lambda \leqq \infty)$ we denote the matrix functions from the representation (1.11) and by $B_1(\lambda)$ their product: $B_1(\lambda) = \overline{R}(\lambda)\, \overline{D}(\lambda)$. Let E_+ be one of the spaces L_+^p $(1 \leqq p < \infty)$ or C_+^0.

We consider the operators $\hat{B}$ defined on the space $(E_+)_n$ by the equation

$$\hat{B}\varphi = c\varphi(t) - \int_0^\infty b(t - s)\, \varphi(s)\, ds . \tag{3.1}$$

Here c is the regular number matrix $c = B_1(\infty)$ and $b(t)$ the inverse Fourier transform of the matrix function $c - B_1(\lambda)$. By $\hat{R}_-$ we denote the corresponding integral operator with symbol $R(\lambda)$. By Theorem 4.2, Chapter 7, $\hat{R}_-$ is invertible in the space $(E_+)_n$. Hence by Lemma 2.1, Chapter 5, we have $\ker \hat{B} = \{0\}$.

The analytic description of the image space im B arises in the same way as in 8.2.1. Here we now have to take the operator $\hat{B}_\alpha^{-1}$ from Corollary 2.4, Chapter 5. We denote the space im $\hat{B}$ equipped with the corresponding norm again by $(\overline{E}_+)_n$.

Remark. By Corollary 2.5, Chapter 5, we have

$$(L_+^{p,\mu})_n \subset (\overline{L}_+^p)_n\,, \qquad 1 \leqq p < \infty\,; \qquad \mu = \max_{j,l} \mu_j^{(l)}\,.$$

2. The space $(\widetilde{E}_+)_n = \hat{G}(E_+)_n$ is also introduced completely analogously to the discrete case (8.2.1, no. 2). In the present case we have to replace (2.3) by the diagonal matrix

$$D(\lambda) = \left\{\left(\frac{\lambda - \alpha_1}{\lambda + i}\right)^{\tilde{\mu}_j^{(1)}} \cdots \left(\frac{\lambda - \alpha_r}{\lambda + i}\right)^{\tilde{\mu}_j^{(r)}} \delta_{jk}\right\}_1^n, \tag{3.2}$$

and $R(\lambda)$ $(-\infty \leqq \lambda \leqq \infty)$ is a polynomial matrix with respect to $\dfrac{1}{\lambda + i}$ with a constant determinant different from zero. By $\hat{D}$ we denote the integral operator of the form (3.1) with the symbol $D_1(\lambda) = D(\lambda)\, R(\lambda)$. Furthermore, now in the formula (2.4_1) $\hat{G}_\alpha$ is the operator introduced in 5.2.1, no. 1, and φ_j $(j = 1, \ldots, n)$ are locally integrable functions on $(0, \infty)$.

Remark. We have (cf. 5.2.1, no. 8)

$$(\widetilde{L}_+^p)_n \subset (L_+^{p,\,-\tilde{\mu}})_n\,, \quad 1 < p < \infty\,, \quad \tilde{\mu} = \max_{j,l} \tilde{\mu}_j^{(l)}\,; \quad (\widetilde{L}_+^1)_n \subset (L_+^{1,\,-(\tilde{\mu}+\varepsilon)})_n \quad (\varepsilon > 0)\,.$$

3. Let now $\overline{D}(\lambda)$ be the diagonal matrix (1.13) and $B_\infty(\lambda) = \overline{R}(\lambda)\, \overline{D}(\lambda)$. The meaning of $\overline{R}(\lambda)$ and $\hat{R}_-$ is the same as in no. 1. By $\hat{B}_\infty$ we denote the Wiener-Hopf integral operator of the first kind defined on the space $(E_+)_n$ by the equation

$$\hat{B}_\infty \varphi = \int_0^\infty b_\infty(t - s)\, \varphi(s)\, ds\,. \tag{3.3}$$

Here $b_\infty(t)$ is the inverse Fourier transform of $B_\infty(\lambda)$. From Lemma 2.1, Chapter 5, it follows that $\ker \hat{B}_\infty = \{0\}$.

The analytic description of the image space im $\hat{B}_\infty$ is easily obtained from Theorem 2.1, Chapter 5: Let $f \in (E_+)_n$ and

$$g = \hat{R}_-^{-1} f = (g_1, g_2, \ldots, g_n)\,.$$

Then $f \in \operatorname{im} \hat{B}_\infty$ holds if and only if the derivative $g_j^{(k)}(t)$ $(j = 1, \ldots, n)$ is absolutely continuous on each finite interval of the positive half-line $(0, +\infty)$ for $k = 0, 1, \ldots, \mu_j - 1$ and if $g_j^{(k)}(t) \in E_+$ for $k = 0, 1, \ldots, \mu_j$.

If the above conditions are satisfied, then we obtain (see Corollary 2.2, Chapter 5):

$$\widetilde{B}_\infty^{-1} f = \left\{ i^{\mu_1} \left(\frac{d}{dt} - 1 \right)^{\mu_1} g_1, \dots, i^{\mu_n} \left(\frac{d}{dt} - 1 \right)^{\mu_n} g_n \right\}.$$

With the norm $\|f\| = \|\hat{B}_\infty^{-1} f\|_{(E_+)_n}$ the space im $\hat{B}_\infty$ becomes a Banach space which in the sequel we denote by $(\overline{E}_+^{(\infty)})_n$.

4. Finally let $\tilde{\mu}_j$ $(j = 1, \dots, n)$ be non-negative integers, $D(\lambda)$ the diagonal matrix function

$$D(\lambda) = \left\{ \left(\frac{1}{\lambda + i} \right)^{\tilde{\mu}_j} \delta_{jk} \right\}_1^n \tag{3.4}$$

and $R(\lambda)$ the polynomial matrix defined in no. 2. By $\hat{D}_\infty$ we denote the integral operator of the form (3.3) with symbol $D_\infty(\lambda) = D(\lambda)\, R(\lambda)$.

On $(E_+)_n$ we consider the operator $\hat{G}$ defined for any vector $f = (f_1, f_2, \dots, f_n) \in (E_+)_n$ by the equation

$$\hat{G} f = \hat{R}_+^{-1} (\hat{G}_\infty^{\tilde{\mu}_1} f_1, \hat{G}_\infty^{\tilde{\mu}_2} f_2, \dots, \hat{G}_\infty^{\tilde{\mu}_n} f_n).$$

Here $\hat{G}_\infty$ is the operator of generalized differentiation

$$\hat{G}_\infty h = i[h(t) + h'(t)] \qquad (h \in E_+)$$

introduced in 5.2.1, no. 1. By relations (1.2), Chapter 5, the operators $\hat{D}_\infty$ and $\hat{G}$ satisfy condition (2.5). Thus on the image set $\hat{G}(E_+)_n$ we can introduce the norm defined in 8.2.1. We denote the Banach space obtained in this way by $(\widetilde{E}_+^{(\infty)})_n$.

8.3.2. Systems of Equations of the Second Kind

Now in the space $(E_+)_n$ (E_+ one of the spaces L_+^p, $1 \leqq p < \infty$ or C_+^0) we consider the system of Wiener-Hopf integral equations of the second kind

$$\hat{A}\varphi = \varphi(t) - \int_0^\infty k(t - s)\, \varphi(s)\, ds = f(t) \qquad (0 \leqq t < \infty), \tag{3.5}$$

where $k(t)$ is a matrix function from $L_{n \times n}$.

We assume that the determinant of the symbol

$$\mathcal{A}(\lambda) = I - \int_{-\infty}^{\infty} e^{i\lambda t}\, k(t)\, dt \qquad (-\infty \leqq \lambda \leqq \infty)$$

(I the unit matrix) has a finite number of zeros of the integral orders m_l at the points $\alpha_l \in (-\infty, \infty)$ $(l = 1, \dots, r)$. Moreover, we assume that $\mathcal{A}(\lambda) \in$

$\in \bigcap_{l=1}^{r} \mathfrak{L}(\alpha_l, m_l)$.[1]) By $m_j^{(l)}$ $(j = 1, \dots, n)$ we denote the partial orders of the zero α_l.

The further considerations are the same as in 8.2.2. We choose arbitrary non-negative integers $\mu_j^{(l)}$ and $\tilde{\mu}_j^{(l)}$ satisfying the equations (2.7). Then by 8.1.4 the matrix function $\mathcal{A}(\lambda)$ can be represented in the form

$$\mathcal{A}(\lambda) = B_1(\lambda)\, \mathcal{A}_0(\lambda)\, D_1(\lambda)\,. \tag{3.6}$$

Here $B_1(\lambda)$, $D_1(\lambda)$ are the matrix functions introduced in 8.3.1, no. 1—2, and $\mathcal{A}_0(\lambda) \in \mathfrak{L}_{n\times n}$, $\det \mathcal{A}_0(\lambda) \neq 0$ $(-\infty \leqq \lambda \leqq \infty)$. If by $\hat{A}_0$ we now denote the Wiener-Hopf integral operator of the second kind with symbol $\mathcal{A}_0(\lambda)$ considered in the space $(E_+)_n$, then the representation (2.9) of the operator (3.5) follows immediately from (3.6).

By Theorem 4.2, Chapter 7, we conclude from this that in the present case the Theorem 2.1 and the subsequent Remarks 1 and 2 retain their complete validity, where now in the formulae (2.10) the numbers $\varkappa_j$ $(j = 1, \dots, n)$ are the right partial indices of the matrix function $\mathcal{A}_0(\lambda) \in \mathfrak{L}_{n\times n}$ (comp. Theorem 2.8, Chapter 7).

Moreover, in the case $(E_+)_n = (L_+^p)_n$ we can represent the solvability conditions for the system (3.5) in a similar form as in the case $n = 1$ (comp. Theorem 2.8, Chapter 5).

8.3.3. Systems of Equations of the First Kind

In the space $(E_+)_n$ we consider the system of Wiener-Hopf integral equations of the first kind

$$\hat{A}\varphi = c\varphi(t) - \int_0^\infty k(t-s)\,\varphi(s)\,ds = f(t) \qquad (0 \leqq t < \infty)\,, \tag{3.7}$$

where $k(t)$ is a matrix function from $L_{n\times n}$ and c a singular number matrix ($\det c = 0$).

We assume that for the determinant of the symbol

$$\mathcal{A}(\lambda) = c - \int_{-\infty}^{\infty} e^{i\lambda t}\, k(t)\, dt \qquad (-\infty \leqq \lambda \leqq \infty) \tag{3.8}$$

the point $\lambda = \infty$ is a zero of the integral order m. For all other λ with $-\infty < < \lambda < \infty$ let first $\det \mathcal{A}(\lambda) \neq 0$. Furthermore, we assume that the elements of the matrix function $\mathcal{A}(\lambda)$ satisfy the conditions (1)—(3) of Theorem 2.10 of Chapter 5. By $m_j^{(\infty)}$ we denote the partial orders of the zero $\lambda = \infty$.

Now let μ_j and $\tilde{\mu}_j$ be arbitrary non-negative integers with $\mu_j + \tilde{\mu}_j = m_j^{(\infty)}$ $(j = 1, 2, \dots, n)$. Then by 8.1.4 we can represent the matrix function $\mathcal{A}(\lambda)$

---

[1]) For notation see 5.2.2, no. 3.

in the form

$$\mathcal{A}(\lambda) = B_\infty(\lambda)\, \mathcal{A}_0(\lambda)\, D_\infty(\lambda)\,.$$

Here $B_\infty(\lambda)$, $D_\infty(\lambda)$ are the matrix functions defined in 8.3.1, no. 3–4 and $\mathcal{A}_0(\lambda)$ has the same meaning as in (3.6). Hence $\hat{A} = \hat{B}_\infty \hat{A}_0 \hat{D}_\infty$.

By means of the same considerations as in 8.2.2 and 8.3.2 we obtain the following theorem.

Theorem 3.1. *The operator* $\hat{A}$ *is a* Φ*-operator from* $(\tilde{E}_+^{(\infty)})$ *into the space* $(\overline{E}_+^{(\infty)})_n$. *The formula* (2.10) *holds, where* $\varkappa_j$ $(j = 1, \ldots, n)$ *are the right partial indices of the matrix function* $\mathcal{A}_0(\lambda) \in \mathfrak{L}_{n\times n}$.

Remark 1. For the index we have the formula

$$-\operatorname{Ind} \hat{A} = \mu + \operatorname{ind}\,[(\lambda + i)^m \det \mathcal{A}(\lambda)]\,, \qquad \mu = \sum_1^n \mu_j\,.$$

Remark 2. If we assume that the determinant of the symbol (3.8) has zeros in the point at infinity and also at the finite points α_j $(j = 1, \ldots, r)$ with the integral orders m and m_j, respectively, then by means of the constructions from 8.1 we can represent the symbol in the form

$$\mathcal{A}(\lambda) = B_\infty(\lambda)\, B_1(\lambda)\, \mathcal{A}_0(\lambda)\, D_\infty(\lambda)\, D_1(\lambda)\,.$$

This yields the following representation for the operator $\hat{A}$:

$$A = \hat{B}_\infty \hat{B} \hat{A}_0 \hat{D}_\infty \hat{D}\,.$$

Carrying out the constructions of 8.3.1, no. 1 and no. 3, now we obtain a Banach space $(\overline{E}_+)_n \subset (E_+)_n$, where the operator $\hat{B}_\infty \hat{B}$ maps the space $(E_+)_n$ isometrically onto $(\overline{E}_+)_n$. By means of the considerations from 8.3.1, no. 2 and no. 4, in a similar way we can construct a space $(\tilde{E}_+)_n \supset (E_+)_n$, where the operator $\hat{D}_\infty \hat{D}$ maps $(\tilde{E}_+)_n$ isometrically onto $(E_+)_n$. Then for the pair of spaces $(\tilde{E}_+)_n$, $(\overline{E}_+)_n$ the Theorem 3.1 remains valid.

Remark 3. Applying the results of 5.2.3 and 5.2.4 using the same scheme we can also treat the case where the determinant of the symbol (3.8) has finitely many zeros of non-integral orders.

8.4. Systems of Paired Wiener-Hopf Equations

Using the same methods we employed to investigate systems of Wiener-Hopf equations of non-normal type in the preceding two sections we can also treat the corresponding paired equations. As an example we shall consider here systems of paired integral equations of the form (4.7) resp. (4.7′) of Chapter 7.

8.4.1. The Spaces Generated by Zeros of the Symbol

In the sequel E is one of the spaces $L^p = L^p(-\infty, \infty)$ $(1 \leqq p < \infty)$ or C°, and P, Q are the projections defined on the space E_n by

$$(P\varphi)(t) = \begin{cases} \varphi(t), & t > 0 \\ 0, & t < 0 \end{cases} \qquad (\varphi \in E_n), \qquad Q = I - P.$$

1. Let $\alpha_l, \beta_l \in (-\infty, \infty)$ be points on the real line and $\mu_j^{(l)}, \nu_j^{(l)}$ $(l = 1, \dots, r;$ $j = 1, \dots, n)$ arbitrary non-negative integers. We introduce the matrix functions

$$B_1(\lambda) = \overline{R}(\lambda)\,\overline{D}(\lambda), \qquad B_2(\lambda) = S(\lambda)\,C(\lambda). \tag{4.1}$$

Here $\overline{R}(\lambda)$, $\overline{D}(\lambda)$ are the matrix functions from the representations (1.11)—(1.11′). $S(\lambda)$ is a polynomial matrix with respect to $\dfrac{1}{\lambda + i}$ with a constant determinant different from zero, and

$$C(\lambda) = \left\{\left(\frac{\lambda - \beta_1}{\lambda + i}\right)^{\nu_j^{(1)}} \cdots \left(\frac{\lambda - \beta_r}{\lambda + i}\right)^{\nu_j^{(r)}} \delta_{jk}\right\}_1^n.$$

By B_k $(k = 1, 2)$ we denote the convolution operator on the space E_n with symbol $B_k(\lambda)$:

$$B_k\varphi = c_k\varphi(t) - \int_{-\infty}^{\infty} b_k(t - s)\,\varphi(s)\,ds \qquad (\varphi \in E_n), \tag{4.2}$$

where $c_k = B_k(\infty)$ and $b_k(t)$ is the inverse Fourier transform of $c_k - B_k(\lambda)$. Then for the paired operator $B = PB_1 + QB_2$ we obtain the representation

$$B = PB_1 + QB_2 = (P\overline{R} + QS)(P\overline{D} + QC). \tag{4.3}$$

Here $\overline{R}$, S, $\overline{D}$, C are the operators of the form (4.2) with the symbols $\overline{R}(\lambda)$, $S(\lambda)$, $\overline{D}(\lambda)$, $C(\lambda)$.

The operator $V = P\overline{R} + QS$ is invertible in E_n, where $V^{-1} = P\overline{R}^{-1} + QS^{-1}$. From Lemma 4.1, Chapter 5, it follows that $\dim \ker (P\overline{D} + QC) = 0$. Thus by (4.3) we have $\ker B = \{0\}$.

Similarily as in the preceding two sections the analytic description of the image space $\operatorname{im} B$ can easily be reduced to the case $n = 1$ considered in Section 5.4. To this end for the functions $h(t) \in E$ we introduce the operator (cf. 5.4.1, no. 3)

$$(B_{l,j}^{(-1)}h)(t) = \begin{cases} [I + (\alpha_l - i)\,R_{\alpha_l}]^{\mu_j^{(l)}}\,h, & t > 0 \\ [I + (\beta_l + i)\,R_{\beta_l}]^{\nu_j^{(l)}}\,h, & t < 0. \end{cases}$$

Here R_α is the operator defined by equations (2.19), Chapter 5. Using relation (4.3) we obtain the following statement from the results of Subsection 5.4.1:

Let $f \in E_n$ and $g = V^{-1}f = (g_1, g_2, \dots, g_n)$. Then $f \in \operatorname{im} B$ holds if and only if

$$h_j = \prod_{l=1}^{r} B_{l,j}^{(-1)} g_j \in E \qquad (j = 1, \dots, n) . \tag{4.4}$$

If the conditions (4.4) are satisfied, then $B^{-1}f = (h_1, h_2, \dots, h_n)$. We denote the image space $\operatorname{im} B$ equipped with the norm $\|f\| = \|B^{-1}f\|_{E_n}$ by $\overline{E}_n$.

Remark. The following formula holds:

$$L_n^{p,\mu} \subset \overline{L}_n^p , \qquad 1 \leq p < \infty ; \qquad \mu = \max_{j,l} \{\mu_j^{(l)}, \nu_j^{(l)}\} .$$

2. Again let $\tilde{\mu}_j^{(l)}, \tilde{\nu}_j^{(l)}$ $(l = 1, \dots, r;\ j = 1, \dots, n)$ be non-negative integers, $D(\lambda)$ the diagonal matrix (3.2), $R(\lambda)$ $(\overline{S}(\lambda))$ a polynomial matrix with respect to $\dfrac{1}{\lambda + i}$ $\left(\dfrac{1}{\lambda - i}\right)$ with a constant determinant different from zero and $\overline{C}(\lambda)$ the diagonal matrix

$$\overline{C}(\lambda) = \left\{\left(\frac{\lambda - \beta_1}{\lambda - i}\right)^{\tilde{\nu}_j^{(1)}} \cdots \left(\frac{\lambda - \beta_r}{\lambda - i}\right)^{\tilde{\nu}_j^{(r)}} \delta_{jk}\right\}_1^n .$$

We form the matrix functions

$$D_1(\lambda) = D(\lambda)\, R(\lambda) , \qquad D_2(\lambda) = \overline{C}(\lambda)\, \overline{S}(\lambda) . \tag{4.5}$$

By D_k $(k = 1, 2)$ we denote the convolution operator of the form (4.2) with the symbol $D_k(\lambda)$ and by D, W the paired operators

$$D = D_1 P + D_2 Q , \qquad W = RP + \overline{S}Q . \tag{4.6}$$

Then $W^{-1} = R^{-1}P + \overline{S}^{-1}Q$.

Now we put

$$G = W^{-1} \prod_{l=1}^{r} G_l .$$

Here G_l is the operator which maps any vector $\varphi = (\varphi_1, \varphi_2, \dots, \varphi_n)$ with locally integrable components on the real line onto the vector $f = (f_1, f_2, \dots, f_n)$ with

$$f_j = P\hat{G}_{\alpha_l}^{\tilde{\mu}_j^{(l)}} P\varphi_j + P''\hat{G}_{-\beta_l}^{\tilde{\nu}_j^{(l)}} P'\varphi_j \qquad (j = 1, \dots, n) .$$

The operators P', P'' and $\hat{G}_\alpha$ were introduced in 5.4.1, no. 1. By the results of Subsection 5.4.1 it is easily seen that the operators D and G satisfy a condition of the form (2.5) for all $\varphi \in E_n$. Hence the space $\widetilde{E}_n = G(E_n)$ can be introduced in the same way as in 8.2.1.

Remark. The following formulae hold:

$$\widetilde{L}_n^p \subset L_n^{p,-\tilde{\mu}} , \quad 1 < p < \infty , \quad \tilde{\mu} = \max_{j,l} \{\tilde{\mu}_j^{(l)}, \tilde{\nu}_j^{(l)}\} ; \quad \widetilde{L}_n^1 \subset L_n^{1,-(\tilde{\mu}+\varepsilon)} \quad (\varepsilon > 0) .$$

3. The above introduced spaces $\overline{E}_n$ and $\widetilde{E}_n$ correspond to systems of the second kind. Analogously we can define the corresponding spaces for systems of the first kind.

Again let μ_j, ν_j $(j = 1, \dots, n)$ be non-negative integers, $\overline{D}(\lambda)$ the diagonal matrix (1.13),

$$C(\lambda) = \left\{ \left(\frac{1}{\lambda + i} \right)^{\nu_j} \delta_{jk} \right\}_1^n$$

and $\overline{R}(\lambda)$, $S(\lambda)$ the polynomial matrices from no. 1. We put

$$B_1^{(\infty)}(\lambda) = \overline{R}(\lambda)\, \overline{D}(\lambda)\,, \qquad B_2^{(\infty)}(\lambda) = S(\lambda)\, C(\lambda) \tag{4.7}$$

and

$$B^{(\infty)} = PB_1^{(\infty)} + QB_2^{(\infty)}\,, \qquad V = P\overline{R} + QS\,. \tag{4.8}$$

In the same way as in no. 1 we obtain $\ker B^{(\infty)} = \{0\}$. We denote the image space $\operatorname{im} B^{(\infty)}$ equipped with the norm $\|f\| = \|[B^{(\infty)}]^{-1} f\|_{E_n}$ by $\overline{E}_n^{(\infty)}$.

The analytic description of the space $\overline{E}_n^{(\infty)}$ can be given in the same way as in no. 1. We have only to replace the operator $\prod_{l=1}^{r} B_{l,j}^{(-1)}$ by the operator

$$(B_{\infty,j}^{(-1)} h)(t) = \begin{cases} i^{\mu_j} \left(\dfrac{d}{dt} - 1 \right)^{\mu_j} h(t)\,, & t > 0 \\[2ex] i^{\nu_j} \left(\dfrac{d}{dt} + 1 \right)^{\nu_j} h(t)\,, & t < 0 \end{cases}$$

in the relations (4.4).

4. In order to define the space $\widetilde{E}_n^{(\infty)}$ by $D(\lambda)$ and $\overline{C}(\lambda)$ we denote the diagonal matrices (3.4) and

$$\overline{C}(\lambda) = \left\{ \left(\frac{1}{\lambda - i} \right)^{\tilde{\nu}_j} \delta_{jk} \right\}_1^n .$$

By $R(\lambda)$ and $\overline{S}(\lambda)$ we denote the polynomial matrices from no. 2. We put

$$D_1^{(\infty)}(\lambda) = D(\lambda)\, R(\lambda)\,, \qquad D_2^{(\infty)}(\lambda) = \overline{C}(\lambda)\, \overline{S}(\lambda) \tag{4.9}$$

and

$$D^{(\infty)} = D_1^{(\infty)} P + D_2^{(\infty)} Q\,, \qquad W = RP + \overline{S}Q\,. \tag{4.10}$$

By $G^{(\infty)}$ we denote the operator which maps any vector $\varphi = (\varphi_1, \varphi_2, \dots, \varphi_n) \in E_n$ onto the vector with the components

$$P\widehat{G}_\infty^{\tilde{\mu}_j} P\varphi_j + P''(-\widehat{G}_\infty)^{\tilde{\nu}_j} P'\varphi_j \qquad (j = 1, \dots, n)$$

(cf. 5.4.1, no. 1). We introduce the space $\widetilde{E}_n^{(\infty)}$ as the image space of the operator $G = W^{-1}G^{(\infty)}$ with the corresponding norm. Then $\widetilde{E}_n^{(\infty)} \supset E_n$, and $D^{(\infty)}$ maps the space $\widetilde{E}_n^{(\infty)}$ isometrically onto E_n.

8.4.2. Systems of Equations of the Second Kind

1. Now we consider the paired operator

$$A = A_1 P + A_2 Q\,, \tag{4.11}$$

where A_1 and A_2 are convolution operators on E_n of the form (4.2) with the symbols[1])

$$\mathcal{A}_1(\lambda) = \mathcal{A}_0(\lambda)\, D_1(\lambda)\,, \qquad \mathcal{A}_2(\lambda) = \mathcal{B}_0(\lambda)\, D_2(\lambda)\,. \tag{4.12}$$

Here $D_1(\lambda)$, $D_2(\lambda)$ are the matrix functions defined by (4.5), $\mathcal{A}_0(\lambda), \mathcal{B}_0(\lambda) \in \mathfrak{L}_{n\times n}$ and

$$\det \mathcal{A}_0(\lambda) \neq 0\,, \qquad \det \mathcal{B}_0(\lambda) \neq 0 \qquad (-\infty \leqq \lambda \leqq \infty)\,. \tag{4.13}$$

Then for the operator A we have the representation

$$A = (A_0 P + B_0 Q)\, D\,,$$

and by the results of Section 4.1 we immediately obtain the following theorem from Theorem 3.3, Chapter 7.

Theorem 4.1. *The operator A is a Φ-operator from $\widetilde{E}_n$ into the space E_n. Here the formulae*

$$\dim\ker A = -\sum_{\varkappa_j<0} \varkappa_j\,, \qquad \dim\operatorname{coker} A = \sum_{\varkappa_j>0} \varkappa_j\,, \tag{4.14}$$

hold, where $\varkappa_j$ $(j = 1, \dots, n)$ are the right partial indices of the matrix function $\mathcal{B}_0^{-1}(\lambda)\, \mathcal{A}_0(\lambda) \in \mathfrak{L}_{n\times n}$.

2. An analogous result holds for the paired operator of the form $PA_1' + QA_2'$, where A_1' and A_2' are the operators of the form (4.2) with the symbols

$$\mathcal{A}_1'(\lambda) = B_1(\lambda)\, \mathcal{A}_0(\lambda)\,, \qquad \mathcal{A}_2'(\lambda) = B_2(\lambda)\, \mathcal{B}_0(\lambda) \tag{4.15}$$

and $B_j(\lambda)$ $(j = 1, 2)$ the matrix functions defined by (4.1). We have:

Theorem 4.2. *The operator $PA_1' + QA_2'$ is a Φ-operator from E_n into the space $\overline{E}_n$. The formulae (4.14) hold for this operator, where $\varkappa_j$ are the right partial indices of the matrix function $\mathcal{A}_0(\lambda)\, \mathcal{B}_0^{-1}(\lambda)$.*

3. Now we consider the paired operator A of the form (4.11) under the general hypothesis that the matrix functions $\mathcal{A}_1(\lambda)$, $\mathcal{A}_2(\lambda)$ admit the representation

$$\mathcal{A}_1(\lambda) = B_1(\lambda)\, \mathcal{A}_0(\lambda)\, D_1(\lambda)\,, \qquad \mathcal{A}_2(\lambda) = B_2(\lambda)\, \mathcal{B}_0(\lambda)\, D_2(\lambda)\,. \tag{4.16}$$

Here the matrix functions $\mathcal{A}_0(\lambda)$, $\mathcal{B}_0(\lambda)$ are to satisfy conditions (4.13). Furthermore, we assume that $B_1(\lambda)\, \mathcal{A}_0(\lambda)$ and $B_2(\lambda)\, \mathcal{B}_0(\lambda)$ belong to the class[1])

$$\bigcap_{l=1}^{r} \mathfrak{L}(\alpha_l, \mu^{(l)}) \cap \mathfrak{L}(\beta_l, \nu^{(l)})$$

---

[1]) Sufficient conditions for the representability of the matrix functions $\mathcal{A}_1(\lambda)$, $\mathcal{A}_2(\lambda) \in \mathfrak{L}_{n\times n}$ in the form (4.12) were mentioned in 8.1.4 (see also 8.3.2).

[1]) For notation see 5.2.2, no. 3.

where

$$\mu^{(l)} = \max_{j=1,\dots,n} \mu_j^{(l)}, \qquad \nu^{(l)} = \max_{j=1,\dots,n} \nu_j^{(l)}. \tag{4.17}$$

Introducing the operators

$$V = B_1 A_0 P + B_2 B_0 Q, \qquad W = P B_1 A_0 + Q B_2 B_0$$

we can represent the operator A in the form

$$A = VD = B(PA_0 + QB_0) D + (V - W) D. \tag{4.18}$$

Using Lemma 4.3, Chapter 5 by means of similar considerations as in 5.4.2 we can easily see that the operator $V - W$ maps the space E_n compactly into the space $\overline{E}_n = \operatorname{im} B$. Thus we obtain the next theorem as an immediate consequence of the relation (4.18) (cf. 4.3.5).

Theorem 4.3. *The operator A is a Φ-operator from $\widetilde{E}_n$ into the space $\overline{E}_n$, and for its index the formula*

$$\operatorname{Ind} A = \frac{1}{2\pi}\left[\arg \frac{\det \mathcal{B}_0(\lambda)}{\det \mathcal{A}_0(\lambda)}\right]_{\lambda=-\infty}^{\infty}$$

holds.

Remark. We assume that the determinants of the matrix functions $\mathcal{A}_1(\lambda)$ and $\mathcal{A}_2(\lambda)$ possess zeros of the integral orders m_l and n_l in the points α_l and $\beta_l \in (-\infty, \infty)$, respectively. Furthermore, let $\mathcal{A}_1(\lambda)$ and $\mathcal{A}_2(\lambda)$ belong to the class

$$\bigcap_{l=1}^{r} \mathfrak{L}(\alpha_l, 2m_l) \cap \mathfrak{L}(\beta_l, 2n_l).$$

By $m_j^{(l)}, n_j^{(l)}$ $(j = 1, \dots, n)$ we denote the partial orders of the zeros α_l, β_l for the functions $\det \mathcal{A}_1(\lambda)$, $\det \mathcal{A}_2(\lambda)$. Choosing arbitrary non-negative integers $\mu_j^{(l)}$, $\nu_j^{(l)}$, $\tilde{\mu}_j^{(l)}$, $\tilde{\nu}_j^{(l)}$ such that

$$\mu_j^{(l)} + \tilde{\mu}_j^{(l)} = m_j^{(l)}, \qquad \nu_j^{(l)} + \tilde{\nu}_j^{(l)} = n_j^{(l)}, \tag{4.19}$$

for all j and l we can represent the matrix functions $\mathcal{A}_1(\lambda)$, $\mathcal{A}_2(\lambda)$ in the form (4.16), where $\mathcal{A}_0(\lambda)$, $\mathcal{B}_0(\lambda)$ have the required properties.

This follows from Theorem 1.2 and some results from 2.3.4.

A theorem analogous to Theorem 4.3 holds for the operator $A = PA_1 + QA_2$ (cf. 4.4.2). Moreover, the results of this subsection can be carried over to the case of zeros of non-integral orders (cf. 4.4.3).

4. If in the representations (4.16) we put in particular $D_1(\lambda) = D_2(\lambda)$ equal to the unit matrix and if in the considerations of the Subsection 4.5.2 we assume that

$$Z = L_n^{p,\mu}, \qquad 1 \leq p < \infty, \qquad \mu = \max_l \{m_l, n_l\}$$

and $R = B^{-1}$, then it follows from Theorem 5.2 and Lemma 5.1 of Chapter 4 that the closed Operator $\underline{A}$ $(D(\underline{A}) \to Z)$ defined by

$$\underline{A}\varphi = A\varphi, \qquad D(\underline{A}) = \{\varphi \in L_n^p : A\varphi \in L_n^{p,\mu}\}$$

is a Φ-operator with

$$\ker \underline{A} = \ker A, \qquad (\operatorname{im} \underline{A})^\perp = (\operatorname{im} A)^\perp.$$

8.4.3. Systems of Equations of the First Kind

Completely analogously we obtain corresponding results for paired operators of the first kind of the form (4.11) resp. $PA_1' + QA_2'$ if in the representations (4.12), (4.15) and (4.16) we replace the matrix functions $D_j(\lambda)$, $B_j(\lambda)$ ($j = 1, 2$) by the matrix functions $D_j^{(\infty)}(\lambda)$, $B_j^{(\infty)}(\lambda)$ from (4.9), (4.7), respectively. In the formulations of Theorems 4.1–4.3 we have to interchange the spaces $\widetilde{E}_n$, $\overline{E}_n$ by $\widetilde{E}_n^{(\infty)}$, $\overline{E}_n^{(\infty)}$, respectively.

8.4.4. Non-Bounded Regularization

In the space L_n^p ($1 \leqq p < \infty$) we consider the paired operator of the second kind $A = A_1P + A_2Q + T$, where the matrix functions $\mathcal{A}_1(\lambda)$, $\mathcal{A}_2(\lambda)$ have the form (4.15) and moreover belong to the class

$$\bigcap_{l=1}^{r} \mathfrak{L}(\alpha_l, m_l) \cap \mathfrak{L}(\beta_l, n_l) .$$

Furthermore, let $T \in \mathcal{K}(L_n^p, \overline{L_n^p})$. The construction of a non-bounded left regularizer of the operator A can be carried out completely analogously to Section 5.5 if we use the inverse operator B^{-1} constructed in 8.4.1. no. 1.

A corresponding statement holds for the paired operator of the first kind.

8.5. Systems of Singular Integral Equations

In this section we consider systems of singular integral equations of non-normal type of the form

$$\mathfrak{A}\varphi = A(t)\,\varphi(t) + \frac{B(t)}{\pi i}\int_{\Gamma} \frac{\varphi(\tau)}{\tau - t}\,d\tau = f(t) \tag{5.1}$$

in the space $L_n^p(\Gamma)$ ($1 < p < \infty$). Here Γ is a sufficiently smooth Ljapunov curve system of the form described in 6.2.1 and $A(t)$, $B(t) \in C_{n\times n}(\Gamma)$. In the sequel we maintain the notation introduced in Section 7.5.

We assume that the determinants of the matrix functions

$$C(t) = A(t) + B(t)\,, \qquad D(t) = A(t) - B(t)$$

possess at most finitely many zeros of integral or fractional orders on Γ.

8.5.1. The Spaces Generated by Zeros of the Symbol

1. Let α_l, β_l ($l = 1, \dots, r$) be points on Γ and $\mu_j^{(l)}, \nu_j^{(l)}$ ($j = 1, \dots, n$) non-negative integers. We form the matrix functions

$$B_1(t) = \overline{R}(t)\,\overline{D}(t)\,, \qquad B_2(t) = S(t)\,\widetilde{C}(t) \qquad (t \in \Gamma)\,. \tag{5.2}$$

Here $\overline{R}(t)$ $(S(t))$ is an arbitrary polynomial matrix in t^{-1} (in t) with a constant determinant different from zero. $\overline{D}(t)$ is the diagonal matrix (1.10), and

$$\widetilde{C}(t) = \{(t - \beta_1)^{\nu_j^{(1)}} \dots (t - \beta_r)^{\nu_j^{(r)}} \delta_{jk}\}_1^n .$$

For the paired operator

$$\mathfrak{B} = PB_1 + QB_2$$

the representation (4.3) again holds. By Lemma 2.2, Chapter 6, this implies $\ker \mathfrak{B} = \{0\}$.

The analytic description of the image space im $\mathfrak{B}$ can be given in the same way as in 8.4.1, no. 1, where now in the relations (4.4) $E = L^p(\Gamma)$ and $B_{l,j}^{(-1)}$ $(l = 1, \dots, r; j = 1, \dots, n)$ is the inverse of the scalar operator

$$P(t^{-1} - \alpha_l^{-1})^{\mu_j^{(l)}} + Q(t - \beta_l)^{\nu_j^{(l)}}$$

given by the formula (2.13), Chapter 6. In the sequel we denote the Banach space im $\mathfrak{B}$ equipped with the norm $\|f\| = \|\mathfrak{B}^{-1}f\|_{L_n^p(\Gamma)}$ by $\overline{L_p^n}(\Gamma)$.

By Corollary 2.2. Chapter 6. we have

$$W_{p,n}(\Gamma; (\alpha_l, \mu^{(l)})_1^r, \ (\beta_l, \nu^{(l)})_1^r) \subset \overline{L_n^p}(\Gamma) , \tag{5.3}$$

where the numbers $\mu^{(l)}, \nu^{(l)}$ are defined by (4.17).

2. Now let $\widetilde{\mu}_j^{(l)}, \widetilde{\nu}_j^{(l)}$ $(l = 1, \dots, r; j = 1, \dots, n)$ be non-negative integers, $R(t)$ $(\overline{S}(t))$ an arbitrary polynomial matrix in t (in t^{-1}) with a constant determinant different from zero and $\widetilde{D}(t)$, $\overline{C}(t)$ the diagonal matrices

$$\widetilde{D}(t) = \left\{\prod_{l=1}^{r} (t - \alpha_l)^{\widetilde{\mu}_j^{(l)}} \delta_{jk}\right\}_1^n , \qquad \overline{C}(t) = \left\{\prod_{l=1}^{r} (t^{-1} - \beta_l^{-1})^{\widetilde{\nu}_j^{(l)}} \delta_{jk}\right\}_1^n .$$

In this number we assume that the points α_l and β_k $(\in \Gamma; l, k = 1, \dots, r)$ are distinct. By $D_1(t)$, $D_2(t)$ we denote the matrix functions

$$D_1(t) = \widetilde{D}(t)\, R(t) , \qquad D_2(t) = \overline{C}(t)\, \overline{S}(t) \tag{5.4}$$

and by $\mathfrak{D}$ the paired operator

$$\mathfrak{D} = D_1(t)\, P + D_2(t)\, Q . \tag{5.5}$$

Furthermore, on $L_n^p(\Gamma)$ we introduce the operator $\mathfrak{D}^{(-1)}$ defined for any $f \in L_n^p(\Gamma)$ by the formula

$$\mathfrak{D}^{(-1)}f = R^{-1}(t)\, \widetilde{D}^{-1}(t)\, Pf + \overline{S}^{-1}(t)\, \overline{C}^{-1}(t)\, Qf . \tag{5.6}$$

By $\widetilde{L_n^p}(\Gamma)$ we denote the collection of all functions of the form (5.6) with $f \in L_n^p(\Gamma)$.

Obviously, we have

$$\mathfrak{D}^{(-1)}\mathfrak{D}f = f \qquad (f \in L_n^p(\Gamma)) ,$$

i.e. $L_n^p(\Gamma) \subset \widetilde{L}_n^p(\Gamma)$. We extend the operator $\mathfrak{D}$ to an operator $\widetilde{\mathfrak{D}}$ defined on the whole space $\widetilde{L}_n^p(\Gamma)$ by putting

$$\widetilde{\mathfrak{D}}\varphi = D_1(t)\,\widetilde{P}\varphi + D_2(t)\,\widetilde{Q}\varphi\,, \qquad \varphi \in \widetilde{L}_n^p(\Gamma)\,.$$

Here $\widetilde{P}, \widetilde{Q}$ are the extensions of the projections P, Q, respectively, defined for the vector $\varphi \in \widetilde{L}_n^p(\Gamma)$ of the form (5.6) by the equations

$$\widetilde{P}\varphi = R^{-1}(t)\,\widetilde{D}^{-1}(t)\,Pf\,, \qquad \widetilde{Q}\varphi = \overline{S}^{-1}(t)\,\overline{C}^{-1}(t)\,Qf\,.$$

It is immediately seen that

$$\widetilde{\mathfrak{D}}\mathfrak{D}^{(-1)}f = f \qquad (f \in L_n^p(\Gamma))\,.$$

Hence the operator $\mathfrak{D}$ satisfies condition (2′) of 4.1.1 so that $\widetilde{L}_n^p(\Gamma)$ with the norm

$$||\varphi|| = ||\widetilde{\mathfrak{D}}\varphi||_{L_n^p(\Gamma)}$$

is a Banach space, where the operator $\widetilde{\mathfrak{D}}$ maps this space isometrically onto $L_n^p(\Gamma)$.

8.5.2. Symbols with Zeros of Integral Orders

1. We consider the operator $\mathfrak{A}$ of the form (5.1):

$$\mathfrak{A} = C(t)\,P + D(t)\,Q$$

under the hypothesis that the matrix functions $C(t)$, $D(t)$ possess the representations[1])

$$C(t) = C_0(t)\,D_1(t)\,, \qquad D(t) = D_0(t)\,D_2(t)\,. \tag{5.7}$$

Here $D_1(t)$, $D_2(t)$ are the matrix functions given by (5.4), $C_0(t), D_0(t) \in C_{n\times n}(\Gamma)$ and

$$\det C_0(t) \neq 0\,, \qquad \det D_0(t) \neq 0 \qquad (t \in \Gamma)\,. \tag{5.8}$$

By the relation

$$\mathfrak{A} = \mathfrak{A}_0\mathfrak{D}\,, \qquad \mathfrak{A}_0 = (C_0P + D_0Q)$$

Theorem 5.1, Chapter 7, now implies the following theorem.

Theorem 5.1. *The operator $\mathfrak{A}$ is a Φ-operator from $\widetilde{L}_n^p(\Gamma)$ into the space $L_n^p(\Gamma)$, and the following formulae hold:*

$$\dim\ker\mathfrak{A} = \dim\ker\mathfrak{A}_0\,, \qquad \dim\operatorname{coker}\mathfrak{A} = \dim\operatorname{coker}\mathfrak{A}_0\,, \tag{5.9$_1$}$$

$$\operatorname{Ind}\mathfrak{A} = \frac{1}{2\pi}\left[\arg\frac{\det D_0(t)}{\det C_0(t)}\right]_\Gamma\,. \tag{5.9$_2$}$$

---

1) Sufficient conditions for the representability of the matrix functions $C(t)$, $D(t) \in C_{n\times n}(\Gamma)$ in the form (5.7) were mentioned in 8.1.3.

2. We obtain an analogous result for the operator $PC_* + QD_*$, where $C_*(t)$ and $D_*(t)$ are matrix functions of the form

$$C_*(t) = B_1(t)\, C_0(t)\,, \qquad D_*(t) = B_2(t)\, D_0(t)$$

and $B_j(t)$ $(j = 1, 2)$ are the matrix functions defined by (5.2).

Theorem 5.2. *The operator $PC_* + QD_*$ is a Φ-operator from $L_n^p(\Gamma)$ into the space $\bar{L}_n^p(\Gamma)$. For this operator the formulae* (5.9) *also hold.*

3. Now we consider the operator $\mathfrak{A}$ of the form (5.1) in the more general case that the matrix functions $C(t)$, $D(t)$ admit the representation

$$C(t) = B_1(t)\, C_0(t)\, D_1(t)\,, \qquad D(t) = B_2(t)\, D_0(t)\, D_2(t)\,. \tag{5.10}$$

Here again the matrix functions $C_0(t)$, $D_0(t)$ shall satisfy the conditions (5.8). Furthermore, we assume that $B_1(t)\, C_0(t)$ and $B_2(t)\, D_0(t)$ belong to the class[1]) $\bigcap_{l=1}^{r} C(\alpha_l, \mu^{(l)}) \cap C(\beta_l, \nu^{(l)})$, where the numbers $\mu^{(l)}$, $\nu^{(l)}$ are defined by (4.17). Then we have:

Theorem 5.3. *The operator $\mathfrak{A} = CP + DQ$ is a Φ-operator from $\tilde{L}_n^p(\Gamma)$ into the space $\bar{L}_n^p(\Gamma)$. For its index the formula* (5.9_2) *holds.*

The proof of Theorem 5.3 can be carried out analogously to the proof of Theorem 4.3 by using Lemma 4.1, Chapter 4.

If we assume that the elements of the matrix functions $C_0(t) \cdot D_1(t)$ and $D_0(t)\, D_2(t)$ belong to the class

$$\bigcap_{l=1}^{r} C(\alpha_l, \tilde{\mu}^{(l)}) \cap C(\beta_l, \tilde{\nu}^{(l)})\,, \qquad \tilde{\mu}^{(l)} = \max_j \tilde{\mu}_j^{(l)}\,, \qquad \tilde{\nu}^{(l)} = \max_j \tilde{\nu}_j^{(l)}\,,$$

then by means of the considerations from Subsection 4.4.2 we see that Theorem 5.3 also holds for the operator $A = PC + QD$.

Remark 1. By the constructions of Section 8.1 the representations (5.10) are possible provided the determinants of the matrix functions $C(t)$ and $D(t)$ possess zeros of the integral orders m_l, $n_l \geqq 0$ in the points α_l, $\beta_l \in \Gamma$ $(l = 1, \ldots, r)$, respectively, and if $C(t)$, $D(t)$ belong to the class

$$\bigcap_{l=1}^{r} C(\alpha_l, 2m_l) \cap C(\beta_l, 2n_l)\,.$$

Here for $\mu_j^{(l)}$, $\nu_j^{(l)}$, $\tilde{\mu}_j^{(l)}$, $\tilde{\nu}_j^{(l)}$ we can choose arbitrary non-negative integers satisfying the equations (4.19), where $m_j^{(l)}$, $n_j^{(l)}$ are the partial orders of the zeros α_l, β_l for the determinants $\det C(t)$, $\det D(t)$, respectively.

Remark 2. Under the hypotheses of the Remark 1 we can especially put

$$\tilde{\mu}_j^{(l)} = \tilde{\nu}_j^{(l)} = 0 \qquad (l = 1, \ldots, r\,; \quad j = 1, \ldots, n)\,.$$

---

[1]) For notation see 2.3.4.

Then we have $\tilde{L}_n^p(\Gamma) = L_n^p(\Gamma)$. Here it is sufficient that the matrix functions $C(t)$, $D(t)$ belong to the class $\bigcap_{l=1}^{r} C(\alpha_l, m_l) \cap C(\beta_l, n_l)$. Furthermore, in this case some of the points α_l and β_l can also coincide since this has no influence on the construction of the space $\overline{L}_n^p(\Gamma) = \text{im B}$.

Remark 3. Using the results of Section 6.4 we can apply the constructions of the spaces $\overline{L}_n^p(\overline{\Gamma})$ and $\tilde{L}_n^p(\Gamma)$ given in 8.5.1 also to the case of non-integral numbers $\mu_j^{(l)}$, $\nu_j^{(l)}$, $\tilde{\mu}_j^{(l)}$, $\tilde{\nu}_j^{(l)}$ without additional difficulties. 4.4.3 shows that in this case also Theorems 5.1 to 5.3 remain valid.

Remark 4. The preceding results of this section can (under the corresponding hypotheses on the coefficients $A(t)$, $B(t)$) easily be carried over to the case of the space $H_n^\lambda(\Gamma)$ $(0 < \lambda < 1)$ (cf. the Chapters 6 and 7).

4. If in particular choose $D_1(t) = D_2(t) = I$ (I the unit matrix) in (5.10) and if we put

$$Z = W_{p,n}(\Gamma; (\alpha_l, \mu^{(l)})_1^r, (\beta_l, \nu^{(l)})_1^r), \qquad R = \mathfrak{B}^{-1},$$

in the considerations of Subsection 4.5.2, then Theorem 5.2 and Lemma 5.1 of Chapter 4 show that the closed operator $\underline{\mathfrak{A}}(D(\underline{\mathfrak{A}}) \to Z)$ defined by

$$\underline{\mathfrak{A}}\varphi = \mathfrak{A}\varphi, \qquad D(\underline{\mathfrak{A}}) = \{\varphi \in L_n^p(\Gamma) : \mathfrak{A}\varphi \in Z\}$$

is a Φ-operator, where

$$\ker \underline{\mathfrak{A}} = \ker \mathfrak{A}, \qquad (\text{im } \underline{\mathfrak{A}})^\perp = (\text{im } \mathfrak{A})^\perp .$$

8.5.3. Symbols with Zeros of Non-Integral Orders

In the sequel we assume that Γ is a simple closed Ljapunov curve enclosing the origin of the coordinate system. In this subsection we consider systems of singular integral equations of non-normal type of the form (5.1) in the space $L_n^p(\Gamma)$ $(1 < p < \infty)$, where the symbol $C(t) = A(t) + B(t)$, $D(t) = A(t) - B(t)$ satisfies the following condition:

$$\text{(I)} \quad C(t) = R_1(t)\, C_0(t), \qquad D(t) = R_2(t)\, D_0(t)$$

with $R_1 = S_{10} R_{10}$, $R_2 = R_{20} S_{20}$ and

$$S_{10} = t^{-\xi_2} F_1 G_1, \qquad R_{10} = L_1 M_1, \qquad R_{20} = t^{\eta_2} L_2 M_2, \qquad S_{20} = F_2 G_2,$$

$$G_1 = \left\{\left(\frac{1}{\alpha} - \frac{1}{t}\right)^{m_{1j}+r_{1j}} \delta_{jk}\right\}_1^n, \qquad M_1 = \{(\alpha - t)^{m_{2j}+r_{2j}}\, \delta_{jk}\}_1^n,$$

$$G_2 = \left\{\left(\frac{1}{\beta} - \frac{1}{t}\right)^{n_{2j}+s_{2j}} \delta_{jk}\right\}_1^n, \qquad M_2 = \{(\beta - t)^{n_{1j}+s_{1j}}\, \delta_{jk}\}_1^n .$$

Here F_1, F_2 (resp. L_1, L_2) are arbitrary polynomial matrices in t^{-1} (resp. in t) with a constant determinant different from zero. α and β are two different points on Γ, m_{1j}, m_{2j}, n_{1j}, n_{2j} are non-negative integers, $0 \leq r_{1j}, r_{2j}, s_{1j}, s_{2j} < 1$ and

$$\xi_l = \max_{j=1,\dots,n} \{m_{lj} - [-r_{lj}]\}, \qquad \eta_l = \max_{j=1,\dots,n} \{n_{lj} - [s_{lj}]\} \qquad (l = 1, 2).$$

Furthermore, let $C_0(t)$, $D_0(t)$ be continuous matrix functions on Γ, where

$$\det C_0(t) \neq 0, \qquad \det D_0(t) \neq 0 \qquad (t \in \Gamma).$$

The curve Γ belongs to the class $C^{\bar{n},\lambda}$ with $0 < \lambda < 1$ and

$$\bar{n} = \max\{\tilde{m}, \tilde{n}\}, \qquad \tilde{m} = \xi_1 + \xi_2, \qquad \tilde{n} = \eta_1 + \eta_2.$$

1. First of all, we begin with the study of the operators

$$\mathfrak{B} = PR_1P + QR_2Q \quad \text{and} \quad \tilde{\mathfrak{B}} = R_1P + R_2Q.$$

Theorem 5.4. *If the condition*

$$\text{(II)} \quad S_{10}R_{20} = R_{20}S_{10}$$

is satisfied, then

$$\dim \ker \tilde{\mathfrak{B}} = n(\xi_2 + \eta_2) - \sum_{j=1}^{n} \left(m_{2j} + n_{2j} + \left[r_{2j} + \frac{1}{p'} \right] + \left[s_{2j} + \frac{1}{p'} \right] \right),$$

where $\dfrac{1}{p'} = 1 - \dfrac{1}{p}$.

Proof. The operator $\tilde{\mathfrak{B}}$ can be written in the form $\tilde{\mathfrak{B}} = t^{-\xi_2}\mathfrak{B}_1\mathfrak{B}_2$, where

$$\mathfrak{B}_1 = F_1G_1P + t^{\xi_2+\eta_2}L_2M_2Q, \qquad \mathfrak{B}_2 = R_{10}P + S_{20}Q.$$

Obviously, we have $\dim \ker \mathfrak{B}_2 = 0$. From Lemma 2.1, Chapter 6, it follows easily that

$$\dim \ker (F_1G_1P + L_2M_2Q) = 0. \tag{5.12}$$

Using condition (II) we can immediately see that each vector φ of the form

$$\varphi = (L_2M_2 - t^{-\xi_2-\eta_2}F_1G_1)\, P_{\xi_2+\eta_2-1} \tag{5.13}$$

belongs to the kernel of the operator $\mathfrak{B}_1$. Here $P_{\xi_2+\eta_2-1}$ is an arbitrary polynomial vector in t whose degree is not greater than the number $\xi_2 + \eta_2 - 1$. (For $\xi_2 + \eta_2 = 0$ we put $P_{\xi_2+\eta_2-1} \equiv 0$).

By X we denote the $n(\xi_2 + \eta_2)$-dimensional vector space formed by the elements of the form (5.13). Then obviously $L_n^p(\Gamma)$ is the direct sum

$$L_n^p(\Gamma) = \operatorname{im}(P + t^{-\xi_2-\eta_2}Q) \dotplus X.$$

Using this representation and the relation (5.12) in the same way as in the proof of Theorem 5.1, Chapter 6, we conclude that $\ker \mathfrak{B}_1$ consists only of elements of the form (5.13).

Now let $\widetilde{\mathfrak{B}}\varphi = 0$. Then $\mathfrak{B}_2\varphi \in \ker \mathfrak{B}_1$, and by (5.13) we have

$$L_2M_2P_{\xi_2+\eta_2-1} = R_{10}P\varphi\,, \qquad -t^{-\xi_2-\eta_2}F_1G_1P_{\xi_2+\eta_2-1} = S_{20}Q\varphi\,. \tag{5.14}$$

We put

$$L_1^{-1}L_2M_2 = M\,, \qquad t^{-\xi_2-\eta_2}F_2^{-1}F_1G_1 = G\,.$$

From (5.14) we obtain that the polynomial vector $P_{\xi_2+\eta_2-1}$ has to satisfy the conditions

$$M_1^{-1}MP_{\xi_2+\eta_2-1} \in L_n^p(\Gamma)\,, \qquad G_2^{-1}GP_{\xi_2+\eta_2-1} \in L_n^p(\Gamma)\,. \tag{5.15}$$

If conversely $P_{\xi_2+\eta_2-1}$ is an arbitrary polynomial vector whose degree is not greater than $\xi_2 + \eta_2 - 1$ and which moreover satisfies the conditions (5.15), then it is easily checked that the vector

$$\varphi = (M_1^{-1}M - G_2^{-1}G)\,P_{\xi_2+\eta_2-1}$$

is a solution (and simultaneously the general solution) of the equation $\overline{\mathfrak{B}}\varphi = 0$. It follows from the preceding considerations that

$$\dim\ker\widetilde{\mathfrak{B}} = n(\xi_2 + \eta_2) - r\,, \tag{5.16}$$

where r is equal to the number of linearly independent conditions which the vector $P_{\xi_2+\eta_2-1}$ has to satisfy in accordance with relations (5.15).

Taking into consideration the concrete form of the matrix functions M_1 and G_2 and using Lemma 1.9 from Chapter 6 we can easily see that the relations (5.15) are valid if and only if the following conditions are satisfied:

$$\left[\sum_{k=1}^{n} a_{jk}(t)\,p_k(t)\right]_{t=\alpha}^{(l)} = 0\,, \qquad l = 0, 1, \ldots, \mu_j - 1\,, \tag{5.17$_1$}$$

$$\left[\sum_{k=1}^{n} b_{jk}(t)\,p_k(t)\right]_{t=\beta}^{(l)} = 0\,, \qquad l = 0, 1, \ldots, \nu_j - 1\,. \tag{5.17$_2$}$$

Here

$$P_{\xi_2+\eta_2-1} = (p_1(t), p_2(t), \ldots, p_n(t))\,,$$
$$M = \{a_{jk}(t)\}_1^n\,, \qquad G = \{b_{jk}(t)\}_1^n\,,$$
$$\mu_j = m_{2j} + \left[r_{2j} + \frac{1}{p'}\right], \qquad \nu_j = n_{2j} + \left[s_{2j} + \frac{1}{p'}\right]. \tag{5.17$_3$}$$

Since $\xi_2 + \eta_2 \geqq \mu_j$, $\xi_2 + \eta_2 \geqq \nu_j$ $(j = 1, \ldots, n)$ and

$$\det|a_{jk}(\alpha)| \neq 0\,, \qquad \det|b_{jk}(\beta)| \neq 0\,,$$

for the polynomial vector $P_{\xi_2+\eta_2-1}$ the number of conditions (5.17$_1$) and (5.17$_2$) is equal to $\sum \mu_j$ and $\sum \nu_j$, respectively. Hence

$$r = \sum_{j=1}^{n} (\mu_j + \nu_j)\,.$$

This together with (5.16) yields the assertion of the Theorem 5.4.

Corollary 5.1. *The following formula holds:*

$$\dim \ker \mathfrak{B} = n(\xi_2 + \eta_2) - \sum_{j=1}^{n} \left(m_{2j} + n_{2j} + \left[r_{2j} + \frac{1}{p'} \right] + \left[s_{2j} + \frac{1}{p'} \right] \right) .$$

Using the relations (cf. formulae (2.4). Chapter 2)

$$\dim \ker (PR_1P \mid \operatorname{im} P) = \dim \ker (R_1P + Q) ,$$

$$\dim \ker (QR_2Q \mid \operatorname{im} Q) = \dim \ker (P + R_2Q)$$

we immediately obtain Corollary 5.1 from Theorem 5.4.

Similarily as in Subsection 6.5.1 we now introduce the image spaces $\hat{L}_n^p(R_1, R_2) = \operatorname{im} (R_1P + R_2Q)$ and $\overline{L}_n^p(R_1, R_2) = \operatorname{im} (PR_1P + QR_2Q)$. We equip both spaces with the corresponding factor norm (comp. Section 4.1.3). For the analytic description of these Banach spaces we still introduce the following notation:

$$S_{11} = \left\{ \left(\frac{1}{\alpha} - \frac{1}{t} \right)^{\xi_1 - m_{1j} - r_{1j}} t^{\xi_1 + \eta_2} \delta_{jk} \right\}_1^n F_1^{-1} ,$$

$$R_{11} = \{ (\alpha - t)^{\xi_2 - m_{2j} - r_{2j}} \delta_{jk} \}_1^n L_1^{-1} ,$$

$$\sigma = \xi_1 + \xi_2 + \eta_1 + \eta_2 .$$

Theorem 5.5. *Let the condition* (II) *be satisfied. Then the vector function* $f \in L_n^p(\Gamma)$ *belongs to the space* $\hat{L}_n^p(R_1, R_2)$ *if and only if there exists a polynomial vector* $P_{\sigma-1}$ *of degree* $\leqq \sigma - 1$ *such that*

$$\varphi_1 = (R_{11}S_{11}S_{10}R_{10})^{-1} (PR_{11}S_{11}f - P_{\sigma-1}) \in L_n^p(\Gamma) ,$$

$$\varphi_2 = (R_{11}S_{11}R_{20}S_{20})^{-1} (QR_{11}S_{11}f + P_{\sigma-1}) \in L_n^p(\Gamma) .$$

If these conditions are satisfied, then

$$\widetilde{\mathfrak{B}}(\varphi_1 + \varphi_2) = f .$$

Proof. It is easily seen that $S_{11}S_{10}R_{20} = R_{20}S_{11}S_{10}$. From this we conclude by the condition (II) that the matrix functions S_{11} and R_{20} commute with each other. Using this fact we can prove Theorem 5.5 in the same way as Theorem 5.3, Chapter 6.

Remark. The analytic description of the space $\overline{L}_n^p(R_1, R_2)$ is easily obtained from Theorem 5.5 by using the equalities

$$\operatorname{im} (PR_1P + Q) = \operatorname{im} (R_1P + Q) ,$$

$$\operatorname{im} (P + QR_2Q) = \operatorname{im} (P + R_2Q)$$

(cf. the formulae (2.2) and (2.3) of Chapter 2 and Remark 2 at the end of Section 2.1.5).

By Theorem 5.4, Chapter 6, we can easily obtain the following theorem.

Theorem 5.6. *The following relation holds:*

$$W_{p,n}(\Gamma; (\alpha, \tilde{m}) , (\beta, \tilde{n})) \subset \overline{L}_n^p(R_1, R_2) .$$

Remark. If the condition (II) is satisfied, then it can be shown that Theorem 5.6 holds also for the space $\hat{L}_n(R_1, R_2)$ instead of $\overline{L}_n^p(R_1, R_2)$ (comp. S. PRÖSSDORF and B. SILBERMANN [2]).

2. Now we study the properties of the singular matrix operator $CP + DQ$ on the pair of spaces $L_n^p(\Gamma)$, $\overline{L}_n^p(R_1, R_2)$ $(1 < p < \infty)$. We assume that the conditions (I) are satisfied with the matrix functions

$$C_0(t) \in C_{n\times n}(\beta, \eta_1 + \eta_2) \cap C_{n\times n}(\alpha, \tilde{\xi}_2)\,,$$
$$D_0(t) \in C_{n\times n}(\alpha, \xi_1 + \xi_2) \cap C_{n\times n}(\beta, \tilde{\eta}_2)\,,$$

where

$$\tilde{\xi}_2 = \max_j \{-[-\xi_2 + m_{2j} + r_{2j}]\}\,, \qquad \tilde{\eta}_2 = \max_j \{-[-\eta_2 + n_{2j} + s_{2j}]\}\,.$$

Under these hypotheses we have:

Theorem 5.7. *The operator $\mathfrak{A} = CP + DQ\colon L_n^p(\Gamma) \to \overline{L}_n^p(R_1, R_2)$ is defined and continuous. $\mathfrak{A}$ is a Φ_+-operator or a Φ_--operator if and only if*

$$\det C_0(t) \neq 0\,, \qquad \det D_0(t) \neq 0 \qquad (t \in \Gamma)\,.$$

If these conditions are satisfied, then $\mathfrak{A} \in \mathscr{L}(L_n^p(\Gamma), \overline{L}_n^p(R_1. R_2))$ is a Φ-operator with the index

$$\operatorname{Ind}\mathfrak{A} = \varkappa + n(\xi_2 + \eta_2) - \sum_{j=1}^{n}\left(m_{2j} + n_{2j} + \left[r_{2j} + \frac{1}{p'}\right] + \left[s_{2j} + \frac{1}{p'}\right]\right),$$

where $\frac{1}{p'} = 1 - \frac{1}{p}$ and

$$\varkappa = \frac{1}{2\pi}\left[\arg\frac{\det D_0(t)}{\det C_0(t)}\right]_\Gamma\,.$$

Proof. We put

$$R_3 = \{(\alpha - t)^{\xi_2 - (m_{2j} + r_{2j})}\,\delta_{ij}\}_1^n\, L_1^{-1}\,,$$
$$R_4 = \left\{\left(\frac{1}{\beta} - \frac{1}{t}\right)^{\eta_2 - (n_{2j} + s_{2j})}\delta_{ij}\right\}_1^n F_2^{-1}\,.$$

Obviously, we have R_3, $R_2R_4 \in C_{n\times n}^+(\Gamma)$ and R_4, $R_1R_3 \in C_{n\times n}^-(\Gamma)$. By hypothesis $\alpha \neq \beta$ holds. Hence

$$C(t) = R_1(t)\,C_0(t) \in C_{n\times n}(\beta, \eta_1 + \eta_2)\,,$$
$$D(t) = R_2(t)\,D_0(t) \in C_{n\times n}(\alpha, \xi_1 + \xi_2)\,.$$

From the results of Section 2.3.4 we conclude immediately that there exists a polynomial matrix $T_2 \in C_{n\times n}^+(\Gamma)$ such that

$$C(t) = \{(t - \beta)^{\eta_1 + \eta_2}\,\delta_{ij}\}_1^n\,\tilde{C}(t) + T_2 = R_2R_4C_2 + T_2\,, \qquad (*)$$

where $\widetilde{C}, C_2 \in C_{n\times n}(\Gamma)$. Analogously we obtain the representations

$$D = R_1R_3D_2 + T_4\,,$$

$$C_0 = R_3C_1 + T_1\,, \qquad D_0 = R_4D_1 + T_3 \qquad (**)$$

with $C_j, D_j \in C_{n\times n}(\Gamma)$ ($j = 1, 2$) and $T_1 \in C^+_{n\times n}(\Gamma)$, $T_3, T_4 \in C^-_{n\times n}(\Gamma)$. Using the relations (*) and (**) for the operator $\mathfrak{A} = CP + DQ$ we easily obtain the representation

$$\mathfrak{A} = \mathfrak{B}\mathfrak{A}_0 \qquad (***)$$

with

$$\mathfrak{A}_0 = PC_0P + QD_0Q + PR_3D_2Q + QR_4C_2P + T$$

and

$$T = -\,PR_3Q(QC_1P + QD_2Q) - QR_4P(PD_1Q + PC_2P)\,.$$

Since T is a compact operator on the space $L^p_n(\Gamma)$, the assertion follows from the representation (***), Theorem 5.1 of Chapter 7, Corollary 5.1 and the results of Section 1.3.

Remark. The proof of Theorem 5.7 shows that the hypothesis $\alpha \neq \beta$ can be omitted if $r_{ji} = s_{ji} = 0$ for $j = 1, 2$ and $i = 1, \dots, n$ (cf. also Subsection 8.5.2). Furthermore, the results of this subsection can easily be generalized for the case that $\det R_1(t)$ and $\det R_2(t)$ have a finite number of zeros.

8.5.4. The Solution of a System of Singular Integral Equations of Non-Normal Type by Means of Factorization

1. In this subsection we deal with the solution of the system (5.1) under the following hypotheses for the coefficients:

$$C(t) = C_0M_1L_1\,, \qquad D(t) = D_0G_2F_2\,. \qquad (5.18)$$

The matrix functions C_0, D_0, M_1, L_1, G_2, F_2 have the same meaning as in 8.5.3. Now in addition let $C_0(t), D_0(t) \in C^\lambda(\Gamma; (\alpha, \widetilde{m}), (\beta, \tilde{n}))$ $(0 < \lambda < 1)$. For the right-hand side of the equation (5.1) we require

$$f \in W_n = W_{p,n}(\Gamma; (\alpha, \widetilde{m}), (\beta, \tilde{n}))$$

with the numbers $\widetilde{m}$, $\tilde{n}$ introduced in 8.5.3. no. 1.

By Theorem 2.5, Chapter 7, the matrix function $C_0^{-1}(t)\,D_0(t) \in C^\lambda(\Gamma; (\alpha, \widetilde{m}), (\beta, \tilde{n}))$ admits a left canonical factorization of the form

$$C_0^{-1}(t)\,D_0(t) = \Delta_+(t)\,\widetilde{D}(t)\,\Delta_-(t) \qquad (t \in \Gamma)$$

with $\Delta_\pm(t) \in C^\lambda(\Gamma; (\alpha, \widetilde{m}), (\beta, \tilde{n}))$ and a diagonal matrix $\widetilde{D}(t) = \{t^{\varkappa_j}\delta_{jk}\}_1^n$, where $\varkappa_1 \geqq \dots \geqq \varkappa_q > 0 \geqq \varkappa_{q+1} \geqq \dots \geqq \varkappa_n$. (If all $\varkappa_j$ are positive, then $q = n$. If all $\varkappa_j \leqq 0$, then we put $q = 0$.)

For the sequel we put

$$\varkappa_+ = \sum_{j=1}^{q} \varkappa_j\,, \qquad \varkappa_- = \sum_{j=q+1}^{n} \varkappa_j\,.$$

Under the above hypotheses the operator $\mathfrak{A} = CP + DQ$ can be represented in the form

$$\mathfrak{A} = C_0\varDelta_+(\varDelta_+^{-1}P + \widetilde{D}\varDelta_-Q)\,(M_1L_1P + G_2F_2Q)\,.$$

It is easily seen that

$$M_1L_1P + G_2F_2Q = \widetilde{\mathfrak{B}}\mathfrak{A}_1$$

(cf. 8.5.3), where $\mathfrak{A}_1$ is an operator of normal type.

In order to find the general solution and the solvability conditions for the equation (5.1) we first assume that a solution $\varphi \in L_n^p(\varGamma)$ of (5.1) exists. We put

$$g = \varDelta_+^{-1}C_0^{-1}f\,, \qquad \psi = M_1L_1P\varphi + G_2F_2Q\varphi\,.$$

Then

$$\varDelta_+^{-1}P\psi + \widetilde{D}\varDelta_-Q\psi = g\,. \tag{5.19}$$

Obviously, $\varDelta_+^{-1}P\psi \in \operatorname{im} P$. Subtracting a suitably chosen polynomial vector $r(t)$ from $\widetilde{D}\varDelta_-Q\psi$ we obtain a vector from im Q. Here the degree of the j-th component $r_j(t)$ must not be greater than $\varkappa_j - 1$ for $j = 1, \ldots, q$ and $r_j(t) \equiv$ $\equiv 0$ for $j = q+1, \ldots, n$. Applying the operators P and Q to both sides of the equation (5.19) we obtain

$$P\psi = M_1L_1P\varphi = \varDelta_+(Pg - r)\,, \tag{5.20}$$

$$Q\psi = G_2F_2Q\varphi = \varDelta_-^{-1}\widetilde{D}^{-1}(Qg + r)\,. \tag{5.21}$$

The last relation implies $\widetilde{D}^{-1}Qg \in \operatorname{im} Q$. Hence the vector $g = (g_1, g_2, \ldots, g_n)$ satisfies the relations

$$t^{-\varkappa_j}Qg_j \in \operatorname{im} Q \qquad (j = 1, \ldots, n)$$

or, what is the same, the conditions

$$\int_\varGamma t^k g_j(t)\,dt = 0 \qquad (k = 0, 1, \ldots, -\varkappa_j - 1,\ j = q+1, \ldots, n)\,. \tag{5.22}$$

Furthermore, using the relation $L_1P\varphi \in L_n^p(\varGamma)$ and Lemma 1.9, Chapter 6, we obtain from (5.20) that the first $\mu_j - 1$ derivatives of the j-th component of the vector $\varDelta_+(Pg - r)$ are equal to zero at the point α:

$$[\varDelta_+(Pg - r)]_j^{(l)}\,(\alpha) = 0 \qquad (l = 0, 1, \ldots, \mu_j - 1,\ j = 1, \ldots, n)\,. \tag{5.23$_1$}$$

Analogously from (5.21) we obtain the conditions

$$[\varDelta_-^{-1}\widetilde{D}^{-1}(Qg + r)]_j^{(l)}\,(\beta) = 0 \qquad (l = 0, 1, \ldots, \nu_j - 1,\ j = 1, \ldots, n)\,. \tag{5.23$_2$}$$

Here μ_j and ν_j are the numbers defined by (5.17$_3$).

Thus we have proved the following theorem.

Theorem 5.8. *Under the hypotheses* (5.18) *for the solvability of the system* (5.1) *with* $f \in W_n$ *it is necessary and sufficient that* f *satisfy the following two conditions:*

a) *The vector* $g = \Delta_+^{-1} C_0^{-1}$ *satisfies the integral relations* (5.22).

b) *There exists a polynomial vector* $r(t)$ *with the degree of* $r_j(t) \leqq \varkappa_j - 1$ $(j = 1, \dots, q)$ *and* $r_j(t) = 0$ $(j = q + 1, \dots, n)$ *such that the relations* (5.23) *hold.*

If the conditions a) *and* b) *are satisfied, then*

$$\varphi = L_1^{-1} M_1^{-1} \Delta_+ (Pg - r) + F_2^{-1} G_2^{-1} \Delta_-^{-1} \widetilde{D}^{-1} (Qg + r)$$

is the general solution of equation (5.1).

The last assertion of the theorem can easily be checked by using Lemma 1.2 and Corollary 2.1 of Chapter 6.

Remark. The preceding proof at once shows that Theorem 5.8 also holds without the hypothesis $f \in W_n$. In the case of an arbitrary vector $f \in L_n^p(\Gamma)$ on the left-hand sides of (5.23_1) and (5.23_2) we have to take the corresponding mean derivatives, and the specified conditions must still be completed by

$$M_1^{-1} \Delta_+ (Pg - r) \in L_n^p(\Gamma), \qquad G_2^{-1} \Delta_-^{-1} \widetilde{D}^{-1} (Qg + r) \in L_n^p(\Gamma).$$

2. Now we still state a formula for the number dim ker $\mathfrak{A}$. It follows immediately from the preceding considerations of this subsection that this number is equal to the number of arbitrary constants in the polynomial vector $r(t) = \{r_1(t), r_2(t), \dots, r_n(t)\}$ satisfying conditions (5.23) with $g = 0$ and the conditions

$$\text{degree } r_j(t) \leqq \varkappa_j - 1 \qquad (j = 1, \dots, q), \qquad r_j(t) \equiv 0 \ (j = q + 1, \dots, n).$$

In order to determine this number we first introduce the following notation. By $d_{ij}(t)$ $(i, j = 1, \dots, n)$ we denote the elements of the matrix function $\Delta_+(t)$ and by K_α the triangle matrix

$$K_\alpha = \begin{pmatrix} K_{11} & 0 & 0 & . & 0 \\ K_{21} & K_{22} & 0 & . & 0 \\ . & . & . & . & . \\ K_{m1} & K_{m2} & . & . & K_{mm} \end{pmatrix}.$$

Here $m = \max\limits_{k=1,\dots,n} \mu_k$, and K_{ij} $(i = 1, \dots, m;\ j \leqq i)$ is the number matrix consisting of the elements $\binom{i-1}{j-1} d_{kl}^{(i-j)}(\alpha)$ for all $k = 1, \dots, n$ with $\mu_k \geqq i$ and all $l = 1, \dots, q$ with $\varkappa_l \geqq j$ (in the case $m = 0$ or $q = 0$ the matrix K_α is the zero matrix). Let ϱ_i be the number of rows and σ_j the number of columns of the matrix K_{ij}. Obviously, K_α is a matrix with

$$\sum_{i=1}^{m} \varrho_i = \sum_{k=1}^{n} \mu_k$$

rows and

$$\sum_{i=1}^{m} \sigma_i = \sum_{l=1}^{q} \min(\varkappa_l, m)$$

columns.[1])

Replacing $\Delta_+(t)$ by $\Delta_-^{-1}(t)\,\widetilde{D}^{-1}(t)$, μ_k by ν_k and α by β in the preceding constructions we obtain a corresponding triangle matrix K_β. We put $\varrho_\alpha =$ = rank of K_α, $\varrho_\beta =$ rank of K_β and

$$\varrho = \varrho_\alpha + \varrho_\beta .$$

Theorem 5.9. *Let $\alpha \neq \beta$. Then*

$$\dim \ker \mathfrak{A} = \begin{cases} \varkappa_+ - \varrho & \text{if} \quad \varkappa_+ > \varrho \\ 0 & \text{if} \quad \varkappa_+ \leqq \varrho . \end{cases}$$

Proof. Putting $g = 0$ in (5.23_1), calculating the derivatives by the product rule and writing down all conditions we obtain a homogenous linear equation system with the unknowns $r_1(\alpha), \dots, r_q(\alpha), \dots, r_1^{(m-1)}(\alpha), \dots, r_{\sigma_m}^{(m-1)}(\alpha)$ and the coefficient matrix K_α. An analogous statement holds for the system (5.23_2). Hence ϱ is equal to the number of linearly independent conditions which the vector $r(t)$ has to satisfy in view of conditions (5.23). Taking into consideration finally that the number of all coefficients in the polynomial vector $r(t)$ is equal to $\varkappa_+$ we obtain the assertion of the theorem.

Finally we consider some special cases in which we can immediately find the ranks ϱ_α and ϱ_β.

1°. $\varrho_\alpha = 0$ if $\mu_j = 0$ $(j = 1, \dots, n)$ $\left(\text{i.e. } m_{2j} = 0 \text{ and } r_{2j} < \dfrac{1}{p}\right)$.

2°. If all subindices $\varkappa_j$ are non-positive, then $\varrho_\alpha = \varrho_\beta = \dim\ker \mathfrak{A} = 0$.

3°. If $\varkappa_j \geqq m$ $(j = 1, \dots, n)$, then $\varrho_\alpha = \sum_{j=1}^{n} \mu_j$. Namely, in this case we have $\sigma_j = q = n$, rank of $K_{ii} = \varrho_i$.

4°. If $\mu_k = \mu_0$ $(k = 1, \dots, n)$, then

$$\varrho_\alpha = \sum_{l=1}^{q} \min(\varkappa_l, \mu_0) .$$

Namely, in this case we have $m = \mu_0$, $\varrho_i = n$, rank of $K_{ii} = \sigma_i$.

Analogous statements hold for the number ϱ_β.

Remark 1. The indices $\varkappa_j$ $(j = 1, \dots, n)$ of the matrix function $C_0^{-1}(t)\,D_0(t)$ are independent of the concrete form of the representations (5.18) in the following sense: Let

$$C(t) = \widetilde{C}_0 M_1 \widetilde{L}_1 , \qquad D(t) = \widetilde{D}_0 G_2 \widetilde{F}_2$$

be another representation of the singular matrix functions $C(t)$, $D(t)$ (cf. Section 8.1). Then after some elementary calculations we obtain

$$C_0^{-1}(t)\,D_0(t) = L\widetilde{C}_0^{-1}\widetilde{D}_0 F ,$$

[1]) The last two equations can be easily checked by means of elementary considerations.

where $L(F)$ is a polynomial matrix in t (in t^{-1}) with a constant determinant different from zero. It follows from this that the left indices of the matrix functions $C_0^{-1}D_0$ and $\widetilde{C}_0^{-1}\widetilde{D}_0$ coincide (comp. Theorem 2.1, Chapter 7).

Following the definition given by N. I. MUSHELIŠVILI [1] in the normal case we could say here that the numbers $\varkappa_j$ are the *indices of the system* (5.1) (of non-normal type).

Remark 2. Let $k = \dim \ker \mathfrak{A}$ and k' the number of solvability conditions (5.22) and (5.23) for the inhomogeneous system (5.1). Then using Theorem 5.7 we conclude easily that

$$k - k' = \varkappa - \sum_{j=1}^{n} \left(m_{2j} + n_{2j} + \left[r_{2j} + \frac{1}{p'} \right] + \left[s_{2j} + \frac{1}{p'} \right] \right).$$

8.6. Non-Bounded Regularization of Systems of Singular Integral Equations

8.6.1. Non-Bounded Regularization

In the space $L_n^p(\Gamma)$ $(1 < p < \infty)$ we consider a system of singular integral equations of the form

$$\mathfrak{A}\varphi = A(t)\, \varphi(t) + \frac{B(t)}{\pi i} \int_\Gamma \frac{\varphi(\tau)}{\tau - t}\, d\tau + \int_\Gamma T(t, \tau)\, \varphi(\tau)\, d\tau = f(t)\,. \quad (6.1)$$

Here Γ is a Ljapunov curve system of the form described in 6.2.1. For the coefficients of the equation (6.1) we state the following hypotheses:

(1) The elements of the matrix functions $A(t)$, $B(t)$ and $T(t, \tau)$ satisfy a Hölder condition on Γ and $\Gamma \times \Gamma$, respectively.

(2) The determinants of the matrix functions

$$C(t) = A(t) + B(t)\,, \qquad D(t) = A(t) - B(t)$$

vanish on Γ at the points $\alpha_1, \alpha_2, \dots, \alpha_r$ and $\beta_1, \beta_2, \dots, \beta_s$, respectively. The positive integers m_j and n_k $(j = 1, \dots, r;\ k = 1, \dots, s)$ are the orders of these zeros.

(3) In a neighborhood (on Γ) of the points α_j and β_k $(j = 1, \dots, r;\ k = 1, \dots, s)$ the elements of the matrix functions A, B and T have derivatives with respect to t up to order m_j resp. n_k satisfying a Hölder condition with exponent $\lambda > \dfrac{p-1}{p}$.

Furthermore, let $\Gamma \in C^{\widetilde{m}, \mu}$ with $\widetilde{m} = \max(m_1, \dots, m_r,\ n_1, \dots, n_s)$ and $0 < \mu \leqq 1$.

In the space $L_n^p(\Gamma)$ we consider the operator $\mathfrak{B}$ given by the equation

$$\mathfrak{B}\varphi = \frac{1}{\det C(t)} (I - H_1)\, \widetilde{C}(t)\, P\varphi + \frac{1}{\det D(t)} (I - H_2)\, \widetilde{D}(t)\, Q\varphi\,.$$

Here H_1, H_2 have the same meaning as in 6.7.1, and $\widetilde{C}(t) = \{C_{ji}\}_1^n$, where C_{ij} is the cofactor of the element c_{ij} in the determinant $\det C(t) = \det |c_{ij}(t)|$. $\widetilde{D}(t)$ is defined analogously. We consider the operator $\mathfrak{B}$ on the set $D(\mathfrak{B}) = \{\varphi \in L_n^p(\Gamma) : \mathfrak{B}\varphi \in L_n^p(\Gamma)\}$ dense in $L_n^p(\Gamma)$.

Repeating the considerations of 6.7.1 and using the relations

$$C\widetilde{C} = \widetilde{C}C = (\det C)\, E\,, \qquad D\widetilde{D} = \widetilde{D}D = (\det D)\, E$$

(E the unit matrix) we can easily see that $\operatorname{im} \mathfrak{A} \subset D(\mathfrak{B})$ and

$$\mathfrak{B}\mathfrak{A}\varphi = \varphi(t) + \int_\Gamma K(t, \tau)\, \varphi(\tau)\, d\tau \qquad (\varphi \in L_n^p(\Gamma))\,. \tag{6.2}$$

Here

$$K(t, \tau) = \frac{1}{\det C(t)} [N_n(t, \tau) - H_1 N_n] + \frac{1}{\det D(t)} [M_n(t, \tau) - H_2 M_n]\,,$$

$$N_n(t, \tau) = \widetilde{C}(t)\, N(t, \tau)\,, \qquad M_n(t, \tau) = \widetilde{D}(t)\, M(t, \tau)\,,$$

and $N(t, \tau)$, $M(t, \tau)$ are defined by the same formulae as in the case $n = 1$ (see 6.7.1). From the results of the Subsection 6.7.1 it follows immediately that the integral operator with the kernel $K(t, \tau)$ from (6.2) is a compact operator on the space $L_n^p(\Gamma)$. Hence $\mathfrak{B}$ is a non-bounded regularizer of the operator $\mathfrak{A}$.

Theorem 7.1 and Corollary 7.1 of Chapter 6 are true also in the present case.

8.6.2. Equivalent Non-Bounded Regularization

Under the hypotheses of the preceding subsection in the sequel we assume in addition that $\alpha_j \neq \beta_k$ ($j = 1, \ldots, r$; $k = 1, \ldots, s$) holds.

In order to construct an equivalent non-bounded regularizer of the operator $\mathfrak{A}$ first we represent the matrix functions $C(t)$, $D(t)$ in the form

$$C(t) = U(t)\, C_0(t)\,, \qquad D(t) = V(t)\, D_0(t)\,. \tag{6.3}$$

Here $U(t)$ ($V(t)$) is a polynomial matrix in t (in t^{-1}), where

$$\det U(t) = R_0(t) = \prod_{j=1}^{r} (t - \alpha_j)^{m_j}\,,$$

$$\det V(t) = R_1(t) = \prod_{k=1}^{s} (t^{-1} - \beta_k^{-1})^{n_k}\,.$$

$C_0(t)$ and $D_0(t)$ are matrix functions regular everywhere on Γ which moreover belong to the class $C^\lambda(\Gamma; (\alpha_j, m_j)_1^r, (\beta_k, n_k)_1^s)$ (cf. Subsection 8.1.3).

We form the operator $\widetilde{\mathfrak{B}} = \mathfrak{B}_1 + \mathfrak{B}_2$, where

$$\mathfrak{B}_1 = R_0^{-1} D_0^{-1} (I - \widetilde{H}_1) \frac{R_0}{\det C} D_0 \widetilde{C} P ,$$

$$\mathfrak{B}_2 = R_1^{-1} C_0^{-1} (I - H_2) \frac{R_1}{\det D} C_0 \widetilde{D} Q .$$

Here H_2 has the same meaning as in 6.7.1 and 8.6.1. $\widetilde{H}_1\varphi = \sum_{k=1}^{\mu} \frac{a_k}{t^k}$ is the polynomial vector defined by the conditions

$$(\widetilde{H}_1\varphi)^{(l)}(\alpha_j) = \varphi^{\{l\}}(\alpha_j) \qquad (l = 0, \ldots, m_j - 1; j = 1, \ldots, r)$$

where

$$\mu = \sum_{j=1}^{r} m_j , \qquad \nu = \sum_{k=1}^{s} n_k .$$

In the same way as in 8.6.1 we can see that $\widetilde{\mathfrak{B}}$ is also a left regularizer of the operator $\mathfrak{A}$. By 1.5.2 $\widetilde{\mathfrak{B}}$ is an equivalent regularizer if and only if dim ker $\widetilde{\mathfrak{B}} = 0$.

Let $\widetilde{\mathfrak{B}}\varphi = 0$. By some elementary calculations we can transform this equation into the form

$$R_1 \frac{R_0}{\det C} \widetilde{C} P\varphi - R_0 C_0^{-1} H_2 \varphi_2 = R_1 D_0^{-1} \widetilde{H}_1 \varphi_1 - R_0 \frac{R_1}{\det D} \widetilde{D} Q\varphi , \qquad (6.4)$$

where $\varphi_1, \varphi_2 \in L_n^p(\Gamma)$. By Theorem 2.5, Chapter 7, the matrix function $C_0(t) \times D_0^{-1}(t)$ admits a left canonical factorization

$$C_0(t) D_0^{-1}(t) = \Delta_+(t) \Delta(t) \Delta_-(t) \qquad (t \in \Gamma) \qquad (6.5)$$

with a diagonal matrix $\Delta(t) = \{t^{\varkappa_j} \delta_{jk}\}_1^n$, where $\varkappa_1 \geqq \varkappa_2 \geqq \ldots \geqq \varkappa_n$ are integers. Furthermore, (6.3) implies

$$R_0 C_0 = \widetilde{U} C , \qquad R_1 \widetilde{D} = (\det D) D_0^{-1} \widetilde{V} , \qquad (6.6)$$

where the symbol $\sim$ over a matrix has the same meaning as in 8.6.1. Multiplying both sides of the equation (6.4) on the left by the matrix function $C_0(t)$ and using the representations (6.5) and (6.6) we obtain

$$\Delta_+^{-1}[R_1 \widetilde{U} P\varphi - R_0 H_2 \varphi_2] = \Delta\Delta_-[R_1 \widetilde{H}_1 \varphi_1 - R_0 \widetilde{V} Q\varphi] . \qquad (6.7)$$

In the sequel we assume that

$$\varkappa_j \leqq -(\mu + \nu) \qquad (j = 1, \ldots, n) . \qquad (6.8)$$

Multiplying both sides of the equation (6.7) by t^ν we obtain that the left-hand side belongs to im P and the right-hand side to im Q. Thus

$$P\varphi = \frac{U(t)}{R_1(t)} \sum_{l=0}^{\nu-1} b_l t^l , \qquad Q\varphi = \frac{V(t)}{R_0(t)} \sum_{k=1}^{\mu} \frac{a_k}{t^k} .$$

Since $\alpha_j \neq \beta_k$ and $P\varphi, Q\varphi \in L_n^p(\Gamma)$ it follows from this that

$$b_l = 0 \qquad (l = 0, \dots, \nu - 1), \qquad a_k = 0 \qquad (k = 1, \dots, \mu)$$

and consequently $\varphi = 0$. Thus we have proved the following theorem.

Theorem 6.1. *If the indices of the canonical factorization* (6.5) *satisfy the relations* (6.8), *then* $\widetilde{\mathfrak{B}}$ *is an equivalent left regularizer of the operator* $\mathfrak{A}$.

Remark. It is easily seen that the results of this subsection coincide with the corresponding results of 6.7.2 for $n = 1$ (cf. especially the Remark 1 from 6.7.2).

8.7. Comments and References

8.1. The constructions of this section generalize the wellknown process of the transformation of polynomial matrices into cononical diagonal matrices (see F. R. Gantmaher [1], Kapitel VI, § 2). The author owes the hint of the fact that each matrix functions of the form (1.7) can be represented in the form (1.1) to I. C. Gohberg.

Similar constructions were first used by F. D. Gahov [2] for the solution of the Riemann-Hilbert boundary problem with a singular coefficient matrix.

8.2.—8.4. Here the considerations from the paper [14] of the author dealing with systems of singular integral equations are carried over to systems of Wiener-Hopf equations. Most of the results of these sections are here published for the first time (cf. also S. Prössdorf [16]). Some results can be found already in the author's paper [11], where at all systems of convolution equations in the non-normal case were first investigated.

We also mention the paper of Š. I. Galiev [3] in which similar results as in 8.2.2 and 8.3.2 were obtained in other spaces than the spaces $\overline{E}_n$ and $\widetilde{E}_n$ considered here. The spaces constructed by this author have a more complicated structure, and the analytic description is more difficult. They have the advantage that they are invariant with respect to the commutation of two arbitrary columns in the symbol matrix. Finally we mention the paper of A. Voigtländer [1] in which the results of Subsection 8.5.3 were carried over to systems of paired Wiener-Hopf equations.

8.5. The results of Subsections 8.5.1—8.5.2 are due to the author [14], [10]. The results of the Subsection 8.5.3 were proved by S. Prössdorf and B. Silbermann [2] under the hypothesis that the condition (II) is satisfied. The here given proof of Theorem 5.7 including the representation (***) is due to B. Silbermann [8]. In Subsection 8.5.4 we follow the paper of the author [12].

8.6. The results of this section were stated by the author [2].

CHAPTER 9

SINGULAR EQUATIONS IN SOME COUNTABLY NORMED SPACES AND SPACES OF DISTRIBUTIONS

The theory of the singular integral equations of non-normal type assumes an especially simple form in the countably normed space $C^{\infty}(\Gamma)$ of the infinitely differentiable functions and in the distribution space $C^{-\infty}(\Gamma) = [C^{\infty}(\Gamma)]^*$. In these spaces we have the following remarkable analogue to the theory in the space $L^p(\Gamma)$: The singular integral operator (with coefficients from $C^{\infty}(\Gamma)$) is a Noether operator if and only if its symbol has at most finitely many zeros of finite orders.

In Chapter 6 we saw that for the homogeneous singular integral equation of non-normal type new solutions appear if the class of solution functions is extended from $L^p(\Gamma)$ to the space $\widetilde{L^p}(\Gamma)$. In this chapter we show among other things that this already yields all possible solutions of this equation in the distribution space $C^{-\infty}(\Gamma)$.

Analogous results hold for the Wiener-Hopf equations in the corresponding countably normed and distribution spaces.

All considerations of this chapter are made for paired equations and for the case $n \geqq 1$. In an introductory section we state the necessary definitions and a general scheme for the investigation of certain operator equations in distribution spaces.

9.1. Operator Equations in Distribution Spaces

9.1.1. Let X be an F-space embedded continuously and densely into a Banach space E. As usual we denote the corresponding conjugate spaces by X^* and E^*. In the sequel we put $H = E^*$.

Each element $\varphi \in H$ can be considered as a functional on X. In fact, for any $x \in X$ we have

$$|\langle x, \varphi \rangle| \leqq ||\varphi||_H \, ||x||_E \,, \tag{1.1}$$

and $x_n \to 0$ (in the sense of the metric of X) implies $||x_n||_E \to 0$ and thus $\langle x_n, \varphi \rangle \to 0$. Since, further, X is dense in the space E, the values of the functional $\langle x, \varphi \rangle$ $(x \in X)$ define the corresponding element $\varphi \in H$ uniquely. Hence

$$H \subset X^* .$$

Lemma 1.1. *Let* $\{\varphi_n\}_1^\infty \subset H$ *and* $\|\varphi_n\|_H \to 0$. *Then* $\varphi_n \to 0$ *in the sense of the strong topology of the space* X^*.

Proof. We recall that the convergence given by the strong topology in X^* is uniform convergence on each bounded set $G \subset X$.

Let $G \subset X$ be an arbitrary bounded set. It is easily seen (cf. 1.1) that G is then also a bounded set in the B-space E. Hence by (1.1) we have

$$|\langle x, \varphi_n \rangle| \leq C\|\varphi_n\|_H \qquad \forall\, x \in G\,,$$

i.e. $\langle x, \varphi_n \rangle$ converges uniformly to zero on G.

Analogously to the theory of distributions under the above hypotheses X is called a *basic space* and the functionals $f \in X^*$ *generalized functions* or *distributions* on the basic space X. The distributions $\varphi \in H$ are called *regular functionals* (cf. I. M. Gelfand and G. E. Šilov [2]).

If E is a Hilbert space, then $E = H = E^*$ and thus

$$X \subset H \subset X^*\,. \tag{1.2}$$

9.1.2. Now let $A \in \mathscr{L}(H)$ be a continuous linear operator on the space H. Furthermore, let $A = (A')^*$ for a certain operator $A' \in \mathscr{L}(E)$. Then obviously we have

$$\langle x, Af \rangle = \langle A'x, f \rangle \qquad (x \in X) \tag{1.3}$$

for any $f \in H$.

Now we assume that $A'(X) \subset X$ and that the restriction A'_X of the operator A' to X is continuous on the space X: $A'_X \in \mathscr{L}(X)$. Under this hypothesis the operator A can be defined for any $f \in X^*$ by means of the formula (1.3). Obviously, this extension of the operator A to the whole space $X^*(\supset H)$ is equal to $(A'_X)^*$ and thus a continuous linear operator on the distribution space X^* (not only with respect to the strong topology but also with respect to the weak topology of X^*).

In the sequel the set of operators A with the above properties is denoted by $\Pi(X; H)$. We make the following definition.

Definition. Let $A \in \Pi(X; H)$ and $f \in X^*$. The element $u \in X^*$ is called a *generalized solution* of the equation

$$Au = f \tag{1.4}$$

if[1])

$$\langle A'x, u \rangle = \langle x, f \rangle \;\forall\; x \in X\,.$$

If $f \in H$ and if the element $u \in H$ satisfies the equation (1.4), then u is called a *classical solution* of this equation.

We remark that by (1.2) in the case of a Hilbert space $H = E$ the right-hand side f of the equation (1.4) can in particular be chosen from X.

---

[1]) We remark that this definition corresponds to the definition of the generalized (or weak) solution of boundary value problems for partial differential equations of elliptic type.

9.1.3. Let $A \in \Pi(X; H)$. Obviously, if the equation (1.4) with $f \in X^*$ possesses a generalized solution, then

$$\langle x, f \rangle = 0 \qquad \forall\, x \in \ker A'_X . \tag{1.5}$$

According to the definitions given in Section 1.2 the operator A (or the equation (1.4)) is said to be *normally solvable in generalized functions* if the condition (1.5) is also sufficient for the solvability (in X^*) of the equation (1.4). A is called a *Φ-operator on the distribution space* X^* if A is normally solvable and if the following two numbers are finite:

$$\dim \ker_* A = \beta(A'_X)\,, \qquad \dim \operatorname{coker}_* A = \alpha(A'_X)\,.$$

The difference of these two numbers[1])

$$\operatorname{Ind}_* A = -\operatorname{Ind} A'_X \tag{1.6}$$

is called the *index* of the operator A on the distribution space X^*.

The results of the Chapter 1 (cf. loc. cit. Theorem 3.5 and the Banach-Hausdorff theorem) immediately show:

Theorem 1.1. *The operator A is a Φ-operator in the distribution space X^* if and only if $A'_X \in \Phi(X)$.*

We remark that in the proof of Theorem 3.5, Chapter 1, a *particular solution* u_0 of the equation (1.4) was also constructed: Let A be a Φ-operator in X^* and let f satisfy condition (1.5). Then

$$u_0 = (A_1^{-1}Q)^* f\,. \tag{1.7}$$

Here Q is a continuous projection of X onto $\operatorname{im} A'_X$ and A_1 the restriction of the operator A'_X to a topological complement of $\ker A'_X$ in X.

9.1.4. Now we recall some results of Section 4.1. Let the operator $A \in \mathcal{L}(H)$ be representable in the fom

$$A = A_0 D \tag{1.8}$$

with $A_0, D \in \mathcal{L}(H)$, where the operator D satisfies condition (2′), i.e. for a certain linear operator $D^{(-1)}$ on H (with values in general outside H) we have

$$D^{(-1)} D x = x\,, \qquad D D^{(-1)} x = x \qquad \forall x \in H\,. \tag{1.9}$$

Then the operator A can be extended to an operator $\tilde{A} \in \mathcal{L}(\tilde{H}, H)$ in a natural way. Here $\tilde{A}$ is a $\Phi_\pm$-operator if $A_0 \in \Phi_\pm(H)$ holds, and $\tilde{H}$ is the Banach space $\tilde{H} = D^{(-1)}(H)$ with the norm defined in 4.1.1.

In the sequel we are interested in the question when the solutions of the homogeneous equation

$$Au = 0 \tag{1.10}$$

[1]) Obviously, the number $\operatorname{Ind}_* A$ defined by (1.6) coincides with the index of the operator $A = (A'_X)^* \in \mathcal{L}(X^*)$ in the sense of Section 1.3.

coincide in the spaces $\tilde{H}$ and X^*. Now we give sufficient conditions for this. All the concrete types of equations considered in the present chapter satisfy these conditions. Thus the general solution $\tilde{u}$ of the equation (1.10) in the distribution space X^* can then be given as follows

$$\tilde{u} = D^{(-1)}\varphi\,, \qquad \varphi \in \ker_H A_0\,.$$

For the general solution of the inhomogeneous equation (1.4) we consequently obtain $u = \tilde{u} + u_0$ (cf. (1.7)).

Thus let $A \in \Pi(X; H)$. We assume that the operator A' admits the representation

$$A' = BC \tag{1.11}$$

where the operators $B, C \in \mathscr{L}(E)$ satisfy the following conditions:

(1) The operator B maps the space X biuniquely and continuously onto itself.

(2) $C \in \Phi(E)$.

(3) $C\varphi \in X (\varphi \in E)$ implies $\varphi \in X$.

First we remark that for the operator $A = (A')^* \in \mathscr{L}(H)$ the relation (1.11) immediately implies (1.8) with

$$A_0 = C^*\,, \qquad D = B^*\,.$$

Since X is a dense subset of E, we obtain from condition (1) that

$$\dim \ker B = \dim \ker D = 0\,.$$

Hence D satisfies condition (1.9) (see 4.1.1). The space $\tilde{H} = D^{(-1)}(H)$ can be embedded into the distribution space X^* as follows: For $f \in \tilde{H}$ we put

$$f(\varphi) = \langle B^{-1}\varphi, Df\rangle \qquad (\varphi \in X)\,. \tag{1.12}$$

Thus we obtain a linear continuous functional on X, i.e. we have

$$H \subset \tilde{H} \subset X^*\,. \tag{1.13}$$

From (1.3) and (1.12) we obtain easily that $\tilde{A}f = Af (f \in \tilde{H})$.

Theorem 1.2. *Let the operator $A \in \Pi(X; H)$ satisfy the conditions* (1.11) *and* (1)—(3). *Then A is a Φ-operator on the distribution space X^*, and the equation* (1.10) *has the same solutions in the spaces $\tilde{H}$ and X^*.*

Proof. We put $\bar{E} = \operatorname{im} B (= B(E))$ and $\tilde{A}' = \bar{B}C$ (comp. 4.1.1). Then $\tilde{A}' \in \Phi(E, \bar{E})$ and $\tilde{A}' \,|X = A'_X$. Using conditions (1)—(3) we obtain by Theorem 3.9, Chapter 1, that $A'_X \in \Phi(X)$. By Theorem 1.1 this implies the first assertion of Theorem 1.2.

By (1.13) we have $\ker_{\tilde{H}} A \subset \ker_{X^*} A$. Contrary to the assertion of the theorem we assume that there exists a functional $f_0 \in \ker_{X^*} A = \ker (A'_X)^*$ which does not belong to $\ker_{\tilde{H}} A$. Then, as is wellknown, there exists an element $y_0 \in X$ such that

$$\langle y_0, f_0 \rangle = 1 , \qquad y_0 \perp \ker_{\tilde{H}} A . \tag{1.14}$$

By (1.12) the second relation (1.14) means that

$$\langle B^{-1} y_0, \psi \rangle = 0 \qquad \forall \psi \in \ker A_0 .$$

By condition (2) this implies the existence of an element $\varphi \in E$ with $C\varphi = B^{-1} y_0$, and the conditions (1), (3) yield $\varphi \in X$. Thus we obtain

$$\langle y_0, f_0 \rangle = \langle A' \varphi, f_0 \rangle = 0 ,$$

but this contradicts the first relation in (1.14). Thus the theorem is proved.

Corollary 1.1. *Let the operator* $A \in \Pi(X; H)$ *satisfy the following conditions* a) $A' \in \Phi(E)$, b) $A'\varphi \in X (\varphi \in E)$ *implies* $\varphi \in X$.

Then each generalized solution $u (\in X^*)$ *of the equation* (1.10) *is a classical solution.*

This follows immediately from Theorem 1.2 since now condition (1) is satisfied with $B = I$ and thus $\tilde{H} = H$ holds.

Remark. Theorem 1.2 shows that under conditions (1)—(3) it is sufficient to extend the operator A to the space $\tilde{H}$ in order to obtain all generalized solutions $u \in X^*$ of the homogeneous equation (1.10). But under the conditions of Corollary 1.1 there appeare no new solutions of equation (1.10) if we go from the space H to the space X^*.

9.1.5. The so-called *countably normed spaces* with pairwise coordinated norms are F-spaces which are especially important for applications. These spaces were first introduced and taken as a basis for a general distribution theory by I. M. Gelfand and G. E. Šilov (see I. M. Gelfand and G. E. Šilov [2]). We give their definition:

In the linear space X let a countable system of norms $\|x\|_1 \leqq \|x\|_2 \leqq \leqq \dots \leqq \|x\|_p \leqq \dots$ be given. Moreover, let these norms be *pairwise coordinated.* This last condition means that each sequence $\{x_n\} \subset X$ which is a Cauchy sequence with respect to the p-th norm and which converges to zero with respect to the $(p-1)$-th norm also converges to zero with respect to the p-th norm.

By X^p $(p = 1, 2, \dots)$ we denote the Banach space obtained by completion of the space X with respect to the p-th norm. By means of the topololgy generated by the above-mentioned norm system X becomes a linear topological space. It is easily seen that the inclusions

$$X^1 \supset X^2 \supset \dots X^p \supset \dots . \supset X$$

hold in the sense of a continuous embedding. The linear topological space X is complete if and only if

$$X = \bigcap_{p=1}^{\infty} X^p , \tag{1.15}$$

(cf. I. M. GELFAND and G. E. ŠILOV [2], p. 14). If condition (1.15) is satisfied, then X is called a *countable normed space*. Obviously, this space is an F-space in the sense of Section 1.1.

For the sequel we need the following lemma.

Lemma 1.2. *Let X be a countably normed space and T a linear operator defined on the whole space X^1 with values in X, where $T \in \mathcal{K}(X^1, X^p)$ for all $p = 1, 2, \ldots$. Then $T_X \in \mathcal{K}(X)$ (T_X restriction of the operator T to X).*

Proof. In the space X we consider the neighborhood $U = \{x \in X: ||x||_1 < < \varepsilon\}$, where $\varepsilon > 0$ is an arbitrary fixed number. It is sufficient to show that $T(U) \subset X$ is a relatively compact set.

Since U is a bounded set in the space X^1, by hypothesis the image set $T(U)$ contains a sequence $\{y_{1n}\}$ $(n = 1, 2, \ldots)$ which is a Cauchy sequence with respect to $||x||_1$. Because of $T \in \mathcal{K}(X^1, X^2)$ this sequence contains a subsequence which is a Cauchy sequence with respect to $||x||_2$. Repeating these considerations we obtain a system of sequences $\{y_{mn}\}$ $(m, n = 1, 2, \ldots)$ for which the m-th is a Cauchy sequence with respect to the norm $||x||_m$. Then the sequence of the diagonal terms $\{y_{mm}\}$ is a Cauchy sequence with respect to each of norms $||x||_p$ $(p = 1, 2, \ldots)$. This means that $\{y_{mm}\}$ is a Cauchy sequence in the space X. Since X is complete, this sequence converges, i.e. $T(U)$ is relatively compact in X. This proves the assertion.

9.2. Discrete Wiener-Hopf Equations

9.2.1. Some Sequence Spaces

Let E be any one of the sequence spaces l^p $(1 \leqq p < \infty)$, m, c or c° introduced in Section 3.1 and k an integer. By E^k we denote the Banach space of all sequences of the form $f = \left\{\frac{\xi_j}{(1 + |j|)^k}\right\}_{j=-\infty}^{\infty}$ with the norm $||f||_k = ||\xi||_E$, where $\xi = \{\xi_j\}_{-\infty}^{\infty} \in E$. By E^∞ we denote the countably normed space $E^\infty = \bigcap_{k=0}^{\infty} E^k$ with the generating system of norms $||f||_k$ $(k = 0, 1, \ldots)$ and by $E^{-\infty}$ the conjugate space of E^∞. In the case of the spaces $E = l^p$ $(1 \leqq p < \infty)$ for any $k \geqq 0$ we have

$$(E^k)^* = E_*^{-k}, \qquad E_* = l^{p'} \qquad \left(p' = \frac{p}{p-1}, l^\infty = m\right)$$

and thus $E^{-\infty} = \bigcup_{k=0}^{\infty} E^{-k}$.

The symbols E_n^∞ and $E_n^{-\infty}$ (n a natural number) have the corresponding meaning (cf. Subsection 7.1.1).

Finally we put $W^\infty = \bigcap_{m=0}^{\infty} W^m$ (comp. 5.1.2, no. 3).

9.2.2. Equations in Basic Spaces

1. Let $a(z) \in W^\infty$ and let a_j ($j = 0, \pm 1, \ldots$) be the Fourier coefficients of the function $a(z)$. By A $(\in \Re(U))$ we denote the operator, which is defined on the space E by the Toeplitz matrix $\{a_{j-k}\}_{j,k=-\infty}^{\infty}$. We show that A is a continuous operator on the space E^∞.

Let $\xi \in E^\infty$ and $\eta = A\xi$. For $l = 1, 2, \ldots$ and $j = 0, \pm 1, \ldots$ we put

$$\alpha_j^{(l)} = (1 + |j|)^l, \qquad \tilde{\xi}_j = \alpha_j^{(l)} \xi_j, \qquad \tilde{\eta}_j = \alpha_j^{(l)} \eta_j .$$

Using the obvious relation $a_{j+k}^{(l)} \leqq \alpha_j^{(l)} \alpha_k^{(l)}$ and the inequality (1.5), Chapter 2, we then obtain the estimate

$$\|\eta\|_l = \|\tilde{\eta}\|_E \leqq c \|\tilde{\xi}\|_E = c \|\xi\|_l$$

with

$$c = \sum_{j=-\infty}^{\infty} \alpha_j^{(l)} |a_j| < \infty .$$

Thus we have proved the assertion $A \in \mathscr{L}(E^\infty)$ for any one of the above-mentioned spaces.

In the sequel by P_1 we denote the projector defined on the space E (resp. E^∞) by the relations (1.2), Chapter 3, and we put $Q_1 = I - P_1$.

Lemma 2.1. *The commutator $T = [P_1, A]$ is a compact operator from E into the space E^l for any $l \geqq 0$ ($E^\circ = E$).*

Proof. Let l be an arbitrary fixed non-negative integer and $\xi \in E$. For real x we define

$$\operatorname{sgn} x = \begin{cases} 1 & \text{for } x \geqq 0 , \\ -1 & \text{for } x < 0 . \end{cases}$$

Then for the terms η_k of the sequence $\eta = T\xi$ we obtain

$$\eta_k = \tfrac{1}{2} \sum_{j=-\infty}^{\infty} a_{k-j} (\operatorname{sgn} k - \operatorname{sgn} j)\, \xi_j \qquad (k = 0, \pm 1, \ldots) .$$

Since for any $m = 0, 1, 2, \ldots$ the obvious relation

$$(|k| + |j|)^m\, |\operatorname{sgn} k - \operatorname{sgn} j| = |k - j|^m\, |\operatorname{sgn} k - \operatorname{sgn} j|$$

holds, we can easily see that the estimate

$$(1 + |k|)^l\, |\eta_k| \leqq \tfrac{1}{2} \sum_{j=-\infty}^{\infty} b_{k-j}\, |\operatorname{sgn} k - \operatorname{sgn} j|\, |\xi_j| \equiv \zeta_k \tag{2.1}$$

holds with $b_k = (1 + |k|)^l\, |a_k|$.

By hypothesis we have $\{b_k\}_{-\infty}^{\infty} \in l^1$. Let B $(\in \Re(U))$ be the operator given on the space E by the Toeplitz matrix $\{b_{k-j}\}_{k,j=-\infty}^{\infty}$. Obviously, for the sequence $\zeta = \{\zeta_k\}_{-\infty}^{\infty}$ the relation

$$\zeta = [P_1, B]\, \varphi , \qquad \varphi = \{|\xi_k|\}_{-\infty}^{\infty} , \tag{2.2}$$

holds. Hence $\zeta \in E$. By (2.1) this implies $\eta \in E^l$.

By Lemma 2.1, Chapter 2, the commutator $[P_1, B]$ is compact on E. From this it follows easily by the estimate (2.1) and the relation (2.2) that the operator $T: E \to E^l$ is compact.

2. Now let $A(z)$ and $B(z)$ be matrix functions from the algebra $W^{\infty}_{n\times n}$ (n a fixed natural number) and a_j, b_j ($j = 0, \pm 1, \ldots$) the matrices formed from their Fourier coefficients. We assume that the determinants $\det A(z)$ and $\det B(z)$ possess finitely many zeros of the integral orders m_l and n_l ($l = 1, \ldots, r$) at the points α_l and β_l, respectively, on the unit circle.

We put $P = \{P_1 \delta_{jk}\}_1^n$, $Q = I - P$. Furthermore, let A_1 and A_2 be the matrix operators with symbols $A(z)$ and $B(z)$, respectively. In the space E_n^{∞} we consider the paired operator

$$A_t = PA_1 + QA_2 \tag{2.3}$$

given by the equation system (4.6′), Chapter 7. By no. 1 we have $A_t \in \mathcal{L}(E_n^{\infty})$.

It was shown in Section 8.1 that the matrix functions $A(z)$ and $B(z)$ can be represented in the form

$$A(z) = B_1(z)\, A_0(z)\,, \qquad B(z) = B_2(z)\, B_0(z)\,. \tag{2.4}$$

Here $B_1(z)$, $B_2(z)$ are the polynomial matrices defined by the relations (5.2), Chapter 8, and $A_0(z)$, $B_0(z) \in W^{\infty}_{n\times n}$, where

$$\det A_0(z) \neq 0\,, \qquad \det B_0(z) \neq 0 \qquad (|z| = 1)\,. \tag{2.5}$$

Theorem 2.1. *Let E be any one of the above-mentioned sequence spaces. The operator A_t is a Φ-operator on the space E_n^{∞}, and the formulae*

$$\dim \ker A_t = -\sum_{\varkappa_j < 0} \varkappa_j\,, \qquad \dim \operatorname{coker} A_t = \sum_{\varkappa_j > 0} \varkappa_j \tag{2.6}$$

hold, where $\varkappa_j$ ($j = 1, \ldots, n$) are the right partial indices of the matrix function $A_0(z)\, B_0^{-1}(z) \in W^{\infty}_{n\times n}$.

Proof. By B_k ($k = 0, 1, 2$) we denote the matrix operator with the symbol $B_k(z)$. A_0 has the corresponding meaning. Introducing the paired operators

$$B = PB_1 + QB_2\,, \qquad A_0' = PA_0 + QB_0$$

for A_t we obtain the representation (cf. 8.4.2)

$$A_t = BA_0'\,. \tag{2.4′}$$

By 7.4.3 we have $A_0' \in \Phi(E_n)$. Taking into consideration Theorem 3.3, Chapter 7, and the remark at the end of the Subsection 2.4.2 we easily see that each solution $\xi \in E_n$ of the equation

$$A_0'\xi = \eta \qquad (\eta \in E_n^{\infty})$$

necessarily belongs to E_n^{∞}. From this it follows by Theorem 3.9, Chapter 1, that $A_0' \in \Phi(E_n^{\infty})$ and that for this operator the formulae (2.6) hold. Now we show that the operator B maps the space E_n^∞ biuniquely onto itself. By (2.4') from this we immediately obtain the assertion of the theorem.

The operator B can be represented as a product of the form (4.3), Chapter 8, and the operator $P\overline{R} + QS$ maps the space E_n^∞ biuniquely onto itself. Since moreover the operator $P\overline{D} + QC$ has a diagonal form, it is sufficient to prove that the scalar operator ($n = 1$) of the form

$$B^0 = P_1 B_\beta + Q_1 D_\alpha \qquad (|\alpha| = |\beta| = 1)$$

(cf. 5.3) maps the space E^∞ biuniquely onto itself.

First we conclude from Lemma 3.1, Chapter 5, that the homogeneous equation $B^0\xi = 0$ has only the trivial solution $\xi = 0$ in the space E^∞. It follows easily from the results of Sections 5.3 and 5.1 that the operator B^0 has a continuous inverse

$$(B^0)^{-1} = P_1 B_1^{-1} + Q_1 B_2^{-1}$$

in E^∞, where B_1^{-1} and B_2^{-1} are defined by formulae (3.8'), Chapter 5. This proves the theorem.

Corollary 2.1. *Under the hypotheses of the Theorem* 2.1 *the formula*

$$\operatorname{Ind} A_t = \varkappa - \sum_{l=1}^{r} m_l \tag{2.7}$$

with

$$\varkappa = \frac{1}{2\pi}\left[\arg\left\{\frac{\det B(z)}{\det A(z)} \prod_{l=1}^{r} \frac{(z-\alpha_l)^{m_l}}{(z-\beta_l)^{n_l}}\right\}\right]_{|z|=1} \tag{2.7'}$$

holds.

This follows immediately from (2.4) and (2.6) since the integers $\mu_j^{(l)}$, $\nu_j^{(l)}$ (cf. (5.2$_2$), Chapter 8) satisfy the equations

$$\sum_{j=1}^{n} \mu_j^{(l)} = m_l\,, \qquad \sum_{j=1}^{n} \nu_j^{(l)} = n_l\,.$$

3. Now we consider the paired operator $A = A_1P + A_2Q$ instead of the operator (2.3). By Lemmas 1.2 and 2.1 the difference $A - A_t = [A_1, P] + [A_2, Q]$ is a compact operator on the space E_n^∞. By Theorem 2.1 and Theorem 3.7, Chapter 1, from this we obtain the following theorem.

Theorem 2.2. *The operator A is a Φ-operator on the space E_n^∞, and its index can be calculated by the formula* (2.7).

4. In the sequel we shall show that the above assumed condition of the finiteness of the number of the zeros of the functions

$$\det A(z)\,, \qquad \det B(z) \tag{2.8}$$

is also necessary for the operator $A = A_1P + A_2Q$ (or, what is the same by no. 3, for the operator A_t) to be a Φ-operator on the space E_n^∞. Thus the following theorem holds.

Theorem 2.3. *Let $A(z)$, $B(z) \in W^{\infty}_{n \times n}$. For the paired operator $A = A_1 P + A_2 Q$ to be a Φ-operator on the space E^{∞}_n it is necessary and sufficient that each of the two functions* (2.8) *have at most finitely many zeros of integral orders on the unit circle.*

By Theorem 2.2 we have only prove the necessity of these conditions. By Theorem 1.1, Chapter 7, and by the remarks made in no. 3 moreover it is sufficient to consider the scalar operator ($n = 1$).

The proof uses the following two lemmas. By Γ we denote a closed rectifiable Jordan curve of class C^{∞} and by $C^{\infty}(\Gamma)$ the countably normed space defined already in Section 1.1 with the generating norm system

$$||f||_p = ||f||_{C^p(\Gamma)} = \sum_{j=0}^{p} \max_{t \in \Gamma} |f^{(j)}(t)| \qquad (p = 0, 1, 2, \ldots). \tag{2.9}$$

Lemma 2.2. *Let $a(t) \in C^{\infty}(\Gamma)$ and for a point $t_0 \in \Gamma$ let $a^{(k)}(t_0) = 0$ for all $k = 0, 1, 2, \ldots$. Then there exists a sequence of real valued functions $f_n \in C^{\infty}(\Gamma)$ ($n = 1, 2, \ldots$) such that the following conditions are satisfied:*

a) $0 \leqq f_n \leqq 1$, $f_n(t_0) = 1$.

b) $\operatorname{supp} f_n \underset{n \to \infty}{\to} t_0$ (i.e. *for any neighborhood $U \subset \Gamma$ of the point t_0 there exists a natural number $N(U)$ with* $\operatorname{supp} f_n \subset U$ *for all* $n > N(U)$)[1]).

c) af_n *converges to zero in the topology of the space $C^{\infty}(\Gamma)$.*

Proof. Let γ be the length of the curve Γ and $t = t(s)$ $(0 \leqq s \leqq \gamma)$ its equation. For the arc abscissa s_0 of the point $t_0 = t(s_0)$ we assume without loss of generality that $0 < s_0 < \gamma$. By $\hat{C}^{\infty}[0, \gamma]$ we denote the following subspace of the space $C^{\infty}[0, \gamma]$:

$$\hat{C}^{\infty}[0, \gamma] = \{f \in C^{\infty}[0, \gamma]: f^{(k)}(0) = f^{(k)}(\gamma)\,,\ k = 0, 1, \ldots\}\,.$$

The mapping

$$\psi: C^{\infty}(\Gamma) \to C^{\infty}[0, \gamma]$$

which associates the function $\hat{f} = f(t(s)) \in \hat{C}^{\infty}[0, \gamma]$ with each function $f \in C^{\infty}(\Gamma)$ obviously defines a topological isomorphism of the algebra $C^{\infty}(\Gamma)$ onto the algebra $\hat{C}^{\infty}[0, \gamma]$. Thus it is sufficient to prove the assertion for $\hat{C}^{\infty}[0, \gamma]$ instead of $C^{\infty}(\Gamma)$.

Let $a(s) \in \hat{C}^{\infty}[0, \gamma]$ and $a^{(k)}(s_0) = 0$ $(k = 0, 1, \ldots)$. By $f(s) \in \hat{C}^{\infty}[0, \gamma]$ we denote an arbitrary real valued function on the interval $[0, \gamma]$ with $0 \leqq f \leqq 1$, $f(s_0) = 1$ and with the property that f vanishes outside a sufficiently small neighborhood of the point s_0, e.g. outside the interval $|s - s_0| \leqq c$.

Now let $p = 0, 1, \ldots$ be an arbitrary fixed number and $A = \sup\limits_{\substack{0 \leqq s \leqq \gamma \\ 0 \leqq k \leqq p}} |f^{(k)}(s)|$. We put $f_n(s) = f(s_0 + n(s - s_0))$. Then we obtain

$$|f_n^{(k)}(s)| \leqq n^k A = O(n^k) \qquad (k = 0, 1, \ldots, p)\,.$$

[1]) By $\operatorname{supp} f$ we denote the support of the function f.

On the other hand. using the Taylor formula (e.g. with the Peano remainder) we easily find the estimates

$$|a^{(r)}(s)| = o(d^{p-r}) \qquad (r = 0, 1, \dots, p) \tag{2.10}$$

for the derivatives of the function $a(s)$ on the interval $|s - s_0| \leqq d$.

Now we estimate the norm $\|af\|_p$. Since the function $f_n(s)$ vanishes outside the interval $|s - s_0| \leqq \frac{c}{n}$, we can put $d = \frac{c}{n}$ in (2.10). Then by the Leibniz rule for any $m = 0, 1, \dots, p$ we find

$$|(af_n)^{(m)}| \leqq \sum_{k=0}^{m} \binom{m}{k} |f_n^{(k)}|\, |a^{(m-k)}| = \sum_{k=0}^{m} \binom{m}{k} O(n^k)\, o\left(\frac{1}{n^{p-m+k}}\right) = o\left(\frac{1}{n^{p-m}}\right).$$

This implies

$$\lim_{n\to\infty} \|af_n\|_p = 0 \qquad (p = 0, 1, \dots).$$

By construction the sequence $\{f_n\}$ satisfies the required conditions a)—c). This proves the lemma.

Remark 1. The sequence $\{f_n\}$ constructed above does not depend on the function $a(t)$ but only on the point $t_0 \in \Gamma$.

Remark 2. The sequence $\tilde{f}_n = \frac{1}{n^3} f_n$ has the properties b), c) and

$$\text{a}') \quad \lim_{n\to\infty} \|\tilde{f}_n\|_2 = 0\,, \qquad \tilde{f}_n^{(3)}(t_0) = 1\,.$$

Lemma 2.3. *Let $a(z) \in W^\infty$ and $A \in \mathfrak{R}(U)$ the operator with the symbol $a(z)$. We assume that for a point t_0 of the unit circle the equations*

$$a^{(k)}(t_0) = 0 \qquad (k = 0, 1, 2, \dots) \tag{2.11}$$

hold. Then there exists a sequence $\{\xi_n\}$ $(n = 1, 2, \dots)$ of elements of the space E^∞ with the following properties:

(1) *$A\xi_n \to 0$ in the space E^∞.*

(2) $\lim_{n\to\infty} \|\xi_n\|_E = 0$.

(3) *None of the sequences $\{\xi_n\}$, $\{P_1\xi_n\}$, $\{Q_1\xi_n\}$ converges to zero in the space E^∞.*

Proof. By K we denote the unit circle. First we remark that the relations

$$a(z) \in C^\infty(K)\,, \qquad C^{m+2}(K) \subset W^m \qquad (m = 0, 1, \dots) \tag{2.12}$$

hold (cf. I. M. Gelfand, D. A. Raikov and G. E. Šilov [1], § 41).

Now let $\tilde{f}_n \in C^\infty(K)$ $(n = 1, 2, \dots)$ be the sequence given in the Remark 2 to Lemma 2.2. We put

$$\tilde{f}_n(z) = \sum_{j=-\infty}^{\infty} \xi_j^{(n)} z^j \qquad (|z| = 1)\,, \qquad \xi_n = \{\xi_j^{(n)}\}_{j=-\infty}^{\infty}\,.$$

Since $\tilde{f}_n \in W^\infty$, we have $\xi_n \in l^{1,\infty}$ and consequently $\xi_n \in E^\infty$. Furthermore, by (2.12) we obtain the estimates

$$||A\xi_n||_{E^m} \leqq ||A\xi_n||_{l^{1,m}} = ||a(z)\,\tilde{f}_n(z)||_{W^m} \leqq M||a(z)\,\tilde{f}_n(z)||_{C^{m+2}(K)}\,.$$

By the condition c) of the Lemma 2.2 it follows from this that the sequence $\{\xi_n\}$ has the property (1).

The property (2) follows immediately from

$$||\xi_n||_E \leqq ||\xi_n||_{l^1} = ||\tilde{f}_n(z)||_W \leqq M||\tilde{f}_n(z)||_2$$

(cf. Remark 2 to Lemma 2.2).

Since the functions $\tilde{f}_n$ are real valued, we have $\bar{\xi}_j^{(n)} = \xi_{-j}^{(n)}$. Thus the sequences $\{\xi_n\}$, $\{P_1\xi_n\}$ and $\{Q_1\xi_n\}$ simultaneously are null sequences of the space E^∞ or not. We assume that $\xi_n \to 0$ in the topology of E^∞. Using the Hölder inequality we can easily show that then $\{\xi_n\}$ is a null sequence in the space $l^{1,3}$. But this contradicts the inequality

$$1 = |\tilde{f}_n^{(3)}(t_0)| \leqq \sum_{j=-\infty}^{\infty} (1 + |j|)^3\,|\xi_j^{(n)}| = ||\xi_n||_{l^{1,3}}\,.$$

Thus the sequence $\{\xi_n\}$ satisfies also the property (3). This completes the proof of the lemma.

Now we come to the proof of Theorem 2.3. We consider the scalar operator

$$C = AP_1 + BQ_1\,,$$

where A and $(B \in \mathfrak{R}(U))$ are the operators with symbols $a(z)$ and $b(z) \in W^\infty$, respectively. Let $C \in \Phi(E^\infty)$. As we remark already it is sufficient to show that each of the functions $a(z)$ and $b(z)$ have at most finitely many zeros of integral orders on the unit circle.

We assume that e.g. the function $a(z)$ does not satisfy this condition. Then either $a(z)$ has only finitely many zeros on the unit circle but one of these zeros is of infinite order or the set of zeros of $a(z)$ is infinite. In the last case it follows easily from Rolle's theorem of differential calculus that $a(z)$ has a zero of an infinite order at each accumulation point of this set. Thus in each case there exists a point $t_0(|t_0| = 1)$ such that the relations (2.11) hold. Let $\{\xi_n\} \subset E^\infty$ be the sequence constructed in Lemma 2.3. We obtain

$$AP_1\xi_n = P_1A\xi_n + [A, P_1]\,\xi_n\,.$$

From the estimate (2.1) it follows that

$$||[A, P_1]\,\xi_n||_{E^l} \leqq c_l||\xi_n||_E \qquad (l = 0, 1, \ldots)\,.$$

Thus by the properties of the sequence $\{\xi_n\}$ mentioned in Lemma 2.3 we obtain the convergence $AP_1\xi_n \to 0$ in the sense of the topology of the space E^∞.

Since by hypothesis C is a Φ_+-operator in E^∞, by Corollary 3.2, Chapter 1, the operator $CP_1 = AP_1$ is normally solvable. That means that the restriction $\hat{A} = A \mid \operatorname{im} P_1 : \operatorname{im} P_1 \to E^\infty$ is normally solvable. Furthermore. we have $\dim\ker \hat{A} = 0$. In fact, let

$$\hat{A}\xi = A\xi = 0\,, \qquad \xi = \{\xi_j\}_0^\infty \in P_1(E^\infty)\,.$$

We put $f(z) = \sum_0^\infty \xi_j z^j$ ($|z| = 1$). Then we obtain $a(z)\,f(z) = 0$, and by the uniqueness theorem for power series we conclude that $f(z) \equiv 0$ ($|z| = 1$), for, since $\dim\ker C < \infty$, the function $a(z)$ cannot be identically equal to zero on the unit circle. This yields $\xi = 0$.

Consequently by Lemma 2.3, Chapter 1, the operator $\hat{A}$ has a continuous inverse, and since $P\hat{A}_1\xi_n \to 0$ we can conclude that $P_1\xi_n \to 0$ (in the sense of the topology of E^∞). But this contradicts the assertion (3) of Lemma 2.3. Thus we have shown that $a(z)$ has at most finitely many zeros of integral orders on the unit circle. In an analogous way we can prove this statement for the function $b(z)$.

This concludes the proof of Theorem 2.3.

Remark. In the preceding proof we have proved even more than the assertion of Theorem 2.3. Namely, the conditions formulated in Theorem 2.3 are necessary for $A \in \Phi_+(E_n^\infty)$.

9.2.3. Equations in Distribution Spaces

Now we apply the considerations from Section 9.1 to the paired operators

$$A = A_1P + A_2Q\,, \qquad A_t = PA_1 + QA_2\,,$$

where A_1 and A_2 have the same meaning as in 9.2.2, no. 2. As the spaces X and E (in the notations of 9.1) we choose $l_n^{p,\infty}$ and l_n^p ($1 \leq p < \infty$), respectively. Then

$$H = l_n^{p'}\,, \qquad X^* = l_n^{p,-\infty} = \bigcup_{k=0}^{\infty} l_n^{p',-k}\,,$$

where $p' = \dfrac{p}{p-1}$ for $p > 1$ and $p' = \infty$ for $p = 1$.

The value of the functional

$$f = (f^{(1)}, f^{(2)}, \ldots, f^{(n)}) \in l_n^{p'}\,, \qquad f^{(m)} = \{f_j^{(m)}\}_{j=-\infty}^{\infty}$$

at the element $\xi = (\xi^{(1)}, \xi^{(2)}, \ldots, \xi^{(n)}) \in l_n^p$ is determined by the formula

$$\langle \xi, f\rangle = \sum_{m=1}^{n} \sum_{j=-\infty}^{\infty} \xi_j^{(m)} f_j^{(m)} \tag{2.13}$$

(cf. also formula (1.12), Chapter 3). Using the commutation formula (1.13), Chapter 3, and a wellknown property of transposed matrices we easily obtain the relation

$$\langle \xi, Af\rangle = \langle A'\xi, f\rangle \qquad (\xi \in l_n^{p,\infty}) \tag{2.14}$$

for any $f \in l_n^{p'}$. Here A' is the operator given by the equation system

$$\left.\begin{aligned} \sum_{k=-\infty}^{\infty} a'_{k-j}\xi_k = \eta_j \qquad (j = 0, 1, \ldots)\,, \\ \sum_{k=-\infty}^{\infty} b'_{k-j}\xi_k = \eta_j \qquad (j = -1, -2, \ldots)\,, \end{aligned}\right\}$$

where a'_k is the transposed matrix of a_k. According to 9.1.2 we extend the operator A by the formula (2.14) to the space $X^* = l_n^{p', -\infty}$.

Remark. Since every functional $f \in X^*$ belongs to one of the spaces $l_n^{p', -k}$ $(k = 0, 1, \ldots)$ and moreover $A(z),\ B(z) \in W_{n\times n}^{\infty}$, the operator A can also be immediately defined by the left-hand side of the equation (4.6), Chapter 7. On the other hand, the value of the functional $f \in X^*$ on $\xi \in X$ can also be defined by the relation (2.13). Then by a simple calculation we can directly check that the relation (2.14) holds for any $f \in X^*$. Thus the operator A defined in this way coincides with the operator $A = (A'_X)^*$ introduced above.

The symbol of the operator A' is the matrix function

$$A'\left(\frac{1}{z}\right)\frac{1+\theta}{2} + B'\left(\frac{1}{z}\right)\frac{1-\theta}{2} \qquad (|z| = 1,\ \theta = \pm 1)\,.$$

By Section 8.1 the matrix functions $A(z)$ and $B(z)$ can be represented in the form

$$A(z) = C_0(z)\,D_1(z)\,, \qquad B(z) = D_0(z)\,D_2(z)\,.$$

Here $D_1(z)$, $D_2(z)$ are the polynomial matrices defined by (5.4), Chapter 8, and $C_0(z),\ D_0(z) \in W_{n\times n}^{\infty}$ satisfy the condition (2.5). Thus we obtain

$$A'\left(\frac{1}{z}\right) = \widetilde{B}_1(z)\,\widetilde{A}_0(z)\,, \qquad B'\left(\frac{1}{z}\right) = \widetilde{B}_2(z)\,\widetilde{B}_0(z)$$

with

$$\widetilde{B}_j(z) = D'_j\left(\frac{1}{z}\right) \qquad (j = 0, 1, 2)\,, \qquad \widetilde{A}_0(z) = C'_0\left(\frac{1}{z}\right).$$

Hence for the operator A' the representation (1.11) holds, where

$$B = P\widetilde{B}_1 + Q\widetilde{B}_2\,, \qquad C = P\widetilde{A}_0 + Q\widetilde{B}_0\,. \tag{2.15}$$

The next theorem follows immediately from Theorems 1.1, 2.1 and 2.3.

Theorem 2.4. *For the operator $A = A_1P + A_2Q$ to be a Φ-operator on the distribution space $l_n^{p,-\infty}$ $(1 \leq p < \infty)$ it is necessary and sufficient that each of the functions (2.8) has at most finitely many zeros of integral orders.*

If this condition is satisfied, then

$$\dim \ker_* A = \sum_{\varkappa_j > 0} \varkappa_j\,, \qquad \dim \operatorname{coker}_* A = -\sum_{\varkappa_j < 0} \varkappa_j\,,$$

where $\varkappa_j$ $(j = 1, 2, \ldots, n)$ are right partial indices of the matrix function $\widetilde{A}_0(z) \times \widetilde{B}_0^{-1}(z)$.

Corollary 2.2. *Under the hypotheses of no. 2 of 9.2.2 for the index of the operator A in $l_n^{p,-\infty}$ the formula*

$$\mathrm{Ind}_* A = \varkappa + \sum_{l=1}^{r} n_l$$

holds with the number $\varkappa$ defined by (2.7′).

Analogous results can be obtained for the operator A_t on the space $l_n^{p,-\infty}$.

It follows from the proof of Theorem 2.1 that the operators (2.15) satisfy conditions (1)—(3) of 9.1.4. If we put

$$D = B' = D_1 P + D_2 Q , \qquad D^{(-1)} = D_1^{-1} P + D_2^{-1} Q ,$$

then moreover the relations (1.9) hold. Thus by Theorem 1.2 we obtain the following theorem.

Theorem 2.5. *The generalized solutions $\varphi \in l_n^{p,-\infty}$ of the homogeneous equation*

$$A\varphi = 0$$

coincide with the solution of this equation in the space $\tilde{l}_n^{p'} = D^{(-1)}(l_n^{p'})$.

Remark. If $n = 1$ and the zeros of the functions $A(z)$ and $B(z)$ are distinct, then the operators A and A_t are at least one-sided invertible in the space E^∞ (resp. in the space $l^{p,-\infty}$, $1 \leqq p < \infty$).

In fact, since the operator A (resp. A_t) has the same index in the spaces E^∞ and E, in these spaces the kernels also coincide. This together with Theorem 3.5, Chapter 5, proves the assertion.

9.3. Wiener-Hopf Integral Equations

9.3.1. Some Function Spaces

Now let E be any one of the spaces

$$L^p \ (1 \leqq p < \infty) , \qquad C^0 \subset C \subset M^u \subset M^c \subset M \tag{3.1}$$

introduced in Section 3.2.

For any integer k by E^k we denote the Banach space of all functions $f(x)$ defined on the real line for which $(x + i)^k f(x) \in E$. We define the norm in E^k by

$$\|f\|_k = \|(x + i)^k f(x)\|_E .$$

The countably normed space E^∞ and the distribution space $E^{-\infty}$ are introduced analogously to 9.2.1. In the case $E = L^p$ $(1 \leqq p < \infty)$ we again obtain $E^{-\infty} = \bigcup_{k=0}^{\infty} E_*^{-k}$ with

$$E_*^{-k} = L^{p',-k} \left(p' = \frac{p}{p-1}, \quad L^\infty = M \right).$$

By $\mathfrak{L}^\infty$ we denote the algebra $\mathfrak{L}^\infty = \bigcap_{m=0}^{\infty} \mathfrak{L}^m$ (cf. 5.2.3. no. 3).

9.3.2. Equations of the Second Kind

1. By P, Q we denote the projections in the space E_n (resp. E_n^∞) defined in 8.4.1 and by A_1, A_2 the matrix operators of convolution type with the symbols $\mathcal{A}_1(\lambda)$, $\mathcal{A}_2(\lambda) \in \mathfrak{L}_{n\times n}^\infty$:

$$A_j\varphi = c_j\varphi(t) - \int_{-\infty}^{\infty} k_j(t-s)\,\varphi(s)\,ds \qquad (j = 1, 2)\,,$$

where $k_j(t) \in L_{n\times n}^{1,\infty}$ and $\det c_j \neq 0$ $(j = 1, 2)$.

In the space E_n^∞ we consider the paired operator A_t of the form (2.3) given by the paired equations (4.7′), Chapter 7. Completely analogously to 9.2.2 we can prove that A_t is continuous on E_n and the commutator $[P, A_j]$ is a compact operator from E_n into the space E_n^l $(l = 0, 1, \ldots)$.

Now we assume that on the real line the determinants

$$\det \mathcal{A}_1(\lambda)\,, \qquad \det \mathcal{A}_2(\lambda) \qquad (-\infty \leqq \lambda \leqq \infty) \tag{3.3}$$

possess finitely many zeros of integral orders m_l and n_l $(l = 1, \ldots, r)$ at the points α_l and $\beta_l \in (-\infty, \infty)$, respectively. Then by 8.1.5 the matrix function $\mathcal{A}_1(\lambda)$, $\mathcal{A}_2(\lambda)$ can be represented in the form

$$\mathcal{A}_1(\lambda) = B_1(\lambda)\,\mathcal{A}_0(\lambda)\,, \qquad \mathcal{A}_2(\lambda) = B_2(\lambda)\,\mathcal{B}_0(\lambda)\,.$$

Here $B_1(\lambda)$, $B_2(\lambda)$ are the matrix functions defined by equations (4.1), Chapter 8, $\mathcal{A}_0(\lambda)$, $\mathcal{B}_0(\lambda) \in \mathfrak{L}_{n\times n}^\infty$ and

$$\det \mathcal{A}_0(\lambda) \neq 0\,, \qquad \det \mathcal{B}_0(\lambda) \neq 0 \qquad (-\infty \leqq \lambda \leqq \infty)\,. \tag{3.4}$$

Using the results of Sections 8.4 and 5.4 by means of the same considerations as in the preceding section we obtain the following theorems.

Theorem 3.1. *The operator $A_t = PA_1 + QA_2$ is a Φ-operator in the space E_n^∞, and the formulae* (2.6) *hold, where $\varkappa_j$ $(j = 1, \ldots, n)$ are the right partial indices of the matrix function $\mathcal{A}_0(\lambda)\,\mathcal{B}_0^{-1}(\lambda) \in \mathfrak{L}_{n\times n}^\infty$.*

Theorem 3.2. *The operator $A = A_1P + A_2Q$ is a Φ-operator on the space E_n^∞, and*

$$\operatorname{Ind} A = \operatorname{Ind} A_t = \varkappa - \sum_{l=1}^{r} m_l\,,$$

where

$$\varkappa = \frac{1}{2\pi}\left[\arg\left\{\frac{\det \mathcal{A}_2(\lambda)}{\det \mathcal{A}_1(\lambda)} \prod_{l=1}^{r} \frac{(\lambda-\alpha_l)^{m_l}}{(\lambda-\beta_l)^{n_l}}\right\}\right]_{\lambda=-\infty}^{\infty}. \tag{3.5}$$

Theorem 2.3 also has a corresponding analogue. We confine ourselves to the case of the space L^p $(1 \leqq p < \infty)$.

Theorem 3.3. *Let $\mathcal{A}_1(\lambda)$, $\mathcal{A}_2(\lambda) \in \mathfrak{L}_{n\times n}^\infty$. For the operator $A = A_1P + A_2Q$ to be a Φ-operator on the space $L_n^{p,\infty}$ $(1 \leqq p < \infty)$ it is necessary and sufficient that each of the two functions* (3.3) *has at most finitely many zeros of integral orders on the real line.*

We have only to prove the necessity of the conditions formulated in the theorem. Furthermore, as in the proof of Theorem 2.3 we can confine ourselves to the case $n = 1$. First we state a lemma analogous to Lemma 2.3.

Lemma 3.1. *Let $k(t) \in L^{1,\infty}$, let be K the scalar convolution operator*

$$K\varphi = \varphi(s) - \int_{-\infty}^{\infty} k(t-s)\, \varphi(s)\, ds$$

and

$$\mathcal{K}(\lambda) = 1 - \int_{-\infty}^{\infty} e^{i\lambda t}\, k(t)\, dt \qquad (-\infty < \lambda < \infty)\,.$$

Assume that for a point t_0 $(-\infty < t_0 < \infty)$ the equations

$$\mathcal{K}^{(l)}(t_0) = 0 \qquad (l = 0, 1, 2, \ldots)$$

hold. Then there exists a sequence of functions $\varphi_n \in L^{p,\infty}$ $(1 \leqq p < \infty; n = 1, 2, \ldots)$ with the following properties:

(1) $K\varphi_n \to 0$ *in the space* $L^{p,\infty}$.

(2) $\lim\limits_{n\to\infty} ||\varphi_n||_{L^p} = 0$.

(3) *None of the sequences* $\{\varphi_n\}$, $\{P\varphi_n\}$, $\{Q\varphi_n\}$ *converges in the space* $L^{p,\infty}$.

Proof. Let $\Phi(t)$ be a real valued function infinitely differentiable on $(-\infty, \infty)$ with $0 \leqq \Phi \leqq 1$ and $||\Phi||_{L^2} = 1$. Furthermore, let $\Phi(t)$ vanish outside a finite interval of the form $|t - t_0| \leqq c$. We put $\Phi_n(t) = \frac{1}{n}\Phi(t_0 + n(t - t_0))$ $(n = 1, 2, \ldots)$. By φ_n, φ we denote the inverse Fourier transforms of the functions Φ_n, Φ, respectively. The function $\Phi_n(t)$ vanishes outside the interval $|t - t_0| \leqq \frac{c}{n}$.

We have

$$\varphi_n(x) = \frac{1}{n^2}\,\varphi\left(\frac{x}{n}\right) e^{-it_0x\left(1-\frac{1}{n}\right)}. \tag{3.6}$$

This immediately implies the property (2). Furthermore, we obtain $||\varphi_n||_{L^p,k} \to \infty$ for $k \geqq 2$ and $n \to \infty$. Hence $\{\varphi_n\}$ is not a convergent sequence of the space $L^{p,\infty}$. The same statement also holds for the sequences $\{P\varphi_n\}$ and $\{Q\varphi_n\}$. For, since the function Φ is real valued, the equality $\overline{\varphi(x)} = \varphi(-x)$ holds. Thus property (3) also holds.

Next we show that for any p, $1 \leqq p < \infty$, and $k = 0, 1, 2, \ldots$ the norm $||K\varphi_n||_{L^p,k}$ can be estimated by the norm $||K\varphi_n||_{L^2,m}$ with $m = m(k)$. Since φ is the inverse Fourier transform of a test function, φ is bounded. Hence by (3.6) we have

$$|\varphi_n(x)| \leqq C\,, \qquad |K\varphi_n| \leqq C \qquad (n = 1, 2, \ldots)$$

simultaneously for all $x \in (-\infty, \infty)$.

Thus for $2 \leqq p < \infty$ we obtain the estimate

$$||K\varphi_n||_{L^p,k} = ||(x+i)^k K\varphi_n||_{L^p} \leqq C^{(p-2)/p} ||K\varphi_n||^{2/p}_{L^2,m},$$

where m can be chosen as the least integer with $m \geqq \frac{kp}{2}$.

Using the Hölder inequality for $1 \leqq p < 2$ and any function $f \in L^{p,k}$ we obtain the estimate

$$||f||_{L^p,k} \leqq C' ||f||_{L^2,m},$$

where C' is a constant independent of f and m the least integer with $m \geqq k + \frac{1}{p}$.

It follows from the preceding considerations that the property (1) and thus also the lemma is proved if we can show that

$$\lim_{n\to\infty} ||K\varphi_n||_{L^2,m} = 0 \qquad (m = 0, 1, 2, \ldots). \tag{3.7}$$

We have

$$||K\varphi_n||_{L^2,m} \leqq C_m \sum_{l=0}^{m} ||(it)^l K\varphi_n||_{L^2}.$$

Using the facts that the Fourier operator F is isometric and that the functions Φ_n are test functions we obtain

$$||(it)^l K\varphi_n||_{L^2} = ||F[(it)^l K\varphi_n]||_{L^2} =$$
$$= ||[\mathcal{K}(\lambda)\, \Phi_n(\lambda)]^{(l)}||_{L^2} \leqq C \max_{\lambda} |[\mathcal{K}(\lambda)\, \Phi_n(\lambda)]^{(l)}|.$$

Repeating the arguments in the proof of Lemma 2.2 we now see that the right-hand side of the last inequality tends to zero for $n \to \infty$. Hence (3.7) holds. This completes the proof of the lemma.

The rest of the argument is the same as in the proof of Theorem 2.3. In the present case the validity of the relation $\dim \ker \hat{A} = 0$ ($\hat{A} = A \mid \operatorname{im} P$) can easily be shown by Fourier transformation. This proves Theorem 3.3.

2. Under the hypotheses of no. 1 we consider the operator $A = A_1 P + A_2 Q$ on the distribution space

$$L_n^{p,-\infty} = \bigcup_{k=0}^{\infty} L_n^{p',-k} \qquad \left(1 \leqq p < \infty;\ p' = \frac{p}{p-1}\right).$$

Using the notations of Section 9.1 we now have to put $E = L_n^p$ and $H = L_n^{p'}$.

The value of the functional $f \in L_n^{p'}$ at the element $\varphi \in L_n^p$ is defined by

$$\langle \varphi, f \rangle = \int_{-\infty}^{\infty} \varphi(t)\, f(t)\, dt, \tag{3.8}$$

where $\varphi(t)\, f(t)$ is the scalar product of the n-component vectors $\varphi(t)$ and $f(t)$ in the sense of Euclidean space. By formula (2.23), Chapter 3, we have

$$\langle \varphi, Af \rangle = \langle A'\varphi, f \rangle \qquad (\varphi \in L_n^{p,\infty}) \tag{3.9}$$

for any $f \in L_n^{p'}$. Here A' is the paired operator[1])

$$A'\varphi = \begin{cases} c_1'\varphi(t) - \int\limits_{-\infty}^{\infty} k_1'(s-t)\, \varphi(s)\, dt\,, & t > 0 \\ c_2'\varphi(t) - \int\limits_{-\infty}^{\infty} k_2'(s-t)\, \varphi(s)\, dt\,, & t < 0\,. \end{cases}$$

By formula (3.9) the operator A is defined for any distribution $f \in L_n^{p',-\infty}$.

All results stated in 9.2.3 can be carried over to the operator A. In particular the corresponding analogues of Theorems 2.4 and 2.5 hold for A in the distribution space $L_n^{p,-\infty}$. Analogous statements hold for the operator $A_t = PA_1 + QA_2$.

Remark. If $n = 1$ and the zeros of the functions $\mathcal{A}_1(\lambda)$ and $\mathcal{A}_2(\lambda)$ are distinct, then the operators A and A_t are at least one-sided invertible in the space E^∞ (resp. in the space $L^{p,-\infty}$, $1 \leqq p < \infty$).

9.3.3. Equations of the First Kind

Now we consider the paired operators

$$A = A_1P + A_2Q\,, \qquad A_t = PA_1 + QA_2$$

under the hypothesis that the operators A_j ($j = 1, 2$) defined by (3.2) satisfy the following conditions

$$\det c_j = 0\,, \qquad \det \mathcal{A}_j(\lambda) \neq 0 \qquad (-\infty < \lambda < \infty)\,.$$

The kernels $k_j(t)$ shall be infinitely differentiable and together with all their derivatives absolutely integrable.

1. Let E be any one of the function spaces (3.1) and $k \geqq 0$ an integer. By E^k we denote the space of all functions on $(-\infty, \infty)$ which together with their derivatives $f^{(l)}(t)$ ($l = 0, 1, \ldots, k-1$) are absolutely continuous on every finite interval, where $f^{(l)} \in E$ ($l = 0, 1, \ldots, k$). Equipped with the norm

$$\|f\|_k = \sum_{l=0}^{k} \|f^{(l)}\|_E$$

E^k becomes a Banach space. By E^∞ we denote the countably normed space $E^\infty = \bigcap\limits_{k=0}^{\infty} E^k$ and by $E^{-\infty}$ the conjugate space of E^∞. In the case of the space $E = L^p$ the space $E^{-\infty}$ consists of distributions representing finite sums of derivatives of different orders of functions in $L^{p'}$ ($p' = p/(p-1)$).

[1]) Here and in the sequel a primed matrix always denotes the transposed matrix.

2. It is easily seen that A and A_t are continuous operators on the space E_n^∞. Moreover, a corresponding analogue of Lemma 2.1 follows from the proof of Lemma 4.3, Chapter 5.

In the sequel we assume that the functions (3.3) have a zero of finite (necessarily integral) order μ, ν, respectively, at the point $\lambda = \infty$. Then by Section 8.1 we can represent the matrix functions $\mathcal{A}_1(\lambda)$, $\mathcal{A}_2(\lambda)$ in the form

$$\mathcal{A}_1(\lambda) = B_1^{(\infty)}(\lambda)\,\mathcal{A}_0(\lambda)\,, \qquad \mathcal{A}_2(\lambda) = B_2^{(\infty)}(\lambda)\,\mathcal{B}_0(\lambda)\,,$$

where $B_1^{(\infty)}(\lambda)$ and $B_2^{(\infty)}(\lambda)$ are defined by formulae (4.7), Chapter 8. The matrix functions $\mathcal{A}_0(\lambda)$, $\mathcal{B}_0(\lambda)$ have the same differentiability properties as $\mathcal{A}_j(\lambda)$ $(j = 1, 2)$ and satisfy conditions (3.4).

In the present case we prove Theorem 3.1 in the same way as in 9.2.2 and 9.3.2. Also in this way we obtain the following theorem.

Theorem 3.2*. *The operators A and A_t are Φ-operators in the space E_n^∞, and the index formula*

$$\operatorname{Ind} A = \operatorname{Ind} A_t = \varkappa - \mu$$

holds with

$$\varkappa = \frac{1}{2\pi}\left[\arg\left\{\frac{\det \mathcal{A}_2(\lambda)\,(\lambda+i)^\nu}{\det \mathcal{A}_1(\lambda)\,(\lambda+i)^\mu}\right\}\right]_{\lambda=-\infty}^{\infty}.$$

By means of similar arguments as in the preceding Subsections we can prove the operator $A(A_t)$ is a Φ-operator on the space $L_n^{p,\infty}$ $(2 \leqq p < \infty)$ *only if* each of the functions (3.3) has a zero of finite order at the point at infinity.

3. The considerations of no. 2 of the preceding Subsection can be carried over word-for-word to equations of the first kind.

9.4. Singular Integral Equations

9.4.1. Lemmas

Now let Γ be a closed curve system of the form described in 6.2.1 consisting of curves of the class C^∞. By $C^\infty(\Gamma)$ we denote the countably normed space of all (complex valued) functions infinitely differentiable on Γ. The topology in the space $C^\infty(\Gamma)$ is defined by the countable norm system (2.9). $C^\infty(\Gamma)$ is a perfect countably normed space and thus reflexive[1]).

---

[1]) A countably normed space is said to be *perfect* if in this space all bounded sets are relatively compact. The perfectness of the space $C^\infty(\Gamma)$ can easily be proved by means of Arzelà's theorem (cf. I. M. Gelfand and G. E. Šilov [2], p. 47).

$C_n^\infty(\Gamma)$ is the countably normed space of all n-dimensional vectors $\varphi(t) = \{\varphi_j(t)\}_1^n$ whose components are elements of the space $C^\infty(\Gamma)$ with the norm system

$$\|\varphi\|_p = \sum_{j=1}^{n} \|\varphi_j\|_p \qquad (p = 0, 1, 2, \ldots)\,.$$

By $C_n^{-\infty}(\Gamma)$ we denote the conjugate space of $C_n^\infty(\Gamma)$.

In the sequel we consider the singular integral operator

$$\mathfrak{A} = C(t)\,P + D(t)\,Q + T \tag{4.1}$$

with the operators P, Q defined in 7.5 and the matrix functions $C(t), D(t) \in C_{n\times n}^\infty(\Gamma)$. First let T be an arbitrary compact operator on the space $C_n^\infty(\Gamma)$. Let $\varphi(t) \in C_n^\infty(\Gamma)$. From

$$S\varphi = \frac{1}{\pi i}\int_\Gamma \frac{\varphi(\tau) - \varphi(t)}{\tau - t}\,d\tau + \varphi(t)$$

(cf. 7.5) it follows immediately that

$$\|S\varphi\|_p \leqq M_p\|\varphi\|_{p+1} \qquad (M_p = \text{const}, p = 0, 1, \ldots)\,.$$

This means that the operator S is continuous on the space $C_n^\infty(\Gamma)$. Hence $\mathfrak{A}$ is also a continuous operator in $C_n^\infty(\Gamma)$.

Lemma 2.1. *Let $T \in \mathscr{L}(L_n^p(\Gamma), C_n^\infty(\Gamma))$ and $f(t) \in C_n^\infty(\Gamma)$. If the operator A is of normal type, then each solution $\varphi(t) \in L_n^p(\Gamma)$ of the equation*

$$\mathfrak{A}\varphi = f \tag{4.2}$$

belongs to $C_n^\infty(\Gamma)$.

Proof. We put $\mathfrak{B} = C^{-1}P + D^{-1}Q$. Applying the operator $\mathfrak{B}$ to both sides of (4.2) we obtain $\varphi + T_1\varphi = \mathfrak{B}f$. It is easily seen that $T_1 \in \mathscr{L}(L_n^p(\Gamma), C_n^\infty(\Gamma))$. Hence $\varphi = \mathfrak{B}f - T_1\varphi \in C_n^\infty(\Gamma)$.

Lemma 2.2. *Let $k(t, \tau) \in C^\infty(\Gamma \times \Gamma)$. The operator K defined on the space $C^\infty(\Gamma)$ by*

$$K\varphi = \int_\Gamma k(t, \tau)\,\varphi(\tau)\,d\tau$$

is compact.

The assertion follows immediately from Lemma 1.2 since by Arzelà's theorem it is easily checked that $K \in \mathscr{K}(C(\Gamma), C^p(\Gamma))$ for all $p = 0, 1, 2, \ldots$.

9.4.2. Equations in the Basic Space

Now we assume that the determinants

$$\det C(t)\,, \qquad \det D(t) \tag{4.3}$$

have finitely many zeros of integral orders $m_l, n_l \geqq 0$ $(l = 1, \ldots, r)$ at the points α_l, β_l, respectively, on Γ. Under these hypotheses the matrix functions

$C(t)$ and $D(t)$ can be represented in the form

$$C(t) = B_1(t)\, A_0(t)\,, \qquad D(t) = B_2(t)\, B_0(t)\,,$$

where the polynomial matrices $B_1(t)$, $B_2(t)$ are given by formulae (5.2), Chapter 8, and $A_0(t)$, $B_0(t) \in C^\infty_{n\times n}(\Gamma)$ satisfy the condition (2.5) (cf. 8.1).

We obtain:

Theorem 4.1. *The operator $\mathfrak{A}_t = PC + QD$ is a Φ-operator on the space $C^\infty_n(\Gamma)$, and formulae* (2.6) *hold, where $\varkappa_j$ $(j = 1, \dots, n)$ are the right partial indices of the matrix function $A_0(t)\, B_0^{-1}(t) \in C^\infty_{n\times n}(\Gamma)$.*

By means of Lemma 2.1 and Corollary 2.4, Chapter 6, the proof is carried out completely analogously to the proof of Theorem 2.1.

The next theorem follows from Theorem 4.1 and Lemma 2.2.

Theorem 4.2. *The operator $\mathfrak{A}$ is a Φ-operator on the space $C^\infty_n(\Gamma)$, and for its index the formula*

$$\operatorname{Ind} \mathfrak{A} = \varkappa - \sum_{l=1}^{r} m_l \tag{4.4}$$

with

$$\varkappa = \frac{1}{2\pi}\left[\arg\left\{\frac{\det D(t)}{\det C(t)} \prod_{l=1}^{r} \frac{(t-\alpha_l)^{m_l}}{(t-\beta_l)^{n_l}}\right\}\right]_\Gamma \tag{4.4'}$$

holds.

Now we prove the corresponding analogue to Theorem 2.3.

Theorem 4.3. *Let $C(t), D(t) \in C^\infty_{n\times n}(\Gamma)$. For the operator* (4.1) *to be a Φ-operator on the space $C^\infty_n(\Gamma)$, it is necessary and sufficient that each of the two functions* (4.3) *has at most finitely many zeros of integral orders on Γ.*

Sufficiency follows from Theorem 4.2. By Theorem 4.1, Chapter 7, and Lemma 2.1 it is sufficient to prove the necessity of the formulated conditions for $n = 1$. Moreover, we can put $T = 0$ (cf. Theorem 3.7, Chapter 1). Thus we assume that the scalar operator

$$A\varphi = a(t)\,\varphi(t) + \frac{b(t)}{\pi i}\int_\Gamma \frac{\varphi(\tau)}{\tau - t}\,d\tau \qquad (t \in \Gamma) \tag{4.5}$$

with $a(t), b(t) \in C^\infty(\Gamma)$ is a Φ-operator on the space $C^\infty(\Gamma)$. We have to prove that the two functions

$$c(t) = a(t) + b(t)\,, \qquad d(t) = a(t) - b(t)$$

have at most finitely many zeros of integral orders on Γ.

Suppose that the function $c(t)$ does not satisfy this condition. Then there exists a point $t_0 \in \Gamma$ such that $c^{(k)}(t_0) = 0$ for all $k = 0, 1, \dots$. By $f_n \in C^\infty(\Gamma)$ we denote the sequence of real valued functions constructed in Lemma 2.2. Then

$$cPf_n = Pcf_n + (cP - Pc)\, f_n\,.$$

Since the kernel of the integral operator $cP - Pc$ is infinitely differentiable, by the properties b) and c) mentioned in Lemma 2.2 we obtain the convergence $cPf_n \to 0$ in the sense of the topology of the space $C^\infty(\Gamma)$. Repeating the arguments used at the end of the proof of Theorem 2.3 we conclude that $Pf_n \to 0$ (in $C^\infty(\Gamma)$).

Now we assume that $\Gamma = K$ is the unit circle. Then we obtain a contradiction to the properties a) and b) of the sequence $\{f_n\}$. In fact, for a real valued function $f(t) \in C^\infty(K)$ the relations

$$f(t) = \sum_{j=-\infty}^{\infty} \xi_j t^j, \qquad Pf = \sum_{j=0}^{\infty} \xi_j t^j, \qquad Qf = \sum_{j=-\infty}^{-1} \xi_j t^j \qquad (|t| = 1)$$

imply

$$\overline{\xi_j} = \xi_{-j}, \qquad Pf = \overline{Qf} + \xi_0 \qquad (\xi_0 \text{ real}).$$

Hence for the sequence $\{f_n\}$ we have

$$f_n - c_n = Pf_n + \overline{Pf_n} \qquad (c_n \text{ real}; n = 1, 2, \ldots). \tag{4.6}$$

Since $Pf_n \to 0$ and moreover $0 \leqq f_n \leqq 1$, we can conclude from (4.6) that there exists a convergent subsequence $\{f_{n_k}\} \subset C^\infty(K)$. But this is not possible because of the specified properties of the sequence $\{f_n\}$. It follows from this contradiction that in the case $\Gamma = K$ the function $c(t)$ can possess at most finitely many zeros of integral orders. Analogously we can prove the same assertion for the function $d(t)$.

Now we consider the case of an arbitrary curve system Γ. By $\Gamma_0 \subset \Gamma$ we denote that closed curve which contains the point t_0. Furthermore, let A_0 be the operator which we obtain if we replace Γ by Γ_0 in (4.5). Obviously, A_0 is a Φ-operator on the space $C^\infty(\Gamma_0)$. There exists a one-to-one mapping between the points $t \subset \Gamma_0$ and the points $s \in K$:

$$t = t(s). \tag{4.7}$$

Here $t(s) \in C^\infty(K)$ and $t'(s) \neq 0$ $(s \in K)$. The mapping (4.7) shows that to the operator A_0 there corresponds the following operator on the space $C^\infty(K)$:

$$\widetilde{A}_0\varphi = a(s)\,\varphi(s) + \frac{b(s)}{\pi i}\int_K \frac{\varphi(\sigma)}{\sigma - s}\,d\sigma + \int_K k(s,\sigma)\,\varphi(\sigma)\,d\sigma \qquad (s \in K).$$

Here $a(s) = a(t(s))$, $b(s) = b(t(s)) \in C^\infty(K)$ and $k(s, \sigma) \in C^\infty(K \times K)$ (cf. 3.4.3). Since A_0 is a Φ-operator, $\widetilde{A}_0$ is a Φ-operator in the space $C^\infty(K)$. But by the earlier part of the proof this contradicts the relations $c^{(k)}(s_0) = 0$ $(k = 0, 1, \ldots;$ $t_0 = t(s_0))$. Hence $c(t)$ has at most finitely many zeros of integral orders on the curve system Γ. The assertion with respect to $d(t)$ is proved simularly.

This completes the proof of Theorem 4.3.

Remark. In the preceding proof there was actually proved that the conditions mentioned in Theorem 4.3 are even necessary for the operator $\mathfrak{A}$ to be a Φ_+-operator in the space $C_n^\infty(\Gamma)$. A corresponding assertion holds with respect to the Φ_--operator (see U. Köhler and B. Silbermann [2]).

9.4.3. Equations in the Distribution Space

Now we apply the results from Section 9.1 to the operator (4.1), where the compact operator T is an integral operator of the form

$$T\varphi = \int_\Gamma T(t, \tau)\, \varphi(\tau)\, d\tau \quad \text{with} \quad T(t, \tau) \in C^\infty_{n\times n}(\Gamma \times \Gamma)$$

and the matrix functions $C(t)$, $D(t)$ satisfy the conditions mentioned in 9.4.2. We put (in the notation of 9.1) $X = C^\infty_n(\Gamma)$. For E we choose the Hilbert space

$$E = H = L^2_n(\Gamma)\,.$$

In this case with each element $f(t) \in L^2_n(\Gamma)$ we associate a regular functional $f \in X = C^{-\infty}_n(\Gamma)$ by the formula

$$\langle \varphi, f\rangle = \int_\Gamma \varphi(t)\, f(t)\, dt \qquad (\varphi \in C^\infty_n(\Gamma))\,. \tag{4.8}$$

Thus the embedding $E \subset X^*$ is fixed. Using the commutation formula (4.22) from Chapter 3 (cf. also 3.4.7) and the formula (4.8) now we obtain

$$\langle \varphi, \mathfrak{A} f\rangle = \langle \mathfrak{A}'\varphi, f\rangle \qquad (\varphi \in C^\infty_n(\Gamma)) \tag{4.9}$$

for any $f(t) \in L^2_n(\Gamma)$. Here $\mathfrak{A}'$ is the integral operator

$$\mathfrak{A}' = PD' + QC' + T'$$

with

$$T'\varphi = \int_\Gamma T'(\tau, t)\, \varphi(\tau)\, d\tau\,.$$

By (4.9) the operator $\mathfrak{A}$ is defined for any $f \in C^{-\infty}_n(\Gamma)$.

The next theorem follows from Theorems 1.1, 4.2 and 4.3.

Theorem 4.4. *For the operator $\mathfrak{A}$ to be a Φ-operator on the distribution space $C^{-\infty}_n(\Gamma)$ it is necessary and sufficient that each of the two functions* (4.3) *has at most finitely many zeros of integral orders on Γ.*

If this condition is satisfied, then for the index of the operator $\mathcal{A}$ the formula

$$\operatorname{Ind}_* \mathfrak{A} = \varkappa + \sum_{l=1}^{r} n_l$$

holds with the number $\varkappa$ defined by (4.4′).

In the sequel we shall consider in particular the operator

$$\mathfrak{A}_0 = C(t)\, P + D(t)\, Q\,.$$

By 8.1 the matrix functions $C(t)$ and $D(t)$ are representable in the form (5.7), Chapter 8. By $\widetilde{L}^2_n(\Gamma)$ we denote the space constructed in 8.5.1, no. 2 (for the case $p = 2$).

Repeating the arguments in 9.2.3 we easily obtain the following theorem from Theorem 4.1.

Theorem 4.5. *For the operator* $\mathfrak{A}_0$ *considered on the distribution space* $C_n^{-\infty}(\Gamma)$ *the formulae*

$$\dim \ker_* \mathfrak{A}_0 = \sum_{\varkappa_j > 0} \varkappa_j , \qquad \dim \operatorname{coker}_* \mathfrak{A}_0 = - \sum_{\varkappa_j < 0} \varkappa_j$$

hold, where $\varkappa_j$ $(j = 1, \dots, n)$ *are right partial indices of the matrix function* $D_0'(t) [C_0'(t)]^{-1}$.

Using Lemma 2.1 and Corollary 2.4, Chapter 6, we obtain the following assertion analogously to Theorem 2.5.

Theorem 4.6. *The generalized solutions* $\varphi \in C_n^{-\infty}(\Gamma)$ *of the homogeneous equation*

$$\mathfrak{A}_0 \varphi = 0$$

coincide with the solutions of this equation in the space $\widetilde{L_n^2}(\Gamma)$.

Remark. If $n = 1$ and the zeros of the functions $C(t)$ and $D(t)$ are distinct, then the operators $\mathfrak{A}_0$ and $\mathfrak{A}_t$ are at least one-sided invertible in the space $C^\infty(\Gamma)$ (resp. in the space $C^{-\infty}(\Gamma)$).

9.5. Comments and References

9.1. Here we essentially follow the paper: S. Prössdorf [6]. The construction of the particular generalized solution (1.7) of a linear equation (1.4) was given simultaneously and independently by A. E. Kosulin [2], S. Prössdorf [4] and V. S. Rogožin [1]. The proof of Lemma 1.2 is in essence the proof of the perfectness criterion for a countably normed space from the book: I. M. Gelfand and G. E. Šilov [2] (p. 46).

9.2 and 9.3. Convolution equations of normal type were investigated in several distribution spaces by J. I. Čerski [2]. Also the Riemann-Hilbert problem was first solved for generalized functions by J. I. Čerski [1].

Theorems 2.1 and 2.2 and Theorem 2.4 (in the sufficiency direction) were stated in the case $n = 1$ by V. B. Dybin and N. K. Karapetjanc [2] (for the Wiener-Hopf operator PAP) and by N. K. Karapetjanc [1] (for the operator $PA + QB$). Their continuous analoga (see Section 9.3) were simultaneously and independently of each other proved for $n = 1$ by V. B. Dybin [3], [4] and for $n \geqq 1$ by S. Prössdorf [6], [11]. Here the proofs of the last mentioned author are essentially stated.

For the solution of the scalar integral equations considered in Section 9.3 in generalized functions see the papers: V. B. Dybin [1], [2], V. B. Dybin and N. K. Karapetjanc [1]. Methods for the solution of scalar Wiener-Hopf integral equations in generalized functions were also stated by M. I. Haikin [2], who investigated for the first time equations of the first kind in distribution spaces.

The basic idea of the proofs of Theorems 2.3 and 3.3 is due to B. Silbermann [3] who for the first proved Theorem 4.3 analogous to these theorems (in the necessity direction). The preliminary Lemma 2.2 is in principle stated in the book: I. M. Gelfand, D. A. Raikov and G. E. Šilov [1] (§ 38).

9.4. Theorems 4.1, 4.2, 4.4 and 4.5 are due to S. Prössdorf [4], [5], [7], [10]. The necessity of the conditions mentioned in Theorem 4.3 was proved by B. Silbermann [3], as we remarked above.

The scalar singular integral equation of non-normal type in the distribution space $C^{-\infty}(\Gamma)$ was first considered by A. E. KOSULIN [2] (under much more restrictive hypotheses on the zeros of the symbol). This author also first stated the theorem about the normal solvability for the specified equation in generalized functions. But the proof of this assertion given in the paper A. E. KOSULIN [3] contains some mistakes. The case of the open curve was also considered by A. E. KOSULIN [4]. In some other classes of distributions and by other methods (use of the integrals in the sense of Hadamard) singular integral equations of non-normal type were investigated by L. A. ČIKIN [1].

Finally we mention some interesting results of N. E. TOVMASJAN [1]. These results concern singular integral operators of the form (4.1) for which the determinants of the matrix functions $C(t)$ and $D(t)$ vanish at all points of the curve Γ. By Theorem 4.3 in this case the operator $\mathfrak{A}$ is not a Φ-operator on $C_n^\infty(\Gamma)$ if T is compact on this space. Tovmasjan proposes the following perturbation T:

$$(T\varphi)(t) = \int_\Gamma K_1(t, \tau) \ln\left(1 - \frac{\tau}{t}\right) \varphi(\tau)\, d\tau + \int_\Gamma K_2(t, \tau) \ln\left(1 - \frac{t}{\tau}\right) \varphi(\tau)\, d\tau + \\ + \int_\Gamma K_3(t, \tau)\, \varphi(\tau)\, d\tau \, .$$

Here $K_j(t, \tau)$ $(j = 1, 2, 3)$ are infinitely differentiable matrix functions. It is assumed that the matrix functions $C(t)$ and $D(t)$ have constant rank (less than n) on Γ and moreover satisfy the following condition: Let $\theta_1(t)$, $\theta_2(t)$ be matrix functions on Γ whose columns are linearly independent and for which $C(t)\, \theta_1(t) = 0$, $D(t)\, \theta_2(t) = 0$ $(t \in \Gamma)$. By $G_1(t)$ $(t \in \Gamma)$ we denote the matrix function whose columns are the columns of the matrix function $C(t)$ linearly independent at the point t and the columns of the matrix function $K_1(t, t)\, \theta_1(t)$. $G_2(t)$ is defined analogously. Then

$$\det G_1(t) \neq 0 \, , \qquad \det G_2(t) \neq 0 \qquad (t \in \Gamma)$$

holds (actually this condition can be somewhat weakened). If these conditions are satisfied, then $\mathfrak{A}$ is a Φ-operator on $C_n^\infty(\Gamma)$, and for the solvability of the equation $\mathfrak{A}\varphi = f$ it is necessary and sufficient that

$$\int_\Gamma f(t)\, \psi(t)\, dt = 0 \qquad \forall\, \psi \in C_n^\infty(\Gamma) : \mathfrak{A}'\psi = 0 \, .$$

The index of the operator $\mathfrak{A}$ can be explicitly calculated. It also depends on $C(t)$, $D(t)$ as well as on the matrix functions $K_1(t, t)$ and $K_2(t, t)$.

Finally we remark that the theory of Wiener-Hopf equations and of the singular integral equations of normal type can be completely carried over from the spaces E_n and $L_n^p(\Gamma)$ (cf. the Sections 7.4 and 7.5) to the spaces E_n^μ $(-\infty < \mu < \infty)$ and $W_p^{(l)}(\Gamma)$ $(l \geqq 0$ integer) respectively, if the symbols belong to the classes $W^{|\mu|}$, $\mathfrak{L}^{|\mu|}$ and $C^{(l)}(\Gamma)$, respectively.

For Wiener-Hopf equations this follows from the fact that the corresponding operators U and P (cf. the Sections 3.1 and 3.2) satisfy conditions (I), (II) and (III) of Chapter 2 in the spaces E^μ. The last statement can easily be obtained by simple norm estimates of the operators U^j $(j = 0, \pm 1, \ldots)$ in E^μ (cf. also 9.2.2, no. 1) or by means of the considerations from 3.2.2. Here we have $W^{|\mu|} \subset \mathfrak{R}(z)$ in the case of discrete equations and $\mathfrak{L}^{|\mu|} \subset \mathfrak{R}(z)$ in the case of integral equations. For singular integral equations this follows from the results of Sections 3.4 and 6.1 (resp. in the case of the unit circle Γ in the same way as for the discrete Wiener-Hopf equations).

CHAPTER 10

SINGULAR EQUATIONS WITH DISCONTINUOUS FUNCTIONS

In this chapter some basic results from the preceding chapters are generalized to the case of paired operators whose coefficients are piecewise continuous functions of an unitary operator on Hilbert space. There is a close connection between such operators and the paired operators of non-normal type with continuous coefficients (see Chapter 4).

The abstract results obtained are applied to paired discrete Wiener-Hopf equations and to singular integral equations with piecewise continuous coefficients. Using the results stated before for degenerate continuous functions we finally succeed in also treating the non-normal case for these equations. Some results can also be extended to the case of measurable bounded functions and to open integration curves. Furthermore, the results on Wiener-Hopf integral equations of the first kind (cf. Chapter 5) are essentially completed.

10.1. Abstract Singular Operators with Piecewise Continuous Coefficients

In the case of an unitary operator U on Hilbert space H the algebra $\mathfrak{R}(U)$ constructed in Section 2.1 can be essentially extended. Namely, with each piecewise continuous function $a(z)$ on the unit circle there can be associated an operator $a(U) \in \mathcal{L}(H)$ in a natural way. In this section we study the abstract singular operators $a(U)\,P + b(U)\,Q$ and $Pa(U) + Qb(U)$ for piecewise continuous functions $a(z)$ and $b(z)$.

10.1.1. By $\Lambda(\Gamma)$ we denote the algebra of all functions piecewise continuous[1]) and continuous on the left on the unit circle Γ. The norm in $\Lambda(\Gamma)$ is given by

$$||a(z)|| = \sup_{|z|=1} |a(z)| \,.$$

With each function $a(z) \in \Gamma$ we associate a function of two variables:

$$a(z, \mu) = \mu a(z + 0) + (1 - \mu)\, a(z)\,, \qquad (|z| = 1, 0 \leqq |\mu| \leqq 1)\,.$$

[1]) A function $a(z)$ is said to be piecewise continuous if at any point $z \in \Gamma$ the one-sided finite limits $a(z \pm 0)$ exist and the number of points of discontinuity is finite.

The set Γ_a of the values of the function $a(z, \mu)$ represents a closed curve in the complex plane. This curve is the union of the range of $a(z)$ ($|z| = 1$) and the line segments

$$\mu a(z_k + 0) + (1 - \mu)\, a(z_k) \qquad (0 \leqq \mu \leqq 1, k = 1, \dots, n),$$

where $t_1, \dots, t_n$ are the points of discontinuity of the function $a(z)$.

The curve Γ_a can be oriented in a natural way. If $0 \notin \Gamma_a$, then the function $a(z)$ is said to be *non-singular*. The winding number of such a curve round the point $z = 0$ is called the *index of the function* $a(z)$ and is denoted by ind $a(z)$.

Lemma 1.1. *If the non-singular functions a and b have no common points of discontinuity, then the function $c = ab$ is also non-singular, and*

$$\operatorname{ind} c = \operatorname{ind} a + \operatorname{ind} b .$$

Proof. It is easy to show that

$$c(z, \mu) - a(z, \mu)\, b(z, \mu) = [a(z + 0) - a(z)]\, [b(z + 0) - b(z)]\, \mu(1 - \mu) .$$

Thus under the hypotheses of the lemma we have $c(z, \mu) = a(z, \mu)\, b(z, \mu)$. This immediately implies the assertion.

Lemma 1.2. *Each non-singular function $a(z) \in \Lambda(\Gamma)$ can be represented in the form*

$$a(z) = r(z)\, g(z) , \tag{1.1}$$

where $r(z)$ is a polynomial in integral powers of z and $g(z)$ is a function in $\Lambda(\Gamma)$ with

$$\sup_{|z|=1} |g(z) - 1| < 1 .$$

Proof. Since the function $a(z)$ is non-singular, $a(z)$ can be represented in the form

$$a(z) = |a(z)|\, e^{i\varphi(z)} ,$$

where the real valued function $\varphi(z)$ is chosen in such a way that in the points of discontinuity $z_1, \dots, z_n$ of the function $a(z)$ the relations

$$|\varphi(z_k) - \varphi(z_k + 0)| < \pi - \delta \qquad (\delta > 0)$$

are satisfied. As the origin for the argument values we choose a point of continuity z_0 of the function $a(z)$. At this point the difference $\varphi(z_0 + 0) - \varphi(z_0)$ is an integral multiple of 2π. At all other points $\varphi(z)$ is continuous.

Now we define real valued functions $b(z)$ and $c(z)$ by

$$b(z_k) = \varphi(z_k) , \qquad b(z_k + 0) = \varphi(z_k + 0) \qquad (k = 0, 1, \dots, n) ,$$

$$c(z_0) = \varphi(z_0) , \qquad c(z_0 + 0) = \varphi(z_0 + 0) ,$$

$$c(z_k) = \tfrac{1}{2}\, [\varphi(z_k) + \varphi(z_k + 0)] \qquad (k = 1, \dots, n) ,$$

at all the other points of the unit circle these functions are defined by linear interpolation with respect to circle arcs. It is easily seen that

$$\sup_{|z|=1} |b(z) - c(z)| < \frac{\pi - \delta}{2}. \tag{1.2}$$

Since the function

$$f(z) = e^{i[\varphi(z) - b(z) + c(z)]}$$

is continuous on Γ, there exists a polynomial $r_1(z)$ such that

$$f(z) = r_1(z)\,(1 + m(z))$$

with $m(z) \in C(\Gamma)$ and

$$\max_{|z|=1} |m(z)| < \frac{1}{2}, \qquad |\arg\,(1 + m(z))| < \frac{\delta}{4}. \tag{1.3}$$

Now we consider the function

$$F(z) = |a(z)|\,(1 + m(z))\,e^{i[b(z) - c(z)]}.$$

We have $\inf |F(z)| > 0$. By (1.2) and (1.3) the relation $|\arg F(z)| < \frac{\pi}{2} - \frac{\delta}{4}$ holds. Hence there exists a number $\gamma > 0$ such that $\sup |1 - \gamma F(z)| < 1$. The functions $g(z) = \gamma F(z)$ and $r(z) = \gamma^{-1} r_1(z)$ satisfy the equations (1.1). This completes the proof of the lemma.

10.1.2. Now let U be an unitary operator and P an orthogonal projection in the Hilbert space H. We suppose that condition (II) of 2.1.2 and condition (III) of 2.2.3 are satisfied:

$$\text{(II)}\ \ UP = PUP,\ UP \neq PU,\ PU^{-1} = PU^{-1}P\,,$$

$$\text{(III)}\ \ \beta(V) = \dim\operatorname{coker}\,(U \mid \operatorname{im} P) < \infty\,.$$

With each function $a(z) \in \Lambda(\Gamma)$ we associate an operator $a(U) \in \mathcal{L}(H)$ in the following way. Let $z_0, z_1, \ldots, z_n$ be the points of discontinuity of the function $a(z)$ and E_z ($|z| = 1$) the partition of unity continuous on the left of the unitary operator U. Then the operator $a(U)$ is defined by the equation

$$a(U) = \sum_{j=0}^{n-1} \int_{z_j+0}^{z_{j+1}} a(z)\,dE_z + \int_{z_n+0}^{z_0} a(z)\,dE_z\,.$$

The set of all these operators forms an algebra isomorphic and isometric to the algebra $\Lambda(\Gamma)$. By reasons which will become clear in the sequel the *symbol of the operator* $a(U)$ but the function $a(z, \mu)$ of the two variables z and μ with $|z| = 1, 0 \leqq \mu \leqq 1$ is called the symbol of the operator $a(U)$ (rather then the function $a(z)$).

As before by V we denote the isometric operator $U \mid \operatorname{im} P$ and by $a(V)$ the restriction $Pa(U)\,P \mid \operatorname{im} P$.

Lemma 1.3. *Let the function* $b(z) \in \Lambda(\Gamma)$ *be of the form*

$$b(z) = a_-(z)\,a(z)\,a_+(z)\,,$$

where $a_+(z)$ *and* $a_-(z)$ *are continuous functions whose Fourier coefficients with non-positive resp. non-negative indices are equal to zero and* $a(z) \in \Lambda(\Gamma)$. *Then*

$$b(V) = a_-(V)\, a(V)\, a_+(V)\,.$$

Proof. The relations (II) imply

$$a_+(U)\, P = Pa_+(U)\, P,\ Pa_-(U) = Pa_-(U)\, P\,.$$

Hence we have

$$b(V) = Pa_-(U)\, a(U)\, a_+(U)\, P = Pa_-(U)\, Pa_+(U)\, Pa_+(U)\, P =$$
$$= a_-(V)\, a(V)\, a_+(V)\,.$$

Theorem 1.1. *Let* $a(z) \in \Lambda(\Gamma)$. *The operator* $A = a(V)$ *is at least one-sided invertible (in the subspace* im $P(H)$) *if and only if*

$$a(z, \mu) \neq 0 \qquad (|z| = 1,\ 0 \leqq \mu \leqq 1)\,.$$

If this condition is satisfied (i.e. if $a(z)$ *is non-singular), then the invertibility of the operator* A *corresponds to the number* $\varkappa =$ ind $a(z)$ *and the index of the operator* A *is equal to* $-\varkappa\beta(V)$.

Proof. First let $a(z)$ be non-singular. Then the polynomial $r(z)$ in the equation (1.1) is not equal to zero on Γ and ind $r =$ ind $a = \varkappa$.

Let the equation

$$r(z) = r_-(z)\, z^\varkappa r_+(z) \tag{1.4}$$

yield the factorization of the polynomial $r(z)$ (see equation (1.12), Chapter 2). Then (1.1) and (1.4) imply

$$a(z) = r_-(z)\, z^\varkappa g(z)\, r_+(z)\,.$$

By Lemma 1.3 we have

$$A = r_-(V)\, g(V)\, V^\varkappa r_+(V) \tag{1.5}$$

for $\varkappa \geqq 0$ and

$$A = r_-(V)\, V^{(\varkappa)} g(V)\, r_+(V) \tag{1.6}$$

for $\varkappa < 0$.

We have frequently remarked that the operators $r_\pm(V)$ are invertible. It follows from the estimates

$$\|g(V) - I\| = \|P(g(U) - I)\, P\| \leqq \sup_{|z|=1} |g(z) - 1| < 1$$

that the operator $g(V)$ is also invertible. Thus we obtain from equations (1.5) and (1.6) that the invertibility of the operator A corresponds to the number $\varkappa$ and that Ind $A = -\varkappa\beta(V)$.

It remains to show that the one-sided invertibility of the operator A implies the non-singularity of the function $a(z)$. Suppose that the operator A is one-sided invertible and $z = 0 \in \Gamma_a$.

First we consider the case that in a certain neighborhood G of the point $z = 0$ the curve Γ_a is a smooth Jordan curve. Let the neighborhood G be so small that for any $\lambda \in G$ the operator $A - \lambda I$ is invertible from the same side as A and its index is equal to the index of the operator A. If $\lambda_1, \lambda_2 \in G$ are on different sides of Γ_a, then $\operatorname{ind}(A(z) - \lambda_1) \neq \operatorname{ind}(a(z) - \lambda_2)$. But this contradicts the proved index formula.

The general case is reduced to the case considered just now in the usual way. Given a function $b(z) \in \Lambda(\Gamma)$ we choose the quantity $\sup |a(z) - b(z)|$ so small that $b(V)$ is at least one-sided invertible and the curve Γ_b contains the point $z = 0$ and is a smooth Jordan curve in a certain neighborhood of this point. This concludes the proof of the theorem.

Remark. It follows easily from the preceding proof that for a singular function $a(z)$ the operator $a(V)$ is neither a Φ_+-operator nor a Φ_--operator.

10.1.3. In the sequel by $Q = I - P$ we denote the complementary projection of the orthogonal projection P. For abstract singular operators we have the following theorem.

Theorem 1.2. *Let $a(z)$, $b(z)$ be functions from $\Lambda(\Gamma)$. The operator $a(U)\,P + b(U)\,Q$ $(Pa(U) + Qb(U))$ is at least one-sided invertible if and only if the condition*

$$a(z)\,b(z+0)\,\mu + a(z+0)\,b(z)\,(1-\mu) \neq 0 \tag{1.7}$$

$$(|z| = 1, 0 \leqq \mu \leqq 1)$$

is satisfied.

If this condition is satisfied, then the invertibility of the operator $a(U)\,P + b(U)\,Q$ $(Pa(U) + Qb(U))$ corresponds to the number $\varkappa = \operatorname{ind}(a(z)/b(z))$ and the index of this operator is equal to $-\varkappa\beta(V)$.

Proof. Here we use the results on paired operators from the Subsection 2.2.1. By $\mathfrak{M}$ we denote the algebra of all operators $c(U)$ with $c(z) \in \Lambda(\Gamma)$. By means of similar arguments as in Subsection 2.1.4 we show that the operator $c(U) \in \mathfrak{M}$ is invertible if and only if the conditions

$$c(z) \neq 0\,, \qquad c(z+0) \neq 0 \qquad (|z| = 1)$$

are satisfied. Here

$$[c(U)]^{-1} = c^{-1}(U)\,.$$

Using the equality

$$\|c(U)\| = \sup_{|z|=1} |c(z)|$$

we obtain easily that the operators invertible in $\mathfrak{M}$ form a set dense in $\mathfrak{M}$ (with respect to the operator norm).

It follows from Theorem 1.1 and Theorem 2.1, Chapter 2, that the operator $a(U)\,P + b(U)\,Q$ is at least one-sided invertible if and only if the following conditions are satisfied:

$$a(z) \neq 0\,, \quad a(z+0) \neq 0\,, \quad b(z) \neq 0\,, \quad b(z+0) \neq 0 \qquad (|z| = 1)$$

and

$$\frac{a(z)}{b(z)}\mu + \frac{a(z+0)}{b(z+0)}(1-\mu) \neq 0 \qquad (|z| = 1, 0 \leqq \mu \leqq 1).$$

Obviously, these conditions are equivalent to the condition (1.7).

The last assertion of the theorem follows immediately from Theorem 1.1 and formula (2.2), Chapter 2. This completes the proof of the theorem.

Remark 1. Using Theorem 2.2, Chapter 2, we obtain from the preceding proof that the operator $a(U)P + b(U)Q$ $(Pa(U) + Qb(U))$ is neither a Φ_+-operator nor Φ_--operator if the condition (1.7) does not hold.

Remark 2. A remarkable property of the paired operator $a(U)P + b(U)Q$ $(Pa(U) + Qb(U))$ in the case of piecewise continuous functions $a(z)$, $b(z)$ consists in the fact that its symbol is a matrix function:

$$\begin{pmatrix} a(z,\mu) & \sqrt{\mu(1-\mu)}\,[b(z+0)-b(z)] \\ \sqrt{\mu(1-\mu)}\,[a(z+0)-a(z)] & \mu b(z) + (1-\mu)\,b(z+0) \end{pmatrix}$$

(see I. C. Gohberg and N. J. Krupnik [3], cf. also Section 10.4).

Remark 3. Obviously, the methods stated in Section 4.2 can also be applied to functions of the form

$$A(z) = \prod_{j=1}^{r} (z - \alpha_j)^{m_j}\, B(z) \qquad (|z| = 1)$$

with $B(z) \in A(\Gamma)$ and $|\alpha_j| = 1$ $(j = 1, \dots, r)$. The same statement holds for the Subsections 4.3.2 to 4.3.4.

If, in particular $B(z)$ is continuous at the points $z \neq \alpha_j$ $(j = 1, \dots, r)$, then $A(z) \in C(\Gamma)$ and thus $A(z) \in \mathfrak{R}(z)$ (cf. Theorem 1.1, Chapter 2). From this it follows already that the theory of the singular equations with degenerate continuous symbols is just as multiform and difficult as the theory of singular equations with discontinuous functions.

10.2. Wiener-Hopf Equations with Piecewise Continuous Coefficients

The abstract results obtained in the preceding section can be applied in a natural way to different paired discrete and integral equations of Wiener-Hopf type. Using the results of Chapter 5 these results can even be extended to equations of non-normal type. As an example we state some results for discrete paired Wiener-Hopf equations.

10.2.1. Theorem 2.1. *Let $a(z)$ and $b(z)$ $(|z| = 1)$ be functions piecewise continuous and continuous on the left and a_j, b_j $(j = 0, \pm 1, \dots)$ its Fourier coefficients. The operator defined in l^2 by the equation system*

$$\sum_{k=0}^{\infty} a_{j-k}\xi_k + \sum_{k=-\infty}^{-1} b_{j-k}\xi_k = \eta_j \qquad (j = 0, \pm 1, \dots) \tag{2.1}$$

is a Φ_+-operator or a Φ_--operator if and only if condition (1.7) holds.

If condition (1.7) *is satisfied, then the invertibility of this operator corresponds to the number* $\varkappa = \operatorname{ind}\big(a(z)/b(z)\big)$ *and the index of this operator is equal to* $-\varkappa$.

An analogous theorem holds for the paired equation system of the form (3.2), Chapter 3.

Theorem 2.1 follows immediately from Theorem 1.2, if the operators U and P are defined on l^2 in the same way as in Section 3.1.1.

10.2.2. Theorem 2.1 shows in particular that the condition

$$\inf_{|z|=1} |a(z)| > 0 \,, \qquad \inf_{|z|=1} |b(z)| > 0 \tag{2.2}$$

is necessary for the operator defined on the space l^2 by (2.1) to be a Φ_+-operator or a Φ_--operator.

Now we assume that condition (2.2) does not hold. Let

$$a(z) = \varrho_-(z)\, c(z) \,, \qquad b(z) = \varrho_+(z)\, d(z) \qquad (|z| = 1) \tag{2.3}$$

with

$$\varrho_-(z) = \prod_{j=1}^{r} (z^{-1} - \alpha_j^{-1})^{m_j} \,, \qquad \varrho_+(z) = \prod_{k=1}^{s} (z - \beta_k)^{n_k} \,.$$

Here α_j $(j = 1, \dots, r)$ and β_k $(k = 1, \dots, s)$ are distinct points of the unit circle and m_j, n_k positive integers. Furthermore, let the functions $c(z)$ and $d(z)$ belong to the subalgebras $\Lambda(\boldsymbol{\beta}, \boldsymbol{n})$ and $\Lambda(\boldsymbol{\alpha}, \boldsymbol{m})$, respectively, of the algebra $\Lambda = \Lambda(\Gamma)$ (cf. 2.3.4):

$$c(z) \in \bigcap_{k=1}^{s} \Lambda(\beta_k, n_k) \,, \qquad d(z) \in \bigcap_{j=1}^{r} \Lambda(\alpha_j, m_j) \,. \tag{2.4}$$

By $\bar{l}^2(\varrho_+, \varrho_-)$ we denote the Banach space $\operatorname{im}(P\varrho_- + Q\varrho_+)$ described in Section 5.3.1.

Theorem 2.2. *Let the preceding conditions be satisfied and let A be the operator defined by the equation system* (2.1) *on* l^2. *Then* $A \in \mathscr{L}(l^2, \bar{l}^2(\varrho_+, \varrho_-))$, *and this operator is a* Φ_+-*operator or a* Φ_--*operator if and only if the condition*

$$c(z)\, d(z + 0)\, \mu + c(z + 0)\, d(z)\, (1 - \mu) \neq 0 \qquad (|z| = 1, 0 \leqq \mu \leqq 1) \tag{2.5}$$

is satisfied.

If this condition holds, then the invertibility of the operator $A \in \mathscr{L}(l^2, \bar{l}^2(\varrho_+, \varrho_-))$ *corresponds to the number* $\varkappa = \operatorname{ind}\big(c(z)/d(z)\big)$ *and the index of this operator is equal to* $-\varkappa$.

Proof. Here we use some considerations from the proof of the Theorem 5.5, Chapter 6. Let $g(z) \in \Lambda(\Gamma)$. Then for simplicity by g we denote the operator defined on l^2 by the Toeplitz matrix $g(U) = \{g_{j-k}\}_{j,k=-\infty}^{\infty}$. Here g_j $(j = 0, \pm 1, \dots)$ are the Fourier coefficients of the function $g(z)$.

By conditions (2.4) from the results of Section 2.3.4 we easily obtain the representations

$$c(z) = \varrho_+(z)\, c_1(z) + h_+(z) \,, \qquad d(z) = \varrho_-(z)\, d_1(z) + h_-(z) \,,$$

where $h_{\pm}(z)$ is a polynomial in $z^{\pm 1}$ and $c_1(z)$, $d_1(z)$ are certain functions in $\Lambda(\Gamma)$. Then analogously to equation (5.14), Chapter 6, the relation

$$A = (\varrho_- P + \varrho_+ Q)\,[cP + dQ + (\varrho_+ - \varrho_-)\,(Pd_1Q - Qc_1P)]\,. \tag{2.6}$$

holds. Completely analogously to the representation (5.7), Chapter 6, the validity of the relation

$$\varrho_- P + \varrho_+ Q = (P\varrho_- + Q\varrho_+)\,(I + T) \tag{2.7}$$

can also be established where T is a compact operator on l^2. By Lemma 4.4, Chapter 4, and Lemma 3.1, Chapter 5, we have $\dim\ker(\varrho_- P + \varrho_+ Q) = \dim\ker(P\varrho_- + Q\varrho_+) = 0$. This together with the equation (2.7) implies that $\varrho_- P + \varrho_+ Q \in \mathcal{L}(l^2, \bar{l}^2(\varrho_+, \varrho_-))$ is an invertible operator. Thus it follows from (2.6) that $A \in \mathcal{L}(l^2, \bar{l}^2(\varrho_+, \varrho_-))$, and this operator is a Φ_+-operator or a Φ_--operator if and only if the operator $A_1 = cP + dQ + (\varrho_+ - \varrho_-) \cdot (Pd_1Q - Qc_1P)$ is such an operator. The operators A_1 and $cP + dQ$ are simultaneously $\Phi_{\pm}$-operators on the space l^2 or not, and we have

$$\operatorname{Ind} A_1 = \operatorname{Ind}(cP + dQ) = -\varkappa \tag{2.8}$$

(cf. I. C. GOHBERG and N. J. KRUPNIK [3], [5], cf. also Section 10.4). By Theorem 2.1 $cP + dQ$ is a Φ_+-operator or a Φ_--operator on l^2 if and only if condition (2.5) is satisfied.

Furthermore, the equation

$$Pa + Qb = (P\varrho_- + Q\varrho_+)\,(Pc + Qd)$$

and Lemma 4.4, Chapter 4[1]), imply the relation

$$\dim\ker A = \dim\ker(Pa + Qb) = \dim\ker(cP + dQ)\,. \tag{2.9}$$

By Theorem 2.1 the assertion of Theorem 2.2 follows from equations (2.8) and (2.9).

Remark 1. Theorem 2.2 also holds in the case of arbitrary real numbers m_j and n_k if instead of (2.4) the following conditions hold:

$$c(z) \in \bigcap_{k=1}^{s} \Lambda(\beta_k, [n_k] + 1)\,, \qquad d(z) \in \bigcap_{j=1}^{r} \Lambda(\alpha_j, [m_j] + 1)\,.$$

Furthermore, the results of Theorem 2.2 can be generalized to the case where in the representations (2.3) the functions $\varrho_-(z)$ and $\varrho_+(z)$ are replaced by more general functions $R_1(z)$ and $R_2(z)$ of the form (5.2) of Chapter 6. The methods considered in Section 6.5 can be carried over to this case in a modified form. The same statement holds for Wiener-Hopf integral equations (see A. VOIGTLÄNDER [1]).

Remark 2. A theorem analogous to Theorem 2.2 also holds for the paired operator $Pa + Qb$ and for the pair of space $\bar{l}^2(\varrho_+', \varrho_-')$, $\bar{l}^2(\varrho_+'', \varrho_-'')$ instead of the space pair l^2, $\bar{l}^2(\varrho_+, \varrho_-)$ (cf. Section 5.3).

[1]) The proof of Lemma 4.4, Chapter 4, is also valid for functions $A(z), B(z) \in \Lambda(\Gamma)$.

10.3. Wiener-Hopf Integro-Difference Equations and Integral Equations of the First Kind

In this section we shall formulate some important results of the theory of the Wiener-Hopf integro-difference equations which was developed by I. C. Gohberg and I. A. Feldman (see I. C. Gohberg and I. A. Feldman [1]). By means of these results the results obtained before about Wiener-Hopf integral equations of the first kind (cf. Subsection 5.2.3) can be completed.

10.3.1. An equation of the form

$$PAP\varphi = f(t) \qquad (0 < t < \infty) \tag{3.1}$$

with

$$(A\varphi)(t) = \sum_{j=-\infty}^{\infty} a_j\varphi(t - \delta_j) + \int_{-\infty}^{\infty} k(t - s)\,\varphi(s)\,ds \qquad (-\infty < t < \infty) \tag{3.2}$$

is called a *Wiener-Hopf integro-difference equation*. Here a_j $(j = 0, \pm 1, \ldots)$ are complex numbers with $\sum_{j=-\infty}^{\infty} |a_j| < \infty$, δ_j $(j = 0, \pm 1, \ldots)$ are arbitrary real numbers and $k(t) \in L^1(-\infty, \infty)$.

The equation (3.1) is considered in the space L^p_+ $(1 \leqq p \leqq \infty)$. (3.1) can also be written in the form

$$\hat{A}\varphi \equiv \sum_{j=-\infty}^{\infty} a_j\varphi_+(t - \delta_j) + \int_0^{\infty} k(t - s)\,\varphi(s)\,ds = f(t) \qquad (0 < t < \infty) \tag{3.1'}$$

with

$$\varphi_+(t - \delta) = \begin{cases} \varphi(t - \delta), & \text{if } t > \delta \\ 0, & \text{if } 0 < t \leqq \delta. \end{cases}$$

The operator A is a linear continuous operator on the space L^p $(1 \leqq p \leqq \infty)$. The function

$$\mathcal{A}(\lambda) = a(\lambda) + K(\lambda) \qquad (-\infty < \lambda < \infty) \tag{3.3}$$

with

$$a(\lambda) = \sum_{j=-\infty}^{\infty} a_j e^{i\delta_j\lambda}, \qquad K(\lambda) = \int_{-\infty}^{\infty} e^{i\lambda t} k(t)\,dt,$$

is called the *symbol* of the operator A resp. of the operator $\hat{A}$ defined in the space L^p_+.

The functions of the form (3.3) form a Banach algebra $\mathfrak{G}$ with the norm

$$\|\mathcal{A}(\lambda)\| = \sum_{-\infty}^{\infty} |a_j| + \int_{-\infty}^{\infty} |k(t)|\,dt.$$

The algebra $\mathfrak{G}$ is the direct sum of the algebra of the almost periodic functions and the algebra $\mathfrak{L}^\circ$ of Fourier transforms of functions in $L^1(-\infty, \infty)$. It is easily seen that $\mathfrak{L}^\circ$ is an ideal of the algebra $\mathfrak{G}$. Let $||A||_p$ be the norm of the operator defined on the space L^p $(1 \leqq p \leqq \infty)$ by the equation (3.2). The operators of the form (3.2) form a commutative Banach algebra $\widetilde{\mathfrak{G}}$ with the norm $||A||_1$. The following equation holds:

$$||A||_1 = ||\hat{A}||_1 = \sum_{-\infty}^{\infty} |a_j| + \int_{-\infty}^{\infty} |k(t)|\, dt\,.$$

Thus the symbol effects an isometric isomorphism between the algebras $\mathfrak{G}$ and $\widetilde{\mathfrak{G}}$. Furthermore, the correspondence between the functions (3.3) and the operators $\hat{A} = PAP$ is one-to-one and linear but not multiplicative (the operators $\hat{A}$ do not form an algebra).

A function $\mathcal{A}(\lambda) \in \mathfrak{G}$ is said to be non-degenerate if

$$\inf_{-\infty<\lambda<\infty} |\mathcal{A}(\lambda)| > 0\,. \tag{3.4}$$

The relation

$$\inf_{-\infty<\lambda<\infty} |a(\lambda) + K(\lambda)| \leqq \inf_{-\infty<\lambda<\infty} |a(\lambda)|$$

which is easily checked shows that together with the function (3.3) its almost periodic component $a(\lambda)$ is non-degenerate.

For each non-degenerate function $\mathcal{A}(\lambda) = a(\lambda) + K(\lambda) \in \mathfrak{G}$ the two numbers

$$\nu(\mathcal{A}) = \lim_{l\to\infty} \frac{1}{2l} [\arg a(\lambda)]_{-l}^{l}\,,$$

$$n(\mathcal{A}) = \frac{1}{2\pi} [\arg (1 + a^{-1}(\lambda)\, K(\lambda))]_{\lambda=-\infty}^{\infty}$$

are defined. The real number $\nu(\mathcal{A})$ and the integer $n(\mathcal{A})$ are called the *indices* of the function $\mathcal{A}(\lambda)$. We remark that the indices of a function do not change with small perturbations (in the sense of the norm of the algebra $\mathfrak{G}$).

The following theorem is the main point of the theory of the Wiener-Hopf integro-difference equations.

Theorem 3.1. (I. C. Gohberg and I. A. Feldman [1]). *The operator $\hat{A} = PAP$ $(A \in \widetilde{\mathfrak{G}})$ is at least one-sided invertible in the space L_+^p $(1 \leqq p \leqq \infty)$ if and only if the condition* (3.4) *is holds.*

If this condition is satisfied, then the operator $\hat{A}$ is only left invertible for $\nu(\mathcal{A}) > 0$ and only right invertible for $\nu(\mathcal{A}) < 0$. For $\nu(\mathcal{A}) = 0$ the invertibility of the operator $\hat{A}$ corresponds to the index $n(\mathcal{A})$, i.e. $\hat{A}$ is invertible, only left invertible or only right invertible, if the index $n(\mathcal{A})$ is equal to zero, positive or *negative, respectively. Moreover, the following formulae hold:*

$$\dim \ker \hat{A} = \begin{cases} \infty, & \text{if } \nu < 0. \\ -n, & \text{if } \nu = 0 \text{ and } n \leqq 0, \end{cases} \tag{3.5}$$

$$\dim \operatorname{coker} \hat{A} = \begin{cases} \infty, & \text{if } \nu > 0, \\ n, & \text{if } \nu = 0 \text{ and } n \geqq 0. \end{cases} \tag{3.6}$$

If the condition (3.4) *is not satisfied, then* $\hat{A}$ *is neither a* Φ_+*-operator nor a* Φ_-*-operator.*

The paired operators $A_1P + A_2Q$ and $PA_1 + QA_2$, where A_1 and A_2 are operators of the form (3.2) with non-degenerate symbols $\mathcal{A}_1(\lambda)$ and $\mathcal{A}_2(\lambda)$, can be reduced in the usual way to the operator $\hat{C} = PCP$, where $C = A_2^{-1}A_1 \in \tilde{\mathfrak{G}}$ is the operator with the symbol $\mathcal{A}_1(\lambda)/\mathcal{A}_2(\lambda)$ (cf. Subsection 2.2.1). Analogous results hold for the paired operators also in the case that A_1 and A_2 belong to the closure $\tilde{\mathfrak{G}}_p$ $(1 \leqq p \leqq \infty)$ of the algebra $\tilde{\mathfrak{G}}$ in the operator norm $\|\cdot\|_p$ (see I. C. Gohberg and I. A. Feldman [1]). Corresponding results hold for the discrete analogue of the equation (3.1′) (see I. C. Gohberg and A. A. Semencul [1]).

10.3.2. There is a close connection between Wiener-Hopf integral equations of the first kind and Wiener-Hopf integro-difference equations. This can already be seen by the following simple example.

We consider the Wiener-Hopf integral equation of the first kind

$$\hat{A}\varphi = \int_0^\infty k(t-s)\,\varphi(s)\,ds = f(t) \qquad (0 < t < \infty) \tag{3.7}$$

with the kernel

$$k(t) = h_\delta(t) + i\int_{-\infty}^{0} e^s k_1(t-s)\,ds \qquad (-\infty < t < \infty),$$

where δ is an arbitrary real number, $k_1(t) \in L^1(-\infty, \infty)$ and

$$h_\delta(t) = \begin{cases} ie^{t-\delta}, & t < \delta, \\ 0, & t > \delta. \end{cases}$$

Obviously, $k(t) \in L^1(-\infty, \infty)$, and for the Fourier transform of the function $k(t)$ (the symbol of the operator $\hat{A}$) we have

$$K(\lambda) = \frac{1}{\lambda - i}\left(e^{i\delta\lambda} + K_1(\lambda)\right) \qquad (-\infty < \lambda < \infty).$$

The case $\delta = 0$ was treated in 5.2.3. Thus we assume $\delta \neq 0$. Then the function $\mathcal{C}(\lambda) = e^{i\delta\lambda} + K_1(\lambda)$ does not belong to the algebra $\mathfrak{L}$ but the algebra $\mathfrak{G}$. The almost periodic function $a(\lambda) = e^{i\delta\lambda}$ is discontinuous at the point $\lambda = \infty$. Obviously, $\nu(\mathcal{C}) = \operatorname{sgn} \delta$.

If the function $\mathscr{C}(\lambda)$ is non-degenerated, then it follows from Theorem 3.1 that the operator (3.7) is only left invertible for $\delta > 0$ and only right invertible for $\delta < 0$ as operator from L^p_+ $(1 \leqq p \leqq \infty)$ into the space $\bar{L}^p_+ \{(\lambda - i^{-1}\}$ (cf. Section 5.3). The cokernel resp. the kernel of this operator is infinite dimensional. This also furnisher an example of a singular operator with a continuous degenerate symbol $K(\lambda) \in \mathfrak{L}^\circ$ for which in the space L^p_+ $(1 \leqq p \leqq \infty)$ one of the numbers dim ker A or dim coker A is infinite.

The next theorem follows immediately from the results of Subsection 5.2.3 and Theorem 3.1.

Theorem 3.2. *Let the symbol of the operator* $\hat{A}$ *defined by equation* (3.7) *have the form*

$$\mathscr{A}(\lambda) = \varrho_-(\lambda)\, \mathscr{C}(\lambda)\, \varrho_+(\lambda)\,, \tag{3.8}$$

where $\varrho_\pm(\lambda)$ *is any one of the functions introduced in Subsection* 5.2.3 *and* $\mathscr{C}(\lambda) \in \mathfrak{G}$.

The operator $\hat{A}$ *is a* Φ_+*-operator or a* Φ_-*-operator from* $\tilde{L}^p_+(\varrho_+)$ *into the space* $\bar{L}^p_+(\varrho_-)$ $(1 \leqq p \leqq \infty)$ *if and only if the function* $\mathscr{C}(\lambda)$ *is non-degenerated.*

If the function $C(\lambda)$ *is non-degenerated, then* $\hat{A} \in \mathscr{L}(\tilde{L}^p_+(\varrho_+), \bar{L}^p_+(\varrho_-))$ *is only left invertible for* $\nu(\mathscr{C}) > 0$ *and only right invertible for* $\nu(\mathscr{C}) < 0$. *For* $\nu(\mathscr{C}) = 0$ *the invertiblity of the operator* $\hat{A}$ *corresponds to the index* $n(\mathscr{C})$. *Moreover, formulae* (3.5) *and* (3.6) *hold.*

An analogous result holds for the Wiener-Hopf integral operator with a symbol of the form (3.8) if $\varrho_\pm(\lambda)$ are the functions introduced in Subsection 5.2.1, no. 2, and $\mathscr{C}(\lambda) \in \mathfrak{G}$ (cf. also Subsection 5.2.2). For the paired Wiener-Hopf integral operator $AP + BQ$ on the pair of spaces $\tilde{L}^p(\varrho_+, \varrho_-)$, L^p $(1 \leqq p < \infty)$ and for the paired operator $PB + QA$ on the pair of spaces L^p, $\bar{L}^p(\varrho_+, \varrho_-)$ $(1 \leqq p < \infty)$ we obtain a result analogous to Theorem 3.2 provided the functions $\mathscr{A}(\lambda)$, $\mathscr{B}(\lambda)$ admit the representations

$$\mathscr{A}(\lambda) = \varrho_+(\lambda)\, \mathscr{A}_0(\lambda)\,, \qquad \mathscr{B}(\lambda) = \varrho_-(\lambda)\, \mathscr{B}_0(\lambda)$$

with the functions $\varrho_\pm(\lambda)$ defined in 5.4.1 and $\mathscr{A}_0(\lambda)$, $\mathscr{B}_0(\lambda) \in \mathfrak{G}$ (cf. Subsection 5.4.2).

10.4. Singular Integral Equations with Piecewise Continuous Coefficients

10.4.1. We consider the singular integral equation

$$(A\varphi)\,(t) \equiv c(t)\,\varphi(t) + \frac{d(t)}{\pi i} \int_\Gamma \frac{\varphi(\tau)}{\tau - t}\, d\tau = f(t) \qquad (t \in \Gamma) \tag{4.1}$$

with piecewise continuous coefficients $c(t)$ and $d(t)$.

First let Γ be the unit circle. Then we can apply Theorem 1.2 to the operator A defined on the Hilbert space $L^2(\Gamma)$ by equation (4.1) if we define the operators U and P on $L^2(\Gamma)$ by the equations

$$(U\varphi)(t) = t\varphi(t),$$

$$(P\varphi)(t) = \frac{1}{2}\left(\varphi(t) + \frac{1}{\pi i}\int_\Gamma \frac{\varphi(\tau)}{\tau - t}\,d\tau\right). \tag{4.2}$$

The operator A can be written in the form

$$A = a(t)\,P + b(t)\,Q \tag{4.3}$$

where

$$a(t) = c(t) + d(t), \qquad b(t) = c(t) - d(t)$$

and $Q = I - P$ (cf. also Section 3.4). The next theorem follows from Theorem 1.2.

Theorem 4.1. *Let Γ be the unit circle and $a(t), b(t) \in \Lambda(\Gamma)$. The operator A defined on $L^2(\Gamma)$ by equation (4.3) is a Φ_+-operator, a Φ_--operator or a Φ-operator if and only if*

$$a(t)\,b(t+0)\,\mu + a(t+0)\,b(t)\,(1-\mu) \neq 0 \qquad (t \in \Gamma, 0 \leqq \mu \leqq 1).$$

If this condition holds, then the invertibility of the operator A corresponds to the number $\varkappa = \operatorname{ind}\big(a(t)/b(t)\big)$, and the index of the operator A is equal to $-\varkappa$.

Theorem 4.1 holds also for the operator $B = Pa + Qb$.

10.4.2. Theorem 4.1 can be generalized to singular integral operators with piecewise continuous matrix coefficients over general (not necessarily closed) curve systems in the space L^p $(1 < p < \infty)$ and in the space L^p with weights (see I. C. Gohberg and N. J. Krupnik [3]).

For simplicity let Γ be a closed Ljapunov curve system of the form described in 6.2.1. By $\Lambda(\Gamma)$ we denote the algebra of all functions piecewise continuous and continuous on the left on Γ and by $\Lambda_{n\times n}(\Gamma)$ the algebra of all matrix functions of order n with elements from $\Lambda(\Gamma)$. Let P be the operator defined on the space $L_n^p(\Gamma)$ $(1 < p < \infty)$ by the equation (4.2), $Q = I - P$ (I the identity operator in $L_n^p(\Gamma)$) and $a(t), b(t) \in \Lambda_{n\times n}(\Gamma)$.

Following to I. C. Gohberg and N. J. Krupnik ([3] we say that the *symbol* of the singular integral operator

$$A = a(t)\,P + b(t)\,Q$$

on the space $L_n^p(\Gamma)$ is the matrix function $\mathcal{A}(t, \mu)$ $(t \in \Gamma, 0 \leqq \mu \leqq 1)$ of order $2n$ defined by

$$\mathcal{A}(t,\mu) = \begin{pmatrix} f_p(\mu)\,a(t+0) + (1 - f_p(\mu)\,a(t)) & h_p(\mu)\,\big(b(t+0) - b(t)\big) \\ h_p(\mu)\,\big(a(t+0) - a(t)\big) & (1 - f_p(\mu)\,b(t+0) + f_p(\mu)\,b(t) \end{pmatrix},$$

where

$$f_p(\mu) = \begin{cases} \dfrac{\sin\theta\mu \exp(i\theta\mu)}{\sin\theta \exp(i\theta)}, & \theta = \pi - \dfrac{2\pi}{p}, \quad p \neq 2 \\ \mu, & p = 2 \end{cases}$$

and $h_p(\mu)$ is a fixed branch of the root $\sqrt{f_p(\mu)(1 - f_p(\mu))}$. In the sequel we write this symbol in the form

$$\mathcal{A}(t, \mu) = \begin{pmatrix} a_{11}(t, \mu) & a_{12}(t, \mu) \\ a_{21}(t, \mu) & a_{22}(t, 1 - \mu) \end{pmatrix}. \tag{4.4}$$

By Γ_0 we denote the curve which we obtain from Γ attaching a loop at each point of discontinuity $t \in \Gamma$ of the matrix functions $a(t)$ and $b(t)$. On each of these loops the parameter μ shall change from 0 to 1. It is easily seen that the function $\det \mathcal{A}(t, \mu)/\det a_{22}(t, 0) \det a_{22}(t, 1)$ is continuous on Γ_0 if $\det a_{22}(t, 0) \det a_{22}(t, 1) \neq 0$.

Theorem 4.2. (I. C. Gohberg and N. I. Krupnik [3]). *The operator $A = aP + bQ$ is a Φ_+-operator, a Φ_--operator or a Φ-operator on the space $L_n^p(\Gamma)$ if and only if*

$$\det \mathcal{A}(t, \mu) \neq 0 \ (t \in \Gamma, 0 \leqq \mu \leqq 1). \tag{4.5}$$

If the condition (4.5) *is satisfied, then A is a Φ-operator with the index*

$$\operatorname{Ind} A = -\frac{1}{2\pi}[\arg \det \mathcal{A}(t, \mu)/\det a_{22}(t, 0) \det a_{22}(t, 1)] \quad (t, \mu) \in \Gamma_0. \tag{4.6}$$

For $n = 1$ A is at least one-sided invertible.

10.4.3. Now we consider operators of the form

$$A = \sum_{j=1}^{k} A_{j1} A_{j2} \dots A_{jr}, \tag{4.7}$$

where $A_{jl} = a_{jl}P + b_{jl}Q$ with $a_{jl}, b_{jl} \in \Lambda_{n\times n}(\Gamma)$. We denote the set of all operators of the form (4.7) by $\mathfrak{R}_n = \mathfrak{R}_n(\Gamma)$.

With the operator A in the representation (4.7) we associate the matrix function

$$\mathcal{A}(t, \mu) = \sum_{j=1}^{k} \mathcal{A}_{j1}(t, \mu) \mathcal{A}_{j2}(t, \mu) \dots \mathcal{A}_{jr}(t, \mu) \tag{4.8}$$

$$(t \in \Gamma, 0 \leqq \mu \leqq 1),$$

where $\mathcal{A}_{jl}(t, \mu)$ is the symbol of the operator A_{jl}. The matrix function $\mathcal{A}(t, \mu)$ is called the *symbol* of the operator (4.7). It is possible to show that the symbol of the operator $A \in \mathfrak{R}_n$ depends only on the operator and not on the kind of representation (4.7).

If $A, B \in \mathfrak{R}_n$ and $\mathcal{A}(t, \mu)$, $\mathcal{B}(t, \mu)$ are their symbols, then the operators $C = A + B$ and $D = AB$ can be represented in the form (4.7) such that for their symbols $\mathscr{C}(t, \mu)$ and $D(t, \mu)$ the equations

$$\mathscr{C}(t, \mu) = \mathscr{A}(t, \mu) + \mathscr{B}(t, \mu),\ D(t, \mu) = \mathscr{A}(t, \mu)\, \mathscr{B}(t, \mu)$$

hold.

Theorem 4.3. (I. C. Gohberg and N. J. Krupnik [3]). *The operator A defined on the space $L_n^p(\Gamma)$ by* (4.7) *is a Φ_+-operator, a Φ_--operator or a Φ-operator if and only if its symbol $\mathscr{A}(t, \mu)$ satisfies condition* (4.5).

If condition (4.5) *holds, then A is a Φ-operator and the index formula* (4.6) *holds with* (4.4)

We remark that the singular integral operator $B = Pa + Qb$ $(a, b \in \Lambda_{n\times n}(\Gamma))$ is an operator of the form (4.7). This follows immediately from the representation

$$Pa + Qb = aP + bQ + PaQ - QaP + QbP - PbQ\,.$$

From this representation we easily obtain that the symbol $\mathscr{B}(t, \mu)$ $(t \in \Gamma, 0 \leqq \mu \leqq 1)$ of the operator B is the transpose matrix of the symbol $\mathscr{A}(t, \mu)$ of the operator $aP + bQ$:

$$\mathscr{B}(t, \mu) = \begin{pmatrix} a_{11}(t, \mu) & a_{21}(t, \mu) \\ a_{12}(t, \mu) & a_{22}(t, \ - \mu) \end{pmatrix}.$$

The functions a_{jk} are the functions from the representation (4.4).

Corollary 4.1. *The operator $B = Pa + Qb$ $(a, b \in \Lambda_{n\times n}(\Gamma))$ is a Φ_+-operator a-Φ_-operator or a Φ-operator on the space $L_n^p(\Gamma)$ if and only if*

$$\det \mathscr{B}(t, \mu) \neq 0 \qquad (t \in \Gamma, 0 \leqq \mu \leqq 1)\,.$$

If this condition holds, then B is a Φ-operator and for its index the formula

$$\operatorname{Ind} B = \operatorname{Ind}\,(aP + bQ)$$

holds. In the case $n = 1$ B is at least one-sided invertible.

This follows immediately from Theorems 4.2 and 4.3 by using relations (2.2) and (2.2′) of Chapter 2.

Theorem 4.4. (I. C. Gohberg and N. J. Krupnik [3]). *An operator A of the form* (4.7) *is compact on the space $L_n^p(\Gamma)$ if and only if its symbol $\mathscr{A}(t, \mu) \equiv 0$ $(t \in \Gamma, 0 \leqq \mu \leqq 1)$.*

Corollary 4.2. *The operators $aP - Pa$, PaQ, QaP $(a \in \Lambda_{n\times n}(\Gamma))$ are compact on $L_n^p(\Gamma)$ if and onl if $a \in C_{n\times n}(\Gamma)$.*

10.5. Singular Integral Equations of Non-Normal Type with Piecewise Continuous Coefficients

Using the results of the preceding section we generalize basic results of Chapter 6 to the case of piecewise continuous coefficients.

10.5.1. Let Γ be a sufficiently smooth closed Ljapunov curve enclosing the origin. In the space $L^p(\Gamma)$ $(1 < p < \infty)$ we consider the singular integral operator

$$A = a(t)\, P + b(t)\, Q$$

with coefficients $a(t), b(t) \in \Lambda(\Gamma)$ satisfying the conditions

$$a(t) = R_1(t)\, c(t),\ b(t) = R_2(t)\, d(t)\,. \tag{5.1}$$

Here $R_1(t)$ and $R_2(t)$ are the continuous functions defined by relations (5.2) and (5.3) of Chapter 6. Let the functions $c(t)$, $d(t)$ belong to the classes

$$\begin{aligned} c(t) &\in \bigcap_{i=1}^{q} \Lambda(\alpha_i, l_{2i} - [-r_{3i}])\,, \\ d(t) &\in \bigcap_{i=1}^{q} \Lambda(\alpha_i, l_{1i} - [-r_{4i}]) \end{aligned} \tag{5.2}$$

with

$$l_{1i} = n_{1i} + n_{3i} - [-r_{1i}] - [-r_{3i}]\,, \qquad l_{2i} = n_{2i} + n_{4i} - [-r_{2i}] - [-r_{4i}]$$
$$(i = 1, \dots, q)\,.$$

In the sequel we maintain the notation introduced in Section 6.5. The following theorem is a generalization of Theorem 5.5, Chapter 6, for the case of piecewise continuous coefficients considered here.

Theorem 5.1. *Let conditions* (5.1) *and* (5.2) *be satisfied. Then* $A = aP + bQ \in \mathcal{L}(L^p(\Gamma), \overline{L}^p(R_1, R_2))$. *This operator is a* Φ_+*-operator, a* Φ_-*-operator or a* Φ*-operator if the symbol of the operator* $C = cP + dQ \in \mathcal{L}(L^p(\Gamma))$ *satisfies the condition*

$$\det \mathscr{C}(t, \mu) \neq 0 \qquad (t \in \Gamma, 0 \leqq \mu \leqq 1)\,. \tag{5.3}$$

If condition (5.3) *holds, then* $A \in \Phi(L^p(\Gamma), \overline{L}^p(R_1, R_2))$ *and*

$$\operatorname{Ind} A = \operatorname{Ind} C - \sum_{i=1}^{q} \left([-r_{3i}] + [-r_{4i}] + \left[r_{3i} + \frac{1}{p'} \right] + \left[r_{4i} + \frac{1}{p'} \right] \right) \tag{5.4}$$

with $\dfrac{1}{p'} = 1 - \dfrac{1}{p}$. *The operator* $A \in \mathcal{L}(L^p(\Gamma), \overline{L}^p(R_1, R_2))$ *is at least one-sided invertible.*

Using the results of Section 10.4 we can carry out the proof analogously to the proof of Theorem 5.5, Chapter 6.

First, the representations (5.13) and (5.14) of Chapter 6 follow from conditions (5.2), where now c_1 and d_1 are functions from $\Lambda(\Gamma)$. Repeating the arguments in the proof of the Theorem 5.5, Chapter 6, and using Theorem 4.2 and the subsequent Lemma 5.1 we obtain the assertion of Theorem 5.1.

Lemma 5.1. *Let (cf. formulae* (5.13), *Chapter* 6*)*

$$c = R_5 c_1 + h_+\,, \qquad d = R_6 d_1 + h_-$$

with $R_5, R_6, h_\pm \in C(\Gamma)$ *and* $c_1, d_1 \in \Lambda(\Gamma)$.

Then in the space $L^p(\Gamma)$ the operators

$$C = cP + dQ \quad and \quad C_1 = C + (R_5 - R_6)(Pd_1Q - Qc_1P)$$

are both $\Phi_\pm$-operators or not. If $C \in \Phi(L^p(\Gamma))$, then Ind $C =$ Ind C_1.

Proof. Obviously C_1 is an operator of the form (4.7). For the symbols of the operators C and C_1 we have

$$\mathscr{C}(t, \mu) = \begin{pmatrix} c_{11}(t, \mu) & h_p(\mu)\, R_5(t)\, (d_1(t+0) - d_1(t)) \\ h_p(\mu)\, R_6(t)\, (c_1(t+0) - c_1(t)) & c_{22}(t, 1-\mu) \end{pmatrix},$$

$$\mathscr{C}_1(t, \mu) = \begin{pmatrix} c_{11}(t, \mu) & h_p(\mu)\, R_6(t)\, (d_1(t+0) - d_1(t)) \\ h_p(\mu)\, R_5(t)\, (c_1(t+0) - c_1(t)) & c_{22}(t, 1-\mu) \end{pmatrix}$$

with

$$c_{11}(t, \mu) = f_p(\mu)\, c(t+0) + (1 - f_p(\mu))\, c(t),$$

$$c_{22}(t, 1-\mu) = (1 - f_p(\mu))\, d(t+0) + f_p(\mu)\, d(t).$$

Obviously, det $C(t, \mu) =$ det $C_1(t, \mu)$. Using the fact that the same elements are in the principal diagonales of both matrices we see that Lemma 5.1 is a direct consequence of Theorems 4.2 and 4.3.

10.5.2. In this subsection we consider the singular integral operator of non-normal type of the form

$$B = Pb + Qa$$

with piecewise continuous coefficients $a(t)$, $b(t) \in \Lambda(\Gamma)$ again satisfying conditions (5.1) and (5.2).

As in Section 6.5 we introduce the functions

$$R_5 = R_2R_3R_4 \quad \text{and} \quad R_6 = R_1R_3R_4,$$

where R_3, R_4 are given by formulae (5.6), Chapter 6. Using the representations (5.13), Chapter 6, by a simple calculation analogous to formula (5.14), Chapter 6, we obtain

$$B = C_1D \tag{5.5}$$

with

$$D = PR_2 + QR_1;$$

$$C_1 = Pd + Qc + (Qc_1P - Pd_1Q)(R_6 - R_5).$$

First we state two simple lemmas.

Lemma 5.2. *In the space $L^p(\Gamma)$ the operators C_1 and $C = Pd + Qc$ are both $\Phi_\pm$-operators or not. If $C \in \Phi(L^p(\Gamma))$, then* Ind $C =$ Ind C_1.

Lemma 5.2 can be proved in the same way as Lemma 5.1 since for the symbols of the operators C_1 and C we obtain the formulae (cf. Section 10.4)

$$\mathscr{C}_1(t,\mu) = \begin{pmatrix} c_{11}(t,\mu) & h_p(\mu)\, R_5(t)\, (d_1(t+0) - d_1(t)) \\ h_p(\mu)\, R_6(t)\, (c_1(t+0) - c_1(t)) & c_{22}(t, 1-\mu) \end{pmatrix},$$

$$\mathscr{C}(t,\mu) = \begin{pmatrix} c_{11}(t,\mu) & h_p(\mu)\, R_6(t)\, (d_1(t+0) - d_1(t)) \\ h_p(\mu)\, R_5(t)\, (c_1(t+0) - c_1(t)) & c_{22}(t, 1-\mu) \end{pmatrix}$$

with

$$c_{11}(t,\mu) = f_p(\mu)\, d(t+0) + (1 - f_p(\mu)\, d(t)\,,$$

$$c_{22}(t, 1-\mu) = (1 - f_p(\mu))\, c(t+0) + f_p(\mu)\, c(t)\,.$$

In the sequel by $\widetilde{L}^p(R_1, R_2)$ we denote the set of all functions of the form

$$\varphi = (R_2 R_6)^{-1}\, (PR_6 f + QR_5 f) \tag{5.6}$$

with $f \in L^p(\Gamma)$. Since by Lemma 2.2, Chapter 6, the formula dim ker $(PR_6 + QR_5) = 0$ holds, by $\|\varphi\| = \|f\|_{L^p(\Gamma)}$ a norm is defined on $\widetilde{L}^p(R_1, R_2)$. With this norm $\widetilde{L}^p(R_1, R_2)$ is a Banach space isometrically isomorphic to $L^p(\Gamma)$. Sometimes we also write $\widetilde{L}^p(R_1, R_2) = \widetilde{L}^p_\Gamma(R_1, R_2)$ in order to indicate that Γ is the integration curve.

Remark 1. If

$$n_{3i} + r_{3i} = n_{4i} + r_{4i} = 0 \qquad (i = 1, \dots, q)\,,$$

then $R_5 = R_2$, $R_6 = R_1$ and Theorem 5.2, Chapter 6, implies that $\widetilde{L}^p(R_1, R_2)$ coincides with the space $\widetilde{L}^p(\varrho_+, \varrho_-)$ introduced in Section 6.2.

Remark 2. Since we already know the analytic description of spaces of the type $\overline{L}^p(R_6, R_5) = \operatorname{im}(PR_6 + QR_5)$ (cf. Section 6.5), we can analytically describe the space $\widetilde{L}^p(R_1, R_2)$. In particular we obtain the following statement from Theorem 5.4, Chapter 6: If the function $g \in L^p(\Gamma)$ is sufficiently smooth in a neighborhood of the points α_i $(i = 1, \dots, q)$, then $(R_2 R_6)^{-1} g \in \widetilde{L}^p(R_1, R_2)$.

Lemma 5.3. *The inclusion $L^p(\Gamma) \subset \widetilde{L}^p(R_1, R_2)$ holds (in the sense of a continuous embedding).*

Proof. For any $\varphi \in L^p(\Gamma)$ we have

$$R_2 R_6 \varphi = (PR_6 + QR_5)\, (PR_2 + QR_1)\, \varphi$$

and thus

$$\varphi = (R_2 R_6)^{-1}\, (PR_6 + QR_5)\, f\,. \tag{5.7}$$

Here $f \in (PR_2 + QR_1)\, \varphi$ and

$$\|\varphi\| = \|f\|_{L^p(\Gamma)} \leqq c \|\varphi\|_{L^p(\Gamma)}\,.$$

This completes the proof of the lemma.

Now we extend the operator $D = PR_2 + QR_1$ to the space $\widetilde{L}^p(R_1, R_2)$. We define

$$\widetilde{D}\varphi = f$$

for any element $\varphi \in \widetilde{L}^p(R_1, R_2)$ of the form (5.6). The relation (5.7) implies immediately that $\widetilde{D} | L^p(\Gamma) = D$.

Hence the operator $\widetilde{B} = C_1\widetilde{D}$ is an extension of the operator $B = C_1 D$ from $L^p(\Gamma)$ to the space $\widetilde{L}^p(R_1, R_2)$. Thus we obtain the following analogue to Theorem 5.1.

Theorem 5.2. *Let conditions* (5.1), (5.2) *be satisfied and* $B = Pb + Qa$. *Then* $\widetilde{B} \in \mathcal{L}(\widetilde{L}^p(R_1, R_2), L^p(\Gamma))$, *and this operator is a* Φ_+*-operator or a* Φ_-*-operator or a* Φ*-operator if and only if the symbol of the operator* $C = Pd + Qc \in \mathcal{L}(L^p(\Gamma))$ *satisfies the condition*

$$\det \mathscr{C}(t, \mu) \neq 0 \qquad (t \in \Gamma, 0 \leqq \mu \leqq 1) . \tag{5.8}$$

If condition (5.8) *holds, then* $\operatorname{Ind} \widetilde{B} = \operatorname{Ind} (Pd + Qc)$ *and the operator* $\widetilde{B}$ *is at least one-sided invertible.*

Proof. Since the operator $\widetilde{D} \in \mathcal{L}(\widetilde{L}^p(R_1, R_2), L^p(\Gamma))$ is continuously invertible, the properties of the operator $\widetilde{B} = C_1\widetilde{D}$ depend only on the properties of the operator C_1. Then Lemma 5.2 yields all the assertions of Theorem 5.2 with the exception of the statement concerning the one-sided invertibility. In order to show this we use the identity

$$(Pd_+ Qc)\,(PR_6 + QR_5) = (PR_6 + QR_5)\, C_1 .$$

From this we see that the operator $C = Pd + Qc$ is defined and continuous on the space $\overline{L}^p(R_6, R_5) = \operatorname{im}(PR_6 + QR_5)$ and has the same properties as the operator $C_1 \in \mathcal{L}(L^p(\Gamma))$. We have

$$\operatorname{Ind} C = \operatorname{Ind} C_1 = \operatorname{Ind} (C | \overline{L}^p(R_6, R_5)) .$$

Since $\overline{L}^p(R_6, R_5)$ is dense in $L^p(\Gamma)$ (cf. Theorem 5.4, Chapter 6), it follows that $\dim\ker C_1 = \dim\ker C$. Thus together with the operator C the operator C_1 is also at least one-sided invertible. This proves the assertion.

10.6. Singular Integral Equations of Non-Normal Type on Non-Closed Curves

In this section the results of the preceding section are carried over to the case of non-closed integration curves. We use a method proposed by F. D. Gahov [1] for the reduction of singular integral equation (of normal type) over non-closed curves to the case of closed curves.

10.6.1. Let $\widetilde{\Gamma}$ be a sufficiently smooth closed Ljapunov curve enclosing the origin and $\Gamma \subset \widetilde{\Gamma}$ a part of $\widetilde{\Gamma}$ consisting of open curves.

We assume that Γ is closed, i.e. the end points of open subcurves of Γ belong to Γ. It is well-known that then the singular integral operator S_Γ defined by the equation

$$(S_\Gamma \varphi)(t) = \frac{1}{\pi i} \int_\Gamma \frac{\varphi(\tau)}{\tau - t} d\tau \qquad (t \in \Gamma)$$

is bounded on the space $L^p(\Gamma)$ $(1 < p < \infty)$. Contrary to the case of a closed curve we now have $S_\Gamma^2 \neq I$. As before we introduce the operators

$$P_\Gamma = \tfrac{1}{2}(I + S_\Gamma), \qquad Q_\Gamma = \tfrac{1}{2}(I - S_\Gamma).$$

But these operators are no longer projections. The operators (projections!) $P_{\tilde{\Gamma}}$ and $Q_{\tilde{\Gamma}}$ are defined analogously.

Now we consider the singular integral operator

$$A = aP_\Gamma + bQ_\Gamma$$

with coefficients $a, b \in \Lambda(\Gamma)$ which we can assume as continuous at the end points of the curve Γ. With the operator A we associate an operator $A_{\tilde{\Gamma}} = \tilde{a}P_{\tilde{\Gamma}} + \tilde{b}Q_{\tilde{\Gamma}}$ acting on the space $L^p(\tilde{\Gamma})$ as follows:

$$\tilde{a}(t) = \begin{cases} a(t), & t \in \Gamma \\ 1, & t \in \tilde{\Gamma} \setminus \Gamma \end{cases}, \qquad \tilde{b}(t) = \begin{cases} b(t), & t \in \Gamma \\ 1, & t \in \tilde{\Gamma} \setminus \Gamma \end{cases}. \tag{6.1}$$

Let $\chi(t)$ and $\tilde{\chi}(t)$ be the characteristic functions of the set Γ and $\tilde{\Gamma} \setminus \Gamma$, respectively. The operators $\chi(t)\, I$, $\tilde{\chi}(t)\, I$ (I the identity operator in $L^p(\tilde{\Gamma})$) are continuous projections in the space $L^p(\tilde{\Gamma})$, where im χI can be identified with $L^p(\Gamma)$, im $\tilde{\chi} I$ with $L^p(\tilde{\Gamma} \setminus \Gamma)$ and thus $L^p(\tilde{\Gamma})$ with the direct sum $L^p(\tilde{\Gamma}) = L^p(\Gamma) \dotplus L^p(\tilde{\Gamma} \setminus \Gamma)$. Using this fact we see at once that the operator A is equal to the restriction of the operator $\chi A_{\tilde{\Gamma}} \chi$ to the space $L^p(\Gamma)$.

Since

$$A_{\tilde{\Gamma}} = (\chi + \tilde{\chi})\, A_{\tilde{\Gamma}}(\chi + \tilde{\chi}) = \tilde{\chi} A_{\tilde{\Gamma}} \tilde{\chi} + \chi A_{\tilde{\Gamma}} \chi + \tilde{\chi} A_{\tilde{\Gamma}} \chi + \chi A_{\tilde{\Gamma}} \tilde{\chi}$$

and since by (6.1) we have $\tilde{\chi} A_{\tilde{\Gamma}} \tilde{\chi} = \tilde{\chi} I$, $\tilde{\chi} A_{\tilde{\Gamma}} \chi = 0$, we obtain

$$A_{\tilde{\Gamma}} \varphi = \begin{pmatrix} A & A_{12} \\ 0 & I \end{pmatrix} \cdot \begin{pmatrix} \chi\, \varphi \\ \tilde{\chi}\, \varphi \end{pmatrix}$$

with $A_{12} = \chi A_{\tilde{\Gamma}} \tilde{\chi}$. Hence $A_{\tilde{\Gamma}}$ can be interpreted as an operator

$$A_{\tilde{\Gamma}} = \begin{pmatrix} A & A_{12} \\ 0 & I \end{pmatrix} : L^p(\Gamma) \dotplus L^p(\tilde{\Gamma} \setminus \Gamma) \to L^p(\tilde{\Gamma}),$$

where $A\colon\ L^p(\Gamma) \to L^p(\Gamma)$, $A_{12}\colon L^p(\tilde{\Gamma} \setminus \Gamma) \to L^p(\tilde{\Gamma})$ and $I\colon\ L^p(\tilde{\Gamma} \setminus \Gamma) \to L^p(\tilde{\Gamma} \setminus \Gamma)$. Since obviously the equation

$$A_{\tilde{\Gamma}} = \begin{pmatrix} I & A_{12} \\ 0 & I \end{pmatrix} \begin{pmatrix} A & 0 \\ 0 & I \end{pmatrix} \tag{6.2}$$

holds, where the first factor possesses the continuous inverse

$$\begin{pmatrix} I & A_{12} \\ 0 & I \end{pmatrix}^{-1} = \begin{pmatrix} I & -A_{12} \\ 0 & I \end{pmatrix}, \tag{6.3}$$

we immediately obtain the following theorem.

Theorem 6.1. *The operator* $A: L^p(\Gamma) \to L^p(\Gamma)$ *is normally solvable, a* Φ_+-*operator, a* Φ_--*operator or a* Φ-*operator, left or right invertible if and only if the operator* $A_{\tilde{\Gamma}}: L^p(\tilde{\Gamma}) \to L^p(\tilde{\Gamma})$ *has the corresponding property. Moreover, the formulae*

$$\dim \ker A = \dim \ker A_{\tilde{\Gamma}}\,, \qquad \dim \operatorname{coker} A = \dim \operatorname{coker} A_{\tilde{\Gamma}}$$

hold.

The coefficients $\tilde{a}$ and $\tilde{b}$ can be modified at finitely many points as belong to the class $\Lambda(\Gamma)$ without changing the operator $A_{\tilde{\Gamma}}$. Consequently the theory of the operators A can be reduced to the theory of the operators $A_{\tilde{\Gamma}}$ already known (in the normal case).

It follows especially from Theorem 6.1 that the condition

$$\inf_{t\in\Gamma} |a(t)| > 0\,, \qquad \inf_{t\in\Gamma} |b(t)| > 0 \tag{6.4}$$

is necessary for the operator $A = aP_\Gamma + bQ_\Gamma$ to be a Φ_+-operator or a Φ_--operator in the space $L^p(\Gamma)$ $(1 < p < \infty)$. Contrary to the case of a closed curve Γ, condition (6.4) is not sufficient for $A \in \Phi_\pm(L^p(\Gamma))$ even in the case of continuous coefficients.

10.6.2. In the sequel we consider the case in which the condition (6.4) does not hold and the functions $a(t), b(t) \in \Lambda(\Gamma)$ again satisfy conditions (5.1) and (5.2), where the functions $R_1(t)$ and $R_2(t)$ are defined by relations (5.2) and (5.3) of Chapter 6. Moreover, we assume that the points α_j $(j = 1, \dots, q)$ are *interior* points of Γ. This hypothesis is essential for the following arguments.

With the operator $A = aP_\Gamma + bQ_\Gamma \in \mathcal{L}(L^p(\Gamma))$ we associate the operator $A_{\tilde{\Gamma}} = \tilde{a}P_{\tilde{\Gamma}} + \tilde{b}Q_{\tilde{\Gamma}} \in \mathcal{L}(L^p(\Gamma))$ as in 10.6.1. Since all points α_j $(j = 1, \dots, q)$ are interior points of Γ, we can write the coefficients $\tilde{a}$ and $\tilde{b}$ in the form

$$\tilde{a}(t) = R_1(t)\,\tilde{c}(t)\,, \qquad \tilde{b}(t) = R_2(t)\,\tilde{d}(t)\,, \tag{6.5}$$

where

$$\tilde{c}(t) = \begin{cases} c(t), & t \in \Gamma \\ \dfrac{1}{R_1(t)}, & t \in \tilde{\Gamma}\setminus\Gamma \end{cases}, \qquad \tilde{d}(t) = \begin{cases} d(t), & t \in \Gamma \\ \dfrac{1}{R_2(t)}, & t \in \tilde{\Gamma}\setminus\Gamma \end{cases} \tag{6.6}$$

are piecewise continuous functions in $\Lambda(\tilde{\Gamma})$. Since obviously the operator $A_{\tilde{\Gamma}}$ satisfies the conditions of Theorem 5.1, we know the theory of this operator in the pair of spaces $L^p(\tilde{\Gamma})$, $\overline{L}^p(R_1, R_2) = \operatorname{im}(P_{\tilde{\Gamma}}R_1P_{\tilde{\Gamma}} + Q_{\tilde{\Gamma}}R_2Q_{\tilde{\Gamma}})$. In order to avoid errors, we put $\overline{L}^p(R_1, R_2) = \overline{L}^p_{\tilde{\Gamma}}(R_1, R_2)$.

Lemma 6.1. *Let χ and $\tilde{\chi}$ be the characteristic functions of the sets Γ and $\tilde{\Gamma} \setminus \Gamma$. The operators χI and $\tilde{\chi} I$ (I the identity operator in $\overline{L}^p_{\tilde{\Gamma}}(R_1, R_2)$ are continuous projections in the space $\overline{L}^p_{\tilde{\Gamma}}(R_1, R_2)$, and the formula*

$$\overline{L}^p_{\tilde{\Gamma}}(R_1, R_2) = \operatorname{im} \chi I \dotplus L(\tilde{\Gamma} \setminus \Gamma)$$

holds.

Proof. Let φ be an arbitrary function from $L^p(\tilde{\Gamma})$. Since obviously $\tilde{\chi}\varphi \in W_p(\tilde{\Gamma}; (\alpha_i, l_i)_{i=1}^q)$ with $l_i = \max\{l_{1i}, l_{2i}\}$, by Theorem 5.4, Chapter 6, we have $\tilde{\chi}\varphi \in \overline{L}^p_{\tilde{\Gamma}}(R_1, R_2)$. Hence for any $\psi \in \overline{L}^p_{\tilde{\Gamma}}(R_1, R_2)$ the relations $\tilde{\chi}\psi \in \overline{L}^p_{\tilde{\Gamma}}(R_1, R_2)$ and $\chi\psi \in \overline{L}^p_{\tilde{\Gamma}}(R_1, R_2)$ hold. Moreover, these arguments show that

$$\operatorname{im} \tilde{\chi} I = L^p(\tilde{\Gamma} \setminus \Gamma) \subset \overline{L}^p_{\tilde{\Gamma}}(R_1, R_2) .$$

Since $\overline{L}^p_{\tilde{\Gamma}}(R_1, R_2)$ is continuously embedded into $L^p(\tilde{\Gamma})$ and thus $\tilde{\chi} I$ is a closed operator on $\overline{L}^p_{\tilde{\Gamma}}(R_1, R_2)$, it follows from the closed graph theorem that the operator $\tilde{\chi} I$ is continuous on $\overline{L}^p_{\tilde{\Gamma}}(R_1, R_2)$. Obviously, this operator is a projection. i.e. $L^p(\tilde{\Gamma} \setminus \Gamma)$ is closed in the norm of $\overline{L}^p_{\tilde{\Gamma}}(R_1, R_2)$. It follows from Banach's theorem on homomorphisms that this norm is equivalent to the usual norm of $L^p(\tilde{\Gamma} \setminus \Gamma)$. Moreover, the equality $\chi I = I - \tilde{\chi} I$ implies $\chi I \in \mathcal{L}(\overline{L}^p_{\tilde{\Gamma}}(R_1, R_2))$ and $(\chi I)^2 = \chi I$. This completes the proof of the lemma.

In the sequel we put

$$\overline{L}^p_{\Gamma}(R_1, R_2) = \operatorname{im} \chi I = \{\chi\varphi : \varphi \in \overline{L}^p_{\tilde{\Gamma}}(R_1, R_2)\} .$$

Then

$$\overline{L}^p_{\tilde{\Gamma}}(R_1, R_2) = \overline{L}^p_{\Gamma}(R_1, R_2) \dotplus L^p(\tilde{\Gamma} \setminus \Gamma) .$$

The operator $A_{\tilde{\Gamma}} = \tilde{a}P_{\tilde{\Gamma}} + \tilde{b}Q_{\tilde{\Gamma}} \in \mathcal{L}(\tilde{L}^p(\tilde{\Gamma}), \overline{L}^p_{\tilde{\Gamma}}(R_1, R_2))$ can be represented in the form

$$A_{\tilde{\Gamma}} = (\chi + \tilde{\chi}) A_{\tilde{\Gamma}}(\chi + \tilde{\chi}) = \chi A_{\tilde{\Gamma}}\chi + \tilde{\chi} A_{\tilde{\Gamma}}\tilde{\chi} + \chi A_{\tilde{\Gamma}}\tilde{\chi} ,$$

where

$$\chi A_{\tilde{\Gamma}}\chi | L(^p\Gamma) = A \in \mathcal{L}(L^p(\Gamma) , \overline{L}^p_{\Gamma}(R_1, R_2)) ,$$

$$\tilde{\chi} A_{\tilde{\Gamma}}\tilde{\chi} | L^p(\tilde{\Gamma} \setminus \Gamma) = I \in \mathcal{L}(L^p(\tilde{\Gamma} \setminus \Gamma)) ,$$

$$A_{12} = \chi A_{\tilde{\Gamma}}\tilde{\chi} | L^p(\tilde{\Gamma} \setminus \Gamma) \in \mathcal{L}(L^p(\tilde{\Gamma} \setminus \Gamma), \overline{L}^p_{\Gamma}(R_1, R_2)) .$$

Consequently we have

$$A_{\tilde{\Gamma}} = \begin{pmatrix} A & A_{12} \\ 0 & I \end{pmatrix} : L^p(\Gamma) \dotplus L^p(\tilde{\Gamma} \setminus \Gamma) \to \overline{L}^p_{\Gamma}(R_1, R_2) \dotplus L^p(\tilde{\Gamma} \setminus \Gamma) .$$

Here I is the identity operator on $L^p(\tilde{\Gamma} \setminus \Gamma)$. Thus by the relations (6.2) and (6.3) and by Theorem 5.1 we obtain:

Theorem 6.2. *Let conditions* (5.1) *and* (5.2) *hold. Then* $A = aP_\Gamma + bQ_\Gamma \in \mathscr{L}(L^p(\Gamma), \overline{L}^p_\Gamma(R_1, R_2))$. *This operator is a* Φ_+*-operator, a* Φ_-*-operator or a* Φ*-operator if and only if the symbol of the operator* $C_{\tilde{\Gamma}} = \tilde{c}P_{\tilde{\Gamma}} + \tilde{d}Q_{\tilde{\Gamma}} \in \mathscr{L}(L^p(\Gamma))$ *satisfies the condition*

$$\det \mathscr{C}_{\tilde{\Gamma}}(t, \mu) \neq 0 \qquad (t \in \tilde{\Gamma}, 0 \leqq \mu \leqq 1) . \tag{6.7}$$

If condition (6.7) *holds, then* $A \in \Phi(L^p(\Gamma), \overline{L}^p_\Gamma(R_1, R_2))$ *and formula* (5.4) *holds with* $C = C_{\tilde{\Gamma}}$. *The operator* $A \in \mathscr{L}(L^p(\Gamma), \overline{L}^p_\Gamma(R_1, R_2))$ *is at least one-sided invertible.*

Remark. Obviously we have

$$W_p(\Gamma; (\alpha_i, l_i)_{i=1}^q) \subset \overline{L}^p_\Gamma(R_1, R_2) .$$

10.6.3. Nȯ we consider the singular integral operator

$$B = P_\Gamma b + Q_\Gamma a$$

with piecewise continuous coefficients on the non-closed curve Γ. Again we assume that the functions $a, b \in \Lambda(\Gamma)$ satisfy conditions (5.1) and (5.2).

With the operator B we associate the operator $B_{\tilde{\Gamma}} = P_{\tilde{\Gamma}}\tilde{b} + Q_{\tilde{\Gamma}}\tilde{a} \in \mathscr{L}(L^p(\tilde{\Gamma}))$, where the coefficients $\tilde{a}, \tilde{b}$ are given by relations (6.1) resp. (6.5) and (6.6). Since $B_{\tilde{\Gamma}}$ satisfies the hypotheses of Theorem 5.2, its theory is known for the pair of spaces $\tilde{L}^p_{\tilde{\Gamma}}(R_1, R_2)$, $L^p(\Gamma)$.

The following analogue to the Lemma 6.1 is holds.

Lemma 6.2. *Let* χ *and* $\tilde{\chi}$ *be the characteristic functions of the sets* Γ *and* $\tilde{\Gamma} \setminus \Gamma$, *respectively. The operators* χI *and* $\tilde{\chi} I$ *(*I *the identity operator in* $\tilde{L}^p_{\tilde{\Gamma}}(R_1, R_2)$*) are continuous projections in the space* $\tilde{L}^p_{\tilde{\Gamma}}(R_1, R_2)$. *The following formula holds:*

$$\tilde{L}^p_{\tilde{\Gamma}}(R_1, R_2) = \operatorname{im} \chi I \dotplus L^p(\tilde{\Gamma} \setminus \Gamma) .$$

Using the structur of the functions R_5 and R_6 (cf. Subsection 10.5.2) of Lemma 6.1 and the remark at the end of the Subsection 10.6.2 we easily obtain the proof of Lemma 6.2.

In the sequel let

$$\tilde{L}^p_\Gamma(R_1, R_2) = \operatorname{im} \chi I = \{\chi\varphi : \varphi \in \tilde{L}^p_{\tilde{\Gamma}}(R_1, R_2)\} .$$

Then

$$\tilde{L}^p_{\tilde{\Gamma}}(R_1, R_2) = \tilde{L}^p_\Gamma(R_1, R_2) \dotplus L^p(\tilde{\Gamma} \setminus \Gamma) .$$

For the operator $\tilde{B}_{\tilde{\Gamma}} \in \mathscr{L}(\tilde{L}^p_{\tilde{\Gamma}}(R_1, R_2), L^p(\tilde{\Gamma}))$ (cf. Theorem 5.2) we have

$$\tilde{B}_{\tilde{\Gamma}} = \chi\tilde{B}_{\tilde{\Gamma}}\chi + \tilde{\chi}\tilde{B}_{\tilde{\Gamma}}\tilde{\chi} + \chi\tilde{B}_{\tilde{\Gamma}}\tilde{\chi}$$

analogously to the operator $A_{\widetilde{\Gamma}}$. Here the restriction

$$\widetilde{B} = \chi B_{\widetilde{\Gamma}}\chi \mid \widetilde{L}^p_\Gamma(R_1, R_2) \tag{6.8}$$

is an extension of the operator $B \in \mathcal{L}(L^p(\Gamma))$ to the space $\widetilde{L}^p_\Gamma(R_1, R_2) \supset L^p(\Gamma)$. Moreover, it is obvious that

$$\widetilde{\chi} B_{\widetilde{\Gamma}} \widetilde{\chi} \mid L^p(\widetilde{\Gamma} \setminus \Gamma) = I \in \mathcal{L}(L^p(\widetilde{\Gamma} \setminus \Gamma)) .$$

$$B_{12} = \chi B_{\widetilde{\Gamma}} \widetilde{\chi} \mid L^p(\widetilde{\Gamma} \setminus \Gamma) \in \mathcal{L}(L^p(\widetilde{\Gamma} \setminus \Gamma), L^p(\Gamma)) .$$

This implies

$$B_{\widetilde{\Gamma}} = \begin{pmatrix} \widetilde{B} & B_{12} \\ 0 & I \end{pmatrix} : \widetilde{L}^p_\Gamma(R_1, R_2) \dotplus L^p(\widetilde{\Gamma} \setminus \Gamma) \to L^p(\widetilde{\Gamma}) .$$

By relations (6.2) and (6.3) and Theorem 5.2 we obtain the following theorem.

Theorem 6.3. *Let conditions* (5.1), (5.2) *be satisfied and* $\widetilde{B} \in \mathcal{L}(\widetilde{L}^p_\Gamma(R_1, R_2), L^p(\Gamma))$ *the extension of the operator* $B = P_\Gamma b + Q_\Gamma a \in \mathcal{L}(L^p(\Gamma))$ *defined by* (6.8). *The operator* $\widetilde{B}$ *is a* Φ_+*-operator, a* Φ_-*-operator or a* Φ*-operator if and only if the symbol of the operator* $C_{\widetilde{\Gamma}} = P_{\widetilde{\Gamma}}\widetilde{d} + Q_{\widetilde{\Gamma}}\widetilde{c} \in \mathcal{L}(L^p(\widetilde{\Gamma}))$ *satisfies the condition*

$$\det \mathscr{C}_{\widetilde{\Gamma}}(t, \mu) \neq 0 \qquad (t \in \widetilde{\Gamma}, 0 \leqq \mu \leqq 1) . \tag{6.9}$$

If condition (6.9) *holds then* Ind $\widetilde{B}$ = Ind $C_{\widetilde{\Gamma}}$ *and the operator* $\widetilde{B}$ *is at least one sided invertible.*

10.7. Singular Integral Equations of Non-Normal Type with Bounded Measurable Coefficients

In this section we show that by the methods considered in this book the investigation of certain classes of singular integral equations of non-normal type with general bounded measurable coefficients can be reduced to the investigation of such equations of normal type. Singular integral equations with bounded measurable coefficients (in the normal case) were first studied by I. B. Simonenko [2], [3] (see also I. C. Gohberg and N. J. Krupnik [4]).

Let Γ be a sufficiently smooth closed Ljapunov curve enclosing the origin. We consider the singular integral operator $A = aP + bQ$ with bounded measurable coefficients a and b on $\Gamma (a, b \in L^\infty(\Gamma))$. Then the condition

$$\operatorname*{ess\,inf}_{t \in \Gamma} |a(t)| > 0 , \qquad \operatorname*{ess\,inf}_{t \in \Gamma} |b(t)| > 0 \tag{7.1}$$

is necessary for the operator A to be a Φ_+-operator or a Φ_--operator on the space $L^p(\Gamma)$ $(1 < p < \infty)$ (I. B. Simonenko [3]).

In the sequel we assume that condition (7.1) does not hold and the functions a and b have the following representation

$$a(t) = R_1(t)\, c(t)\,, \qquad b(t) = R_2(t)\, d(t)\,. \tag{7.2}$$

Here $R_1(t)$ and $R_2(t)$ are the continuous functions defined by relations (5.2) and (5.3) of Chapter 6. The functions c and d satisfy the conditions

$$c, d, c^{-1}, d^{-1} \in L^\infty(\Gamma) \quad \text{and}$$

$$c, d \in \bigcap_{i=1}^{q} L^\infty(\alpha_i, l_{2i} - [-r_{3i}])\,, \tag{7.3}$$

$$d \in \bigcap_{i=1}^{q} L^\infty(\alpha_i, l_{1i} - [-r_{4i}])\,.$$

The numbers l_{ji} ($j = 1, 2$; $i = 1, \dots, q$) were defined in 10.5.1 (cf. also Section 6.5).

From conditions (7.3) and the results of Subsection 2.3.4 we obtain the representations

$$d = R_5 d_1 + h_1, \quad d = R_6 d_2 + h_2\,, \quad d^{-1} = R_5 d_3 + h_3\,, \quad d^{-1} = R_6 d_4 + h_4\,,$$
$$g = d^{-1}c = R_5 g_1 + q_1\,, \qquad g^{-1} = dc^{-1} = R_5 g_2 + q_2 \tag{7.4}$$

with

$$d_j, h_j \in L^\infty(\Gamma),\ g_i, q_i \in L^\infty(\Gamma) \qquad (j = 1, \dots, 4,\ i = 1, 2)\,,$$
$$Ph_jP = h_jP \qquad (j = 1, 3)\,, \qquad Qh_jQ = h_jQ \qquad (j = 2, 4)\,,$$
$$Pq_iP = q_iP \qquad (i = 1, 2)\,.$$

The functions $R_5 \in C^+(\Gamma)$ and $R_6 \in C^-(\Gamma)$ were defined in Subsection 10.5.2 (see also 6.5.2).

The operator $A = aP + bQ$ can be written in the form

$$A = d(gR_1P + R_2Q) \tag{7.5}$$

Lemma 7.1. *The operator of multiplication by the function d is defined, continuous and invertible on the space $\overline{L}^p(R_1, R_2)$.*

Proof. Since the spaces $\overline{L}^p(R_1, R_2)$ and $\hat{L}^p(R_1, R_2)$ are topologically equivalent (see Theorem 5.2, Chapter 6) it is sufficient to prove this assertion for the space $\hat{L}^p(R_1, R_2)$.

From (7.4) we easily obtain the representations

$$d(R_1P + R_2Q) = (R_1P + R_2Q)\,[dI + (R_5 - R_6)\,(Pd_2Q - Qd_1P)]\,,$$
$$d^{-1}(R_1P + R_2Q) = (R_1P + R_2Q)\,[d^{-1}I + (R_5 - R_6)\,(Pd_3Q - Qd_4P)]\,.$$

Since $R_1P + R_2Q \in \Phi(L^p(\Gamma), \overline{L}^p(R_1, R_2))$ and the operators enclosed in square brackets are continuous on the space $L^p(\Gamma)$, we obtain from the last equations that the operators dI and $d^{-1}I$ are defined and continuous on $\overline{L}^p(R_1, R_2)$.

Finally the relations $dd^{-1}\varphi = d^{-1}d\varphi = \varphi$ $(\varphi \in \overline{L}^p(R_1, R_2))$ imply that $d^{-1}I$ is the inverse of dI. This completes the proof of the lemma.

For the second factor in (7.5) the representation

$$gR_1P + R_2Q = (R_1P + R_2Q)\,[gP + Q + (R_6 - R_5)\,Qg_1P] \tag{7.6}$$

holds. Here the operator

$$G = gP + Q + (R_6 - R_5)\,Qg_1P$$

is normally solvable, a $\Phi_\pm$-operator or one-sided invertible on $L^p(\Gamma)$, if and only if the operator $gP + Q$ has this property. This follows immediately from the next lemma:

Lemma 7.2. *The following equality holds:*

$$G = G_1(gP + Q)\,G_2\,, \tag{7.7}$$

where G_1, G_2 are invertible operators on the space $L^p(\Gamma)$.

Proof. It is easily seen that relation (7.7) holds with the factors

$$G_1 = [g + (R_5 - I)\,(PgQ - Qg_1P)]\,g^{-1}\,,$$
$$G_2 = (I - Pg^{-1}Q + R_5Pg^{-1}Q)\,[I + (R_6 - I)\,Qg_1P]\,.$$

By $R_5 \in C^+(\Gamma)$, $R_6 \in C^-(\Gamma)$ the first factor of the operator G_2 has the form $I + PBQ$ and the second factor the form $I + QCP$. Hence G_2 is invertible (cf. Section 2.2). We show that also G_1 is invertible.

The relations (7.4) imply

$$g(P + R_5Q) = (P + R_5Q)\,[g + (R_5 - I)\,(PgQ - Qg_1P)]\,,$$
$$g^{-1}(P + R_5Q) = (P + R_5Q)\,[g^{-1} + (R_5 - I)\,(Pg^{-1}Q - Qg_2P)]\,.$$

By Theorem 5.1, Chapter 6, we have $\dim\ker(P + R_5Q) = 0$. We equip the space $X = \operatorname{im}(P + R_5Q)$ with the usual norm (cf. Section 4.1). Then the last equations show that gI, $g^{-1}I \in \mathscr{L}(X)$ and thus $gI \in \mathscr{L}(X)$ is continuously invertible. Hence also $g + (R_5 - I)\,(PgQ - Qg_1P)$ and thus G_1 is continuously invertible in $\mathscr{L}(L^p(\Gamma))$. This completes the proof of the lemma.

The next theorem follows immediately from equations (7.5)–(7.7).

Theorem 7.1. *Let the singular integral operator $A = aP + bQ$ $(a, b \in L^\infty(\Gamma))$ satisfy conditions* (7.2) *and* (7.3). *Then $A \in \mathscr{L}(L^p(\Gamma), \overline{L}^p(R_1, R_2))$ and this operator is a Φ_+-operator, a Φ_--operator or a Φ-operator if and only if the operator $C = cP + dQ \in \mathscr{L}(L^p(\Gamma))$ has this property.*

If $A \in \Phi(L^p(\Gamma), \overline{L}^p(R_1, R_2))$, then

$$\operatorname{Ind} A = \operatorname{Ind} C + K(R_1, R_2)$$

with

$$K(R_1, R_2) = -\sum_{i=1}^{q}\left([-r_{3i}] + [-r_{4i}] + \left[r_{3i} + \frac{1}{p'}\right] + \left[r_{4i} + \frac{1}{p'}\right]\right).$$

In the case $K(R_1, R_2) = 0$ *the operators*

$$A \in \mathcal{L}(L^p(\Gamma), \overline{L}^p(R_1, R_2)) \quad \textit{and} \quad C \in \mathcal{L}(L^p(\Gamma))$$

are both one-sided invertible or not.

Remark 1. Using the results from 10.5.2 analogously we can investigate the operator $B = Pb + Qa$ $(a, b \in L^\infty(\Gamma))$.

Remark 2. Theorem 7.1 can in particular be applied to such coefficients as are continuous on Γ with the exception of finitely many points of discontinuity of the second kind of almost periodic type. (The isolated point of discontinuity z_0 of the function $f(z)$ defined on the unit circle is called a *point of discontinuity of almost periodic type* if there exists a uniformly almost periodic function $p(\lambda)$, $-\infty < \lambda < \infty$, such that

$$\lim_{z \to z_0} \left[[f(z) - p\left(-i \frac{z + z_0}{z - z_0}\right) \right] = 0 .$$

For such coefficients c, d the theory of the singular integral operators $cP + dQ$ $(Pc + Qd)$ was developed by I. C. Gohberg and A. A. Semencul [1] (cf. also A. A. Semencul [1]). For these operators analogous results hold as for the Wiener-Hopf integro-difference equations in the space $L^p(\Gamma)$ (cf. 10.3.1). Thus the results of this section yield the possibility of constructing examples of singular integral equations with continuous and pointwise degenerate coefficients for which the kernel resp. the cokernel is infinite dimensional in the space $L^p(\Gamma)$. (In this connection cf. also the paper A. A. Semencul [1], in which the method described in Section 4.1 is used for the investigation of such operators).

We remark that the methods of this section can also be applied to paired Wiener-Hopf integro-difference equations of non-normal type (cf. 10.3.2).

10.8. Systems of Singular Integral Equations of Non-Normal Type with Piecewise Continuous Coefficients

Let Γ be a sufficiently small closed Ljapunov curve enclosing the coordinate origin. In this section we consider singular integral operators of the form

$$\mathfrak{A} = CP + DQ$$

on the space $L_n^p(\Gamma)$ $(1 < p < \infty)$. Here P and Q are the projection in the space $L_n^p(\Gamma)$ introduced in Section 7.5. The matrix functions $C, D \in \Lambda_{n \times n}(\Gamma)$ satisfy the condition

$$C(t) = R_1(t)\, C_0(t)\,, \qquad D(t) = R_2(t)\, D_0(t)\,. \tag{8.1}$$

The continuous matrix functions $R_1(t)$ and $R_2(t)$ whose determinants vanish at the points α and β, respectively, $(\alpha, \beta \in \Gamma, \alpha \neq \beta)$ are defined in Subsection 8.5.3, no. 1. The matrix functions $C_0, D_0 \in \Lambda_{n \times n}(\Gamma)$ will belong to the following classes[1])

$$C_0 \in \Lambda_{n\times n}(\beta, \eta_1 + \eta_2) \cap \Lambda_{n\times n}(\alpha, \tilde{\xi}_2)\,,$$
$$D_0 \in \Lambda_{n\times n}(\alpha, \xi_1 + \xi_2) \cap \Lambda_{n\times n}(\beta, \tilde{\eta}_2)\,. \tag{8.2}$$

Analogously to the arguments in Subsection 8.5.3 we obtain the representations (*) and (**) from conditions (8.2), where now C_j, $D_j \in \Lambda_{n\times n}(\Gamma)$ ($j = 1, 2$). Hence the relation (***) of Subsection 8.5.3 remains true, i.e.

$$\mathfrak{A} = \mathfrak{B}\mathfrak{A}_0 \tag{8.3}$$

with $\mathfrak{B} = PR_1P + QR_2Q$ and

$$\mathfrak{A}_0 = PC_0P + QD_0Q + PR_3D_2Q + QR_4C_2P + T\,,$$

where

$$T = -PR_3Q(QC_1P + QD_2Q) - QR_4P(PD_1Q + PC_2P)$$

is a compact operator on the space $L_n^p(\Gamma)$. By Section 10.4 for the symbol $\mathcal{A}_0(t, \mu)$ $(t \in \Gamma)$ $0 \leqq \mu \leqq 1)$ of the operator $\mathfrak{A}_0 \in \mathcal{L}(L_n^p(\Gamma))$ we obtain

$$\mathcal{A}_0(t, \mu) = \begin{pmatrix} C_{11}(t, \mu) & D_{12}(t, \mu) \\ C_{21}(t, \mu) & D_{22}(t, \mu) \end{pmatrix}$$

with

$$C_{11}(t, \mu) = f_p(\mu)\, C_0(t + 0) + (1 - f_p(\mu))\, C_0(t)\,,$$
$$C_{21}(t, \mu) = h_p(\mu)\, R_4(t)\, (C_2(t + 0) - C_2(t))\,,$$
$$D_{12}(t, \mu) = h_p(\mu)\, R_3(t)\, (D_2(t + 0) - D_2(t))\,,$$
$$D_{22}(t, \mu) = (1 - f_p(\mu))\, D_0(t + 0) + f_p(\mu)\, D_0(t)\,.$$

Theorem 8.1. *Let conditions* (8.1) *and* (8.2) *hold.*

(a) *The operator* $\mathfrak{A} = CP + DQ\colon L_n^p(\Gamma) \to \overline{L}_n^p(R_1, R_2)$ *is defined and continuous.* $\mathfrak{A}$ *is a* Φ_+*-operator or a* Φ_-*-operator if and only if*

$$\det \mathcal{A}_0(t, \mu) \neq 0 \qquad (t \in \Gamma,\ 0 \leqq \mu \leqq 1)\,. \tag{8.4}$$

If condition (8.4) *holds then* $\mathfrak{A} \in \Phi(L_n^p(\Gamma), \overline{L}_n^p(R_1, R_2))$ *and*

$$\operatorname{Ind} \mathfrak{A} = \operatorname{Ind} \mathfrak{A}_0 + K\,,$$

$$K = n(\xi_2 + \eta_2) - \sum_{j=1}^{n} \left(m_{2j} + n_{2j} + \left[r_{2j} + \frac{1}{p'} \right] + \left[s_{2j} + \frac{1}{p'} \right] \right),$$

where $\dfrac{1}{p'} = 1 - \dfrac{1}{p}$.

(b) *If the matrix functions* $C_0(t)$ *and* $D_0(t)$ *possess no common points of discontinuity on* Γ, *then the operators* $\mathfrak{A} \in \mathcal{L}(L_n^p(\Gamma), \overline{L}_n^p(R_1, R_2))$ *and* $C_0P + D_0Q \in \mathcal{L}(L_n^p(\Gamma))$ *are both* $\Phi_\pm$*-operators or not. If* $C_0P + D_0Q \in \Phi(L_n^p(\Gamma))$ *then*

$$\operatorname{Ind} \mathfrak{A} = \operatorname{Ind} (C_0P + D_0Q) + K\,.$$

[1]) For notation see Subsection 8.5.3.

Proof. (a) The relation (8.3) shows that $\mathfrak{A}$ is a linear continuous operator from $L_n^p(\Gamma)$ into the space $\overline{L_n^p}(R_1, R_2)$. Using Theorem 5.4, Chapter 8, and Theorem 4.2 of Section 10.4 we obtain the assertion (a) by means of the same arguments as in the proof of Theorem 5.1.

(b) Since in the representations (**) of 8.5.3 the matrix functions T_1 and T_3 are continuous, the symbol $\mathcal{A}_1(t, \mu)$ of the operator $\mathfrak{A}_1 = C_0P + D_0Q$ has the form

$$\mathcal{A}_1(t, \mu) = \begin{pmatrix} C_{11}(t, \mu) & h_p(\mu)\, R_4(t)\, (D_1(t+0) - D_1(t)) \\ h_p(\mu)\, R_3(t)\, (C_1(t+0) - C_1(t)) & D_{22}(t, \mu) \end{pmatrix}.$$

If $C_0(t)$ and $D_0(t)$ have no common points of discontinuity on Γ, then it is easily seen that the same statement holds also for the matrix functions $R_3C_1 + R_4D_1$ and for R_4C_2 and R_3D_2. Hence $\mathcal{A}_0(t, \mu)$ and $\mathcal{A}_1(t, \mu)$ are quasi-diagonal matrices. But this means

$$\det \mathcal{A}_0(t, \mu) = \det C_{11}(t, \mu)\, D_{22}(t, \mu) = \det \mathcal{A}_1(t, \mu)\,.$$

Now the assertion (b) follows immediately from the results of Section 10.4.2.

Remark 1. The remark made at the end of Subsection 8.5.3 remains valid also in the present case of piecewise continuous matrix functions $C(t)$ and $D(t)$.

Remark 2. If the matrix functions $C_0(t)$ and $D_0(t)$ possess common points of discontinuity on Γ, then it is possible that one of the operators $\mathfrak{A}_0$ or $\mathfrak{A}_1 = C_0P + D_0Q$ is a Φ-operator on $L_n^p(\Gamma)$ but the other operator is not even normally solvable (for $n = 1$ this is not possible, see Theorem 5.1). Situations of this kind can occur when singular integral operators of non-normal type with matrix coefficients on non-closed curves are investigated (by the methods considered in Section 10.6 and in this section).

10.9. Comments and References

10.1. Lemma 1.3 is a special case of a theorem of I. B. Simonenko [2]. All the other results of this section are due to I. C. Gohberg and I. A. Feldman. In our statements we essentially follow the monograph I. C. Gohberg and I. A. Feldman [1] (comp. Kapitel IV, § 1, and Kapitel V, § 5).

10.2. The results of 10.2.1 were stated by I. C. Gohberg and N. J. Krupnik [1], [2]. The results of 10.2.2 were first proved by B. Silbermann in his thesis [8].

Theorem 2.1 was generalized by R. V. Dudučava [1] to l^p spaces $(1 < p < \infty)$. By using these generalizations the results from 10.2.2 can also be carried over to l^p spaces.

10.3. The theory of Wiener-Hopf integro-difference equations and of the corresponding paired equations was developed by I. C. Gohberg and I. A. Feldman (see I. C. Gohberg and I. A. Feldman [1]).

The thesis of Š. I. Galiev [3] deals with questions similar to those of the Subsection 10.3.2. In particular this author made a detailed study of Wiener Hopf integral equations of the first kind with a symbol of the form $\varrho_-(\lambda)\, \mathcal{A}(\lambda)$, where $\mathcal{A}(\lambda)$ is a non-degenerate symbol of a Wiener-Hopf integro-difference operator and $\varrho_-(\lambda)$ has the form (2.26_1) resp. (2.28), Chapter 5, or is the reciprocal of an entire Polya function[1]) (and thus has a zero of infinite order at the point $\lambda = \infty$).

Finally we mention at a paper of V. N. GAPONENKO and W. B. DYBIN [1], where Wiener-Hopf integro-difference equations are considered under the assumption that the symbol has the form

$$\mathcal{A}(\lambda) = (\xi - e^{-i\alpha\lambda})^{\gamma} \mathcal{B}(\lambda) \qquad (-\infty < \lambda < \infty)$$

with $|\xi| = 1$, $\operatorname{Im} \alpha = 0$, $\gamma > 0$ (arbitrary real) and $\mathcal{B}(\lambda)$ non-degenerated. The authors apply the method of the normalization equivalent to the method of Section 4.1 (in the case $D = I$) to such equations (cf. also 1.5.3 and 4.5.3).

10.4. All results of this section were proved by I. C. GOHBERG and N. J. KRUPNIK [3] (comp. also [1], [2], [4]).

Similar results were stated by R. V. DUDUČAVA [2] for paired integral operators of convolution type with discontinuous matrix symbols. More general classes of integral operators with discontinuous coefficients containing the singular integral operators over the real line as well as paired convolution operators were also recently studied by R. V. DUDUČAVA [3], [4].

10.5.—10.8. All results of these sections are due to B. SILBERMANN [5]—[8]. In our statements we essentially follow the thesis of B. SILBERMANN [8]. We remark that in the paper of B. SILBERMANN [5] singular integral operators of non-normal type of the kind considered in Section 10.5 are also considered on some other pairs of spaces.

Finally we remark: It is possible to show that the last assertion of Theorem 7.1 also holds in the case $K(R_1, R_2) \neq 0$.

By means of the results of Section 10.8 singular integral equations with Carleman shift and the Tricomi integral equation were recently studied in non-normal cases by Ch. MEYER [1].

In the papers of M. I. HAIKIN [4], [5] singular operators of the form $A = aP + bQ$ $(a, b \in \Lambda(\Gamma))$ were investigated in the space $L^p(\Gamma)$. Here the symbol degenerates in another sense: Condition (6.4) is satisfied but the function $a(t)/b(t)$ is singular (cf. Section 10.4). The method proposed by M. I. HAIKIN is based on the approximation of the operator A by singular operators whose symbol does not degenerate. By means of the concept of one-sided approximative inverse introduced by this author $\ker A$ and $\operatorname{im} A$ are described (comp. also M. I. HAIKIN [6]).

---

[1]) Cf. I. I. HIRSCHMAN and D. V. WIDDER ([1], III. 3).

CHAPTER 11

APPROXIMATION METHODS FOR THE SOLUTION OF SINGULAR EQUATIONS

In the present chapter we consider projection methods with projections unbounded in general. First, for linear continuous operators on Banach spaces, we prove theorems on the stability of the manifold of convergence of projection methods with respect to compact or small perturbations. We state error estimates and criterions for the numerical stability of these methods. Then we establish two stable projection methods for abstract singular equations. From this we show that the reduction method resp. the collocation method can be applied to system of discrete Wiener-Hopf equations, Wiener-Hopf integral equations and singular integral equations in the normal case as well as in the non-normal case. For various concrete equation types and spaces we determine the rate of convergence of these methods.

11.1. General Theorems on Projection Methods

Let X and Y be Banach spaces and $\{P_n\}$, $\{Q_n\}$ $(n = 1, 2, ...)$ two sequences of (in general unbounded) projections with domains $D(P_n) \subset X$, $D(Q_n) \subset Y$ and the closed image spaces $\operatorname{im} P_n \subset X$, $\operatorname{im} Q_n \subset Y$.

Let $A \in \mathcal{L}(X, Y)$. In general, the transition from the operator equation

$$Ax = y \tag{1.1}$$

to the projection equation

$$Q_n A P_n x_n = Q_n y \qquad (y \in D(Q_n)) \tag{1.2}$$

for which a solution $x_n \in \operatorname{im} P_n$ is found is called a *projection method* $\{P_n, Q_n\}$ for an approximate solution of equation (1.1).

In the case of bounded projections P_n and Q_n the projection method $\{P_n, Q_n\}$ is said to be *applicable* to the operator A if there exists an n_0 such that for all $n \geqq n_0$ and for all elements $y \in Y$ the equation (1.2) possesses one and only one solution x_n and if the sequence of these solutions converges to the solution of equation (1.1) for $n \to \infty$. In this case we write $A \in \Pi\{P_n, Q_n)$. We say in this case also that for the operator A the projection method $\{P_n, Q_n\}$ converges for *all* $y \in Y$. If in particular $A \in \mathcal{L}(X)$ and if the projection method $\{P_n, P_n\}$ converges for all $y \in \operatorname{im} A$, then we speak of the convergence of the *natural projection method* $\{P_n, P_n\}$.

If for $n \to \infty$ the projections P_n and Q_n converge strongly to the identity operator (in X resp. Y), then $A \in \Pi\{P_n, Q_n\}$ if and only if the operators $A\colon X \to Y$ and $Q_nAP_n\colon \operatorname{im} P_n \to \operatorname{im} Q_n$ $(n \geqq n_0)$ are invertible and the operators $(Q_nAP_n)^{-1} Q_n$ converge strongly to the inverse A^{-1} for $n \to \infty$.[1]) This follows easily from the Banach-Steinhaus theorem and the estimate

$$||P_nx|| = ||(Q_nAP_n)^{-1} Q_nAP_nx|| \leqq c||Q_nAP_nx|| \, .$$

But in many applications the projection method does not converge for any element $y \in Y$ $(y \in \operatorname{im} A)$ (see Section 11.4 and the following sections of this chapter). Obviously, we have such a case if the space X is not separable and the projections P_n are finite dimensional or if the projections Q_n are unbounded. Therefore it is reasonable to give the following definition.

Definition. In the sequel we assume that for $n \geqq n_0$ the relation $A(\operatorname{im} P_n) \subset \operatorname{im} Q_n$ holds and the operators $Q_nAP_n\colon \operatorname{im} P_n \to \operatorname{im} Q_n$ are continuous and (continuously) invertible. By $\mathfrak{K}(A; \{P_n, Q_n\})$ we denote the set of all $y \in Y$ with the following properties:

1. $y \in D(Q_n)$ for $n \geqq n'(y)$.
2. The sequence of elements $x_n = (Q_nAP_n)^{-1} Q_ny$ converges in the norm of X to an element $x \in X$ for $n \to \infty$.
3. $Ax = y$.

The set $\mathfrak{K}(A; \{P_n, Q_n\})$ is called the *manifold of convergence* of the operator A with respect to the projection method $\{P_n, Q_n\}$.

Obviously, the manifold of convergence is a linear set contained in the image $\operatorname{im} A$. When A is invertible the set $\mathfrak{K}(A; \{P_n, Q_n\})$ is the maximal linear set on which the operators $(Q_nAP_n)^{-1} Q_n$ converge strongly to A^{-1} for $n \to \infty$. Furthermore, we have $\mathfrak{K}(A; \{P_n, Q_n\}) = Y$ if and only if $A \in \Pi\{P_n, Q_n\}$.

In this section we investigate the relations between the manifolds of convergence of the operators A and $A + T$, where T is a compact or a small (in the operator norm) operator. Furthermore, we state estimates for the error $||x - x_n||$ and formulate criteria for the stability of a projection method.

Theorem 1.1. *Let $A \in \mathscr{L}(X, Y)$ be an invertible operator, $Z \subset Y$ a Banach space continuously embedded into Y*[2]) *with $Z \subset \mathfrak{K}(A; \{P_n, Q_n\})$ and $T \in \mathscr{K}(X, Z)$. Let the following two conditions hold*

$$1.\ \dim \ker (A + T) = 0 \, ,$$

$$2.\ Q_n|Z \in \mathscr{L}(Z, Y) \, .$$

[1]) The same assertion holds in the case $\dim \operatorname{im} P_n = \dim \operatorname{im} Q_n < \infty$ (see I. C. Gohberg and V. I. Levčenko [1]).

[2]) In the sequel an inclusion of two Banach spaces $Z \subset Y$ shall always be understood in the sense of a continuous embedding.

Then the operators $Q_n(A + T) P_n$: im $P_n \to$ im Q_n *are invertible for* $n \geqq n_1$ *and*

$$\mathfrak{R}(A; \{P_n, Q_n\}) = \mathfrak{R}(A + T; \{P_n, Q_n\}) .$$

Proof. The invertibility of the operator A and condition 1 of the theorem show that the operator $\tilde{A} = A + T$ and thus also the operator $I + A^{-1}T$ is invertible. Furthermore, we have $\tilde{A}^{-1} = (I + A^{-1}T)^{-1} A^{-1}$.

Condition 2 of the theorem yields im $T \subset D(Q_n)$ and $Q_nT \in \mathcal{L}(X, Y)$ ($n = 1, 2, \ldots$). We put

$$A_n = Q_nAP_n, \tilde{A}_n = Q_n\tilde{A}P_n = A_n + Q_nTP_n \qquad (n \geqq n_0) .$$

By the hypotheses of the theorem the sequence of operators $A_n^{-1}Q|Z \in \mathcal{L}(Z, X)$ converges strongly to the operator $A^{-1}|Z$ on the Banach space Z for $n \to \infty$. As is well-known, this convergence then holds uniformly on each relatively compact subset of the space Z, and since $T \in \mathcal{K}(X, Z)$, we obtain

$$\delta_n = ||A^{-1}T - A_n^{-1}Q_nT|| \to 0 \quad \text{for} \quad n \to \infty . \tag{1.3}$$

Hence for $n \geqq n_1$ the operator $I + A_n^{-1}Q_nT \in \mathcal{L}(X)$ is invertible. Thus for $n \geqq n_1$ the operator $\tilde{A}_n$: im $P_n \to$ im Q_n is also invertible and

$$\tilde{A}_n^{-1} = (I + A_n^{-1}Q_nT)^{-1} A_n^{-1} . \tag{1.4}$$

Now let y be an arbitrary element of $\mathfrak{R}(A; \{P_n, Q_n\})$. We will show that the sequence $\tilde{A}_n^{-1}Q_ny$ converges in the norm of X to the element $\tilde{A}^{-1}y$ for $n \to \infty$. Since $A_n^{-1}Q_ny \to A^{-1}y$ for $n \to \infty$ and the equation (1.4) holds, it is sufficient to show that the sequence

$$z_n = [(I + A^{-1}T)^{-1} - (I + A_n^{-1}Q_nT)^{-1}] A_n^{-1}Q_ny$$

converges to zero. But this follows immediately from the well-known estimate

$$||z_n|| \leqq ||(I + A^{-1}T)^{-1}||^2 \frac{\delta_n ||A_n^{-1}Q_ny||}{1 - \delta_n ||(I + A^{-1}T)^{-1}||}$$

together with the relation (1.3). Hence

$$\mathfrak{R}(A; \{P_n, Q_n\}) \subset \mathfrak{R}(\tilde{A}; \{P_n, Q_n\}) .$$

Since $A = \tilde{A} - T$ and $Z \subset \mathfrak{R}(\tilde{A}; \{P_n, Q_n\})$, interchanging the roles of A and $\tilde{A}$ and repeating the preceding considerations we obtain

$$\mathfrak{R}(\tilde{A}; \{P_n, Q_n\}) \subset \mathfrak{R}(A; \{P_n, Q_n\}) .$$

This completes the proof of the theorem.

Corollary 1.1. *Let* $Q_n \in \mathcal{L}(Y)$ $(n = 1, 2, \ldots)$. *Let* $A \in \Pi\{P_n, Q_n\}$ *be an invertible operator and* $T \in \mathcal{K}(X, Y)$ *a compact operator. If the operator* $A + T$ *is invertible, then this operator also belongs to the class* $\Pi\{P_n, Q_n\}$.

This assertion follows from Theorem 1.1 for $Z = Y$.

There is an analogue of Theorem 1.1 for perturbations of the operator A by operators small with respect to the norm.

Theorem 1.2. *Let $A \in \mathcal{L}(X, Y)$ be an invertible operator and $Z \subset Y$ a Banach space with $Z \subset \mathfrak{R}(A; \{P_n, Q_n\})$. Furthermore, let $Q_n | Z \in \mathcal{L}(Z, Y)$, $P_n \in \mathcal{L}(X)$ and $P_n x \to x$ $(n \to \infty)$ for all $x \in X$. Then there exists a constant $\gamma > 0$ such that for any operator $T \in \mathcal{L}(X, Z)$ with $\|T\|_{X\to Z} < \gamma$ the operators $Q_n(A + T) P_n$: $\operatorname{im} P_n \to \operatorname{im} Q_n$ $(n \geqq n_0)$ are invertible and*

$$\mathfrak{R}(A; \{P_n, Q_n\}) \subset \mathfrak{R}(A + T; \{P_n, Q_n\}) .$$

Proof. We maintain the notation introduced in the proof of Theorem 1.1.

Since the continuous operators $A_n^{-1}Q_n | Z$ converge strongly on Z to $A^{-1} | Z$ and the operators $P_n \in \mathcal{L}(X)$ converge strongly on X to the identity operator, by the Banach-Steinhaus theorem for certain positive constants γ_1 and γ_2 we have the estimates

$$\|A_n^{-1}Q_n\|_{Z\to X} \leqq \gamma_1 , \ \|P_n\|_X \leqq \gamma_2 \qquad (n = 1, 2, \ldots).$$

We put $\gamma = 1/\gamma_1\gamma_2$.

Let $T \in \mathcal{L}(X, Z)$ be an arbitrary operator with the norm $\|T\|_{X\to Z} < \gamma$. For $\beta = \gamma_1\gamma_2\|T\|_{X\to Z}$ we then obtain $\beta < 1$. The operator $\widetilde{A}_n = A_n + Q_n T P_n$ can be written in the form

$$\widetilde{A}_n = A_n(I + B_n) , \qquad B_n = A_n^{-1}Q_n T P_n \qquad (n \geqq n_0) ,$$

where

$$\|B_n\|_X \leqq \beta < 1 . \tag{1.5}$$

Thus the operator $I + B_n$ $(n \geqq n_0)$ is invertible on X as well as on the subspace $\operatorname{im} P_n$, and we have

$$\widetilde{A}_n^{-1} = \left(\sum_{k=0}^{\infty} (-1)^k B_n^k\right) A_n^{-1} \qquad (n \geqq n_0) .$$

Since by the hypotheses of the theorem B_n converges strongly to $A^{-1}T$, the relation (1.5) shows that $\|A^{-1}T\|_X \leqq \beta$. This implies the existence of the continuous operator

$$(I + A^{-1}T)^{-1} = \sum_{k=0}^{\infty} (-1)^k (A^{-1}T)^k .$$

Now we show that for $n \to \infty$ the operator

$$C_n = \sum_{k=0}^{\infty} (-1)^k B_n^k$$

converges strongly on the space X to $(I + A^{-1}T)^{-1}$.

Let ε be an arbitrary positive number and $x \in X$. We choose an integer $m = m(x) > 0$ such that $\frac{\beta^{m+1}}{1-\beta}\|x\| < \frac{\varepsilon}{4}$. Then

$$\left\|\sum_{k=m+1}^{\infty} (-1)^k (A^{-1}T)^k x - \sum_{k=m+1}^{\infty} (-1)^k B_n^k x\right\| < \frac{\varepsilon}{2}.$$

Since obviously for $0 \leqq k \leqq m$ and $n \to \infty$ B_n^k converges strongly to $(A^{-1}T)^k$, for $n > N(\varepsilon)$ we obtain the inequality

$$\left\|\sum_{k=0}^{m} (-1)^k (A^{-1}T)^k x - \sum_{k=0}^{m} (-1)^k B_n^k x\right\| < \frac{\varepsilon}{2}.$$

Hence C_n converges strongly to $(I + A^{-1}T)^{-1}$.

Now let $y \in \mathfrak{R}(A; \{P_n, Q_n\})$. Then $y \in D(Q_n)$ for $n \geqq n'(y)$ and $A_n^{-1}Q_n y$ converges to $A^{-1}y$ in X. Thus we obtain

$$\tilde{A}_n^{-1}Q_n y = C_n A_n^{-1} Q_n y \to (I + A^{-1}T)^{-1} A^{-1} y = \tilde{A}^{-1} y,$$

i.e. $y \in \mathfrak{R}(A + T; \{P_n, Q_n\})$. This completes the proof of the theorem.

Corollary 1.2. *Let $P_n \in \mathcal{L}(X)$, $Q_n \in \mathcal{L}(Y)$ $(n = 1, 2, \ldots)$ and $P_n x \to x (\forall x \in X)$. Let $A \in \Pi\{P_n, Q_n\}$ be an invertible operator. Then there exists a constant $\gamma > 0$ such that for all $T \in \mathcal{L}(X, Y)$ with $\|T\| < \gamma$ the relation $A + T \in \Pi\{P_n, Q_n\}$ holds.*

The next theorem yields an error estimate for a certain class of operators A.

Theorem 1.3. *Let $A = CB$, where $B \in \mathcal{L}(X, Y)$ and $C \in \mathcal{L}(Y)$ are invertible operators satisfying the following two conditions*

$$\text{(a)}\ Q_n B P_n x = B P_n x, \qquad \forall x \in D(P_n), \qquad n = 1, 2, \ldots;$$

$$\text{(b)}\ Q_n C^{\pm 1} Q_n y = Q_n C^{\pm 1} y, \qquad \forall y \in D(Q_n).$$

Furthermore, let the conditions of Theorem 1.1 hold.

If $y \in \mathfrak{R}(A; \{P_n, Q_n\}$ and $x \in X$, $x_n \in P_n$ are the solutions of the equations

$$Ax + Tx = y, \qquad Q_n(A + T)\, x_n = Q_n y,$$

respectively, then we have the estimate

$$c_1\|Bx - Q_n Bx\|_Y \leqq \|x - x_n\|_X \leqq c_2\|Bx - Q_n Bx\|_Y \tag{1.6}$$

and in the case of continuous projections Q_n the estimate

$$\|x - x_n\|_X \leqq c\|Q_n\|\, E_n(Y, Bx). \tag{1.7}$$

Here c_1, c_2, c are positive constants and

$$E_n(Y, f) = \inf_{y_n \in \operatorname{im} Q_n} \|f - y_n\|_Y.$$

Proof. We put $T_1 = C^{-1}T$. Obviously, the equations

$$Ax + Tx = y \quad \text{and} \quad Bx + T_1 x = C^{-1} y$$

and by condition (b) also the equations

$$Q_n(A + T)\, x_n = Q_n y \quad \text{and} \quad Q_n(B + T_1)\, x_n = Q_n C^{-1} y$$

are equivalent. Using the condition (a) we obtain

$$(B + Q_n T_1)\,(x - x_n) = Bx - Q_n Bx\,. \tag{1.8}$$

The inequality (1.6) holds if

$$\|T_1 - Q_n T_1\|_{X\to Y} \to 0 \quad \text{for} \quad n \to \infty\,. \tag{1.9}$$

Namely, from this we get the uniform convergence

$$B + Q_n T_1 \to B + T_1\,,\ (B + Q_n T_1)^{-1} \to (B + T_1)^{-1}\,,$$

and (1.8) at once implies inequality (1.6).

We prove relation (1.9). We put $A_n = Q_n A P_n$. It follows from (a) and (b) that

$$A_n = Q_n C Q_n B P_n\,,$$

$$B^{-1} Q_n C^{-1} A_n = B^{-1} Q_n B P_n = P_n$$

and $Q_n C Q_n$: $\operatorname{im} Q_n \to \operatorname{im} Q_n$ is invertible. Thus it follows from the invertibility of A_n: $\operatorname{im} P_n \to Q_n$ that the operator BP_n: $\operatorname{im} P_n \to \operatorname{im} Q_n$ is invertible and

$$A_n^{-1} Q_n = B^{-1} Q_n C^{-1}\,.$$

Consequently we obtain

$$\|T_1 - Q_n T_1\| \leqq \|B\|\, \|B^{-1}C^{-1}T - B^{-1}Q_n C^{-1} T\| = \|B\|\, \|A^{-1}T - A_n^{-1} Q_n T\|,$$

and this together with (1.3) implies (1.9).

It remains to prove inequality (1.7). For arbitrary elements $z \in D(Q_n)$ and $z_n \in \operatorname{im} Q_n$ we have

$$\|z - Q_n z\| \leqq \|z - z_n\|_Y + \|Q_n(z - z_n)\|_Y \leqq (1 + \|Q_n\|)\, \|z - z_n\|_Y\,.$$

From this it follows at once for $z = Bx$ that

$$\|Bx - Q_n Bx\|_Y \leqq 2\|Q_n\|\, E_n(Y, Bx)\,.$$

This together with the second inequality (1.6) implies the estimate (1.7). The theorem is proved.

Corollary 1.3. *Let the conditions of the Theorem* 1.3 *hold. Furthermore, let* $Bx \in Z$, *where* $Z \subset Y$ *are Banach spaces satisfying the condition*

$$\text{(c)}\ \|y - Q_n y\|_Y \to 0\,, \qquad \forall y \in Z\,, \qquad n \to \infty\,.$$

Then the estimate

$$\|x - x_n\|_X \leqq c E_n(Z, Bx)$$

holds.

Proof. In view of the continuous embedding $Z \subset Y$ for any $z \in Z$, $z_n \in \operatorname{im} Q_n$ we have the estimate

$$\|z - Q_n z\|_Y \leqq (c' + \|Q_n\|_{Z\to Y})\, \|z - z_n\|_Z\,.$$

From this we obtain the assertion using condition (c) and the Banach-Steinhaus theorem.

The following theorem yields a sufficient condition for a certain set Z to belong to the manifold of convergence of an operator.

Theorem 1.4. *Let $A = CB$, where $B \in \mathcal{L}(X, Y)$ and $C \in \mathcal{L}(Y)$ are invertible operators satisfying the following two conditions:*

(a') im $BP_n =$ im Q_n $(n = 1, 2, \ldots)$;
(b') *There exists a set $Z \subset Y$ with $C^{-1}(Z) \subset Z$.*

Furthermore, let the above-mentioned conditions (b) *and* (c) *be satisfied.*

Then $Z \subset \mathfrak{K}(A; \{P_n, Q_n\})$.

Proof. Condition (b) implies

$$Q_n C^{-1} Q_n C = Q_n = Q_n C Q_n C^{-1} .$$

This together with (a') shows that the operator $A_n = Q_n A P_n$: im $P_n \to$ im Q_n is invertible, where

$$A_n^{-1} Q_n = B^{-1} Q_n C^{-1} .$$

Let $z \in Z$ be an arbitrary element. By condition (d) we have $C^{-1} z \in Z$, and using condition (c) we obtain

$$A_n^{-1} Q_n z \to B^{-1} C^{-1} z = A^{-1} z ,$$

i.e. $z \in \mathfrak{K}(A; \{P_n, Q_n\})$. This completes the proof of the theorem.

Remark. It is easy to see from the preceding arguments that Theorems 1.3 and 1.4 and Corollary 1.3 can be modified in the following way:

Let $Y \subset X$ and $A = CB$, where $B \in \mathcal{L}(X)$ and $C \in \mathcal{L}(X, Y)$ are invertible operators. We replace the conditions specified in these theorems by the following conditions:

(a) $\quad P_n B P_n = B P_n$ (resp. (a') im $BP_n =$ im P_n) ;

(b) $\quad Q_n C P_n = Q_n C , \qquad P_n C^{-1} Q_n = P_n C^{-1}$;

(c) $\quad \|y - P_n y\|_X \to 0 , \qquad \forall\, y \in Z$.

Then the statements of Theorems 1.3 and 1.4 and Corollary 1.3 remain valid, where in the estimates (1.6) and (1.7) we have to put $Q_n = P_n$ and $Y = X$.

Because of the errors appearing in the projection equation (1.2), in calculating of the operator $A_n = Q_n A P_n$ and the right-hand side $Q_n y$ (e.g. rounding off errors in the calculation of scalar products, integrals etc.) in fact we have nor to solve the equation (1.2) but a certain perturbed equation of the form

$$(A_n + \Gamma_n)\, \tilde{x}_n = Q_n y + \delta_n . \tag{1.10}$$

Of course we shall demand that $\tilde{x} \in$ im P_n, $\delta_n \in$ im Q_n and Γ_n: im $P_n \to$ $\to$ im Q_n. The projection method $\{P_n, Q_n\}$ is stable[1]) in the following sense.

[1]) For the definition of stability cf. S. G. Mihlin [5].

Theorem 1.5. *Let $A \in \mathcal{L}(X, Y)$ be an invertible operator and $Z \subset Y$ a Banach space with $Z \subset \mathfrak{R}(A; \{P_n, Q_n\})$. Furthermore, let* $\operatorname{im} Q_n \subset Z$, $Q_n | Z \in \mathcal{L}(Z, Y)$ *and $y \in Z$.*

Then there exist positive constants p, q, γ independent of n and y such that for $\|\Gamma_n\|_{X\to Z} < \gamma$ the following statements hold:

1. *For any $\delta_n \in \operatorname{im} Q_n$ $(n \geqq n')$ the equation* (1.10) *is uniquely solvable.*
2. $\|\tilde{x}_n - x_n\|_X \leqq p\,\|y\|_Z\,\|\Gamma_n\|_{X\to Z} + q\,\|\delta_n\|_Z$.

Proof. Since the continuous operators $A_n^{-1}Q_n | Z$ converge strongly to $A^{-1}|Z$ on the Banach space Z, by the Banach-Steinhaus theorem there exists a constant $\gamma_1 > 0$ independent of n such that

$$\|A_n^{-1}Q_n\|_{Z\to X} < \gamma_1 \qquad (n = 1, 2, \ldots)\,.$$

Let γ be chosen such that $\gamma\gamma_1 = \beta < 1$. Then the operator $\tilde{A}_n = A_n + \Gamma_n$ is invertible for $\|\Gamma_n\|_{X\to Z} < \gamma$ and

$$\tilde{A}_n^{-1} = \left[\sum_{k=0}^{\infty} (-1)^k (A_n^{-1}Q_n\Gamma_n)^k\right] A_n^{-1}\,. \tag{1.11}$$

The relations

$$x_n = A_n^{-1}Q_n y\,, \qquad \tilde{x}_n = \tilde{A}_n^{-1}Q_n y + \tilde{A}_n^{-1}Q_n\delta_n$$

yield

$$\tilde{x}_n - x_n = (-A_n^{-1}Q_n\Gamma_n)\left[\sum_{k=0}^{\infty}(-1)^k (A_n^{-1}Q_n\Gamma_n)^k\right] A_n^{-1}Q_n y +$$
$$+ \tilde{A}_n^{-1}Q_n\delta_n = -A_n^{-1}Q_n\Gamma_n\tilde{A}_n^{-1}Q_n y + \tilde{A}_n^{-1}Q_n\delta_n\,.$$

For the operator $\tilde{A}_n^{-1}Q_n$ we obtain the norm estimate

$$\|\tilde{A}_n^{-1}Q_n\|_{Z\to X} < \frac{\gamma_1}{1-\beta}$$

from the relation (1.11). Hence

$$\|\tilde{x}_n - x_n\|_X \leqq \frac{\gamma_1^2}{1-\beta}\|\Gamma_n\|_{X\to Z}\,\|y\|_Z + \frac{\gamma_1}{1-\beta}\|\delta_n\|_Z\,.$$

This completes the proof of the theorem.

Remark 1. If the conditions of the Theorem 1.5 are satisfied but y is an *arbitrary* element from the manifold of convergence $\mathfrak{R}(A; \{P_n, Q_n\})$, then the estimate

$$\|\tilde{x}_n - x_n\|_X \leqq \frac{\gamma_1}{1-\beta}[\|\Gamma_n\|_{X\to Z}\,\|A_n^{-1}Q_n y\|_X + \|\delta_n\|_Z]\,.$$

holds.

Remark 2. All results of this section remain valid if the index set for the index n is not the set of the natural numbers but an arbitrary non-bounded set of positive real numbers.

11.2. Projection Methods for the Solution of Abstract Singular Equations

Let X be a Banach space, P a continuous projection in X and $Q = I - P$. Furthermore, let $\mathfrak{M}$ be a subalgebra of the algebra $\mathcal{L}(X)$ of all continuous linear operators in X with the property that each operator $a \in \mathfrak{M}$ invertible in $\mathcal{L}(X)$ has an inverse $a^{-1} \in \mathfrak{M}$.

We consider the abstract singular operator of the form

$$A = aP + bQ + T \tag{2.1}$$

with $a, b \in \mathfrak{M}$ and $T \in \mathcal{K}(X)$. Under certain additional hypotheses on the operator a, b and T in this section we establish two different projection methods for the approximate solution of the abstract singular equation

$$Ax = y\,. \tag{2.2}$$

Similarily as in Section 2.2 by $\mathfrak{M}^+(\mathfrak{M}^-)$ we denote the subalgebra of the algebra $\mathfrak{M}$ consisting of all operators $a \in \mathfrak{M}$ with the property $aP = PaP$ $(Pa = PaP)$. By $\mathfrak{M}_l(\mathfrak{M}_r)$ we denote the collection of all operators $a \in \mathfrak{M}$ representable in the form

$$a = a_+a_- \qquad (a = a_-a_+)$$

with $a_+^{\pm 1} \in \mathfrak{M}^+$ and $a_-^{\pm 1} \in \mathfrak{M}^-$.

11.2.1. First we consider operators of the form (2.1) with coefficients

$$a = \varrho_+ c\,, \qquad b = \varrho_- d\,, \tag{2.3}$$

where $c, d \in \mathfrak{M}$ are invertible operators and $\varrho_\pm$ operators from $\mathfrak{M}^\pm$.

We consider a sequence of projections $P_n \in \mathcal{L}(X)$ $(n = 1, 2, \ldots)$. We assume that the following conditions hold:

Ia. The operators P_n converge strongly to the identity operator in X for $n \to \infty$.

Ib. For any $c_\pm \in \mathfrak{M}^\pm$ we have the relations

$$P_n(Pc_- + Qc_+)\, P_n = (Pc_- + Qc_+)\, P_n\,,$$

$$P_n(c_+P + c_-Q)\, P_n = P_n(c_+P + c_-Q)\,.$$

Ic. Let $D = \varrho_+P + \varrho_-Q$. Then

$$\dim \ker (P_nD \mid \operatorname{im} P_n) = 0\,.$$

In particular, it follows from these conditions that $\dim \ker D = 0$. Thus the image space $X^D = \operatorname{im} D$ is a Banach space in the norm $\|y\| = \|x\|_X$ $(y = Dx)$, and X^D is continuously embedded into X (cf. Section 4.1). In the sequel we assume that there exists a Banach space $Z \subset X^D$ such that the following conditions hold:

Id. $a \mid Z \in \mathcal{L}(Z)$ and $Pa - aP \in \mathcal{K}(X, Z)$ for any $a \in \mathfrak{M}$.

Ie. $\operatorname{im} P_n \subset Z$.

It follows from Ie by the closed graph theorem that the restriction $P_n^D = P_n|X^D$ is a continuous projection in the space X^D.

Remark. If the projections P_n ($n = 1, 2, \ldots$) are finite dimensional, then Ic is a consequence of condition Ie.

In fact, it follows from Ie that

$$\operatorname{im} P_n D P_n = \operatorname{im} P_n .$$

But in a finite dimensional space this is only possible if Ic holds.

Theorem 2.1. *Let the hypotheses* Ia *to* Ie *hold. Let A be an operator of the form* (2.1) *with $T \in \mathcal{K}(X, Z)$ and the coefficients* (2.3), *where $c \in \mathfrak{M}_l$ and $d \in \mathfrak{M}_r$. Furthermore, let* $\dim \ker A = 0$. *Then $A \in \mathcal{L}(X, X^D)$ is an invertible operator and*

$$A \in \Pi\{P_n, P_n^D\} .$$

If $x \in X$ and $x_n \in \operatorname{im} P_n$ are the solutions of the equations (2.2) *and*

$$P_n A x_n = P_n y ,$$

respectively, then the estimate

$$\|x - x_n\|_X \leqq \gamma \|y_0 - P_n y_0\|_X \tag{2.4}$$

holds. Here $\gamma > 0$ is a constant and $y_0 = (Pc_- + Qd_+)\, x$, where

$$c = c_+ c_- , \qquad d = d_- d_+ . \tag{2.5}$$

Proof. Using relations (2.5) we easily obtain the representation

$$A = D(c_+ P + d_- Q)\,(Pc_- + Qd_+) + T' , \tag{2.6}$$

where

$$T' = \varrho_+ c_+ Q c_- P + \varrho_- d_- P d_+ Q + T . \tag{2.7}$$

The condition Id implies $T' \in \mathcal{K}(X, Z)$, and equation (2.6) yields $A \in \mathcal{L}(X, X^D)$. The operator

$$A_0 = D(c_+ P + d_- Q)\,(Pc_- + Qd_+) \in \mathcal{L}(X, X^D)$$

has the continuous inverse

$$A_0^{-1} = (Pc_-^{-1} + Qd_+^{-1})\,(c_+^{-1} P + d_-^{-1})\, D^{-1} .$$

Since $\dim \ker A = 0$, the operator $A = A_0 + T' \in \mathcal{L}(X, X^D)$ is also invertible.

By the conditions Ib and Ic the operator

$$P_n^D D P_n = P_n^D D \colon \operatorname{im} P_n \to \operatorname{im} P_n^D$$

is invertible, where

$$(P_n^D D P_n)^{-1} P_n^D = P_n D^{-1} .$$

Hence also the operator $P_n^D A_0 P_n \colon \operatorname{im} P_n \to \operatorname{im} P_n^D$ is invertible and

$$(P_n^D A_0 P_n)^{-1} P_n^D = (Pc_-^{-1} + Qd_+^{-1})\, P_n (c_+^{-1} P + d_-^{-1} Q)\, D^{-1} .$$

By Ia it follows immediately from this that $A_0 \in \Pi\{P_n, P_n^D\}$. Finally, Corollary 1.1 shows that $A \in \Pi\{P_n, P_n^D\}$.

From Theorem 1.3 we obtain the estimate (2.4) taking into consideration the remark after Theorem 1.4, where in the present case we have to put

$$C = D(c_+P + d_-Q)\,, \qquad B = (Pc_- + Qd_+)$$

(see equation (2.6)).

Remark 1. Theorem 2.1 is also true if in its hypotheses we put $Z = X^D$ everywhere.

Remark 2. Under the hypotheses of Theorem 2.1 the statements of Theorem 1.5 hold with $Z = X^D$, i.e. the projection method $\{P_n, P_n^D\}$ is stable.

11.2.2. In this subsection we consider abstract singular operators of the form (2.1) with coefficients of the form

$$a = c\varrho_-\,, \qquad b = d\varrho_+\,, \tag{2.8}$$

where $c, d \in \mathfrak{M}$ are invertible operators and $\varrho_\pm \in \mathfrak{M}^\pm$.

Let the following conditions hold:

IIa. $\dim \ker B = 0$ for $B = P\varrho_- + Q\varrho_+$.

Again we equip the image space $X^B = \operatorname{im} B \subset X$ with the norm $\|y\| = \|x\|_X$ $(y = Bx)$. In the sequel we assume that there exist two Banach spaces $Z_0 \subset Z \subset X^B$ with the following properties:

IIb. $P\,|\,Z \in \mathcal{L}(Z)$ and $a\,|\,Z \in \mathcal{L}(Z)$ $\quad (\forall a \in \mathfrak{M})$.

IIc. $a\,|\,Z_0 \in \mathcal{L}(Z_0)$ and $Pa - aP \in \mathcal{K}(X, Z_0)$ $\quad (\forall a \in \mathfrak{M})$.

Let $\{P_n\}$ and $\{Q_n\}$ be two sequences of projections in the space X with $P_n \in \mathcal{L}(X)$. Let the domain $D(Q_n)$ of the projection Q_n be a linear subset of X. Furthermore, let the following conditions be satisfied (for $n \geqq n_0$):

IId. $P_n(Pc_- + Qc_+)\,P_n = (Pc_- + Qc_+)\,P_n \quad (\forall c_\pm \in \mathfrak{M}^\pm)$.

IIe. $\operatorname{im} P_n \subset B\,(\operatorname{im} P_n)$.

IIf. $\operatorname{im} P_n = \dim Q_n \subset Z$.

IIg. $Q_n\,|\,Z_0 \in \mathcal{L}(Z_0, Z)$ and for any $z \in Z_1$ the sequence $Q_n z$ converges to $z (n \to \infty)$ in the norm of Z, where $Z_0 \subset Z_1 \subset D(Q_n)$. Finally, let one of the two following conditions hold:

IIh. $Q_n a Q_n = Q_n a$ and $a(Z_1) \subset Z_1 \quad (\forall a \in \mathfrak{M})$.

IIi. For any operator of the form $C = c_+P + c_-Q$ with $c_\pm \in \mathfrak{M}^\pm$ we have

$$Q_n C Q_n = Q_n C \quad \text{and} \quad C(Z_j) \subset Z_j \qquad (j = 0, 1)\,.$$

Remark. If the projectors P_n are finite dimensional, then IIe is a consequence of conditions IIa and IId.

Namely, by IId we have

$$P_n B P_n = B P_n\,,$$

i.e. $B(\operatorname{im} P_n) \subset \operatorname{im} P_n$. Thus condition IIe means that B maps the subspace $\operatorname{im} P_n$ onto itself. But in the case of a finite dimensional space $\operatorname{im} P_n$ this follows already from IIa.

Theorem 2.2. *Let the conditions* IIa *to* IIh *hold. Let A be an operator of the form* (2.1) *with $T \in \mathcal{K}(X, Z_0)$ and the coefficients* (2.8), *where $c, d \in \mathfrak{M}$ are invertible operators and $d^{-1}c = b_+b_- \in \mathfrak{M}_l$. Furthermore, let* $\dim \ker A = 0$.

Then for $n \geqq n_1$ and for any $y \in Z_1$ the equation

$$Q_n A P_n x_n = Q_n y$$

possesses a unique solution $x_n \in \operatorname{im} P_n$ and x_n converges in the norm of X to the (unique) solution $x \in X$ of the equation $Ax = y$ for $n \to \infty$. We have the estimate

$$\|x - x_n\|_X \leqq \gamma \|y_0 - Q_n y_0\|_Z \,, \tag{2.9}$$

where $\gamma > 0$ is a constant (independent of n and y) and

$$y_0 = (Pb_- + Qb_+^{-1})\, Bx \in Z_1 \,.$$

If $y \in Z_0$, then $y_0 \in Z_0$ and

$$\|x - x_n\| \leqq \gamma' E_n(Z_0, y_0) \tag{2.10}$$

with

$$E_n(Z_0, y_0) = \inf_{y_n \in \operatorname{im} Q_n} \|y_0 - y_n\|_{Z_0} \,.$$

Proof. Let $\bar{X} = \{x \in X\colon Bx \in Z\}$. $\bar{X}$ with the norm $|x| = \|Bx\|_Z$ $(x \in \bar{X})$ is a Banach space, where $\bar{X} \subset X$. IIb implies $Z \subset \bar{X}$. Obviously the restriction $\bar{B} = B|\bar{X}$ maps the space $\bar{X}$ continuously and biuniquely onto Z.

Let $E\colon \bar{X} \to X$ be the embedding operator. Because of $d^{-1}c = b_+b_-$ for the restriction $\bar{A} = A\,|\,\bar{X}$ we obtain

$$\bar{A} = db_+(Pb_- + Qb_+^{-1})\, \bar{B} + T' \,, \tag{2.11}$$

where

$$T' = db_+(Qb_-\varrho_-P + Pb_+^{-1}\varrho_+Q)\, E + TE \,. \tag{2.12}$$

IIb implies the continuous invertibility of the operator $A_0 = db_+(Pb_- + Qb_+^{-1})\, \bar{B} \in \mathcal{L}(\bar{X}, Z)$, and the condition IIc yields $T' \in \mathcal{K}(\bar{X}, Z_0)$. In view of $\dim \ker A = 0$ then $\bar{A} \in \mathcal{L}(\bar{X}, Z)$ is invertible. By the closed graph theorem it follows from IIf that the restriction $\bar{P}_n = P_n|\bar{X}$ is a continuous projection in the space $\bar{X}$.

Using conditions IId to IIh we conclude easily from Theorem 1.4 that

$$Z_1 \subset \mathfrak{R}(A_0; \{\bar{P}_n, Q_n\}) \,.$$

Theorem 1.1 yields $\mathfrak{R}(\bar{A}; \{\bar{P}_n, Q_n\}) = \mathfrak{R}(A_0; \{\bar{P}_n, Q_n\})$. From this the first assertion of the theorem follows immediately. Finally, Theorem 1.3 and Corollary 1.3 imply the estimates (2.9) and (2.10) with

$$y_0 = (Pb_- + Qb_+^{-1})\, \bar{B}x = b_+^{-1} d^{-1}(y - T'x)$$

(see equation (2.11)). This completes the proof of the theorem.

Theorem 2.3. *Let the conditions* IIa *to* IIg *and* IIi *hold. Let A be an operator of the form* (2.1) *with $T \in \mathcal{K}(X, Z_0)$ and coefficients a, b of the form* (2.8), *where $c = c_+c_- \in \mathfrak{M}_l$ and $d = d_-d_+ \in M_r$. Furthermore, let* $\dim \ker A = 0$.

Then the assertions of Theorem 2.2 *hold with $y_0 = (Pc_- + Qd_+)\, Bx$.*

Proof. We use the notation introduced in the preceding proof. Under the hypotheses of the Theorem 2.3 we obtain $\overline{A} = A_0 + T'$, where

$$A_0 = (c_+P + d_-Q)\,(Pc_- + Qd_+)\,\overline{B}\,, \tag{2.13}$$

$$T' = (c_+Qc_-\varrho_-P + d_-Pd_+\varrho_+Q)\,E + TE\,. \tag{2.14}$$

The remainder of the arguments is the same as in the proof of Theorem 2.2 if we put

$$y_0 = (Pc_- + Qd_+)\,\overline{B}x = (c_+^{-1}P + d_-^{-1}Q)\quad(y - T'x)\,.$$

Remark 1. If $d^{-1}c = a_-a_+ \in \mathfrak{M}_r$ $(cd^{-1} = a_-a_+ \in \mathfrak{M}_r)$, then the operator $cP + dQ$ $(Pc + Qd)$ is invertible on the space X.

This follows immediately from the relations

$$cP + dQ = d(a_-P + Q)\,(a_+P + Q)\,,$$

$$Pc + Qd = (Pa_- + Q)\,(Pa_+ + Q)\,d\,.$$

Remark 2. Under the hypotheses of Theorem 2.2 (resp. of Theorem 2.3) the statements of Theorem 1.5 hold with $Z = Z_0$, i.e. the projection method $\{P_n, Q_n\}$ is stable.

Remark 3. If under the hypotheses of the Subsections 11.2.1 resp. 11.2.2 we still require that the projections $P_nP = PP_n$ and $Q_nP = PQ_n$ commute, then we obtain a projection method for the abstract Wiener-Hopf operator $PaP \mid \operatorname{im} P$ with respect to the projection systems P_nP resp. P_nP and Q_nP changing over from the operator $A = aP + Q$ to the operator $PAP = PaP$.

11.3. Discrete Wiener-Hopf Equations of Normal Type

Now we shall apply the general theorems. In this and in the next section let E be any one of the sequence spaces l^p $(1 \leq p < \infty)$ or c^0, E^μ $(-\infty < \mu < \infty)$ the space defined in 5.1.1, no. 5, and W^μ $(0 \leq \mu < \infty)$ the algebra of all functions in W whose Fourier coefficients form a sequence of the space $l^{1,\mu}$ (cf. 5.1.2, no. 3). E_m and E_m^μ $(m = 1, 2, \ldots)$ are the corresponding spaces of the m-dimensional vectors (cf. 7.1.1).

Theorem 3.1. *Let $a(t)$, $b(t)$ be matrix functions from the algebra $W_{m\times m}$ and $\{a_j\}_{-\infty}^{\infty}$, $\{b_j\}_{-\infty}^{\infty}$ their matrix Fourier coefficients. Furthermore, let the following conditions hold:*

1. $\det a(t) \neq 0$, $\det b(t) \neq 0$ $(|t| = 1)$.
2. *The left indices of $a(t)$ are equal to zero.*
3. *The right indices of the matrix functions $b(t)$ and $b^{-1}(t)\,a(t)$ are equal to zero.*

Then for all sufficiently great n and for any sequence $\eta = \{\eta_j\}_{-\infty}^{\infty} \in E_m$ *the system*

$$\sum_{k=0}^{n} a_{j-k}\xi_k + \sum_{k=-n}^{-1} b_{j-k}\xi_k = \eta_j \qquad (j = 0, \pm 1, \dots, \pm n) \tag{3.1}$$

has precisely one solution $\{\xi_k^{(n)}\}_{k=-n}^{n}$ *and for* $n \to \infty$ *the sequence*

$$\xi^{(n)} = \{\dots, 0, \xi_{-n}^{(n)}, \dots, \xi_n^{(n)}, 0, \dots\} \tag{3.1'}$$

converges in the norm of E_m *to the solution* $\xi \in E_m$ *of the system*

$$\sum_{k=0}^{\infty} a_{j-k}\xi_k + \sum_{k=-\infty}^{-1} b_{j-k}\xi_k = \eta_j \qquad (j = 0, \pm 1, \dots) . \tag{3.2}$$

Moreover, in the case $a(t), b(t) \in W_{m \times m}^{\mu}$ $(0 < \mu < \infty)$ *and* $\eta \in E_m^{\mu}$ *the estimate*

$$||\xi - \xi^{(n)}||_{E_m} = O(n^{-\mu})$$

holds.

Proof. Theorem 3.1 follows from Theorem 2.1 under the following hypotheses: $X = Z = E_m$, $\varrho_+ = \varrho_- = I$ (I the unit matrix), $\mathfrak{M} = W_{m \times m}$ (we identify the matrix function $a(t) \in W_{m \times m}$ with the Toeplitz operator $\{a_{j-k}\}_{j,k=-\infty}^{\infty}$). We define the projections P and P_n in the space E_m by the equations

$$P\{\xi_j\}_{-\infty}^{\infty} = \{\dots, 0, \xi_0, \xi_1, \dots\} ,$$

$$P_n\{\xi_j\}_{-\infty}^{\infty} = \{\dots, 0, \xi_{-n}, \dots, \xi_n, 0, \dots\} .$$

Then the paired operator defined in E_m by the left-hand side of the system (3.2) can be written in the form $A = aP + bQ$ and the system (3.1) represent the projection equation $P_nAP_n\xi = P_n\eta$. The conditions Ia and Ib can easily be checked. The conditions Ic to Ie are obviously satisfied. Moreover, the hypotheses 1 to 3 of the Theorem 3.1 yield $a \in \mathfrak{M}_l$, $b \in \mathfrak{M}_r$ and the invertibility of the operator A (cf. Theorem 2.3 and Corollary 3.1 of Chapter 7). Thus Theorem 2.1 at once implies the first assertion of the theorem and also the estimate

$$||\xi - \xi^{(n)}||_{E_m} \leqq \gamma ||y_0 - P_n y_0||_{E_m}$$

with $y_0 = (Pc_- + Qd_+)\, \xi$, where

$$a = c_+c_- , \qquad b = d_-d_+$$

are the factorizations of the matrix functions a and b.

Now let $a, b \in W_{m \times m}^{\mu}$ $(0 < \mu < \infty)$ and $\eta \in E_m^{\mu}$. Then the matrix functions $c_\pm$, $d_\pm$ and their inverses belong to $W_{m \times m}^{\mu}$ (cf. the remark after Theorem 2.3, Chapter 7) and the equations (2.6) and (2.7) yield

$$y_0 = (c_+^{-1}P + d_-^{-1}Q)\,(\eta - T'\xi)$$

with $T'\xi = (c_+Qc_-P + d_-Pd_+Q)\,\xi \in E_m^\mu$ (cf. the proof of Lemma 2.1 of Chapter 9). Hence also $y_0 \in E_m^\mu$. A simple calculation now yields

$$||y_0 - P_n y_0||_{E_m} = O(n^{-\mu}) .$$

This completes the proof of the theorem.

The projection method for the solution of the system (3.2) is called a *reduction method*. In an analogous way it can be applied to the paired system

$$\left.\begin{aligned} \sum_{k=-\infty}^{\infty} a_{j-k}\xi_k &= \eta_j \qquad (j = 0, 1, \ldots) , \\ \sum_{k=-\infty}^{\infty} b_{j-k}\xi_k &= \eta_j \qquad (j = -1, -2, \ldots) . \end{aligned}\right\} \tag{3.3}$$

In this case the projection equation is the finite system

$$\left.\begin{aligned} \sum_{k=-n}^{n} a_{j-k}\xi_k &= \eta_j \qquad (j = 0, 1, \ldots, n) , \\ \sum_{k=-n}^{n} b_{j-k}\xi_k &= \eta_j \qquad (j = -1, \ldots, -n) . \end{aligned}\right\} \tag{3.4}$$

All assertions of Theorem 3.1 remain valid if in the formulation of condition 3 the matrix function $b^{-1}(t)\,a(t)$ is replaced by $a(t)\,b^{-1}(t)$.

Remark 1. In the case $m = 1$ (scalar equation) conditions 2 and 3 assume the simple form

$$\operatorname{ind} a(t) = \operatorname{ind} b(t) = 0 . \tag{3.5}$$

Then the case of non-zero indices can be reduced to case (3.5) by the socalled *index deletion* (for this comp I. C. Gohberg and I. A. Feldman [1]).

Remark 2. An analogous theorem holds for the space E_m^ν $(-\infty < \nu < \infty)$ instead of E_m if we require that $a(t), b(t) \in W_{m\times m}^{(\nu)}$ is demanded.

Remark 3. Conditions 1 to 3 of Theorem 3.1 are also necessary for the convergence of the reduction method in the space E_m (see A. V. Kozak [1]).

11.4. Discrete Wiener-Hopf Equations of Non-Normal Type

The reduction method cannot be applied to Wiener-Hopf equations of non-normal type in the space E_m or in the space E_m^ν $(-\infty < \nu < \infty)$ (see I. C. Gohberg and I. A. Feldman [1], cf. also Remark 3 at the end of the preceding chapter). In this section we investigate the manifold of convergence for the reduction method in the spaces E_m^ν $(-\infty < \nu < \infty)$. Here again E is any one of the sequence spaces l^p $(1 \leq p < \infty)$ or c° and E_+ is the corresponding space of the one-sided sequences $\{\xi_j\}_0^\infty$.

11.4.1. First of all, as a simple example in the case $m = 1$ we consider the (scalar) discrete Wiener-Hopf equation

$$\sum_{k=0}^{\infty} a_{j-k}\xi_k = \eta_j \qquad (j = 0, 1, \ldots) . \tag{4.1}$$

Under the hypothesis that the symbol

$$a(t) = \sum_{j=-\infty}^{\infty} a_j t^j (|t| = 1)$$

possesses the system of zeros $\boldsymbol{\alpha} = \{\alpha_1, \dots, \alpha_r\}$ of integral multiplicity $\boldsymbol{m} = \{m_1, \dots, m_r\}$ on the unit circle we investigate the convergence of the *natural reduction method*

$$\sum_{k=0}^{n} a_{j-k}\xi_k = \eta_j \qquad (j = 0, 1, \dots, n) \tag{4.2}$$

in the space E_+. According to the definition from 11.1 we say that for the system (4.1) the natural reduction method converges in the space E_+ if for any sequence $\eta = \{\eta_j\}_0^\infty \in E_+$ for which (4.1) has a solution $\xi \in E_+$ and for all sufficiently great n the system (4.2) has a unique solution $\{\xi_j^{(n)}\}_{j=0}^n$ and if $\xi^{(n)} = \{\xi_0^{(n)}, \dots, \xi_n^{(n)}, 0, \dots\}$ converges in the norm of E_+ to the solution of (4.1).

Theorem 4.1. *Let*

$$a(t) = \prod_{j=1}^{r} (t^{-1} - \alpha_j^{-1})^{m_j} c(t) , \tag{4.3}$$

where $c(t) \in W$, $c(t) \neq 0$ $(|t| = 1)$ *and* $\operatorname{ind} c(t) = 0$.

In the space $\overset{0}{c}_+$ *the natural reduction method for the system* (4.1) *converges if and only if all zeros* α_j *are simple* (*i.e.* $m_j \leqq 1$, $j = 1, 2, \dots, r$).

In the space l_+^p $(1 \leqq p < \infty)$ *the natural reduction method does not converge if* $\sum_1^r m_j > 0$.

Before we give the proof of this theorem we make the following remark. If by $\hat{A}$, $\hat{B}$ and $\hat{C}$ we denote the discrete Wiener-Hopf operators in E_+ with the symbols $a(t)$, $\varrho_-(t) = \prod_{j=1}^{r} (t^{-1} - \alpha_j^{-1})^{m_j}$ and $c(t)$, respectively, then (4.3) implies the representation

$$\hat{A} = \hat{B}\hat{C} . \tag{4.4}$$

The operator $\hat{C}$ is invertible in E_+, and $\hat{B}$ maps the space E_+ isometrically onto the space $\overline{E}_+(\varrho_-) = \overline{E}(\boldsymbol{\alpha}, \boldsymbol{m})$ (cf. Subsection 5.1.1). Hence $\hat{A} \in \mathcal{L}(E_+, \overline{E}_+(\varrho_-))$ is an invertible operator. In the sequel by P_n we denote the projection which is defined in E_+ by the equation

$$P_n\{\xi_j\}_{j=0}^\infty = \{\xi_0, \xi_1, \dots, \xi_n, 0, \dots\}$$

and by $\overline{P}_n \in \mathcal{L}(\overline{E}_+(\varrho_-))$ the restriction $P_n \mid \overline{E}_+(\varrho_-)$. The convergence of the natural reduction method then means that $\hat{A} \in \Pi\{P_n, \overline{P}_n\}$.

The proof of Theorem 4.1 is based on the following lemma.

Lemma 4.1. *The projectors $\overline{P}_n$ converge to the identity operator on the space $\bar{c}^0_+(\varrho_-)$ if and only if*

$$m_j \leqq 1\,, \qquad j = 1, \dots, r\,. \tag{4.5}$$

If condition (4.5) *is not satisfied, then the sequence $\overline{P}_n$ is unbounded in $\bar{c}^0_+(\varrho_-)$. In the space $\bar{l}^p_+(\varrho_-)$ $(1 \leqq p < \infty)$ the sequence of projections $\overline{P}_n$ is unbounded if $\sum_1^r m_j > 0$.*

Proof. By Theorem 2.2, Chapter 4, without loss of generality we can assume that $r = 1$. Then $\hat{B} = \hat{B}^m_\alpha$, where $\hat{B}_\alpha$ is the operator defined in 5.1.1, and the norm in $\overline{E}_+(\varrho_-) = \overline{E}_+(\alpha, m)$ is defined by the equation $|\eta| = \|\xi\|_{E_+}$ $(\eta = \hat{B}^m_\alpha \xi)$. We consider the projections $Q_n = I - P_n$.

Let $m = 1$. We prove that the equation

$$\lim_{n\to\infty} |Q_n \eta| = \lim_{n\to\infty} \|\hat{B}^{-1}_\alpha Q_n \hat{B}_\alpha \xi\|_{c^0_+} = 0 \tag{4.6}$$

holds for any $\xi \in c^0_+$. This immediately shows the sufficiency of the conditions (4.5). Since $\hat{B}_\alpha = V^{(-1)} - \alpha^{-1} I$, we see easily that the equation $Q_n \hat{B}_\alpha = \hat{B}_\alpha Q_n - V^{(-1)}(P_{n+1} - P_n)$ holds. Using Corollary 1.1, Chapter 5, we obtain from this that

$$|Q_n \eta| \leqq \|Q_n \xi\|_{c^0_+} + |\xi_{n+1}|\, \|f\|_{c^0_+}\,, \tag{4.7}$$

where ξ_{n+1} is the $(n+1)$-th coordinate of the vector $\xi \in c^0_+$ and $f = \{\alpha^{n+1}, \alpha^n, \dots, \alpha, 0, 0, \dots\}$. Since $\|f\|_{c^0_+} = 1$ and $\|Q_n \xi\|_{c^0_+} \to 0$, (4.7) implies (4.6).

Now we prove the necessity of conditions (4.5). Obviously, we have $Q_n = V^{(n)} V^{(-n)}$, where $V^{(-n)} = [V^{(-1)}]^n$ and V, $V^{(-1)}$ are the shift operators defined in Section 3.1. Hence

$$Q_n \hat{B}^m_\alpha = V^n \hat{B}^m_\alpha V^{(-n)}\,. \tag{4.8}$$

For the unit vectors $e_n = \{0, \dots, 0, 1, 0, \dots\}$ $(n = 0, 1, \dots)$ the equation

$$V^n \hat{B}^m_\alpha V^{(-n)} e_{n+1} = \frac{(-1)^m}{\alpha^m} e_{n+1}\,. \tag{4.9}$$

can easily be checked. Now let $m \geqq 2$. Then

$$\|\hat{B}^{-m}_\alpha Q_n \hat{B}^m_\alpha e_{n+1}\|_{c^0_+} = \|\hat{B}^{-m+2}_\alpha h_n\|_{c^0_+}\,,$$

where the sequence

$$h_n \overset{\text{def}}{=} \hat{B}^{-2}_\alpha e_{n+1} = \alpha\{(n+2)\,\alpha^{n+2}, (n+1)\,\alpha^{n+1}, \dots, \alpha, 0, \dots\}$$

is unbounded in the space c^0_+. Hence also the sequence $\hat{B}^{-m+2}_\alpha h_n$ is unbounded in c^0_+ since otherwise the sequence $h_n = \hat{B}^{m-2}_\alpha (B^{-m+2}_\alpha h_n)$ would be bounded. With this we have proved that for $m \geqq 2$ the sequence of projections Q_n (and thus also P_n) is unbounded in the space $\bar{c}^0_+(\varrho_-)$.

Now we prove the last assertion of the lemma. The obvious equation

$$\|\hat{B}_\alpha^{-1} e_n\|_{l_+^p} = (n+1)^{1/p} \qquad (n = 0, 1, \ldots)$$

implies that the sequence $\hat{B}_\alpha^{-m} e_n$ is unbounded in the space l^p $(1 \leqq p < \infty)$ if $m \geqq 1$. This together with equations (4.8) and (4.9) yields the assertion. The lemma is proved.

Proof of Theorem 4.1. Let $c(t) = c_-(t)\, c_+(t)$ be the factorization of the function $c(t) \in W$. Then for the operator $\hat{A}$ we obtain the representation $\hat{A} = \hat{B}\hat{C}_-\hat{C}_+ = \hat{C}_-\hat{B}\hat{C}_+$.

We consider the operator $\hat{A}_1 = \hat{C}_+\hat{C}_-\hat{B}$. Obviously, $\dim\ker \hat{A}_1 = 0$, and Lemma 4.1, Chapter 4, implies that $\hat{A} - \hat{A}_1 \in \mathcal{K}(E_+, \overline{E}_+(\varrho_-))$. Hence also the operator $\hat{A}_1 \in \mathcal{L}(E_+, \overline{E}_+(\varrho_-))$ is invertible. By Corollary 1.1 we have $\hat{A} \in \Pi\{P_n, \overline{P}_n\}$ if and only if $\hat{A}_1 \in \Pi\{P_n, \overline{P}_n\}$.

The following relations are easily checked:

$$P_n V P_n = P_n V,\; P_n V^{(-1)} P_n = V^{(-1)} P_n \qquad (n = 0, 1, \ldots)\,.$$

These relations imply the equation

$$\overline{P}_n \hat{A}_1 P_n = \overline{P}_n \hat{C}_+ \overline{P}_n \cdot \overline{P}_n \hat{C}_- \overline{P}_n \cdot \overline{P}_n \hat{B} P_n$$

and moreover

$$(\overline{P}_n \hat{A}_1 P_n)^{-1}\, \overline{P}_n = \hat{B}^{-1} \hat{C}_-^{-1} \overline{P}_n \hat{C}_+^{-1}\,. \tag{4.10}$$

The assertion of the theorem follows immediately from equation (4.10) and Lemma 4.1.

Remark. The results of this subsection can be generalized to the matrix case $(m > 1)$ (for this cf. I. C. Gohberg and S. Prössdorf [1]).

11.4.2. In connection with Theorem 4.1 there arises the question of which vectors belong to the manifold of convergence of the reduction method for the system (4.1) in the space l_+^p resp. in the space c_+^0 if the symbol has zeros resp. not simple zeros. In the present subsection we shall give an answer to this question for the more general case of paired systems and $m \geqq 1$. We maintain the notation of Section 11.3.

Let $a(t)$, $b(t)$ $(\in W_{m \times m})$ be two matrix functions and $\{a_j\}_{-\infty}^{\infty}$ $\{b_j\}_{-\infty}^{\infty}$ their matrix Fourier coefficients. We assume that on the unit circle the function $\det a(t)$ possesses the system of zeros $\boldsymbol{\alpha} = \{\alpha_1, \ldots, \alpha_r\}$ of integral multiplity $\boldsymbol{m} = \{m_1, \ldots, m_r\}$ and $\det b(t)$ the system of zeros $\boldsymbol{\beta} = \{\beta_1, \ldots, \beta_s\}$ of integral multiplicity $\boldsymbol{n} = \{n_1, \ldots, n_s\}$. We put

$$l = \max(m_1, \ldots, m_r, n_1, \ldots, n_s)\,. \tag{4.11}$$

If $a(t) \in W_{m\times m}(\boldsymbol{\alpha}, \boldsymbol{m})$ and $b(t) \in W_{m\times m}(\boldsymbol{\beta}, \boldsymbol{n})$ then by Theorem 2.1, Chapter 8, these matrix functions admit the representations

$$a(t) = A_1(t)\, D_-(t)\, S_-(t) = R_-(t)\, D_-(t)\, A_2(t)\,. \tag{4.12}$$

$$b(t) = B_1(t)\, D_+(t)\, S_+(t) = R_+(t)\, D_+(t)\, B_2(t)\,. \tag{4.13}$$

Here $D_-(t)$ and $D_+(t)$ are diagonal matrices of the form

$$D_-(t) = \left\{\prod_{k=1}^{r} (t^{-1} - \alpha_k^{-1})^{\mu_j^{(k)}}\, \delta_{jl}\right\}_{j,l=1}^{m},$$

$$D_+(t) = \left\{\prod_{k=1}^{s} (t - \beta_k)^{\nu_j^{(k)}}\, \delta_{jl}\right\}_{j,l=1}^{m}.$$

$\mu_1^{(k)} \geqq \mu_2^{(k)} \geqq \dots \geqq \mu_m^{(k)} \geqq 0$ $(k = 1, \dots, r)$ and $\nu_1^{(k)} \geqq \nu_2^{(k)} \geqq \dots \geqq \nu_m^{(k)} \geqq 0$ $(k = 1, 2, \dots, s)$ are integers, $R_\pm(t)$ and $S_\pm(t)$ polynomial matrices in $t^{\pm 1}$ with a constant determinant different from zero and $A_j(t), B_j(t)$ $(j = 1, 2)$ non-singular matrix functions of the class $W_{m\times m}$. Let

$$k = \max\,(\mu_1^{(1)}, \dots, \mu_1^{(r)}, \nu_1^{(1)}, \dots, \nu_1^{(s)})\,. \tag{4.14}$$

Obviously, we have $k \leqq l$.

Theorem 4.2. *Let the representations* (4.12) *and* (4.13) *with matrix functions $A_1(t)$ and $B_1(t)$ in $W_{m\times m}^{\bar{k}}$ be given*[1]), *where*

$$\bar{k} = k, \quad \text{if} \quad E = l^p\ (1 \leqq p < \infty), \quad \text{and} \quad \bar{k} > k, \quad \text{if} \quad E = c^\circ\,.$$

Let the left indices of $A_1(t)$ and the right indices of $B_1(t)$ be equal to zero and let the homogeneous system belonging to (3.2) *possess only the zero solution.*

Then for all sufficiently great n and for any $\eta \in E_m^{\bar{k}}$ the system (3.1) *has precisely one solution $\{\xi_j^{(n)}\}_{j=-n}^{n}$ and the sequence* (3.1′) *converges in the norm of E_m to the solution $\xi \in E_m$ of the system* (3.2) *for $n \to \infty$.*

If $A_1(t), B_1(t) \in W_{m\times m}^{\bar{k}+\mu}$ $(0 < \mu < \infty)$ and $\eta \in E_m^{\bar{k}+\mu}$, then the estimate

$$\|\xi - \xi^{(n)}\|_{E_m} = O(n^{-\mu})$$

holds.

Proof. Theorem 4.2 follows from Theorem 2.3 under the following hypotheses: $X = E_m$, $Z = Z_1 = Z_0 = E_m^{\bar{k}}$, $\varrho_- = D_-S_-$, $\varrho_+ = D_+S_+$, $c = A_1$, $d = B_1$, $\mathfrak{M} = W_{m\times m}^{\bar{k}}$. The projection $P_n = Q_n$ is defined as in Section 11.3. The results of Subsection 8.2.1 and 9.2.2, no. 1, show that the conditions IIa to IIc hold. The validity of the other conditions of the Theorem 2.3 can easily be checked. Thus from Theorem 2.3 we at once obtain the first assertion of the theorem and the estimate

$$\|\xi - \xi^{(n)}\|_{E_m} \leqq \gamma \|y_0 - P_n y_0\|_{E_m^{\bar{k}}}$$

[1]) By Theorem 2.1, Chapter 8, for this it is sufficient that $a(t) \in W_{m\times m}^{\bar{k}}(\boldsymbol{\alpha}, \boldsymbol{m})$ and $b(t) \in W_{m\times m}^{\bar{k}}(\boldsymbol{\beta}, \boldsymbol{n})$.

with $y_0 = (Pc_- + Qd_+) B\xi$, where

$$A_1 = c_+c_- , \qquad B_1 = d_-d_+$$

are the factorizations of the matrix functions $A_1(t)$ and $B_1(t)$.

Now let $A_1(t), B_1(t) \in W^{\bar{k}+\mu}_{m\times m}$ and $\eta \in E^{\bar{k}+\mu}_m$. Then $c_\pm$, $d_\pm$ and their inverses are matrix functions from $W^{\bar{k}+\mu}_{m\times m}$ (cf. the remark after Theorem 2.3, Chapter 7), and the equations (2.13) and (2.14) yield

$$y_0 = (c_+^{-1}P + d_-^{-1}Q)(\eta - T'\xi)$$

with $T'\xi = (c_+Qc_-\varrho_-P + d_-Pd_+\varrho_+Q)\,\xi \in E^{\bar{k}+\mu}_m$ (cf. the proof of Lemma 2.1 from Chapter 9). Hence $y_0 \in E^{\bar{k}+\mu}_m$, and a simple calculation yields

$$\|y_0 - P_n y_0\|_{E^{\bar{k}}_m} = O(n^{-\mu}) .$$

This completes the proof of the theorem.

Remark 1. An analogous theorem holds for the paired system (3.3). Here the homogeneous system belonging to (3.3) has only the zero solution if and only if the right indices of the matrix function $A_2(t)\, B_2^{-1}(t)$ are equal to zero.

Remark 2. If $A_1(t), B_1(t) \in W^{\bar{k}+h}_{m\times m}$ $(0 \leq h < \infty)$, then Theorem 4.2 holds for any of the spaces $E = l^{p,h}$ and $E = c^{0,h}$.

Remark 3. In the case $m = 1$ the representations (4.12) and (4.13) assume the simple form

$$a(t) = \prod_{k=1}^{r} (t^{-1} - \alpha_k^{-1})^{m_k} A_1(t) , \qquad b(t) = \prod_{k=1}^{s} (t - \beta_k)^{n_k} B_1(t) ,$$

where $A_1(t)$ and $B_1(t)$ are non-vanishing functions of the class $W^{\bar{k}}$. In this case the conditions of Theorem 4.2 are satisfied if

$$\operatorname{ind} A_1(t) = \operatorname{ind} B_1(t) = 0 \tag{4.15}$$

and if the points α_j and β_k $(j = 1, \dots, r,\ k = 1, \dots, s)$ are distinct. This follows easily from Lemma 4.4, Chapter 4.

The case $\operatorname{ind} A_1(t) \neq 0$ and $\operatorname{ind} B_1(t) \neq 0$ can be reduced to the case (4.15) (for this cf. S. Prössdorf and B. Silbermann [4], § 4).

11.4.3. In this subsection, for discrete Wiener-Hopf equations of non-normal type (3.2) and (3.3) we shall prove the convergence of the natural reduction method in the spaces E^ν_m for $\nu < 0$.

By Theorem 2.1, Chapter 8, in addition to the representations (4.12) and (4.13) the matrix functions $a(t) \in W_{m\times m}(\boldsymbol{\alpha}, \boldsymbol{m})$ and $b(t) \in W_{m\times m}(\boldsymbol{\beta}, \boldsymbol{n})$ also admit the representations

$$a(t) = R'_+(t)\, D'_+(t)\, A_3(t) = A_4(t)\, D'_+(t)\, S'_+(t) , \tag{4.16}$$

$$b(t) = R'_-(t)\, D'_-(t)\, B_3(t) = B_4(t)\, D'_-(t)\, S'_-(t) . \tag{4.17}$$

Here $D'_+(t)$ and $D'_-(t)$ are diagonal matrices of the form

$$D'_+(t) = \left\{ \prod_{k=1}^{r} (t - \alpha_k)^{\mu_j^{(k)}} \delta_{jl} \right\}_{j,l=1}^{m},$$

$$D'_-(t) = \left\{ \prod_{k=1}^{s} (t^{-1} - \beta_k^{-1})^{\nu_j^{(k)}} \delta_{jl} \right\}_{j,l=1}^{m}$$

and $A_j(t)$, $B_j(t)$ ($j = 3, 4$) non-singular matrix functions in the algebra $W_{m\times m}$. $R'_\pm(t)$ and $S'_\pm(t)$ have the same form as $R_\pm(t)$ and $S_\pm(t)$.

Theorem 4.3. *Let E be any one of the sequence spaces $l^{p,\nu}$ $\left(1 \leqq p < \infty, \nu \leqq 1 - \frac{1}{p}\right)$ or $c^{0,\nu}$ $(\nu \leqq 1)$ and $A_3(t)$, $B_3(t) \in W_{m\times m}^{\tilde{k}+|\nu|}$, where*[1])

$$\tilde{k} = k\,, \quad \text{if} \quad E = c^{0,\nu}(\nu < 1)\,, \quad \text{and}$$

$$\tilde{k} > k\,, \quad \text{if} \quad E = l^{p,\nu}\left(1 \leqq p < \infty, \nu \leqq 1 - \frac{1}{p}\right) \quad \text{or} \quad E = c^{0,1}\,.$$

Let the left indices of $A_3(t)$ and the right indices of the matrix functions $B_3(t)$ and $B_4^{-1}(t)\,A_4(t)$ be equal to zero.

Then the natural reduction method converges for the system (3.2) in the space $E_m^{-\tilde{k}}$, i.e. for all sufficiently great n the system (3.1) has a unique solution and for $n \to \infty$ the sequence of these solutions (3.1′) converges in the norm of the space $E_m^{-\tilde{k}}$ to the solution $\xi \in E_m^{-\tilde{k}}$ of the system (3.2) for any vector $\eta \in E_m^{-\tilde{k}}$ for which (3.2) is solvable in $E_m^{-\tilde{k}}$. In particular this holds for any $\eta \in E_m$.

If $\xi \in E_m^{-\tilde{k}+\mu}$ and $A_3(t)$, $B_3(t) \in W_{m\times m}^{\tilde{k}+|\nu|+\mu}$ $(0 < \mu < \infty)$ then the estimate

$$\|\xi - \xi^{(n)}\|_{E_m^{-\tilde{k}}} = O(n^{-\mu})\,. \tag{4.18}$$

holds.

Proof. Theorem 4.3 follows from Theorem 2.1 under the following hypotheses: $X = E_m^{-\tilde{k}}$, $Z = E_m$, $\varrho_+ = R'_+D'_+$, $\varrho_- = R'_-D'_-$, $c = A_3$, $d = B_3$, $\mathfrak{M} = W_{m\times m}^{\tilde{k}+|\nu|}$. The projector P_n is defined in the space X by the same formula as in Section 11.3. Then conditions Ia and Ie are obviously satisfied. The same holds for condition Ic (cf. the remark before Theorem 2.1). Conditions Ib hold in $E_m^{-\tilde{k}}$ as in E_m. The validity of condition Id follows easily from the results of Subsection 9.2.2, no. 1. We obtain the continuous embedding

$$E_m \subset D(E_m^{-\tilde{k}}) \tag{4.19}$$

as follows.

---

[1]) The number k is to be taken from the equation (4.14).

Let $\hat{G}_\alpha$ ($|\alpha| = 1$) be the operator which is defined by the equation (1.1′), Chapter 5. Simple estimates yield the embedding[1])

$$\hat{G}_\alpha(E_+) \subset E_+^{-\mu}\,, \tag{4.20}$$

where

$$\mu = \begin{cases} 1, & \text{if} \quad E_+ = c_+^{0,\nu} \qquad (\nu < 1)\,, \\ \mu > 1, & \text{if} \quad E_+ = c_+^{0,1} \quad \text{or} \quad E_+ = l_+^{p,\nu} \qquad (1 \leqq p < \infty)\,. \end{cases}$$

Using the relations $(E_+^\mu)^\nu = E_+^{\mu+\nu}$ (μ, ν arbitrary real) and

$$\hat{G}(E_+) = \bigcap_{j=1}^{r} \hat{G}_{\alpha_j}^{m_j}(E_+) \quad \text{with} \quad \hat{G} = \prod_{j=1}^{r} \hat{G}_{\alpha_j}^{m_j}$$

from (4.20) we obtain the embedding

$$\hat{G}(E_+) \subset E_+^{-\tilde{k}}\,, \tag{4.21}$$

where the number $\tilde{k}$ is defined as in Theorem 4.3 with $k = \max(m_1, m_2, \dots, m_r)$. (4.21) together with the formulae for the formal inverse of the operator $D = R'_+D'_+P + R'_-D'_-Q$ (cf. Section 5.3.1 and Chapter 8) show that relation (4.19) holds.

Finally the equation

$$A = (A_4P + B_4Q)\,(D'_+P + D'_-Q)\,(S'_+P + S'_-Q)$$

yields dim ker $A = 0$ (comp. Corollary 3.1, Chapter 7). Thus all the conditions of Theorem 2.1 are satisfied, and this theorem at once yields the first assertion of Theorem 4.3 and the estimate

$$\|\xi - \xi^{(n)}\|_{E_m^{-\tilde{k}}} \leqq \gamma \|y_0 - P_n y_0\|_{E_m^{-\tilde{k}}} \tag{4.22}$$

with $y_0 = (Pc_- + Qd_+)\,\xi$. Here

$$A_3 = c_+c_-\,, \qquad B_3 = d_-d_+$$

are the factorizations of the matrix functions $A_3(t)$ and $B_3(t)$.

If $\xi \in E_m^{-\tilde{k}+\mu}$ and A_3, $B_3 \in W_{m\times m}^{\tilde{k}+|\nu|+\mu}$, then $c_\pm$, $d_\pm$ are matrix functions from $W_{m\times m}^{\tilde{k}+|\nu|+\mu}$ and consequently we have $y_0 \in E_m^{-\tilde{k}+\mu}$. The estimate (4.18) follows from (4.22).

Remark 1. If $\eta \in E_m^\mu$ and A_3, $B_3 \in W_{m\times m}^{\tilde{k}+|\nu|+\mu}$, where

$$0 < \mu \leqq \begin{cases} 1 - 1/p - \nu\,, & \text{if} \quad E = l^{p,\nu}\,, \\ 1 - \nu\,, & \text{if} \quad E = c^{0,\nu}\,, \end{cases}$$

then the estimate (4.18) holds.

In fact, using the embedding (4.19) in the present case we easily obtain the relation $y_0 \in E_m^{-\tilde{k}+\mu}$ from equations (2.6) and (2.7).

---

[1]) For this cf. S. Prössdorf and B. Silbermann [6].

Remark 2. An analogous theorem holds for the system (3.3) if the corresponding homogeneous system has only the zero solution in $E_m^{-\tilde{k}}$ and if the left indices of $A_3(t)$ and the right indices of $B_3(t)$ are equal to zero.

Remark 3. In the case $m = 1$ the representations (4.16) and (4.17) assume the form

$$a(t) = \prod_{k=1}^{r} (t - \alpha_k)^{m_k} A_3(t) , \qquad b(t) = \prod_{k=1}^{s} (t^{-1} - \beta_k^{-1})^{n_k} B_3(t) ,$$

where $A_3(t)$ and $B_3(t)$ are non-vanishing functions of the class $W^{\tilde{k}+|\nu|}$. In this case the conditions of Theorem 4.3 are satisfied if

$$\operatorname{ind} A_3(t) = \operatorname{ind} B_3(t) = 0 . \tag{4.23}$$

If moreover α_j and β_k ($j = 1, \ldots, r$, $k = 1, \ldots, s$) are distinct points, then the homogeneous system belonging to (3.3) has only the zero solution.

The case $\operatorname{ind} A_3(t) \neq 0$, $\operatorname{ind} B_3(t) \neq 0$ can be reduced to the case (4.23) (for this cf. S. Prössdorf and B. Silbermann [5], § 4).

11.5. Wiener-Hopf Integral Equations of Normal Type

All the results of the preceding two sections can be carried over to the case of Wiener-Hopf integral equations. By $L^{p,\mu}$ $(1 \leqq p < \infty, -\infty < \mu < \infty)$ we denote the function spaces introduced in Subsection 5.2.1, no. 6, and by $\mathfrak{L}^\mu$ $(0 \leqq \mu < \infty)$ the function algebra defined in Subsection 5.2.2, no. 3.

Theorem 5.1. *Let $k_j(t) \in L_{m \times m}$ $(j = 1, 2)$ and*

$$\mathcal{A}(\lambda) = 1 - \int_{-\infty}^{\infty} e^{i\lambda t} k_1(t)\, dt , \qquad \mathcal{B}(\lambda) = 1 - \int_{-\infty}^{\infty} e^{i\lambda t} k_2(t)\, dt .$$

Furthermore, let the following conditions be satisfied:

1. $\det \mathcal{A}(\lambda) \neq 0$, $\det \mathcal{B}(\lambda) \neq 0$ $(-\infty < \lambda < \infty)$.
2. *The left indices of $\mathcal{A}(\lambda)$ are equal to zero.*
3. *The right indices of the matrix functions $\mathcal{B}(\lambda)$ and $\mathcal{B}^{-1}(\lambda)\, \mathcal{A}(\lambda)$ are equal to zero.*

Then for all sufficiently great $\tau > 0$ and for any vector $f(t) \in L_m^p$ $(1 \leqq p < \infty)$ the integral equation

$$\varphi(t) - \int_0^\tau k_1(t - s)\, \varphi(s)\, ds - \int_{-\tau}^0 k_2(t - s)\, \varphi(s)\, ds = f(t) \tag{5.1}$$

has precisely one solution $\varphi_\tau(t) \in L_m^p(-\tau, \tau)$ and for $\tau \to \infty$ the functions

$$\tilde{\varphi}_\tau(t) = \begin{cases} \varphi_\tau(t) , & |t| < \tau , \\ 0 , & |t| > \tau \end{cases} \tag{5.1'}$$

converge in the norm of L_m^p to the solution $\varphi \in L_m^p$ of the equation

$$\varphi(t) - \int_0^\infty k_1(t-s)\,\varphi(s)\,ds - \int_{-\infty}^0 k_2(t-s)\,\varphi(s)\,ds = f(t) \qquad (5.2)$$

$$(-\infty < t < \infty)\,.$$

Moreover, in the case $\mathcal{A}(\lambda)$, $\mathcal{B}(\lambda) \in L_{m\times m}^\mu$ $(0 < \mu < \infty)$ and $f(t) \in L_m^{p,\mu}$ the estimate

$$\|\varphi - \tilde{\varphi}_\tau\|_{L_m^p} = O(\tau^{-\mu})\,.$$

holds.

The proof is carried out completely analogously to the proof of Theorem 5.1 if the projections P and P_τ $(0 < \tau < \infty)$ are defined in the space L_m^p by the equations

$$(P\varphi)(t) = \begin{cases} \varphi(t)\,, & t > 0\,, \\ 0\,, & t < 0\,; \end{cases} \qquad (P_\tau\varphi)(t) = \begin{cases} \varphi(t)\,, & |t| < \tau\,, \\ 0\,, & |t| > \tau\,. \end{cases}$$

An analogous theorem holds for the paired system

$$\begin{aligned} \varphi(t) - \int_{-\infty}^\infty k_1(t-s)\,\varphi(s)\,ds &= f(t) \qquad (0 < t < \infty)\,, \\ \varphi(t) - \int_{-\infty}^\infty k_2(t-s)\,\varphi(s)\,ds &= f(t) \qquad (-\infty < t < 0)\,. \end{aligned} \qquad (5.3)$$

Then the projection equation has the form

$$\begin{aligned} \varphi(t) - \int_{-\tau}^\tau k_1(t-s)\,\varphi(s)\,ds &= f(t) \qquad (0 < t < \tau)\,, \\ \varphi(t) - \int_{-\tau}^\tau k_2(t-s)\,\varphi(s)\,ds &= f(t) \qquad (-\tau < t < 0)\,. \end{aligned} \qquad (5.4)$$

Remark 1. An analogous theorem holds for the space $L_m^{p,\nu}$ $(1 \leqq p < \infty$, $-\infty < \nu < \infty)$ instead of L_m^p if $\mathcal{A}(\lambda)$, $\mathcal{B}(\lambda) \in \mathfrak{L}_{m\times m}^{|\nu|}$.

Remark 2. It follows also from Theorem 2.1 that in the case of the Hilbert space L^2 under the hypotheses of Theorem 5.1 for the equation system (5.2) (resp. (5.4)) the Galerkin method with respect to the system of the Laguerre functions $\psi_n(t)$ $(n = 0, \pm 1, \pm 2, \ldots)$ converges. These functions are defined by the equations

$$\psi_n(t) = (U^n\psi_0)(t) \qquad (n = \pm 1, \pm 2, \ldots)\,, \qquad \psi_0(t) = \begin{cases} \sqrt{2}\,e^{-t}\,, & t > 0\,, \\ 0\,, & t < 0 \end{cases}$$

or

$$\psi_n(t) = \begin{cases} \sqrt{2}\,e^{-t}\Lambda_n(2t)\,, & t > 0\,, \\ 0\,, & t < 0 \end{cases} \qquad (n = 0, 1, 2, \ldots)$$

and $\psi_n(t) = -\psi_{-n-1}(-t)$ $(n = -1, -2, \ldots)$. Here U is the operator defined by the equation (2.12), Chapter 3, and Λ_n are the normed Laguerre polynomials.

Moreover, it is possible to show that equations (5.2) and (5.3) in the space L^2 are equivalent to the discrete systems (3.2) and (3.3), respectively, considered in l^2. Thus the convergence of the mentioned Galerkin method follows also from Theorem 3.1 (for this cf. I. C. GOHBERG and I. A. FELDMAN [1]). The same statement holds for the space L^p with $4/3 < p < 4$. For all other p, $1 \leqq p < \infty$, in L^p a manifold of convergence can be determined (see A. POMP [1]).

11.6. Wiener-Hopf Integral Equations of Non-Normal Type

All the results of Subsections 11.4.2 and 11.4.3 can be carried over to such equations. We assume that the determinants of the matrix functions $\mathcal{A}(\lambda)$ and $\mathcal{B}(\lambda)$ mentioned in Theorem 5.1 have the system of zeros $\boldsymbol{\alpha} = \{\alpha_1, \dots, \alpha_r\}$ of integral multiplicity $\boldsymbol{m} = \{m_1, \dots, m_r\}$ and the system of zeros $\boldsymbol{\beta} = \{\beta_1, \dots, \beta_s\}$ of integral multiplicity $\boldsymbol{n} = \{n_1, \dots, n_s\}$, respectively, on the real line.

11.6.1. First we consider the systems (5.2) and (5.3) in the space L_m^p $(1 \leqq p < \infty)$. We assume that the matrix functions

$$a(t) = \mathcal{A}\left(i\frac{1+t}{1-t}\right), \qquad b(t) = \mathcal{B}\left(i\frac{1+t}{1-t}\right) \qquad (|t| = 1) \tag{6.1}$$

defined on the unit circle possess the representations (4.12), (4.13), respectively, where the matrix functions

$$\mathcal{A}_j(\lambda) = A_j\left(\frac{\lambda - i}{\lambda + i}\right), \qquad \mathcal{B}_j(\lambda) = B_j\left(\frac{\lambda - i}{\lambda + i}\right) \qquad (-\infty < \lambda < \infty) \tag{6.2}$$

$(j = 1, 2)$ are non-singular on the real line and moreover belong to the algebra $\mathfrak{L}^k_{m\times m}$ with the number k defined by the equation (4.14).[1])

Theorem 6.1. *Let the left indices of $\mathcal{A}_1(\lambda)$ and the right indices of $\mathcal{B}_1(\lambda)$ be equal to zero, and let the homogeneous system belonging to* (5.2) *have only the zero solution in L_m^p $(1 \leqq p < \infty)$.*

Then for all sufficiently great $\tau > 0$ and for any vector $f(t) \in L_m^{p,k}$ the equation (5.1) *has precisely one solution $\varphi_\tau(t) \in L_m^p(-\tau, \tau)$ and for $\tau \to \infty$ the functions* (5.1′) *converge in the norm of L_m^p to the solution $\varphi \in L_m^p$ of the equation* (5.2).

If $\mathcal{A}_1(\lambda), \mathcal{B}_1(\lambda) \in \mathfrak{L}^{k+\mu}_{m\times m}$ $(0 < \mu < \infty)$ and $f(t) \in L_m^{p,k+\mu}$, then the estimate

$$\|\varphi - \tilde{\varphi}_\tau\|_{L_m^p} = O(\tau^{-\mu})$$

holds.

An analogous theorem holds for the equation (5.3). Furthermore, the remarks made in 11.4.2 can be carried over to the case considered here.

---

1) By Theorem 2.1, Chapter 8, for this it is sufficient that $\mathcal{A}(\lambda) \in \mathfrak{L}^k_{m\times m}(\alpha, m)$ and $\mathcal{B}(\lambda) \in \mathfrak{L}^k_{m\times m}(\beta, n)$.

The proof of Theorem 6.1 can be carried out completely analogously to the proof of Theorem 4.2.

11.6.2. Now we consider the equations (5.2) and (5.3) in the spaces $L_m^{p,\nu}$ for $\nu < 0$. We assume that the matrix functions (6.1) admit the representation (4.16) resp. (4.17), where the corresponding matrix functions $\mathcal{A}_j(\lambda)$ and $\mathcal{B}_j(\lambda)$ ($j = 3, 4$) (cf. (6.2)) are non-singular and belong to the algebra $\mathfrak{L}_{m\times m}^{\tilde{k}+|\nu|}$. Here $\tilde{k} > k$, and k is the number defined by (4.14).

Theorem 6.2. *Let E be any one of the function spaces $L^{p,\nu}$ $\left(1 \leqq p < \infty,\ \nu \leqq 1 - \frac{1}{p}\right)$. Let the left indices of $\mathcal{A}_3(\lambda)$ and the right indices of the matrix functions $\mathcal{B}_3(\lambda)$ and $\mathcal{B}_4^{-1}(\lambda)\ \mathcal{A}_4(\lambda)$ be equal to zero.*

Then for all sufficiently great τ the equation (5.1) has precisely one solution $\varphi_\tau(t) \in L_m^p(-\tau, \tau)$ and for $\tau \to \infty$ the functions (5.1′) converge in the norm of the space $E_m^{-\tilde{k}}$ to the solution $\varphi(t) \in E_m^{-\tilde{k}}$ of equation (5.2) for any vector $f(t) \in E_m^{-\tilde{k}}$ for which (5.2) is solvable in $E_m^{-\tilde{k}}$. In particular this holds for any $f(t) \in E_m$.

If $\varphi(t) \in E_m^{-\tilde{k}+\mu}$ and $\mathcal{A}_3(\lambda), \mathcal{B}_3(\lambda) \in \mathfrak{L}_{m\times m}^{k+|\nu|+\mu}$ $(0 < \mu < \infty)$, then the estimate

$$\|\varphi - \tilde{\varphi}_\tau\|_{E_m^{-\tilde{k}}} = O(\tau^{-\mu}) .$$

holds.

An analogous theorem holds for equation (5.3). Furthermore, all the other remarks made in 11.4.3 can be carried over to the case considered here.

Using the results of Subsection 5.4.1 we obtain the Theorem 6.2 from Theorem 2.1 by means of the same arguments as in the proof of Theorem 4.3. In the present case condition Ic of Section 11.2 follows from the relation $P_\tau D^{(-1)} = P_\tau D^{(-1)} P_\tau$ for the formal inverse $D^{(-1)}$, and the operator $\hat{G}_\alpha$ is defined in Subsection 5.2.1, no. 1.[1])

Remark. By means of the results of the Section 11.4 it is possible to make assertions about the convergence of the Galerkin method mentioned at the end of Section 11.5 for the integral equations (5.2) and (5.3) of non-normal type, especially for equations of the first kind (see A. Pomp [1]).

11.7. Singular Integral Equations of Normal Type

In the present section using the results of Section 11.2 we shall establish the convergence of the reduction method and the collocation method for systems of singular integral equations over the unit circle Γ of the form

$$A\varphi = c(t)\,\varphi(t) + \frac{d(t)}{\pi i} \int_\Gamma \frac{\varphi(\tau)}{\tau - t}\, d\tau + T\varphi = f(t)\ (t \in \Gamma) \tag{7.1}$$

[1]) Fore more detail see S. Prössdorf and B. Silbermann [6], § 4.

in the spaces $L^p_m(\Gamma)$ $(1 < p < \infty)$ and $H^\mu_m(\Gamma)$ $(0 < \mu < 1)$. Here $c(t)$ and $d(t)$ are matrix functions of the class $H^\mu_{m\times m}(\Gamma)$ and T is a linear compact operator on the spcified spaces. We put

$$a(t) = c(t) + d(t)\,, \qquad b(t) = c(t) - d(t)\,. \tag{7.2}$$

Moreover, we use the following notation: $C^{l,\lambda}(\Gamma)$ $(l \geqq 0$ integer, $0 < \lambda \leqq 1)$ is the function algebra introduced in 6.1.1, no. 2. For $\lambda = 0$ we put $C^m(\Gamma) = C^{m,0}(\Gamma)$, and for $l = 0$ we write $C(\Gamma) = C^\circ(\Gamma)$, $H^\lambda(\Gamma) = C^{0,\lambda}(\Gamma)$. By $W^l_p(\Gamma)$ $(1 < p < \infty)$ we denote the Sobolev space of 2π-periodic functions on the real line which together with their derivatives up to order $l - 1$ are absolutely continuous and whose l-th derivative is absolutely integrable to the p-th power (cf. also 6.1.1, no. 3). Furthermore, let $W^{l,\lambda}_p(\Gamma)$ $(0 < \lambda \leqq 1)$ be the set of all functions $f \in W^{(l)}_p(\Gamma)$ whose l-th derivative satisfies a Hölder condition with respect to the L^p-norm, i.e.

$$||f^{(l)}(t + h) - f^{(l)}(t)||_{L^p(\Gamma)} \leqq c|h|^\lambda \qquad (c = \text{const.}).$$

For $\lambda = 0$ we put $W^{l,0}_p(\Gamma) = W^{(l)}_p(\Gamma)$. Obviously, then we have $C^{l,\lambda}(\Gamma) \subset \subset W^{l,\lambda}_p(\Gamma)$. At last let $\mathfrak{R}(\Gamma)$ be the class of all 2π-periodic functions which are bounded and Riemann integrable.

11.7.1. The Reduction Method

In this subsection by f_j, a_j, b_j, c_{jk} $(j, k = 0, \pm 1, \ldots)$ we denote the Fourier coefficients of the vector function $f(t)$ $(\in \mathrm{L}^m_p(\Gamma))$ and the matrix functions $a(t)$, $b(t)$, Tt^k, respectively.

Theorem 7.1. *Let $a(t)$ and $b(t)$ be matrix functions of the class $H^\mu_{m\times m}(\Gamma)$ $(0 < \mu < 1)$ satisfying the following conditions:*

1. $\det a(t) \neq 0$, $\det b(t) \neq 0$ $(|t| = 1)$.
2. *The left indices of $a(t)$ and the right indices of $b(t)$ are equal to zero.*

If T is a compact operator on the space $L^p_m(\Gamma)$ $(1 < p < \infty)$ and $\dim \ker A = 0$, *then for all sufficiently great n and for any vector function $f \in L^p_m(\Gamma)$ the system*

$$\sum_{k=0}^{n} a_{j-k}\xi_k + \sum_{k=-n}^{-1} b_{j-k}\xi_k + \sum_{k=-n}^{n} c_{jk}\xi_k = f_j \tag{7.3}$$
$$(j = 0, \pm 1, \ldots, \pm n)$$

has precisely one solution $\{\xi^{(n)}_k\}^n_{k=-n}$ and for $n \to \infty$ the vector functions

$$\varphi_n(t) = \sum_{k=-n}^{n} t^k \xi^{(n)}_k \tag{7.4}$$

converge in the norm of the space $L^p_m(\Gamma)$ to the solution $\varphi \in L^p_m(\Gamma)$ of the equation (7.1).

If moreover $a(t), b(t) \in C^{r,\mu}_{m\times m}(\Gamma)$, $f \in [W^{r,\nu}_p(\Gamma)]_m$ $(r \geqq 0,\ 0 \leqq \nu < 1)$ *and* $T: L^p_m(\Gamma) \to [W^{r,\nu}_p(\Gamma)]_m$, *then the estimate*

$$||\varphi - \varphi_n||_{L^p_m} = O(n^{-r-\lambda}), \qquad \lambda = \min(\mu, \nu) \tag{7.5}$$

holds.

Proof. We introduce the projections P and $Q = I - P$ in the same way as in Section 7.5. Then the operator A can be written in the form

$$A = aP + bQ + T,$$

where a and b are operators of multiplication by the matrix functions (7.2).

By $P^1_n (n = 1, 2, \ldots)$ we denote the projection defined on the space $L^p(\Gamma)$ by the equation

$$(P^1_n\varphi)(t) = \sum_{j=-n}^{n} \xi_j t^j \qquad (t \in \Gamma) \tag{7.6}$$

for any function $\varphi(t) = \sum_{j=-\infty}^{\infty} \xi_j t^j \in L^p(\Gamma)$. By P_n we denote the corresponding projection in $L^p_m(\Gamma)$, i.e. $P_n = \{\delta_{jk}P^1_n\}^m_{j,k=1}$. As is well-known, for $n \to \infty$ P^1_n converges strongly to the identity operator on $L^p(\Gamma)$ $(1 < p < \infty)$ (see e.g A. Zygmund [1]).

Now Theorem 7.1 follows from Theorem 2.1 under the following hypotheses: $X = Z = L^p_m(\Gamma)$, $\varrho_+ = \varrho_- = I$, $\mathfrak{M} = H^\mu_{m\times m}(\Gamma)$, (we identify the matrix function $a(t) \in H^\mu_{m\times m}(\Gamma)$ with the corresponding multiplication operator a). It is easily seen that the system (7.3) represents the projection equation $P_nAP_n\varphi = P_nf$. Conditions *Ia* and *Ic* to *Ie* are satisfied. The condition *Ib* can easily be checked. Moreover, hypotheses 1 and 2 of Theorem 7.1 yield $a \in \mathfrak{M}_l$ and $b \in \mathfrak{M}_r$. Thus from Theorem 2.1 we immediately obtain the first assertion of the theorem and the estimate

$$||\varphi - \varphi_n||_{L^p_m} \leqq \gamma\, ||y_0 - P_ny_0||_{L^p_m} \leqq \gamma' E_n(L^p_m, y_0) \tag{7.7}$$

(cf. the proof of Theorem 1.3), where

$$y_0 = (Pc_- + Qd_+)\,\varphi, \qquad a = c_+c_-, \qquad b = d_-d_+ .$$

In order to prove the second part of the theorem we first remark that under the additional hypotheses of the theorem we have $y_0 \in [W^{r,\lambda}_p(\Gamma)]_m$. In fact, for any function $g \in W^{0,\nu}_p(\Gamma)$ also the Cauchy singular integral Sg is an element of $W^{0,\nu}_p(\Gamma)$ (see P. L. Butzer and R. J. Nessel [1], Proposition 8.2.9, p. 322). Here by Lemma 1.5, Chapter 6, together with $g \in W^{r,\nu}_p(\Gamma)$ the inclusion $Sg \in W^{r,\nu}_p(\Gamma)$ also holds. Since the matrix functions $c_\pm$, $d_\pm$ and their inverses belong to the algebra $C^{r,\mu}_{m\times m}(\Gamma)$ (Theorem 2.6, Chapter 7), using Theorem 5.2 and Corollary 5.1 of Chapter 7 we conclude from the equation $A\varphi = f - T\varphi \in [W^{r,\nu}_p(\Gamma)]_m$ that $\varphi \in [W^{r,\lambda}_p(\Gamma)]_m$. Then from equations (2.6) and (2.7) we easily obtain $y_0 \in [W^{r,\lambda}_p(\Gamma)]_m$. Hence

$$E_n(L^p_m, y_0) = O(n^{-r-\lambda})$$

(see P. L. Butzer and R. J. Nessel [1], Corollary 2.2.4, p. 99). This together with (7.7) implies the assertion. The theorem is proved.

Remark. For $T = 0$ there holds $\dim \ker A = 0$ if and only if the right indices of the matrix function $b^{-1}(t)\, a(t)$ are equal to zero.

Now we prove the analogue of Theorem 7.1 for the space $H^\mu_m(\Gamma)$.

Theorem 7.2. *Let $a(t), b(t) \in H^\mu_{m\times m}(\Gamma)$ $(0 < \mu < 1)$ be matrix functions satisfying conditions* 1 *and* 2 *of Theorem* 7.1.

If T is a compact operator from $H^\nu_m(\Gamma)$ into the space $H^\mu_m(\Gamma)$ $(0 < \nu < \mu < 1)$ and $\dim\ker A = 0$, *then for all sufficiently great n and for any $f(t) \in H^\mu_m(\Gamma)$ the system* (7.3) *has precisely one solution $\{\xi_k^{(n)}\}_{k=-n}^n$ and for $n \to \infty$ the vector functions* (7.4) *converge in the norm of $H^\nu_m(\Gamma)$ to the solution $\varphi \in H^\nu_m(\Gamma)$ of the equation* (7.1). *Here the estimate*

$$||\varphi - \varphi_n||_{H^\nu_m} = O\left(\frac{\ln n}{n^{\mu-\nu}}\right). \tag{7.8}$$

holds.

If moreover $a(t), b(t) \in C^{r,\mu}_{m\times m}(\Gamma)$, $f(t) \in C^{r,\mu}_m(\Gamma)$ and $T\colon H^\nu_m(\Gamma) \to C^{r,\mu}_m(\Gamma)$, then

$$||\varphi - \varphi_n||_{H^\nu_m} = O\left(\frac{\ln n}{n^{r+\mu-\nu}}\right). \tag{7.9}$$

Proof. We obtain Theorem 7.2 from Theorem 2.3 under the following hypothesis: $X = Z = H^\nu_m(\Gamma)$, $Z_0 = Z_1 = H^\mu_m(\Gamma)$, $\varrho_+ = \varrho_- = I$, $\mathfrak{M} = H^\mu_{m\times m}(\Gamma)$, $Q_n = P_n$. Obviously, conditions IIa, IIb, IIe and IIf are satisfied. By Corollary 4.4, Chapter 3, condition IIc is also satisfied. Conditions IId and IIi can easily be checked. Finally condition IIg follows from the relation[1])

$$||f - P_n f||_{H^\nu_m} \leqq \gamma_1 \frac{\ln n}{n^{\mu-\nu}} ||f||_{H^\mu_m}, \tag{7.10}$$

which holds for any $f(t) \in H^\mu_m(\Gamma)$. Thus from Theorem 2.3 we immediately obtain the first assertion of the theorem and the estimate

$$||\varphi - \varphi_n||_{H^\nu_m} \leqq \gamma\, ||y_0 - P_n y_0||_{H^\nu_m}, \tag{7.11}$$

where $y_0 \in H^\mu_m(\Gamma)$ has the same meaning as in the proof of Theorem 7.1.

Now let $a(t), b(t) \in C^{r,\mu}_{m\times m}(\Gamma)$, $f(t) \in C^{r,\mu}_m(\Gamma)$ and $T\colon H^\nu_m(\Gamma) \to C^{r,\mu}_m(\Gamma)$. Then by means of the same arguments as in the end of the preceding proof we obtain $y_0 \in C^{r,\mu}_m(\Gamma)$. Without more difficulty from (7.10) we arrive at the estimate[1])

$$||y_0 - P_n y_0||_{H^\nu_m} \leqq \gamma_2 \frac{\ln n}{n^{r+\mu-\nu}} ||y_0^{(r)}||_{H^\mu_m}$$

from which together with (7.11) the relation (7.9) follows. This completes the proof of the theorem.

[1]) See S. Prössdorf [16], [17] or V. A. Zolotarevski [1].

11.7.2. The Collocation Method

Now we consider the collocation method for the approximate solution of equation (7.1) with the following collocation points on the unit circle:

$$t_j = t_j^{(n)} = e^{\frac{2\pi i}{2n+1} j} \qquad (j = 0, \pm 1, \dots, \pm n)\,. \tag{7.12}$$

This method can be described as follows. We search for the approximate solution of the equation (7.1) in the form (7.4), where the coefficients $\xi_k = \xi_k^{(n)}$ are to be determined such that the equation $A\varphi_n = f$ is satisfied at the collocation points (7.12), i.e.

$$(A\varphi_n)(t_j) = f(t_j) \qquad (j = 0, \pm 1, \dots, \pm n)\,. \tag{7.13}$$

It is easily seen that equation (7.13) has the form

$$a(t_j) \sum_{k=0}^{n} t_j^k \xi_k + b(t_j) \sum_{k=-n}^{-1} t_j^k \xi_k + \sum_{k=-n}^{n} d_{jk}\xi_k = f(t_j) \qquad (j = 0, \pm 1, \dots, \pm n)\,. \tag{7.14}$$

Here $d_{jk} = (Tt^k)(t_j)$ $(j, k = 0, \pm 1, \dots, \pm n)$.

Theorem 7.3. *Let $a(t)$ and $b(t)$ be matrix functions of the class $H^{\mu}_{m\times m}(\Gamma)$ $(0 < \mu < 1)$ satisfying the following conditions:*

1. $\det a(t) \neq 0$, $\det b(t) \neq 0$ $(|t| = 1)$.
2. *The left indices of the matrix function $b^{-1}(t)\, a(t)$ are equal to zero.*

If T is a compact operator from $L^p_m(\Gamma)$ $\left(\frac{1}{\mu} < p < \infty\right)$ into the space $C_m(\Gamma)$ and $\dim \ker A = 0$, *then for all sufficiently great n and for any vector function $f(t)$ with components from $\mathfrak{R}(\Gamma)$ the system (7.14) has precisely one solution $\{\xi_k^{(n)}\}_{k=-n}^{n}$ and for $n \to \infty$ the vector functions (7.4) converge in the norm of the space $L^p_m(\Gamma)$ to the solution $\varphi \in L^{p}_m(\Gamma)$ of equation (7.1).*

If $a(t)$, $b(t) \in C^{r,\mu}_{m\times m}(\Gamma)$, $f(t) \in C^{r,\nu}_m(\Gamma)$ and T: $L^p_m(\Gamma) \to C^{r,\nu}_m(\Gamma)$ $(r \geqq 0$, $0 \leqq \nu < 1)$, *then the estimate* (7.5) *holds.*

Proof. We obtain Theorem 7.3 from Theorem 2.2 ander the following hypotheses: $X = Z = L^p_m(\Gamma)$, $\varrho_+ = \varrho_- = I$, $Z_0 = C_m(\Gamma)$, $Z_1 = \mathfrak{R}_m(\Gamma)$, $\mathfrak{M} = H^{\mu}_{m\times m}(\Gamma)$. We define the projection P_n in the same way as in 11.7.1 (cf. equation (7.6)). Let $Q_n f$ be the interpolation polynomial of the function $f \in \mathfrak{R}_m(\Gamma)$ at the collocation points (7.12), i.e.

$$(Q_n f)(t) = \sum_{k=-n}^{n} a_k t^k\,, \qquad a_k = \frac{1}{2n+1} \sum_{j=-n}^{n} f(t_j)\, t_j^{-k}\,. \tag{7.15}$$

It is easy to see that the system (7.14) is equivalent to the projection equation $Q_n A P_n \varphi = Q_n f$. Obviously, conditions IIa, IIb and IId to IIf from Section 11.2 are satisfied. Condition IIh can easily be checked, and condition IIc is a consequence of a well-known theorem on the compactness of integral operators with kernels of potential type (see e.g. L. V. KANTOROVIČ and G. P. AKILOV [1], Satz 7 (2.X)). Finally, condition IIg is a well-known property of the convergence of interpolation polynomials in the L^p-norm for functions from $\mathcal{R}(\Gamma)$ (see A. ZYGMUND [1], Chapter X.7). Thus all the hypotheses of Theorem 2.2 are satisfied, and we obtain the first assertion of the theorem and the estimate

$$||\varphi - \varphi_n||_{L^p_m} \leqq \gamma E_n(C_m, y_0) \tag{7.16}$$

with $y_0 \in C_m(\Gamma)$.

If the conditions mentioned at the end of Theorem 7.3 are satisfied, then by means of similar arguments as in the proof of Theorem 7.1 and using equations (2.11) and (2.12) we obtain that $y_0 \in C^{r,\lambda}_m(\Gamma)$ with $\lambda = \min(\mu, \nu)$. Thus

$$E_n(C_m, y_0) = O(n^{-r-\lambda})$$

(see P. L. BUTZER and R. J. NESSEL [1], Corollary 2.2.4, p. 99). From this together with (7.16) we get the estimate (7.5). This completes the proof of the theorem.

The following theorem is the analogue to Theorem 7.2.

Theorem 7.4. *Let $a(t)$, $b(t) \in H^\mu_{m\times m}(\Gamma)$ $(0 < \mu < 1)$ be matrix functions satisfying conditions* 1 *and* 2 *of Theorem* 7.3. *Then the assertions of Theorem* 7.2 *hold with respect to the system* (7.14) (instead of (7.3)).

Proof. This theorem follows from Theorem 2.2 under the following hypotheses: $X = Z = H^\nu_m(\Gamma)$, $Z_0 = Z_1 = H^\mu_m(\Gamma)$, $\varrho_+ = \varrho_- = I$, $\mathfrak{M} = H^\mu_{m\times m}(\Gamma)$. The projections P_n and Q_n are defined by the equations (7.6) and (7.15), respectively. The rest of the argument is essentially the same as in the proofs of Theorems 7.2 and 7.3 if we use the estimate[1])

$$||\varphi - Q_n\varphi||_{H^\nu_m} = O\left(\frac{\ln n}{n^{r+\mu-\nu}}\right)$$

valid for any function $\varphi \in C^{r,\mu}_m(\Gamma)$.

Remark. Theorems 7.3 and 7.4 remain valid if instead of the collocation knots (7.12) we choose the knots

$$t_j = e^{\frac{\pi i}{n} j} \quad (j = 0, 1, \dots, 2n-1).$$

This method is usually called the Multhopp method (see H. MULTHOPP [1]).

11.8. Singular Integral Equations of Non-Normal Type

In the present section we investigate the convergence of the reduction method and of the collocation method for the equation (7.1) in the case where the determinants of the matrix functions (7.2) vanish on the unit

[1]) See V. A. ZOLOTAREVSKI [1].

circle at finitely many points. On the unit circle Γ let the function $\det a(t)$ have the system of zeros $\boldsymbol{\alpha} = \{\alpha_1, \dots, \alpha_r\}$ of integral multiplicity $\boldsymbol{m} = \{m_1, \dots, m_r\}$ and the function $\det b(t)$ the system of zeros $\boldsymbol{\beta} = \{\beta_1, \dots, \beta_s\}$ of integral multiplicity $\boldsymbol{n} = \{n_1, \dots, n_s\}$.

11.8.1. The Reduction Method

Theorem 8.1. *Let the representations* (4.12) *and* (4.13) *be given, where* $A_1(t)$ *and* $B_1(t)$ *are non-singular matrix functions of the class* $C^{k,\mu}_{m\times m}(\Gamma)$ $(0 < \mu < 1)$ *and the number* k *is defined by the equation* (4.14).[1]) *Let the left indices of* $A_1(t)$ *and the right indices of* $B_1(t)$ *be equal to zero. Furthermore, let* T *be a compact operator from* $L^p_m(\Gamma)$ $(1 < p < \infty)$ *into the space* $[W^{(k)}_p(\Gamma)]_m$ *and* $\dim \ker A = 0$.

Then for all sufficiently great n *and for any* $f \in [W^{(k)}_p(\Gamma)]_m$ *the system* (7.3) *has precisely one solution* $\{\xi^{(n)}_k\}^n_{k=-n}$ *and for* $n \to \infty$ *the vector functions* (7.4) *converge in the norm of* $L^p_m(\Gamma)$ *to the solution* $\varphi \in L^p_m(\Gamma)$ *of equation* (7.1).

If $A_1(t), B_1(t) \in C^{k+r,\mu}_{m\times m}(\Gamma)$ $(r > 0)$, $f \in [W^{(k+r)}_p(\Gamma)]_m$ *and* $T\colon L^p_m(\Gamma) \to [W^{(k+r)}_p(\Gamma)]_m$, *then the estimate*

$$\|\varphi - \varphi_n\|_{L^p_m} = O(n^{-r}) .$$

holds.

Proof. Theorem 8.1 follows from Theorem 2.3 under the following hypotheses: $X = L^p_m(\Gamma)$, $Z = Z_1 = Z_0 = [W^{(k)}_p(\Gamma)]_m$, $\varrho_- = D_-S_-$, $\varrho_+ = D_+S_+$, $c = A_1$, $d = B_1$. $\mathfrak{M} = C^{k,\mu}_{m\times m}(\Gamma)$. The projection $P_n = Q_n$ is defined by equation (7.6). The rest of the argument is the same as in the proof of Theorem 4.2 if we use Lemma 1.6, Chapter 6, the results of Subsections 8.5.1 and 8.5.2 and the estimate

$$\|f - P_n f\|_{[W^{(k)}_p]_m} = \sum_{j=0}^{k} \|f^{(j)} - P_n(f^{(j)})\|_{L^p_m} = O(n^{-r}) , \tag{8.1}$$

where $f \in [W^{(k+r)}_p(\Gamma)]_m$ is an arbitrary vector function (see P. L. Butzer and R. J. Nessel [1], Corollary 2.2.4, p. 99). This completes the proof of the theorem.

Remark. In the case $T = 0$ and $m = 1$ for Theorem 8.1 Remark 3 of Subsection 11.4.2 holds.

11.8.2. The Collocation Method

Theorem 8.2. *Let the representations* (4.12) *and* (4.13) *be given, where* $A_1(t)$ *and* $B_1(t)$ *are non-singular matrix functions of the class* $C^{k+1}_{m\times m}(\Gamma)$ *and the number* k *is defined by the equation* (4.14). *Let the left indices of the matrix*

---

[1]) By Theorem 2.1, Chapter 8, for this it is sufficient that $a(t) \in C^{k,\mu}_{m\times m}(\boldsymbol{\alpha}, \boldsymbol{m})$ and $b(t) \in C^{k,\mu}_{m\times m}(\boldsymbol{\beta}, \boldsymbol{n})$.

function $B_1^{-1}(t)\ A_1(t)$ *be equal to zero. Furthermore, let* T *be a compact operator from* $L_m^p(\Gamma)$ $(1 < p < \infty)$ *into the space* $C_{m\times m}^{k,\mu}(\Gamma)$ $(0 < \mu < 1)$ *and* $\dim \ker A = 0$.

Then for all sufficiently great n *and for any* $f \in C_{m\times m}^{k,\mu}(\Gamma)$ *the system* (7.14) *has precisely one solution* $\{\xi_k^{(n)}\}_{k=-n}^{n}$ *and for* $n \to \infty$ *the vector functions* (7.4) *converge in the norm of* $L_m^p(\Gamma)$ *to the solution* $\varphi \in L_m^p(\Gamma)$ *of equation* (7.1). *Furthermore*

$$\|\varphi - \varphi_n\|_{L_m^p} = O(n^{-\mu}) . \tag{8.2}$$

If $A_1(t),\ B_1(t) \in C_{m\times m}^{k+r,\mu}(\Gamma)$ $(r > 0)$, $f \in C_m^{k+r,\mu}(\Gamma)$ *and* $T\colon\ H_m^\nu(\Gamma) \to C_m^{k+r,\mu}(\Gamma)$ $(0 < \nu < 1)$, *then the estimate*

$$\|\varphi - \varphi_n\|_{L_m^p} = O(n^{-r-\lambda}) \tag{8.3}$$

holds for any λ, $0 < \lambda < \mu$.

For the proof of this theorem we need the following lemma.

Lemma 8.1. *Let* $f \in C_m^{k+r,\mu}(\Gamma)$ $(k \geqq 0,\ r \geqq 0,\ 0 < \mu < 1)$ *and* $Q_n f$ *the interpolation polynomial defined by the equation* (7.15). *Then*

$$\|f - Q_n f\|_{[W_p^{(k)}]_m} = O(n^{-r-\mu}) . \tag{8.4}$$

Proof. Obviously, it is sufficient to prove the lemma for the case $m = 1$. We have

$$\|f - Q_n f\|_{W_p^{(k)}} \leqq \|f - P_n f\|_{W_p^{(k)}} + \|P_n f - Q_n f\|_{W_p^{(k)}} . \tag{8.5}$$

Equation (8.1) shows that

$$\|f - P_n f\|_{W_p^{(k)}} = O(n^{-r-\mu}) \tag{8.6}$$

(see P. L. Butzer and R. J. Nessel [1], Corollary 2.2.4, p. 99). Furthermore, by a wellknown theorem of Bernstein (see A. Zygmund [1]) we have

$$\|(P_n f - Q_n f)^{(j)}\|_{L^p} \leqq c_j n^j \|P_n f - Q_n f\|_{L^p} \leqq c_j n^j (\|f - P_n f\|_{L^p} + \|f - Q_n f\|_{L^p})$$
$$(j = 0, 1, \ldots, k)$$

with $c_j = \text{const}$. From this together with the relation (8.6) (for $k = 0$) and the inequality

$$\|f - Q_n f\|_{L^p} \leqq (\gamma_1 + \|Q_n\|_{C \to L^p})\, E_n(C, f) \leqq \gamma n^{-k-r-\mu}$$

we get the estimate

$$\|P_n f - Q_n f\|_{W_p^{(k)}} = O(n^{-r-\mu}) \tag{8.7}$$

The relations (8.5) to (8.7) immediately yield the estimate (8.4).

Proof of Theorem 8.2. We obtain Theorem 8.2 from Theorem 2.2 under the following hypotheses: $X = L_m^p(\Gamma)$, $Z = [W_p^{(k)}(\Gamma)]_m$, $Z_0 = Z_1 = C_m^{k,\nu}(\Gamma)$, $\nu = \min\left(\mu, 1 - \dfrac{1}{p}\right)$, $\varrho_- = D_- S_-$, $\varrho_+ = D_+ S_+$, $c = A_1$, $d = B_1$, $\mathfrak{M} = C_{m\times m}^{k+1}(\Gamma)$.

The projections P_n and Q_n are defined by equations (7.6) and (7.15), respectively.

It follows from the results of Subsection 8.5.1, that condition IIa is satisfied. Conditions IIb, IId, IIf, IIh and thus also IIe are obviously satisfied. Condition IIg follows from the estimate (8.4) for $r = 0$. Using the continuous embedding $W_p^{(k+1)}(\Gamma) \subset C^{k,1-1/p}(\Gamma)$ we obtain from Lemma 1.6, Chapter 6, and the remark following this lemma that condition IIc is also satisfied. Thus it follows immediately from Theorem 2.2 that the first assertion of Theorem 8.2 and the estimate

$$\|\varphi - \varphi_n\|_{L_m^p} \leqq \gamma \|y_0 - Q_n y_0\|_{[W_p^{(k)}]^m} \tag{8.8}$$

hold. From equations (2.11) and (2.12) we conclude that in the present case $y_0 = (Pb_- + Qb_+^{-1}) B\varphi \in C_m^{k,\nu}(\Gamma)$. Using the formulae for the inverse B^{-1} (see Sections 6.2.4 and 8.5), the Lemma 1.1 and Corollary 1.2 of Chapter 6 we then obtain $\varphi \in H_m^\nu(\Gamma)$. Using again formulae (2.11) and (2.12) and Lemma 1.7, Chapter 6, we now obtain $T'\varphi \in C_m^{k,1-\varepsilon}(\Gamma)$ for any ε, $0 < \varepsilon < 1$, and $y_0 = b_+^{-1} d^{-1}(f - T'\varphi) \in C_m^{k,\mu}(\Gamma)$. Then the estimate (8.2) follows from (8.4) and (8.8).

If the conditions mentioned at the end of Theorem 8.2 are satisfied, then we conclude from the preceding considerations that $y_0 \in C_m^{k+r,\lambda}(\Gamma)$ for any λ, $0 < \lambda < \mu$. Thus (8.3) is also a consequence of the estimates (8.4) and (8.8). This completes the proof of the theorem.

11.9. Comments and References

11.1. All results of this section with the exception of Corollaries 1.1 and 1.2 were proved by S. Prössdorf and B. Silbermann in [9], [4]. They are formulated without proofs in the paper [8] of these authors. Corollaries 1.1 and 1.2 are proved in I. C. Gohberg and I. A. Feldman [1], Kapitel II, § 3, for the case of continuous projections P_n and Q_n converging strongly to the identity operator for $n \to \infty$ and by I. C. Gohberg and V. I. Levčenko [1] for the case of finite dimensional projectors with dim im P_n = dim im Q_n. The concept of the manifold of convergence was probably first introduced by I. C. Gohberg and V. I. Levčenko [2]. Also these authors already proved Theorem 1.1 in the case of a finite dimensional space Z and continuous projections P_n, Q_q. In the case $X = Y$ and $A = I$ (identity operator) Theorems 1.1 and 1.3 yield well-known results on the Galerkin method for Riesz-Schauder equations (see M. A. Krasnosielski, G. M. Vainikko and others [1], §§ 15.4—15.5). Theorem 1.1 admits another generalization (see S. Prössdorf and B. Silbermann [8], [9], here also some applications of the results of this section to integro-differential equations can be found).

11.2. The results of Subsection 11.2.1 and 11.2.2 were stated by S. Prössdorf and B. Silbermann in the papers [6] and [10]. Moreover, in [6] a more general projection method for abstract singular equations was established. In a certain sense this method contains the two methods described here. Several special cases resp. modifications of these projection methods can be found in the papers: S. Prössdorf and B. Silbermann [3], [4], [5], [6], S. Prössdorf [16].

11.3. Under the additional hypothesis that the left indices as well as the right indices of the two matrix functions $a(t)$ and $b(t)$ are equal to zero Theorem 3.1 is proved without the error estimate in I. C. Gohberg and I. A. Feldman [1]. (For weaker conditions than the specified ones comp. A. V. Kozak [1], I. C. Gohberg and S. Prössdorf [1] and S. Prössdorf [16]). The error estimate is probably new. The same holds for Theorems 5.1 and 7.1. In the case $m = 1$ the convergence of the reduction method for the discrete Wiener-Hopf equation in the space l^1_+ was first proved by G. Baxter [1].

11.4. The results of Subsection 11.4.1 are due to I. C. Gohberg and V. I. Levčenko [1]. They were generalized in I. C. Gohberg and S. Prössdorf [1] for the matrix case and in S. Prössdorf and B. Silbermann [5], V. I. Levčenko [2] for the case of paired equations.

Theorem 4.2 was proved in the papers: S. Prössdorf and B. Silbermann [4], [6], S. Prössdorf [16] (cf. also I. C. Gohberg and S. Prössdorf [1]). For the special case of a scalar Wiener-Hopf equation ($m = 1$) in the space c^0_+ with a polynomial symbol this theorem was first proved in a weakened form and without the error estimate by I. C. Gohberg and V. I. Levčenko [2]. For $m = 1$ Theorem 4.2 can also be carried over to the case of non-integral orders of the zeros of the symbol (see S. Prössdorf and B. Silbermann [4], B. Silbermann [10]).

Theorem 4.3 was proved by S. Prössdorf and B. Silbermann in the papers [5], [6] (comp. also I. C. Gohberg and S. Prössdorf [1]). For the special case of the scalar equation (3.2) with a polynomial symbol and $E = c^{0,1}$ Theorem 4.3 was first proved in a somewhat weakened form and without the error estimate (4.18) by V. I. Levčenko [2] (cf. also V. I. Levčenko [1]). For $m = 1$ Theorem 4.3 can be carried over to the case of non-integral orders of the zeros of the symbol and to some cases of more general spaces (see S. Prössdorf and B. Silbermann [5]).

11.6. The results of this section are due to S. Prössdorf and B. Silbermann [4], [5], [6] (cf. also S. Prössdorf [16]). For $m = 1$ they can also be carried over to the case of non-integral orders of the zeros of the symbol (see S. Prössdorf and B. Silbermann [4], [5]).

11.7. Theorem 7.1 is proved under additional restrictions and without the error estimate in I. C. Gohberg and I. A. Feldman [1] for $m \geqq 1$. For $m = 1$ it was also proved by V. V. Ivanov [2]. The error estimate (7.5) is possibly new.

Independent of each other and with different methods V. A. Zolotarevski [2], [3] and S. Prössdorf [16] proved the first assertion of Theorem 7.2. V. A. Zolotarevski stated error estimates which are a little rougher than (7.8) resp. (7.9) ($\ln^4 n$ instead of $\ln n$).

Theorem 7.4 was obtained with a rougher error estimate ($\ln^3 n$ instead of $\ln n$) also by V. A. Zolotarevski [1], [3]. In the case $m = 1$ under the hypotheses of Theorem 7.4 the convergence of the collocation method was first proved by V. V. Ivanov [2] with respect to the weaker norm

$$\|\varphi\| = \max_{|t|=1} |(P\varphi)(t)| + \max_{|t|=1} |(Q\varphi)(t)|$$

and then by B. G. Gabdulhaev [1] with respect to the norm of $H^\nu(\Gamma)$ $(0 < \nu < \mu)$.

In the formulation given here the results of Section 11.7 were stated by S. Prössdorf and B. Silbermann in [7].

11.8. All the results of this section are due to S. Prössdorf and B. Silbermann [4], [5], [6] (cf. also S. Prössdorf [16]). For $m = 1$ Theorem 8.1 can be generalized to some cases of non-integral orders of the zeros of the symbol (see S. Prössdorf and B. Silbermann [4], [5] and a paper of J. Steinmüller [1]). By B. Silbermann [9] for the scalar singular integral equation in the space $L^p(\Gamma)$ $(1 < p < \infty)$ the convergence of the natural reduction method was proved in the case where all orders of the zeros of the symbol are less than $1/p$ (cf. Theorem 4.1).

BIBLIOGRAPHY

ATKINSON, F. V. (Аткинсон, Ф. В.)

[1] Нормальная разрешимость линейных уравнений в нормированных пространствах, Матем. сб. **28**, 1 (1951), 3—14.

BANACH, S.

[1] Théorie des opérations linéaires, Monografie Matematyczne, Warszawa 1932.

BAXTER, G.

[1] A norm inequality for a "finite-section" Wiener-Hopf equations, Illinois J. Math. **7** (1963), 97—103.

BERESIN, I. S., and SHIDKOV, N. P. (Березин, И. С., и Жидков, Н. П.)

[1] Методы вычисления I, Физматгиз, Москва 1966 (germ. transl.: Numerische Methoden 1, VEB Dtsch. Verl. d. Wiss., Berlin 1970).

BREUER, M.

[1] Banachalgebren mit Anwendungen auf Fredholmoperatoren und singuläre Integralgleichungen, Bonn. math. Schr. **24** (1965), 108 S.

BROWDER, F. E.

[1] Functional analysis and partial differential equations. I, Math. Ann. **138** (1959), 55—79.

BUDJANU, M. S., and GOHBERG, I. C. (Будяну, М. С., и Гохберг, И. Ц.)

[1] Общие теоремы о факторизации матриц-функций. I. Основная теорема, Матем. исслед., Кишинёв, III: 2 (1968), 87—103.

[2] Общие теоремы о факторизации матриц-функций. II. Некоторые признаки и их следствия, Матем. исслед., Кишинёв, III: 3 (1968), 3—18.

BUTZER, P. L., and NESSEL, R. J.

[1] Fourier Analysis and Approximation. Vol. I, Birkhäuser Verlag, Basel and Stuttgart, 1971.

CALDERON, A. P.

[1] Algebras of singular integral operators, Amer. Math. Soc., Proc. Symp. Pure Math. **X** (1967), 18—55.

CARLEMAN, T.

[1] Sur la résolution de certaines équations intégrales, Arkiv for mat., astr. och fys **16**, Nr. 26 (1922).

ČEBOTAREV, G. N. (Чеботарёв, Г. Н.)

[1] О кольцах функций, интегрируемых с весом, Изв. вузов, Матем., № 5 (1963), 133—145.

[2] Об одном уравнении типа свёртки первого рода, Изв. вузов, Матем., № 2 (1967), 80—92.

[3] Об одном особом случае уравнения Винера-Хопфа в пространстве ограниченных функций, Изв. вузов, матем., № 10 (1967), 92—101.

[4] О нормальной разрешимости уравнений Винера-Хопфа в некоторых особых случаях, Изв. вузов, матем., № 3 (1968), 113—118.

Čerski, J. I. (Черский, Ю. И.)
[1] К решению краевой задачи Римана в классах обобщённых функций, ДАНСССР **125**, 3 (1959), 500—503.
[2] Интегральные уравнения типа свёртки и некоторые их приложения, Докторская диссертация, Тбилиси 1964.
[3] Общее сингулярное уравнение и уравнения типа свёртки, Матем. сб. **41** (**83**), 3 (1957), 277—296.
Čikin, L. A. (Чикин, Л. А.)
[1] Особые случаи краевой задачи Римана и сингулярных интегральных уравнений, Уч. Зап. Казанск. ун-та **113**, 10 (1953), 57—105.
Douglas, R. G.
[1] Banach Algebra Techniques in Operator Theory, Academic Press, New York 1972.
Dudučava, R. V. (Дудучава, Р. В.)
[1] Дискретные уравнения Винера-Хопфа, составленные из коэффициентов Фурье кусочно-винеровских функций, ДАН СССР **207**, № 6 (1972), 1273—1276.
[2] Об интегральных операторах Винера-Хопфа, Math. Nachr. **65** (1975), 59—82.
[3] Об интегральных операторах в свёртках с разрывными коэффициентами, ДАН СССР **218**, № 2 (1974), 264—267.
[4] Об интегральных операторах типа свёртки с разрывными коэффициентами, Math. Nachr. **79** (1977), 75—98.
Dunford, N., and Schwartz, J. T.
[1] Linear operators. Part II. Spectral theory self adjoint operators in Hilbert spaces, IP, New York, London 1963.
Dybin, V. B. (Дыбин, В. Б.)
[1] Исключительный случай интегральных уравнений типа свёртки в классе обобщённых функций, ДАН СССР **161**, 4 (1965), 753—756.
[2] Исключительный случай интегральных уравнений типа свёртки, Изв. АН БССР, сер. физ.-мат. наук, № 3 (1966), 37—45.
[3] Исключительный случай парного интегрального уравнения типа свёртки, ДАН СССР **176**, 2 (1967), 251—254.
[4] Интегральный оператор Винера-Хопфа в классах функций со степенным характером поведения на бесконечности, Изв. АН Арм. ССР, Матем. **2**, 4 (1967), 250—270.
[5] Нормализация оператора Винера-Хопфа, ДАН СССР **191**, 4 (1970), 759—762.
Dybin, V. B., and Karapetjanc, N. K. (Дыбин, В. Б., и Карапетянц, Н. К.)
[1] Об интегральных уравнениях типа свёртки в классе обобщённых функций, Сиб. матем. журнал **7**, 3 (1966), 531—545.
[2] Применение метода нормализации к одному классу бесконечных систем линейных алгебраических уравнений, Изв. вузов, Матем. № 10 (1967), 39—49.
Fichtenholz, G. M. (Фихтенгольц, Г. М.)
[1] Курс дифференциального и интегрального исчисления, т. I—III, Физматгиз, Москва 1960 (germ. transl.: Differential- und Integralrechnung, Bd. I—III, VEB Dtsch. Verl. d. Wiss., Berlin 1964).
Fok, V. A. (Фок, В. А.)
[1] О некоторых интегральных уравнениях математической физики, Мат. сб. **14 (56)**, 1—2 (1944), 3—50.

GABDULHAEV, B. G. (Габдулхаев, Б. Г.)
[1] Приближенное решение сингулярных интегральных уравнений методом механических квадратур, ДАН СССР **179**, № 2 (1968), 260—263.
GAHOV, F. D. (Гахов, Ф. Д.)
[1] Краевые задачи, Физматгиз, М. 1963 (engl. transl.: Pergamon Press, Bd. 85, 1966).
[2] Краевая задача Римана для системы n пар функций, УМН, VII, вып. **4** (1952), 3—54.
[3] Вырожденные случаи особых интегральных уравнений с ядром Коши, ДУ **2**, 4 (1966), 533—543.
GAHOV, F. D., and SMAGINA, V. I. (Гахов, Ф. Д., и Смагина В. И.)
[1] Исключительные случаи интегральных уравнений типа свётрки и уравнения первого рода, Изв. АН СССР **26**, 3 (1962), 361—390.
GALIEV, Š. I. (Галиев, Ш. И.)
[1] Некоторые особые случаи интегральных и интегро-разностных уравнений, Труды КАИ, сер. матем. и мех., вып. 144 (1972).
[2] Особый случай системы уравнений Винера-Хопфа, Изв. вузов, Матем. **3** (1974), 32—42.
[3] Некоторые особые случаи уравнений и систем уравнений в свёртках, Диссертация, Казань 1972.
GALIEV, Š, I., and ČEBOTAREV, G. N. (Галиев, Ш. И., и Чеботарёв, Г. Н.)
[1] Об одной бесконечной системе линейных алгебраических уравнений и её континуальном аналоге, Труды КАИ, сер. матем. и мех., вып. **109** (1969).
GANTMAHER, F. R. (Гантмахер, Ф. Р.)
[1] Теория матриц, Гостехиздат, Москва 1955 (germ. transl.: Matrizenrechnung I, VEB Dtsch. Verl. d. Wiss., Berlin 1958).
GAPONENKO, V. N., and DYBIN, V. B. (Гапоненко, В. Н., и Дыбин, В. Б.)
[1] Интегрально-разностное уравнение Винера-Хопфа с аннулирующимся символом, Матем. исслед., Кишинёв, **VII**, вып. 4 (1972), 50—59.
GELFAND, I. M., RAIKOV, D. A., and ŠILOV, G. E. (Гельфанд, I. M., Райков, Д. А., и Шилов, Г. Е.)
[1] Коммутативные нормированные кольца, Физматгиз, Москва 1960 (germ. transl.: Kommutative normierte Algebren, VEB Dtsch. Verl. d. Wiss., Berlin 1966).
GELFAND, I. M., and ŠILOV, G. E. (Гельфанд, I. M., и Шилов, Г. Е.)
[1] Обобщённые функции и действия над ними I, Физматгиз, Москва, 1958 (germ. transl.: Verallgemeinerte Funktionen (Distributionen), Bd. I, VEB Dtsch. Verl. d. Wiss., Berlin 1960).
[2] Обобщенные функции и действсвия над ними II, Физматгиз, Москва, 1958 (germ. transl.: Verallgemeinerte Funktionen (Distributionen), Bd. II, VEB Dtsch. Verl. d. Wiss., Berlin 1962).
GOHBERG, I. C. (Гохберг, И. Ц.)
[1] Об одном применении теории нормированных колец к сингулярным интегральным уравнениям, УМН **7**, вып. 2 (1952), 149—156.
[2] Задача факторизации в нормированных кольцах, функции от изометрических и симметрических операторов и сингулярные интегральные уравнения, УМН **19**, вып. 1 (1964), 71—124.
GOHBERG, I. C., and FELDMAN, I. A. (Гохберг, И. Ц., и Фельдман, И. А.)
[1] Уравнения в свертках и проекционные методы их решения, Наука, Москва 1971 (germ. transl.: Faltungsgleichungen und Projektionsverfahren zu ihrer Lösung, Akademie-Verlag, Berlin 1974).

GOHBERG, I. C., and KREIN, M. G. (Гохберг, И. Ц., и Крейн, М. Г.)
[1] Основные положения о дефектных числах, корневых числах и индексах линейных операторов, УМН **12**, вып. 2 (1957), 44—118.
[2] Введение в теорию линейных несамосопряжённых операторов, Наука, Москва 1965.
[3] Системы интегральных уравнений на полупрямой с ядрами, зависящими от разности аргументов, УМН **13**, вып. 2 (1958), 3—72.
[4] Парное интегральное уравнение и его транспонированное, Теоретическая и прикладная математика, Львов, № 1 (1958), 58—81.

GOHBERG, I. C., and KRUPNIK, N. Ja. (Гохберг, И. Ц., и Крупник, Н. Я.)
[1] О спектре сингулярных интегральных операторов в пространствах L_p, Studia Mathem. **XXXI** (1968), **347**—362.
[2] О спектре сингулярных интегральных операторов в пространстве L_p с весом, ДАН СССР **185**, 4 (1969), 745—748.
[3] Сингулярные интегральные операторы с кусочно-непрерывными коэффициентами и их символы, Изв. АН СССР **35**, 4 (1971), 940—964.
[4] Введение в теорию одномерных сингулярных интегральных операторов, Штиинца, Кишинёв, 1973.
[5] Об алгебре, порожденной теплицевыми матрицами, Функц. анализ и его приложения **3**, вып. 2 (1969), 46—56.

GOHBERG, I. C., and LEVČENKO, V. I. (Гохберг, И. Ц., и Левченко, В. И.)
[1] О сходимости проекционного метода решения вырожденного дискретного уравнения Винера-Хопфа, Матем. исслед., Кишинёв, VI: 4 (1971), 20—36.
[2] О проекционном методе для вырожденного дискретного уравнения Винера-Хопфа, Матем. исслед., Кишинёв, VII: 3 (1972), 238—253.

GOHBERG, I. C., MARKUS, A. S., and FELDMAN, I. A. (Гохберг, И. Ц., Маркус, А. С., и Фельдман, И. А.)
[1] О нормально разрешимых операторах и связанных с ними идеалах, Изв. Молд. филиала АН СССР **10**, 76 (1960), 51—70.

GOHBERG, I. C., and PRÖSSDORF, S.
[1] Ein Projektionsverfahren zur Lösung entarteter Systeme von diskreten Wiener-Hopf-Gleichungen, Math. Nachr. **65** (1975), 19—45.

GOHBERG, I. C., and SEMENCUL, A. A. (Гохберг, И. Ц., и Семенцул, А. А.)
[1] Тёплицевы матрицы, составленные из коэффициентов Фурье функций с разрывами почти-периодического типа, Матем. исслед., Кишинёв, V, вып. 4 (1970), 63—83.

GOLDBERG, S.
[1] Unbounded linear operators. Theory and applications, Mc. Graw-Hill Book Company 1966.

GRADSTEIN, I. S., and RYSHIK, I. M. (Градштейн, И. С.., и Рыжик И. М.)
[1] Таблицы интегралов, сумм, рядов и произведений, Физматгиз, Москва 1962. (engl. transl.: Pergamon Press 1964).

HAIKIN, M. I. (Хайкин, М. И.)
[1] Об интегральном уравнении типа свёртки первого рода, Изв. вузов, Матем., № 3 (1967), 105—116.
[2] Уравнение Винера-Хопфа в пространствах основных и обобщённых функций, Изв. вузов, Матем., № 10 (1967), 83—91.
[3] О регуляризации операторов с незамкнутой областью значений, Изв. вузов, Матем., № 8 (1970), 118—123.

[4] Об особом случае сингулярного интегрального оператора с кусочно-непрерывными коэффициентами, Матем. исслед., Кишинёв, VII: 3 (1972), 194—209.

[5] О нулях сингулярного интегрального оператора с кусочно-непрерывными коэффициентами в особом случае, Матем. исслед., Кишинёв, VIII: **1** (1973), 171—179.

[6] Сингулярное интегральное уравнение с непрерывными коэффициентами в особом случае, Труды Семинара по краевым задачам. Казанск. университет, вып. **10** (1973), 152—162.

Halilov, Z. I. (Халилов, З. И.)

[1] Линейные сингулярные уравнения в нормированном кольце, Изв. АН СССР, Сер. матем. **13**, 2 (1949), 163—176.

Hardy, G. H., Littlewood, J. E., and Polya, G.

[1] Inequalities, University-Press, London-Cambrigde, 1951.

Hausdorff, F.

[1] Zur Theorie der linearen metrischen Räume, Journal für die reine und angewandte Mathematik **167** (1932), 294—311.

Hirschman, I. I., and Widder, D. V.

[1] The convolution transform, Princeton University-Press, Princeton, New Jersey, 1955.

Hoffman, K.

[1] Banach spaces of analytic functions, Prentice-Hall, INC, 1962.

Hvedelidze, B. V. (Хведелидзе, Б. В.)

[1] Линейные разрывные граничные задачи теории функций, сингулярные интегральные уравнения и некоторые их приложения, Тр. Тбилисск. матем. ин-та АН Груз. ССР **23** (1956), 3—158.

[2] Замечание к моей работе „Линейные разрывные граничные задачи...", Сообщения АН ГрузССР **21**, 2 (1958), 129—138.

Ivanov, V. V. (Иванов, В. В.)

[1] Об уравнении Винера-Хопфа первого рода, ДАН СССР **151**, 3 (1963), 489—492.

[2] Теория приближенных методов и её применение к численному решению сингулярных интегральных уравнений, „Наукова Думка", Киев 1968.

Kantorovič, L. V., and Akilov, G. P. (Канторович, Л. В., и Акилов, Г. П.)

[1] Функциональный анализ в нормированных пространствах. Физматгиз, Москва 1959 (germ. transl.: Funktionalanalysis in normierten Räumen, Akademie-Verlag Berlin 1977).

Karapetjanc, N. K. (Карапетянц, Н. К.)

[1] О нормализации дискретных уравнений типа свёртки, Изв. вузов, Матем. **12** (1968), 45—52.

[2] Дискретное уравнение типа свёртки в одном исключительном случае, Сиб. Мат. журнал XI, 1 (1970), 80—90.

Kato, T.

[1] Perturbation theory for linear operators, Springer-Verlag, Berlin, Heidelberg, New York 1966.

Köhler, U., and Silbermann, B.

[1] Einige Ergebnisse über Φ_+-Operatoren in lokalkonvexen topologischen Vektorräumen, Math. Nachr. **56**, 1—6 (1973), 145—153.

[2] Über algebraische Eigenschaften einer Klasse von Operatorenmatrizen und eine Anwendung auf singuläre Integraloperatoren, Math. Nachr. **57**, 1—6 (1973), 245—258.

Köthe, G.

[1] Zur Theorie der kompakten Operatoren in lokalkonvexen Räumen, Portugaliae Math. **13** (1954), 97—104.

Kosulin, A. E. (Косулин, А. Е.)

[1] Особый случай в теории сингулярных уравнений, Вестн. ЛГУ **19** (1962), 142—148.

[2] Одномерные сингулярные уравнения в обобщённых функциях, ДАН СССР **163**, 5 (1965), 1054—1057.

[3] Одномерные сингулярные уравнения в обобщённых функциях, Матем. сб. **70**, 2 (1966), 180—197.

[4] Одномерные сингулярные уравнения в обобщённых функциях. Случай разомкнутого контура, Вестн. ЛГУ **7** (1966), 157—160.

Kozak, A. V. (Козак, А. В.)

[1] О методе редукции для многомерных дискретных свёрток, Матем. исслед., Кишинёв, VIII: 3 1973, 157—160.

Kračkovski, S. N., and Dikanski, A. S. (Крачковский, С. Н., и Диканский, А. С.)

[1] Фредгольмовы операторы и их обобщения, Итоги науки. Матем. анализ (1968), М., ВИНИТИ 1969, 39—71.

Krasnosielski, M. A., Vainikko, G. M., Zabreiko, P. P., Ruticki, J. B., and Stecenko, V. J. (Красносельский, М. А., Вайникко, Г. М., Забрейко, П. П., Рутицкий, Я. Б., и Стеценко, В. Я.)

[1] Приближенное решение операторных уравнений, Физматгиз, Москва 1971 (germ. transl.: Näherungsverfahren zur Lösung von Operatorgleichungen, Akademie-Verlag, Berlin 1973; engl. transl.: Approximate Solution of operator equation, Wolters Nordhoff, Gronningen 1972).

Krein, M. G. (Крейн, М. Г.)

[1] Интегральные уравнения на полупрямой с ядрами, зависящими от разности аргументов, УМН **13**, вып. 5 (1958), 3—120.

Krein, S. G. (Крейн, С. Г.)

[1] Линейные уравнения в банаховом пространстве, Наука, М. 1971.

Krupnik, N. J. (Крупник, Н. Я.)

[1] К вопросу о нормальной разрешимости и индексе сингулярных интегральных уравнений, Уч. зап. Кишинёвского ун-та **82** (1965), 3—7.

Leiterer, J.

[1] Критерии нормальной разрешимости систем сингулярных интегральных уравнений и уравнений Винера-Хопфа, Матем. сб. **83 (125)**, № 3 (11) (1970), 390—406.

[2] Zur normalen Auflösbarkeit singulärer Integraloperatoren, Math. Nachr. **51**, 1—6 (1971), 197—230.

Levčenko, V. I. (Левченко, В. И.)

[1] О проекционном методе для операторов, опеределяемых вырожденными тёплицевыми матрицами, Матем. исслед., Кишинёв, VII: 4 (1972), 123—140.

[2] О проеционном методе решения вырожденных парных дискретных уравненний Винера-Хопфа, Матем. исслед., Кишинёв, VIII: 3 (1973), 26—45.

Lindenstrauss, J., and Tzafriri, L.

[1] On the complemented subspaces problem, Israel J. Math., **9** (1971), 263—269.

Mandžavidze, G. F., and Hvedelidze, B. V. (Манджавидзе, Г. Ф., и Хведелидзе, Б. В.)

[1] О задаче Римана-Привалова с непрерывными коэффициентами, ДАН СССР **123**, 5 (1958), 791—794.

Meyer, Ch.

[1] Über einige Klassen von singulären Operatoren und ihre Beziehungen zu singulären Integraloperatoren, Math. Nachr. **73** (1976), 171—183.

Meyer, S.

[1] Über eine Klasse singulärer Integralgleichungen, deren Symbol auf einer Menge positiven Maßes entartet. Wiss. Zeitschr. der TH Karl-Marx-Stadt **XIV**, 6 (1972), 681—695.

[2] Über singuläre Integralgleichungen, deren Symbol Entartungen vom logarithmischen Typ besitzt, Wiss. Zeitschr. der TH Karl-Marx-Stadt **XVI**, 3 (1974), 399—418.

Mihlin, S. G. (Михлин, С. Г.)

[1] Сингулярные интегральные уравнения, УМН **3**, 3 (1948), 29—112.

[2] Vorlesungen über lineare Integralgleichungen, VEB Dtsch. Verl. d. Wiss., Berlin 1962.

[3] Многомерные сингулярные интегралы и интегральные уравнения, Физматгиз, М. 1962. (engl. transl.: Pergamon Press, Bd. 83, 1965).

[4] Сингулярные интегральные уравнения с непрерывными коэффициентами, ДАН СССР **59**, 3 (1948), 435—438.

[5] Численная реализация вариационных методов, Физматгиз, Москва, 1967 (germ. transl.: Numerische Realisierung von Variationsmethoden, Akademie Verlag, Berlin 1969).

Multhopp, H.

[1] Die Berechnung der Auftriebsverteilung von Tragflügeln, Luftfahrt-Forschung, **XV**, Nr. 4 (1938), 153—169.

Musheli̇švili, N. I.

[1] Singuläre Integralgleichungen, Akademie-Verlag, Berlin 1965.

Noether, F.

[1] Über eine Klasse singulärer Integralgleichungen, Math. Ann. **82** (1921), 42—63.

Palais, R. S.

[1] Seminar on the Atiyah-Singer index theorem. Chapter VI and Chapter VII, Princeton, New Jersey, Princeton University Press 1965.

Pietsch, A.

[1] Zur Theorie der σ-Transformationen in lokalkonvexen Vektorräumen, Math. Nachr. **21**, 6 (1960), 347—369.

[2] Homomorphismen in lokalkonvexen Vektorräumen, Math. Nachr. **22**, 3—4 (1960), 162—174.

[3] Nukleare lokalkonvexe Räume, Akademie-Verlag, Berlin 1965 (engl. transl.: Nuclear Locally Convex Spaces, Berlin 1972).

Pogorzelski, W.

[1] Integral equations and their applications, Volume 1, Pergamon Press, PWN-Polish Scientific Publishers, Warszawa 1966.

Pomp, A.

[1] Über die Konvergenz des Galerkinschen Verfahrens für Wiener-Hopfsche Integralgleichungen in den Räumen L^p, Math. Nachr. (in press).

Privalov, I. I. (Привалов, И. И.)

[1] Введение в теорию финкций комплексного переменного, Гостехиздат, Москва 1954 (germ. transl.: Einführung in die Funktionentheorie, Teil I—III, Leipzig 1958—1959).

[2] Граничные свойства аналитических функций, Гостехиздат, Москва 1950 (germ. transl.: Randeigenschaften analytischer Funktionen, VEB Dtsch. Verl. d. Wiss., Berlin 1956).

Prössdorf, S.

[1] Операторы, допускающие неограниченную регуляризацию, Вестн. ЛГУ **13** (1965), 59—67.

[2] К теории систем сингулярных интегральных уравнений с вырождающейся символической матрицей I, II, Вестн. ЛГУ **19** (1965), 58—73, **7** (1966), 68—75.

[3] Об устойчивости индекса одномерного сингулярного оператора с обращающимся в нуль символом, Пробл. матем. анализа, изд-во ЛГУ, (1966), 70—79.

[4] О линейных уравнениях в пространствах основных и обобщённых функций, ДАН СССР **166**, 4 (1966), 802—805.

[5] Индекс одномерного сингулярного оператора с обращающимся в нуль символом в пространстве $C^\infty(\Gamma)$, Вестн. ЛГУ **7** (1966), 154—156.

[6] Eindimensionale singuläre Integralgleichungen und Faltungsgleichungen nicht normalen Typs in lokalkonvexen Räumen, Habil.-Schrift, TH Karl-Marx-Stadt 1967, 1—143.

[7] Die Indexformel für ein System eindimensionaler singulärer Integralgleichungen nicht normalen Typs, Math. Zeitschrift **106** (1968), 73—80.

[8] Über eine Verallgemeinerung des Ableitungsbegriffes und die Regularisierung singulärer Integralgleichungen, Wiss. Zeitschrift d. TH Karl-Marx-Stadt **X**, 5 (1968), 551—554.

[9] Ein Satz über die äquivalente Regularisierung abgeschlossener Operatoren, Wiss. Zeitschrift d. TH Karl-Marx-Stadt **X**, 5 (1968), 555—556.

[10] Über eine Klasse singulärer Integralgleichungen nicht normalen Typs, Math. Ann. **183** (1969), 130—150.

[11] Zur Theorie der Faltungsgleichungen nicht normalen Typs, Math. Nachr. **42**, 1—3 (1969), 103—131.

[12] Zur Lösung eines Systems singulärer Integralgleichungen mit entartetem Symbol, Ellipt. Dgl., Bd. I, 111—118, Akademie-Verlag, Berlin 1970.

[13] Сингулярное интегральное уравнение с символом, обращающимся в нуль в конечном числе точек, Матем. исслед., Кишинёв, VII: 1 (1972), 116—132.

[14] О системах сингулярных интегральных уравнений с обращающимся в нуль символом, Матем. исслед., Кишинёв, VII: 2 (1972), 129—142.

[15] Über koerzitive und nichtkoerzitive Probleme bei singulären Integralgleichungen, Mitteil. der Math. Ges. der DDR, H. 3—4 (1970), 61—78.

[16] Systeme einiger singulärer Gleichungen vom nicht normalen Typ und Projektionsverfahren zu ihrer Lösung, Studia math. **LIII** (1975), 225—252.

[17] Zur Konvergenz der Fourierreihen hölderstetiger Funktionen, Math. Nachr. **69** (1975), 7—14.

Prössdorf, S., and Silbermann, B.

[1] Über die normale Auflösbarkeit des singulären Integraloperators vom nicht normalen Typ, Math. Nachr. **55**, 1—6 (1973), 73—88.

[2] Über die normale Auflösbarkeit von Systemen singulärer Integralgleichungen vom nicht normalen Typ, Math. Nachr. **56**, 1—6 (1973), 131—144.

[3] Ein Projektionsverfahren zur Lösung singulärer Gleichungen vom nicht normalen Typ, Wiss. Zeitschr. der TH Karl-Marx-Stadt **XVI**, 3 (1974), 367—376.

[4] Ein Projektionsverfahren zur Lösung abstrakter singulärer Gleichungen vom nicht normalen Typ und einige seiner Anwendungen, Math. Nachr., **61** (1974), 133–155.
[5] Verallgemeinerte Projektionsverfahren zur Lösung singulärer Gleichungen vom nicht normalen Typ, Math. Nachr. **68** (1975), 7–28.
[6] Projektionsverfahren zur Lösung von Systemen singulärer Gleichungen vom nicht normalen Typ, Revue Roum. Math. Pures et Appl. **XXII**, № 7 (1977), 965–991.
[7] О сходимости методов редукции и коллокации для систем сингулярных интегральных уравнений, ДАН СССР **226**, № 3 (1976), 516–519.
[8] Общие теоремы сходимости проекционных методов для операторных уравнений в банаховых пространствах, ДАН СССР **230**. № 3 (1976), 527–529.
[9] Einige allgemeine Sätze zur Theorie der Projektionsverfahren für lineare Operatorgleichungen in Banachräumen, Math. Nachr. **75** (1976), 61–72.
[10] Zur Kollokations- und Reduktionsmethode für Systeme singulärer Integralgleichungen. Vortrag auf dem VII. Internationalen Kongreß über Anwendung der Mathematik in den Ingenieurwissenschaften, Weimar 1975. Kongreßberichte S. 289–293.

PRÖSSDORF, S., and TEICHMANN, E.
[1] Über die äquivalente rechte Regularisierung einer singulären Integralgleichung nicht normalen Typs, Wiss. Zeitschrift d. TH Karl-Marx-Stadt **XI**, 3, (1969), 361–365.

PRÖSSDORF, S., and UNGER, G.
[1] Zur Faktorisierung von Matrixfunktionen in Algebren mit zwei Normen, Math. Nachr. **79** (1977), 37–47.

PRÖSSDORF, S., and v. WOLFERSDORF, L.
[1] Zur Theorie der Noetherschen Operatoren und einiger singulärer Integral- und Integrodifferentialgleichungen, Wiss. Zeitschrift d. TH Karl-Marx-Stadt **13**, 1 (1971), 103–116.

PRZEWORSKA-ROLEWICZ, D.
[1] Equations with transformed argument. An algebraic approach, Elsevier Scientific Publishing Company, Amsterdam 1973.

PRZEWORSKA-ROLEWICZ, D., and ROLEWICZ, S.
[1] Equations in linear spaces, Warszawa 1968.

RIESZ, F., and SZ.-NAGY, B.
[1] Lecons d'analyse fonctionelle, Budapest 1952 (germ. transl.: Vorlesungen über Funktionalanalysis, VEB Dtsch. Verl. d. Wiss., Berlin 1956).

ROBERTSON, A. P., and ROBERTSON, W.
[1] Topological vector spaces, Cambridge University Press, 1964.

ROGOŽIN, V. S. (Рогожин, В. С.)
[1] Общая схема решения краевых задач в пространстве обобщённых функций, ДАН СССР **164**, 2 (1965), 277–280.

SCHAEFER, H. H.
[1] Über singuläre Integralgleichungen und eine Klasse von Homomorphismen in lokalkonvexen Räumen, Math. Zeitschrift **66** (1956), 147–163.
[2] Topological Vector Spaces, Springer-Verlag, New York, Heidelberg, Berlin 1971.

SCHECHTER, M.
[1] Principles of Functional Analysis, Academic Press, New York 1971.

SEMENCUL, A. A. (Семенцул, А. А.)

[1] О сингулярных интегральных уравнениях с коэффициентами, имеющими разрывы почти-периодического типа, Матем. исслед., Кишинёв, VI: 3 (1971), 92—114.

ŠERMAN, D. I. (Шерман, Д. И.)

[1] Об одном случае регуляризации сингулярных интегральных уравнений, Прикл. матем. и мех. **15** (1951), 75—83.

SILBERMANN, B.

[1] Über eine Klasse von singulären Integralgleichungen, deren Symbol nicht mehr als eine endliche Anzahl von Nullstellen ganzzahliger und gebrochener Ordnung aufweist, Math. Nachr. **47**, 1—6 (1970), 245—260.

[2] Über eine Klasse einseitig invertierbarer singulärer Integraloperatoren nicht normalen Typs in gewissen Paaren von Banach-Räumen I—II, Math. Nachr. **51** (1971), 327—342, **52** (1972), 297—313.

[3] О сингулярных интегральных операторах в пространствах бесконечно дифференцируемых и обобщённых функций, Матем. исслед., Кишинёв, VI: 3 (1971), 168—179.

[4] Über einen Zugang zu einer Klasse eindimensionaler singulärer Integraloperatoren nicht normalen Typs, Wiss. Zeitschrift d. TH Karl-Marx-Stadt **13**, 1 (1971), 135—142.

[5] Zur Theorie eindimensionaler singulärer Integraloperatoren nicht normalen Typs mit stückweise stetigen Koeffizienten, Math. Nachr. **57**, 1—6 (1973), 371—384.

[6] Über paarige Operatoren nicht normalen Typs, Math. Nachr. **60** (1974), 79—95.

[7] Über die einseitige Invertierbarkeit gewisser Klassen paariger Operatoren nicht normalen Typs, Wiss. Zeitschr. der TH Karl-Marx-Stadt **XVI**, **3** (1974), 377—381.

[8] Singuläre Integralgleichungen vom nicht normalen Typ mit unstetigen Koeffizienten, Dissertation B, TH Karl-Marx-Stadt, 1974.

[9] Ein Projektionsverfahren für schwach ausgeartete singuläre Integralgleichungen, ZAMM **55** (1975), 525—527.

[10] Ein Projektionsverfahren für einen diskreten Wiener-Hopfschen Operator, dessen Koeffizientensymbole Nullstellen nicht ganzzahliger Ordnungen besitzen, Math. Nachr. **74** (1976), 191—199.

ŠILOV, G. E. (Шилов, Г. Е.)

[1] О локально аналитических функциях, УМН **21**, вып. 6 (1966), 177—182.

SIMONENKO, I. B. (Симоненко, И. Б.)

[1] Краевая задача Римана для n пар функций с непрерывными коэффициентами, Изв. вузов Матем., № 1 (1961), 140—145.

[2] Краевая задача Римана для n пар функций с измеримыми коэффициентами и её применение к исследованию сингулярных интегралов в пространствах L_p с весами, Изв. АН СССР, сер. матем., **28**, 2 (1964), 277—306.

[3] Некоторые общие вопросы теории краевой задачи Римана, Изв. АН СССР, сер. матем., **32**, 5 (1968), 1138—1146.

SMIRNOV, V. I. (Смирнов, В. И.)

[1] Курс высшей математики, Физматгиз, Москва 1959 (germ. transl.: Lehrgang der höheren Mathematik, Bd. V, VEB Dtsch. Verl. d. Wiss., Berlin 1962).

SOBOLEV, S. L. (Соболев, С. Л.)

[1] Некоторые применения функционального анализа в математической физике, Физматгиз, Москва 1961 (germ. transl.: Einige Anwendungen der Funktionalanalysis auf Gleichungen der mathematischen Physik, Akademie-Verlag, Berlin 1964).

SPITZER, F.

[1] The Wiener-Hopf equation whose kernel is a probability density I, II, Duke math. J. **24** (1957), 327—343, **27** (1960), 363—372.

STEINMÜLLER, J.

[1] Ein Projektionsverfahren zur Lösung einer Klasse singulärer Integralgleichungen, deren Symbol Nullstellen nicht ganzzahliger Ordnung besitzt, Beiträge zur Analysis (in press).

TITCHMARSH, E. C.

[1] Theory of Fourier integrals, Oxford 1957.

TOVMASJAN, N. E. (Товмасян, Н. Е.)

[1] К теории сингулярных интегральных уравнений, ДУ **III**, 1 (1967), 69—80.

VEKUA, N. P. (Векуа, Н. П.)

[1] Об одной задаче Гильберта с разрывными коэффициентами и её применении к сингулярным интегральным уравнениям, Тр. Тбилисск. матем. ин-та АН Груз. ССР **XVIII** (1951), 307—313.

[2] Системы сингулярных интегральных уравнений, Наука, М., 1970.

VLADIMIRSKI, J. N. (Владимирский, Ю. Н.)

[1] Φ_--операторы в локально выпуклых пространствах, ДАН СССР **184**, 3 (1969), 514—517.

[2] Строго сингулярные операторы в пространствах Фреше, Матем. исслед., Кишинёв, V: 2 (1970), 79—88.

VOIGTLÄNDER, A.

[1] Über paarige Wiener-Hopf-Operatoren nicht normalen Typs, Wiss. Zeitschrift d. TH Karl-Marx-Stadt **XVI**, 3 (1974), 383—398.

WIDOM, H.

[1] Equations of Wiener-Hopf type, Illinois J. of Math. **2**, 2 (1958), 261—270.

WIENER, N., and HOPF, E.

[1] Über eine Klasse singulärer Integralgleichungen, Sitzungsber. Preuss. Akad. d. Wiss. (1931), 696—706.

YOOD, B.

[1] Properties of linear transformations preserved under addition of a completely continuous transformation, Duke math. J. **18**, 3 (1951), 599—612.

YOSIDA, K.

[1] Functional analysis, Springer-Verlag, Berlin-Göttingen-Heidelberg 1965.

ZOLOTAREVSKI, V. A. (Золотаревский, В. А.)

[1] О сходимости коллокационного метода для систем сингулярных интегральных уравнений, Матем. исслед., Кишинёв, **IX**: 1 (1974), 56—69.

[2] Решение сингулярных интегральных уравнений методом редукции, Матем. исслед., Кишинёв, **IX**: 2 (1974), 38—52.

[3] О приближенном решении сингулярных интегральных уравнений, Матем. исслед., Кишинёв, **IX**: 3 (1974), 82—94.

ZYGMUND, A.

[1] Trigonometric series, Vol. I—II, Univ. Press, Cambridge 1959.

SYMBOL INDEX

AUTHOR INDEX

SUBJECT INDEX